For Marjorie's and my
good friends at Saint
Francis College -

Sincerely,
Harold S. Sharp

March 28, 1977 -

FOOTNOTES TO AMERICAN HISTORY

A Bibliographic Source Book

by

HAROLD S. SHARP

The Scarecrow Press, Inc.
Metuchen, N.J. 1977

Library of Congress Cataloging in Publication Data

Sharp, Harold S
Footnotes to American history.

Includes index.
1. United States--History--Bibliography. 2. United States--History--Miscellanea. I. Title.
Z1236.S48 [E179] 016.973 76-49967
ISBN 0-8108-0944-X

Manufactured in the United States of America

Lovingly dedicated to the memory of
my wife Marjorie (1914-1976)
who helped with other books
and with this one while she could.

TABLE OF CONTENTS

FOREWORD

What was the "Hatfield-McCoy Feud"? The "Dred Scott Decision"? The "Teapot Dome Scandal"? The "Hall-Mills Murder Mystery"? The "Aaron Burr Conspiracy"? The "Underground Railroad"? These and many more historical happenings are vaguely familiar to most people, yet it is often difficult to identify them. Who and what were involved in these incidents? Where and when did they occur? What sources of information are available concerning them? Of what larger events are they segments?

The purpose of this book is to answer these questions. Three hundred and thirteen events of American history are described in short narrative form and significant dates and the names of the principal persons involved are indicated. Some are major events, others are comparatively minor but are significant in terms of human interest. Each historical event is identified by its generally accepted name, such as the "Salem Witchcraft Trials," the "Whiskey Ring Affair," the "Charley Ross Kidnaping," or the "Battle of the Ironclads." An index at the back of the book guides the user to the names of specific persons, to places and to subjects.

For the reader who wishes only a quick identification of a particular event, the descriptive paragraphs should be sufficient. If, however, he wishes additional information, each entry is followed by a list of sources which may be consulted for further information concerning that entry. The reference department of the public library can be of great assistance in helping the user locate these and other sources.

The appended lists (bibliographies) are by no means all-inclusive. Many other sources of information exist. Standard history textbooks, biographies of persons involved and in-depth accounts of particular historical events in books, periodicals and newspapers can supply additional detailed information if desired. These may be located in the public library by means of the card catalog, which lists the books and other materials to be found there.

Aside from these specific sources, there are many general reference books which may be consulted for guidance to information of the historical events covered. Assistance in using these tools may also be had from the reference department of the public library where many such tools constitute a part of the permanent reference collection. Some of these, which suggest information sources, are indicated below.

Cumulative Book Index: a World List of Books in the English Language. New York: H. W. Wilson, 1898--. This index lists all books included by author, title, subject and editor and gives a complete bibliographic citation for each book covered in the author entry.

Subject Guide to Books in Print; an index to the Publisher's Trade List Annual. New York: R. R. Bowker, 1957--. This annual publication indexes books in print by subject heading.

Paperbound Books in Print. New York: R. R. Bowker, 1955--. This monthly publication lists new titles in paperback as well as titles still in print by author, title and subject.

Poole's Index to Periodical Literature, 1820-1881. Rev. ed. Boston: Houghton Mifflin, 1891. 2 vols.

________. ________, Supplements, January, 1882-January, 1907. Boston: Houghton Mifflin, 1887-1908. 5 vols. This is a subject index to 590,100 articles in 12,241 volumes of 470 American and English periodicals.

Nineteenth Century Readers' Guide to Periodical Literature, 1890-1899. With supplementary indexing, 1900-1922. New York: H. W. Wilson, 1944. 2 vols. This author, subject and illustrator index to 51 periodicals (1003 volumes) covers the period from 1890 through 1899.

Readers' Guide to Periodical Literature, 1900--. New York: H. W. Wilson, 1905--. Indexes more than 130 well-known general American and Canadian magazines. Each article is entered by author, as many subjects as needed, and by title.

Biography Index. New York: H. W. Wilson, 1956--. This quarterly, cumulated in annual and three-year volumes, gives birth date and, where applicable, death date of each person listed, together with his nationality and profession or occupation. It also cites sources of biographical information concerning each person.

Book Review Digest. New York: H. W. Wilson, 1905--. This monthly, except February and July, indexes reviews of current books which appear in more than 70 periodicals, showing these reviews in digest form. It has a title and subject index in each issue. A cumulated subject and title index is published every five years.

New York "Times" Index. New York: New York "Times," 1851--. (Semi-monthly, annual cumulation.) This newspaper index gives the exact references to date, page and column. Summaries of articles may answer questions making it unnecessary to refer to the paper itself.

"Times" (London) Index to the "Times." London: London "Times," 1907--. (Bimonthly since 1957.) This indexes the final edition of the London "Times."

Book Review Index. Detroit: Gale Research Company, 1965--. (Bimonthly, quarterly cumulations, annual index.) This publication indexes reviews appearing in more than 200 periodicals. Excerpts are not given.

Choice: Books for College Libraries. Chicago: American Library Association, 1964--. (Monthly except August.) This publication is designed to meet the needs of college, junior college and other libraries concerned with acquiring academic materials. It makes evaluative comparisons.

Webster's Biographical Dictionary. Springfield, Mass.: G. & C. Merriam Co., 1971. This pronouncing biographical dictionary is not restricted by period, nationality, race, religion, or occupation. It gives brief, condensed biographical sketches of prominent persons living and dead.

Essay and General Literature Index. New York: H. W. Wilson, 1900--. This is a cumulative author and subject index to articles which are published as parts of books.

Dictionary of American History. James Truslow Adams, editor. New York: Scribner's, 1940-44, 1961. 6 vols. This dictionary contains brief articles on American history with selected bibliographies. No biographies are given.

Dictionary of American Biography. New York: Scribner's, 1928-1958. 20 vols. Index, 2 supplements. This publication contains long articles on deceased Americans who have made significant contributions to American life. Brief bibliographies are found at the end of most articles.

United States Catalog. New York: H. W. Wilson, 1928. 4th edition. Lists books in print January 1, 1928. Earlier editions: 1st: Books in print, 1899. Minneapolis: H. W. Wilson, 1900. 2nd: Books in print, 1902. Minneapolis: H. W. Wilson, 1903. 3rd: Books in print, 1912. New York: H. W. Wilson, 1912. This publication was designed to provide a consolidated publishers' catalog showing all books available for purchase. The first edition was an author catalog with a title index. Subsequent editions contain author, title and subject entries in one alphabet, with author entries the most complete. It was replaced by the Cumulative Book Index following the publication of the 4th edition in 1928.

Some users of this book may wish to locate stories or novels concerning events in American history. Three guides to fiction of this type are:

Dickinson, A. T. *American Historical Fiction.* 3rd ed. Metuchen, N. J.: Scarecrow Press, 1971.

Logasa, Hannah. *Historical Fiction: Guide for Junior and Senior High Schools and Colleges, Also for General Reader.* Philadelphia: McKinley Publishing Co., 1964.

McGarry, Daniel D., and Sarah Harriman White. *World Historical Fiction Guide: An Annotated Chronological, Geographical, and Topical List of Selected Historical Novels.* 2d ed. Metuchen, N. J.: Scarecrow Press, 1973.

In addition to the above information sources, the user of this book may often profit by consulting a standard encyclopedia for information concerning an historical event. There are a number of good encyclopedias available. It is important to realize that many of the larger encyclopedias have cross-references or indexes which guide the user to information contained in articles of larger scope. For example, information on the destruction of the U.S. battleship "Maine" in Havana Harbor, 1898, might be given in an article on the Spanish-American war. The index would indicate this.

Three standard multi-volume encyclopedias which are found in most public library reference collections are:

Encyclopaedia Britannica: A New Survey of Universal Knowledge. Chicago: Encyclopaedia Britannica, Inc., c. 1974. 24 vols. Continuous revision. This is the oldest and most famous encyclopedia in the English language.

Encyclopedia Americana. New York: Americana Corporation, c. 1976. 30 vols. Continuous revision. This is the oldest major American encyclopedia still in print.

Collier's Encyclopedia. New York: Crowell-Collier Educational Corporation, c. 1976. 24 vols. Continuous revision. This is the newest of the general encyclopedias.

All these encyclopedias contain bibliographies directing the user to additional information on most subjects covered. They also publish annual yearbooks which consider happenings during the previous calendar year.

There are, needless to say, many other encyclopedias and sources of historical information. These are treated in several publications, two of the best of which are:

Winchell, Constance M. *Guide to Reference Books.* 8th edition. Chicago: American Library Association, 1967 (plus supplements). This annotated list of reference books is well indexed, brings books together by subject, and indicates their comparative value.

Murphey, Robert W. *How and Where to Look It Up.* New York:

McGraw-Hill, 1958. This source book contains much information on using a library reference collection and also cites specific publications containing information on many subjects.

While this book was being prepared many persons gave willing assistance. Particularly helpful were Karen Allred, Linda Chapman, Helen Colchin, Geri Cubbal, Catherine Datzman, Janet Gerber, Mary Grigsby, John Hall, Helen Mustin, Susan Pallone, Pegi Ritchie, Martha Rogers, Peter Scott, Richard Seagly, Earlene Shoemaker and Theodore Thieme, all members of the staff of Fred Reynolds, librarian of the Fort Wayne and Allen County Library, Fort Wayne, Indiana. Reference librarians at the Indiana University Library, Bloomington, and the staff of the Indianapolis Public Library also rendered willing and helpful aid.

My wife Marjorie was especially helpful. By reading both manuscript and proof as her health permitted and alphabetizing the 3" x 5" cards on which the bibliographies and index were first typed, she contributed many hours of valuable assistance. I wish to express my deep appreciation to all these persons at this time.

Harold S. Sharp

Fort Wayne, Indiana
June, 1976

THE DISCOVERY OF NORTH AMERICA BY THE NORSEMEN (990-1015)

Not a great deal is known about the Norsemen discovering the North American continent. Much of the information we have comes from Icelandic sagas which were written to glorify and preserve the traditions of a particular family rather than give historically accurate information. The location of the Norsemen's landfall in North America and the extent of their exploration and colonization are quite open to question. Adam of Bremen, a German monk, in approximately 1075, wrote an account of some Norse voyages which furnishes a degree of verification of the saga material.

Of the sagas which have come down to the present day, the Greenlander's Saga, which describes six Norse voyages, is considered to be comparatively authentic in content. It tells of one Bjarne Herjulfsson, the son of an Icelander who had migrated to Greenland about 985 along with Eric the Red. Bjarne set off from Iceland to follow his father, eventually arriving at Greenland after sighting several other islands.

Sometime after 1000 A.D. Leif Ericson, called "Leif the Lucky," the son of Eric the Red, left Greenland for a voyage of exploration. He and his men spent the winter on the mainland of what was probably modern-day North America. This place, where he and his men remained a year, he called "Vinland."

Vinland's exact location is unknown. The sagas describing it mention wine and grapes and some historians have reasoned that it must be situated south of the northern limits of wild grapes. New England, New York and other areas have been suggested. Other historians have considered that the "Vin" in Vinland referred to the old Norse syllable meaning "pasture" rather than "wine," and that Vinland was probably in northern Newfoundland. Archaeological research in Newfoundland, made in 1960, seems to confirm this theory.

In 1965 the so-called "Vinland Map," published by Yale University, apparently indicated Vinland as being located north of the limit of wild grapes. The "Vinland Map" is a world map, supposedly drawn in Switzerland about the year 1440, thus predating the 1492 voyage of Christopher Columbus.

Suggested Readings

Bakeless, John, compiler. The Eyes of Discovery: The Pageant of North America as Seen by the First Explorers. Philadelphia: Lippincott, 1950.

Gathorne-Hardy, G. M. The Norse Discoveries of America. Gloucester, Mass.: Peter Smith, 1970.
Haugen, Einar A., editor and translator. Voyages to Vinland. Chicago: Rand-McNally, 1941.
Holand, Hjalmar R. Explorations in America Before Columbus. New York: Twayne Publishers, 1956.
______. Norse Discoveries and Explorations in North America: Leif Ericson to the Kensington Stone. New York: Dover, 1969.
______. Westward from Vinland: An Account of Norse Discoveries and Explorations in America, 982-1362. New York: Duell, Sloan, 1940.
Ingstad, Helge. Westward to Vinland: The Discovery of Pre-Columbian Norse House-Sites in North America. New York: St. Martin's Press, 1969.
Jones, Gwyn. The Norse Atlantic Saga: Being the Norse Voyages of Discovery and Settlement to Iceland, Greenland, America. London: Oxford University Press, 1964.
Magnusson, Magnus, and Hermann Palsson. The Vinland Sagas: The Norse Discovery of America. New York: New York University Press, 1966.
Morison, Samuel Eliot. The European Discovery of America: The Northern Voyages, A.D. 500-1600. New York: Oxford University Press, 1971.
Oleson, Tryggvij. Early Voyages and Northern Approaches, 1000-1632. Vol. I: Canadian Centenary Series. London: Oxford University Press, 1964.
Reman, Edward. The Norse Discoveries and Explorations in North America. Berkeley, Calif.: University of California Press, 1950.
Skelton, R. A., et al. The Vinland Map and the Tartar Relation. New Haven, Conn.: Yale University Press, 1965.
Wahlgren, Erik. "Fact and Fancy in the Vinland Sagas" in Polome, Edgar G., editor. Old Norse Literature and Mythology: A Symposium. Austin, Texas: University of Texas Press, 1969.

THE DISCOVERY OF AMERICA BY COLUMBUS (1492)

The discovery of America on October 12, 1492, by Christopher Columbus, an Italian navigator serving the Spanish court, has been called by historians "the most spectacular and most far-reaching geographical discovery in recorded human history."

Columbus, born in Genoa, Italy, in 1451, was the son of a wool weaver. As a young boy he went to sea, making voyages to various ports, and in about the year 1477 settled in Lisbon. There he married Doña Felipa Perestrello e Moniz, the daughter of a Portuguese navigator. Columbus had access to the logbooks and charts of his late father-in-law. He became convinced from these and from conversations with various ship captains that the world was spherical and that it was possible, by sailing due west from Europe, to reach Asia.

He submitted proposals to John II, King of Portugal, that the latter finance a voyage of discovery. The King, however, rejected the idea and in 1485, Columbus' wife having passed away, he traveled to Spain. There he met the Count of Medina Celi, who eventually arranged an audience for him with Queen Isabella in 1486. He was, however, unable to obtain royal patronage for his project at that time.

For the next six years Columbus sought in vain to obtain support for his scheme, being rejected at the courts of Portugal, England and France. Finally, in April, 1492, when he had departed from the Spanish court, hoping to discuss his project with King Charles VIII of France once more, Queen Isabella sent a messenger to bring him back. Again at the Spanish court, he was told that Isabella and Ferdinand had agreed to his terms. On April 30, 1492, a document, "Privileges and Prerogatives Granted to Columbus," was signed by their majesties. This document provided that Columbus was to become viceroy of all territories discovered by him, and other rewards were also provided for. These included an hereditary peerage and one-tenth of all precious metals found within his jurisdiction.

Columbus sailed from Palos, Spain, on the first of his four voyages, on August 3, 1492. The three small ships comprising his "fleet" were the "Santa María," commanded by Columbus, with Juan de la Cosa as pilot; the "Pinta," captained by Martín Alonso Pinzón; and the "Niña," with Martin Pinzón's brother Vicente Yáñez Pinzón in command. The three vessels carried a total of 88 men.

Three days out of Palos the "Pinta" was damaged, and a brief stop for repairs was made at Tenerife in the Canary Islands. Then, on September 6, the three vessels weighed anchor, heading due west. This course was maintained until October 7, when it was altered to a few points south of west. The crews became restless and fearful and demanded that the expedition return to Spain. Mutiny was averted only by stringent disciplinary action.

On the same day great flocks of birds were seen flying over the ships. Mutiny seemed imminent, but floating branches gave evidence that land was nearby and the threats of mutiny disappeared. Columbus promised his captains that if land were not sighted within three days he would abandon the voyage and return home.

At 2:00 A.M., October 12, 1492, Rodrigo de Triana, watchman on the "Pinta," sighted land. The ships hove to until daylight and then a landing party disembarked on what is now usually identified as Watlings Island in the Bahamas. Columbus named it San Salvador.

During the next few weeks additional discoveries were made, including Cuba and Haiti. In December the "Santa María" was wrecked on the Cuban coast. A fort was constructed there and garrisoned with 44 men.

Early in January, 1493, the "Niña" and the "Pinta" returned to Spain, after brief stops in the Azores and at Lisbon, arriving at Palos on March 15. The Spanish rulers and people gave him an enthusiastic welcome and the honors conferred by his contract were confirmed. The news of his discoveries was promptly communicated to Pope Alexander VI and King John II of Portugal. King John believed that the lands discovered by Columbus had really been in the region south and west of Guinea. Portugal had control of Guinea and the region south of Cape Bojador, based on discoveries, a 1480 treaty with Spain and Papal bulls. The king was concerned with the effect which Columbus' discoveries would have on the territorial claims of Portugal and Spain.

The matter was submitted to Pope Alexander VI who, by the Bulls of May 3 and May 4, 1493 (the Papal Bull Inter Caetera) drew an imaginary line of demarcation one hundred leagues west of the Cape Verde Islands. Portugal was to have all the rights and possessions she already held east of this line, and Spain was to have the right to explore, colonize and trade in the area west of it. The Bull of May 4 was amended by another Bull of September, 1493. This granted Spain the right to hold lands to the "eastern regions and to India."

On June 7, 1494, King John II, who was not satisfied with the provisions of the Papal Bulls, negotiated the Treaty of Tordesillas, by which Spain agreed to move the line of demarcation 370 leagues west from the Cape Verde Islands, thus giving Portugal a claim to Brazil.

Columbus made three other voyages to the New World. He made further discoveries but failed to gain reinstatement of his honors and privileges, his enemies having succeeded in putting him in a bad light with the Spanish authorities. He died in poverty at Valladolid in 1506.

Suggested Readings

Barclay, Isabel. Worlds Without End. Garden City, N.Y.: Doubleday, 1956.

Carrison, Daniel Jordan. Christopher Columbus: Navigator to the New World. New York: Watts, 1967.

Clark, William R. Explorers of the World. New York: Natural History Press, 1964.

Columbus, Christopher. The Journal of Christopher Columbus (During His First Voyage, 1492-1493) and Documents Relating to the Voyages of John Cabot and Gaspar Corte Real. Translated by Cecil Jane. London: The Hakluyt Society, 1960.

Commager, Henry Steele, editor. "The Papal Bull Inter Caetera (Alexander VI)," (Doc. No. 2) in his Documents of American History, 8th edition. New York: Appleton, 1968.

_______. "Privileges and Prerogatives Granted to Columbus," (Doc. No. 1) in his Documents of American History, 8th edition. New York: Appleton, 1968.

_______. "The Treaty of Tordesillas," (Doc. No. 3) in his Documents of American History, 8th edition. New York: Appleton, 1968.

Crone, G. R. The Discovery of America. London: Hamish Hamilton, 1969.

Cumming, William Patterson, R. A. Skelton and D. B. Quinn. The Discovery of North America. New York: American Heritage Press, 1972.

Elliott, J. H. The Old World and the New. Cambridge: Cambridge University Press, 1970.
Foster, Genevieve. The World of Columbus and Sons. New York: Scribner's, 1965.
Franklin, Walter. Famous American Ships. New York: Simon & Schuster, 1958.
Giardini, Cesare. The Life and Times of Columbus. Translated by Frances Lanza. Philadelphia: Curtis, 1967.
Hodges, C. W. Columbus Sails. New York: Coward-McCann, 1949.
Lamb, Harold. New Found World: How North America Was Discovered and Explored. New York: Doubleday, 1955.
McKendrick, Melveena. Ferdinand and Isabella. New York: American Heritage Press, 1968.
Meredith, Robert K., and E. Brooks Smith, editors. The Quest of Columbus. Boston: Little, Brown, 1966.
Morison, Samuel Eliot. Admiral of the Open Sea: A Life of Christopher Columbus. Boston: Little, Brown, 1951.
________. Christopher Columbus, Mariner. Boston: Little, Brown, 1955.
________, and Mauricio Obregon. The Caribbean as Columbus Saw It. Boston: Little, Brown, 1964.
Oleson, Tryggvij. Early Voyages and Northern Approaches. Toronto: McClelland & Stewart, 1963.
Parry, J. H. The Age of Renaissance: Discovery, Exploration and Settlement, 1450-1650. London: Weidenfeld & Nicholson, 1963.
Penrose, Boies. Travel and Discovery in the Renaissance. Cambridge, Mass.: Harvard University Press, 1955.
Priestley, Herbert Ingram. The Coming of the White Man. New York: Macmillan, 1929.
Wright, Louis B. Gold, Glory, and the Gospel: The Adventurous Lives and Times of the Renaissance Explorers. New York: Atheneum, 1970.

THE LOST COLONIES (1585)

The first English settlement in America was that founded on Roanoke Island, now North Carolina, in 1585. In the previous year Queen Elizabeth I had granted Sir Walter Raleigh, the English soldier, explorer, courtier and author, a charter to explore and colonize the eastern coast of North America. In 1587 he sent a group of 121 colonists headed by governor John White to establish a settlement in Virginia in accordance with the charter. When this group arrived on July 2, 1587, the members of the first settlement had completely disappeared.

White's group established itself on Roanoke Island. Approximately a month later a child, Virginia, was born to Ananias Dare, a colonist, and his wife Elinor, daughter of White. She was the first child born of English parents in America.

Nine days later White sailed for England. War was raging between England and Spain, and for four years communications between England and Roanoke Island were cut off.

In 1591 a ship from England arrived at the site of the settlement but all of its former inhabitants had vanished without trace. Nothing definite concerning their ultimate fate was ever learned and the only clues left behind were the word "Croatoan" carved on a doorpost and the word "Cro" carved on the trunk of a tree.

It is supposed that the colonists took refuge with a tribe of friendly Indians on the island and eventually became absorbed into the tribe. In the latter part of the nineteenth century a group of mixed-blood Indians living in Robeson County, North Carolina, claimed to have descended from the vanished colonists. This claim cannot be substantiated, but the state government has officially recognized these people as "Croatan Indians."

Suggested Readings

Andrews, Charles McLean. Our Earliest Colonial Settlements. Ithaca: Cornell University Press, 1964.

Bothwell, Jean. The Lost Colony: The Mystery of Roanoke Island. New York: Winston, 1953.

Commager, Henry Steele, editor. "Charter to Sir Walter Raleigh," (Doc. No. 5) in his Documents of American History, 8th edition. New York: Appleton, 1968.

Craven, Wesley F. The Southern Colonies in the Seventeenth Century, 1607-1689. Baton Rouge, La.: Louisiana State University Press, 1949.

Green, Paul. The Lost Colony (symphonic drama in two acts). First produced at Roanoke Island Waterside Theater, Roanoke Island, N.C., July 4, 1937. Chapel Hill, N.C.: University of North Carolina Press, 1954.

Henkle, Henrietta. Walter Raleigh: Man of Two Worlds. New York: Random House, 1964.

Johnston, Mary. Croatan (fiction). Folcroft, Pa.: Folcroft Library Editions, 1923.

Parks, George B. Richard Hakluyt and the English Voyages. Edited by James A. Williamson. New York: Frederick Ungar, 1961.

Quinn, David B. Raleigh and the British Empire. London: Hodder & Stoughton, 1947.

Rowse, A. L. The Elizabethans and America. London: Macmillan, 1959.

Stebbing, William. Sir Walter Raleigh. Philadelphia: Richard West, 1973.

Stevenson, Augusta. Virginia Dare: Mystery Girl. Indianapolis: Bobbs-Merrill, 1957.

Syme, Ronald. Walter Raleigh. New York: Morrow, 1962.

Tarbox, Increase N., editor. Sir Walter Raleigh and His Colony in America. New York: Burt Franklin, 1966.

Thompson, Edward. Sir Walter Ralegh: Last of the Elizabethans.

New Haven: Yale University Press, 1935.
Trease, Geoffrey. Sir Walter Raleigh, Captain and Adventurer. New York: Vanguard Press, 1950.
Tyler, Lyon G. England in America: Fifteen Eighty to Sixteen Fifty Two. New York: Cooper Square, 1968.
Wallace, Willard M. Sir Walter Raleigh. Princeton, N.J.: Princeton University Press, 1959.
Wallechinsky, David, and Irving Wallace. "Enigmatic Lands (The Lost Colony of Roanoke)," in their The People's Almanac. Garden City, N.Y.: Doubleday, 1975.
Williamson, James A. "England and the Opening of the Atlantic," in The Cambridge History of the British Empire. Cambridge: Cambridge University Press, 1929-1930.
Winton, John. Sir Walter Raleigh. New York: Coward-McCann, 1975.

THE FOUNDING OF THE JAMESTOWN COLONY (1607)

The first permanent English colony in America was the one established on Jamestown Island, Virginia. This colony was authorized by King James I of England by the First Charter of Virginia, dated April 10, 1606. The document provided for the incorporation of two companies: the London Company and the Plymouth Company. It was the London Company which established the colony at Jamestown.

The expedition of 120 settlers left London in December, 1606, made its first landfall at Cape Henry, Virginia, on April 29, 1607, and established the Jamestown colony on May 14, 1607. Captain Christopher Newport was the expedition's leader and the group included Captain John Smith, an English adventurer, and the Reverend Robert Hunt, an English divine. Smith, although only a member of the governing council of the settlement, was actually the leader in all but name. He is said to have ruled that "he who does not work does not eat," and to have directed the activities of the group with a firm hand.

Smith led exploring and mapping expeditions in the area, on one of which he was captured by the Chickahominy Indians and condemned to death. According to legend, Chief Powahatan's daughter Matoaka (Pocahontas) saved his life by holding his head in her arms as he was about to be clubbed to death by Indian warriors. From 1608 to 1609 he was president of the colony.

In spite of Smith's strong leadership the colony suffered from famine and disease, and in 1608 a fire destroyed the original buildings. On June 7, 1610, the settlement was abandoned. However, at the mouth of the James River the settlers met the ship of Lord Delaware bringing supplies and about 150 new immigrants. Encouraged, the settlers returned to Jamestown Island.

From then on the colony prospered, although it was burned in 1676 during Bacon's Rebellion and another fire destroyed several public buildings in 1698. Jamestown was the capital of Virginia until 1699 when the seat of government was moved to Middle Plantations, now Williamsburg. At this time the establishment on Jamestown Island was abandoned. Today it is a part of the Colonial National Park, with a 21-acre area on the island separate from the Park and owned by the Association for the Preservation of Virginia Antiquities.

Suggested Readings

Andrews, Charles McLean. The Colonial Period in American History. New Haven, Conn.: Yale University Press, 1964.

_______. Our Earliest Colonial Settlements. Ithaca: Cornell University Press, 1964.

Barbour, Philip L. The Three Worlds of Captain John Smith. Boston: Houghton Mifflin, 1964.

Brown, Alexander. English Politics in Early Virginia History. New York: Russell & Russell, 1968.

_______. First Republic in America. New York: Russell & Russell, 1969.

_______, editor. Genesis of the United States. New York: Russell & Russell, 1959.

Chatterton, E. Keble. Seed of Liberty: The Story of the American Colonies. Indianapolis: Bobbs-Merrill, 1929.

Commager, Henry Steele, editor. "First Charter of Virginia," (Doc. No. 6) in his Documents of American History, 8th edition. New York: Appleton, 1968.

_______. "Ordinance for Virginia," (Doc. No. 9) in his Documents of American History, 8th edition. New York: Appleton, 1968.

_______. "Second Charter of Virginia," (Doc. No. 7) in his Documents of American History, 8th edition. New York: Appleton, 1968.

_______. "The Third Charter of Virginia," (Doc. No. 8) in his Documents of American History, 8th edition. New York: Appleton, 1968.

Craven, Wesley F. The Dissolution of the Virginia Company: The Failure of a Colonial Experiment. Gloucester, Mass.: Peter Smith, 1964.

_______. The Southern Colonies in the Seventeenth Century, 1607-1689. Baton Rouge, La.: Louisiana State University Press, 1949.

Foster, Genevieve. The World of Captain John Smith. New York: Scribner's, 1959.

Gallman, Robert E. Developing the American Colonies, 1607-1783. Chicago: Scott, Foresman, 1964.

Lawson, Marie. Pocahontas and Captain John Smith: The Story of the Virginia Colony. New York: Random House, 1950.

McCary, Ben. Indians in Seventeenth Century Virginia. Williamsburg, Va.: 350th Anniversary Celebration Corp., 1957.

Miers, Earl Schenck. Blood of Freedom: The Story of Jamestown, Williamsburg and Yorktown. New York: Holt, 1958.

Morton, Richard L. Colonial Virginia: The Tidewater Periods.

Chapel Hill, N.C.: University of North Carolina Press, 1960.
Notestein, Wallace. The English People on the Eve of Colonization. New York: Harper, 1954.
Osgood, Herbert L. American Colonies in the Seventeenth Century. Gloucester, Mass.: Peter Smith, 1970.
Rouse, Parke, Jr. Virginia: A Pictorial History. New York: Scribner's, 1975.
Seymour, Flora W. Pocahontas: Brave Girl. Indianapolis: Bobbs-Merrill, 1961.
Smith, Bradford. Captain John Smith, His Life and Legend. Philadelphia: Lippincott, 1953.
Smith, John. The Generall Historie of Virginia, New-England, and the Summer Isles. Ann Arbor, Mich.: Xerox University Microfilms, 1966.
Stith, William. History of the First Discovery and Settlement of Virginia. New York: Johnson Reprint, 1969.
Thane, Elswyth. The Virginia Colony. New York: Crowell-Collier, 1969.
Tyler, Lyon G. England in America: Fifteen Eighty to Sixteen Fifty Two. New York: Cooper Square, 1968.
Wertenbaker, Thomas J. Virginia Under the Stuarts. Princeton, N.J.: Princeton University Press, 1914.

THE "MAYFLOWER" COMPACT (1620)

In 1607 William Brewster of Scrooby, England, was discharged from his position of village postmaster because of his religious beliefs and activities. He and his followers, known as Separatists, went to Leyden, Holland, where they lived, worshiping as they saw fit.

After twelve years in Holland they realized that, partly through intermarriage, they were losing their identity. They decided to travel to America and start a new colony in the unsettled part of northern Virginia.

The Separatists, or Pilgrims, as they came to be called, boarded the vessel "Mayflower," Captain Christopher Jones commanding, at Southampton on September 6, 1620. Included in the company of 103 men, women and children were some 70 non-Separatists, two of whom were John Alden and Captain Myles Standish.

Because of stormy weather and navigational errors the "Mayflower" reached, on November 11, not the mouth of the Hudson River as Captain Jones intended, but Cape Cod, near the future site of Provincetown, Massachusetts.

Prior to going ashore at Plymouth, where the Pilgrims decided to establish their new homes, on November 11, 1620, they drew up a document known as the "Mayflower Compact." In 1619 they had received a patent for a private plantation from the Virginia Company. According to William Bradford, some of the non-Separatists were "an undesirable lot," who stated that they were not under the jurisdiction of the Virginia Company and "would use their own libertie."

Because of this it was necessary to establish some form of government. To accomplish this the above-mentioned "Mayflower Compact" was executed. This, which was not intended as a constitution but as an extension of the customary church covenant to civil circumstances, was signed by the 41 adult males aboard. In the Compact the signers bound themselves to create a civil body politic, to make laws, and to elect officers to administer the government of the new colony. As the Plymouth settlers were never able to secure a charter, the "Mayflower Compact" remained the only form of constitution of the colony.

The Compact came to light in 1802 when John Quincy Adams described it as "the first example in modern times of a social compact or system of government instituted by voluntary agreement conformable to the laws of nature, by men of equal rights and about to establish their community in a new country."

Suggested Readings

Adams, James Truslow. The Founding of New England. Gloucester, Mass.: Peter Smith, 1921.

Anderson, Anita Melva. Squanto and the Pilgrims. Edited by Emmett A. Betts. Chicago: Wheeler, 1949.

Andrews, Charles McLean. The Colonial Period in American History. New Haven, Conn.: Yale University Press, 1964.

Arber, Edward. The History of the Pilgrim Fathers, 1606-1623. New York: Kraus Reprint, 1897.

Bradford, William. Bradford's History of the Plymouth Plantation. Edited by William T. Davies. New York: Scribner's, 1908. (Also reprinted in MacDonald, H. Malcolm, Wilfred D. Webb and others. Outside Readings in American Government. New York: Crowell, 1949.)

Commager, Henry Steele, editor. "The Mayflower Compact," (Doc. No. 11) in his Documents of American History, 8th edition. New York: Appleton, 1968.

Dillon, Francis. The Pilgrims. Garden City, N.Y.: Doubleday, 1975.

Doyle, John A. The English Colonies in America. New York: AMS Press, 1969.

Eggleston, Edward. The Beginnings of a Nation. Folcroft, Pa.: Folcroft Library Editions, 1897.

Goodwin, J. A. The Pilgrim Republic. Millwood, N.Y.: Kraus Reprint, 1888.

Langdon, George D., Jr. The Pilgrim Colony: A History of New Plymouth. New Haven, Conn.: Yale University Press, 1966.

McLaughlin, Andrew C. Foundations of American Constitutionalism. Gloucester, Mass.: Peter Smith, 1933.

Morison, Samuel Eliot. "The Pilgrim Fathers: Their Significance in History," in his By Land and By Sea. New York: Knopf, 1953.

Osgood, Herbert L. American Colonies in the Seventeenth Century.

Gloucester, Mass.: Peter Smith, 1970.
Palfrey, John G. History of New England. New York: AMS Press, 1858-1890.
Parrington, Vernon. The Colonial Mind. New York: Harcourt, Brace, 1955.
Rutman, Darrett B. Husbandmen of Plymouth. Boston: Beacon Press, 1967.
Smith, Bradford. Bradford of Plymouth. Philadelphia: Lippincott, 1951.
Tyler, Lyon G. England in America: Fifteen Eighty to Sixteen Fifty Two. New York: Cooper Square, 1968.
Webb, Robert N. We Were There with the Mayflower Pilgrims. New York: Grosset and Dunlap, 1956.
Willison, George F. Saints and Strangers. New York: Reynal and Hitchcock, 1945.
Young, Alexander. Chronicles of the Pilgrim Fathers and the Colony of Plymouth, 1602-1625. New York: Da Capo Press, 1971.
Ziner, Feenie. The Pilgrims and Plymouth Colony. New York: American Heritage Press, 1961.

THE TRIAL OF ANNE HUTCHINSON (1637)

Born in England in 1591, Anne Hutchinson had been greatly influenced by the preaching of John Cotton, a prominent Puritan clergyman. When Cotton was denounced to the British Court of High Commission for his heterodox theological ideas in 1632 he fled to Boston. In 1634 Anne Hutchinson and her family followed him to the colony on Massachusetts Bay where she held religious meetings in her home. At these meetings she preached "a doctrine of salvation through intuitive apprehension of grace rather than by works." She criticized the sermons of the Boston ministers, with the exception of Cotton, thereby making powerful enemies. The fact that she attracted such supporters as Sir Harry Vane, governor of Massachusetts, and some clergymen including her brother-in-law the Reverend John Wheelright, increased the animosity of those she criticized.

In March, 1637, Wheelright was brought to trial for sedition and contempt. In spite of Sir Harry Vane's support he was convicted, but his sentence was postponed. In May of that year John Winthrop was elected governor vice Vane who shortly thereafter departed for England.

Anne Hutchinson was then arrested and tried at what is now Cambridge, Massachusetts. As was customary in the seventeenth century, the accused conducted her own defense. Winthrop presided, with Thomas Dudley and John Endecott also sitting. Anne Hutchinson defended herself ably but before a court of enemies had no chance of acquittal whatsoever. Winthrop delivered the sentence of the court: banishment "as being a woman not fit for our society." The entire trial took just two days.

Separated from her family, Anne Hutchinson was committed to the home of Joseph Weld, a church elder, who attempted in vain to make her change her religious views. In March, 1638, she was given an ecclesiastical trial. She signed a retraction of her views and, shortly after, recanted. She was then excommunicated. Banished from Boston, she made her new home in Aquidneck, now Rhode Island. Her husband passed away in 1642 and she moved to Pelham Bay, then in Dutch New Netherland. In 1643 she and all but one member of her family were killed in an Indian massacre.

Suggested Readings

Adams, Brooks. The Emancipation of Massachusetts: The Dream and the Reality. Clifton, N.J.: Augustus M. Kelley, 1919.

Adams, Charles Francis. Antinomianism in the Colony of Massachusetts Bay. New York: Da Capo Press, 1894.

_______. Three Episodes of Massachusetts History. New York: Russell & Russell, 1965.

Battis, Emery. Saints and Sectaries: Anne Hutchinson and the Antinomian Controversy in the Massachusetts Bay Colony. Chapel Hill, N.C.: University of North Carolina Press, 1962.

Ellis, George E. The Puritan Age and Rule in the Colony of Massachusetts Bay, 1629-1685. New York: Burt Franklin, 1970.

Hosmer, James K. The Life of Thomas Hutchinson. New York: Da Capo Press, 1970.

_______, editor. Winthrop's Journal: History of New England, 1630 to 1649. New York: Barnes & Noble, 1908.

Howe, Daniel W. The Puritan Republic. Indianapolis: Bobbs Merrill, 1899.

Miller, Perry. The New England Mind: From Colony to Province. Cambridge, Mass.: Harvard University Press, 1953.

_______. Orthodoxy in Massachusetts, 1630-1650. Gloucester, Mass.: Peter Smith, 1970.

Morgan, Edmund S. "The Case Against Anne Hutchinson," New England Quarterly, December, 1937.

_______. Puritan Dilemma: The Story of John Winthrop. Edited by Oscar Handlin. Boston: Little, Brown, 1958.

Morris, Richard B. "Jezebel Before the Judges," in his Fair Trial. New York: Knopf, 1952.

Rugg, Winifred K. Unafraid: The Life of Anne Hutchinson. Plainview, N.Y.: Books for Libraries, 1930.

Torpey, William George. Judicial Doctrines of Religious Rights in America. Chapel Hill, N.C.: University of North Carolina Press, 1948.

BACON'S REBELLION (1676)

The 1676 rebellion led by Nathaniel Bacon against the colonial authorities headed by Sir William Berkeley, Governor of Virginia, is considered by many historians as the forerunner of the American Revolutionary War. This rebellion was an early manifestation of the American colonists' protests of Berkeley's arbitrary self-serving policies and such English legislation as the Navigation Acts of 1651 and 1660. These Acts required American colonists to trade only with English firms and individuals at prices established in England. The export duties levied by government officials were considered exorbitantly high.

Governor Berkeley was not only high-handed in his methods of dealing with the colonists but he also enjoyed a lucrative personal monopoly of the fur trade with the Indians. In 1675 the Indians made a series of attacks on the frontier plantations of Virginia. Many colonists were captured, tortured, and murdered but Berkeley, afraid of jeopardizing his trade in furs, made only a token attempt to fight the redskins and refused to permit the colonists to march against them. He authorized the building of a few frontier forts, but these proved inadequate against Indian attacks, although the governor taxed the colonists two million pounds of tobacco in order to raise funds to build them.

Early in 1676, after the poorly constructed forts had proved worthless against Indian forays, Nathaniel Bacon, a colonial plantation owner, was chosen to lead a 300-man army organized to attack the Indians despite Berkeley's orders. Bacon requested Berkeley to grant him a commission. Upon Berkeley's refusal, Bacon led his army against the Indians on his own authority, defeating them in battle. He was declared a rebel by the governor, which aroused the colonists to the point that Berkeley was forced to dissolve the Virginia Assembly of 1662. This body had supported him strongly and for that reason was kept in permanent session.

Bacon was arrested but released on parole and promised a commission, which promise was kept only after he and his followers occupied Jamestown, the capital of the colony. Bacon, commissioned a major general, defeated the Indians once more, this time at the battle of Bloody Run.

Berkeley set about organizing a force with which to fight Bacon and his army, whereupon the latter captured Jamestown a second time, burning it on September 19, 1676. When, in the following month, Bacon died of malaria, his rebellion, without adequate leadership, collapsed. Berkeley promptly took a bloody revenge on Bacon's followers, and Charles II of England, on being informed of it, is said to have remarked, "the old fool has put to death more people in that naked country than I did here for the murder of my father."

Berkeley's policy in the colonies was condemned by a royal commission and he was forced to resign the governorship. He returned to England in 1676 and died the following year.

Suggested Readings

Andrews, Charles McLean. Narratives of the Insurrections, 1675 to 1690. New York: Scribner's, 1915.
Ashley, W. J. "The Commercial Legislation of England and the American Colonies, 1660-1760," Quarterly Journal of Economics. Vol. XIV.
Beverley, Robert. The History and Present State of Virginia. Edited by Louis B. Wright. Chapel Hill, N.C.: University of North Carolina Press, 1947.
The Cambridge History of the British Empire. Cambridge: Cambridge University Press, 1929-1930.
Craven, Wesley F. The Southern Colonies in the Seventeenth Century, 1607-1689. Baton Rouge, La.: Louisiana State University Press, 1949.
Gallman, Robert E. Developing the American Colonies, 1607-1783. Chicago: Scott, Foresman, 1964.
Harper, Lawrence A. The English Navigation Laws. New York: Octagon Books, 1964.
Heaps, Willard Allison. Riots, U.S.A., 1765-1970. New York: Seabury Press, 1970.
Morton, Richard L. Colonial Virginia: The Tidewater Periods. Chapel Hill, N.C.: University of North Carolina Press, 1960.
Washburn, Wilcomb E. The Governor and the Rebel. Chapel Hill, N.C.: University of North Carolina Press, 1957.
Wertenbaker, Thomas J. Torchbearer of the Revolution: The Story of Bacon's Rebellion and Its Leader. Princeton, N.J.: Princeton University Press, 1940.
________. Virginia Under the Stuarts. Princeton, N.J.: Princeton University Press, 1914.

THE CHARTER OAK (1687)

In 1674 Sir Edmund Andros was sent from England by King James II to the then province of New York, in the capacity of governor. By the terms of his commission he had jurisdiction over Long Island, Pemaquid and the region between the Connecticut and Delaware Rivers. In 1686 all of New England was combined into a single province and Andros was designated governor general of the expanded territories.

The colony of Connecticut refused to recognize Andros' authority as governor and in October, 1687, he appeared in the council chamber at Hartford with an armed guard. Here he demanded that the colony's charter be surrendered to him.

The colonists were reluctant to comply with Andros' order but pretended to submit. The charter was carried into the council chamber and, during the ensuing debate, the lights were suddenly extinguished. In the confusion which followed the charter was carried from the room and hidden in a gigantic oak tree which came to be known as the "Charter Oak."

In 1688 James II was deposed by the "Glorious Revolution" and replaced on the English throne by William III. The citizens of Boston imprisoned Andros and some of his officers on April 18, 1689. Jacob Leisler, an American official, set up a rebel government in New York, and in July Andros was sent to England for trial. However, no charges were formally pressed against him.

With the removal of Andros from American soil, concealment of the charter was no longer necessary. The document was retrieved from its hiding place in 1689.

The Charter Oak stood in Hartford until 1856, in which year it was blown down. It was estimated to be about one thousand years old.

Suggested Readings

Andrews, Charles McLean, ed. Narratives of the Insurrections, 1675-1690. New York: Scribner's, 1915.

Andros Tracts: Being a Collection of Pamphlets and Official Papers Issued During the Period Between the Overthrow of the Andros Government and the Establishment of the Second Charter of Massachusetts. Boston: Prince Society, 1868.

Bates, Albert Carlos. The Expedition of Sir Edmund Andros to Connecticut in 1687. Worcester, Mass.: Antiquarian Society, 1939.

Brodhead, John Romeyn. Government of Sir Edmund Andros Over New England in 1688 and 1689. Morrisana, N.Y.: Bradstreet Press, 1867.

Choate, Washington. Oration by Rev. Washington Choate and the Poem by Rev. Edgar F. Davis on the 200th Anniversary of the Resistance to the Andros Tax at Ipswich, July 4, 1887. Salem, Mass.: Salem Observer Print, 1894.

Gallman, Robert E. Developing the American Colonies, 1607-1783. Chicago: Scott, Foresman, 1964.

Gocher, W. H. Wadsworth: or, The Charter Oak. Hartford, Conn.: Gocher, 1904.

Johnston, Edward F. Charter Oak (musical play). Book by Edith M. Burrows. New York: Carl Fischer, 1916.

Van Dusen, Albert. Connecticut. New York: Random House, 1961.

Whitmore, William Henry. A Memoir of Sir Edmund Andros. Boston: T. R. Marvin & Son, 1868.

THE SALEM WITCHCRAFT TRIALS (1692)

The word "witch" is defined by Webster as "one who practices the black art, or magic; one regarded as possessing supernatural or magical power by compact with an evil spirit, especially with the Devil." "Witchcraft," in turn, is defined by the same authority as "the practices or art of witches; the practice of black magic; sorcery; intercourse with evil spirits."

The Puritans of Salem Township, in what is now Massachusetts, were, for the most part, ardent believers in witchcraft, as were most people of the seventeenth century. Deeply religious, they feared the Devil, and even such enlightened men as the Reverend Cotton Mather, a Harvard graduate and prominent Boston minister, believed in and preached against him.

This background of superstition, religion and fear furnished the ideal setting for the witchcraft trials and executions which occurred in Salem, Massachusetts, in 1692. Several young girls, ranging in age from nine to twenty years, were in the habit of visiting one Tituba, a West Indian servant of Samuel Parris. Bored and frustrated, these girls learned and practiced the occult arts taught them by the West Indian: fortune-telling, spiritualism, palmistry and the like.

At first this amounted to little more than a pastime, but as the girls became more proficient in these arts they began to show off before other members of the community. They would writhe, make strange exclamations and wild gestures, apparently faint, and utter loud screams. Eventually they were examined by the local physician who pronounced them "bewitched."

The girls so "afflicted" were Elizabeth Parris, Sarah Churchill, Abigail Williams, Ann Putnam, Susannah Sheldon, Elizabeth Booth, Elizabeth Hubbard, Mercy Lewis, Mary Walcot and Mary Warren. Their "affliction" soon became the talk of the community, which pitied them. The local minister stated that "the Devil was working his wrath on these innocent victims." Urged to name the witches or wizards who had bewitched them, the girls accused Tituba, Sarah Good and Sarah Osburn.

Complaints against these three women were sworn out and on March 1, 1692, magistrates John Hathorne and Jonathan Corwin arrived to conduct preliminary examinations of the accused. The examinations were based on a primary assumption of guilt. The two Sarahs denied being witches; Tituba confessed to witchcraft and implicated the other two. The three women were committed to jail. During the examinations the "bewitched" girls went into their "torments and tantrums" whenever one of the accused protested her innocence. These manifestations ceased when Tituba admitted her "guilt."

Charges were brought against other residents of Salem. Martha Corey maintained her innocence and was ultimately convicted and hanged. Her husband Giles, charged and refusing to plead either "guilty" or "innocent," was executed by being pressed to death.

Still other victims were interrogated and bound over for trial. These included Rebecca Nurse, Sarah Cloyse, Elizabeth Proctor, John Proctor, Abigail Hobbs, George Jacobs, Bridget Bishop, George Burroughs, Ann Hibbens, and Mary Warren. The latter, formerly one of the afflicted, was later charged as an afflicter.

Finally, early in June, 1692, the actual trials began, the court proceeding under an act of James I, passed in 1603 and prescribing the death penalty for those convicted of witchcraft. Bridget Bishop was the first to be tried. She was found guilty and hanged. Before the court reconvened it revived an old colony law making witchcraft a capital offense and the other defendants were tried under this law.

In subsequent trials those who were convicted and executed including Sarah Good, Sarah Wildes, Elizabeth How, Rebecca Nurse, Susanna Martin, George Burroughs, John Proctor, George Jacobs and ten others. In addition to those hanged, over 250 persons were arrested on charges of witchcraft.

The final executions took place on September 22, 1692. By this time men of prominence in the community were beginning to regard evidence based on spectral manifestations with skepticism. Increase Mather, father of Cotton Mather, was not so soundly convinced of the validity of spectral evidence as was his son, and he wrote to this effect. When the wife of Sir William Phips, royal governor of Massachusetts, expressed sympathy for the unfortunate victims of the witchcraft mania and was "cried out upon by the 'afflicted' girls," Phips dismissed the court of oyer and terminer. He also requested and obtained petitions of release for the 150 accused witches still in prison.

On January 3, 1693, 52 "witches" were tried. Spectral evidence was no longer admissible in such cases and without it 49 of the cases resulted in acquittal. Governor Phips issued reprieves for the three who were convicted, and in May discharged all remaining "witches." Since then no one in the American colonies has ever been condemned to death for witchcraft.

Suggested Readings

Alderman, Clifford Lindsey. Cauldron of Witches: The Story of Witchcraft. New York: Messner, 1971.

Aymar, Brandt, and Edward Sagarin. "The Salem Witchcraft Trials," in their A Pictorial History of the World's Great Trials. New York: Crown Publishers, 1967.

Beard, George M. The Psychology of the Salem Witchcraft Excitement of 1692. New York: Putnam, 1882.

Burr, George L. Narratives of the Witchcraft Cases, 1648 to 1706. New York: Barnes & Noble, 1952.

Hansen, Chadwick. Witchcraft at Salem. New York: George Braziller, 1969.

Kimball, Henrietta D. Witchcraft Illustrated. Boston: George A. Kimball, 1892.

Kittredge, George Lyman. Witchcraft in Old and New England. Cambridge, Mass.: Harvard University Press, 1929.

Levin, David. What Happened in Salem? New York: Harcourt, Brace, 1960.

Longfellow, Henry Wadsworth. Giles Corey of the Salem Farms (play). New York: Scribner's, 1867.

Mather, Cotton. Wonders of the Invisible World. London: John Russell Smith, 1862.

Miller, Arthur. The Crucible (play). New York: Viking Press, 1952.

Miller, Perry. The New England Mind: From Colony to Province.

Cambridge, Mass.: Harvard University Press, 1953.
Perley, M. V. B. A Short History of the Salem Witchcraft Trials. M. V. B. Perley, 1911.
Records of Salem Witchcraft. Copied from the original documents and privately printed for W. Elliott Woodward. Roxbury, Mass.: 1864.
Robbins, Russell Hope. The Encyclopedia of Witchcraft and Demonology. New York: Crown Publishers, 1959.
Starkey, Marion L. The Devil in Massachusetts. New York: Knopf, 1949.
Upham, Charles W. Salem Witchcraft. Boston: Wiggin and Lunt, 1867.
Wendell, Barrett. Cotton Mather: The Puritan Priest. New York: Harcourt, Brace, 1963.
Wilkins, Mary E. Giles Corey, Yeoman (play). New York: Scribner's, 1893.

CAPTAIN KIDD'S TREASURE (1700)

William Kidd, better remembered as "Captain" Kidd, was a Scottish shipowner turned pirate and privateer. In 1690, having spent several years at sea, he was established in New York City as a shipowner. From that year until 1695 he served as captain of one of his own ships in the British Colonial Service against French privateers in the West Indies. The Council of New York Colony voted him a reward of 150 pounds for his loyal service in connection with the insurrection instigated by Jacob Leisler.

In 1695, while in London on business, Kidd received a commission as privateer from William III, with orders to act against pirates operating in the Indian Ocean. He sailed from England in April, 1696, but in the year that followed captured no prizes.

The crew mutinied at Madagascar and Kidd struck and fatally injured one of the sailors. He and his men then turned to piracy and in the next two years attacked and looted a number of legitimate merchant ships. In January, 1698, he captured the rich American vessel "Quedagh Merchant" which he took for his own ship and with which he captured other prizes.

Arriving in the West Indies in 1699, Kidd learned that he had been proclaimed a pirate and was wanted by the authorities. He immediately set sail for New England, landing at Oyster Bay, Long Island. There he wrote Lord Bellomont, governor of New England, stating that he could justify his piracy and sending the governor some plunder. On Bellomont's promise of a pardon he proceeded to Boston where he and several members of his crew were promptly arrested. They were sent to London where they were imprisoned at Newgate.

Kidd was tried and found guilty on three piracy indictments. He was also found guilty of fatally wounding one of his sailors during the mutiny at Madagascar. On May 23, 1701, he was hanged.

A small amount of booty was taken from Kidd's ship and other booty was recovered from Gardener's Island, off the east coast of Long Island. With the passing of the years the legend spread that Captain Kidd had buried the bulk of his treasure somewhere on the Atlantic coast near New York City. Edgar Allan Poe's famous story, "The Gold Bug," was inspired by the legend.

Suggested Readings

Bonner, W. H. Pirate Laureate. New Brunswick, N.J.: Rutgers University Press, 1947.

Boyd, Mildred. Black Flags and Pieces of Eight. New York: Criterion, 1965.

Burney, James. History of the Buccaneers of America. London: G. Allen & Company, 1912.

Esquemeling, Alexandre Olivier. The Buccaneers of America. London: S. Sonnenschein, 1893.

Gerhard, Peter. Pirates on the West Coast of New Spain, 1575-1742. Glendale, Calif.: A. H. Clark, 1960.

Grosse, Philip. The Pirate's Who's Who. London: Dulany, 1924.

Hughson, Shirley Carter. The Carolina Pirates and Colonial Commerce. Baltimore: Johns Hopkins University Press, 1894.

Jameson, John Franklin. Privateering and Piracy in the Colonial Period. New York: Macmillan, 1923.

Johnson, Captain Charles. A General History of the Pirates. London: Routledge and Kegan Paul, 1926.

Karraker, Cyrus H. Piracy Was a Business. Durham, N.H.: R. R. Smith, 1953.

Lonsdale, Adrian L., and H. R. Kaplan. A Guide to Sunken Ships in American Waters. Arlington, Va.: Compass Publications, 1964.

Morris, Richard B. "The Politicians and the Pirate," in his Fair Trial. New York: Knopf, 1952.

Nesmith, Robert I. Dig for Pirate Treasure. New York: Devin-Adair, 1958.

Paine, Ralph D. The Book of Buried Treasure. New York: Sturgis and Walton, 1911.

Poe, Edgar Allan. "The Gold Bug" (fiction), in Complete Stories and Poems of Edgar Allan Poe. Garden City, N.Y.: Doubleday, 1966.

Pond, Seymour Gates. True Adventures of Pirates. Boston: Little, Brown, 1954.

Pringle, Patrick. Jolly Roger. New York: Norton, 1953.

Pyle, Howard. Book of Pirates. New York: Harper, 1921.

Reisberg, Harry E. Adventures in Underwater Treasure Hunting. New York: F. Fell, 1965.

________. Treasure! New York: Holt, 1957.

Snow, Edward Rowe. Pirates and Buccaneers of the Atlantic Coast. Boston: Yankee, 1944.

________. True Tales of Buried Treasure. New York: Dodd, Mead, 1951.

Stockton, Frank Richard. Buccaneers and Pirates of Our Coasts.

New York: Grosset and Dunlap, 1926.
Verrill, A. Hyatt. The Real Story of the Pirate. New York: Appleton, 1923.
Whipple, Addison Beecher Colvin. Famous Pirates of the New World. New York: Random House, 1958.
_____. Pirate. New York: Doubleday, 1957.
Wilkins, Harold T. Hunting Hidden Treasure. New York: Dutton, 1929.
Williams, Lloyd H. Pirates of Colonial Virginia. Richmond, Va.: Dietz Press, 1937.

THE ZENGER LIBEL TRIAL (1735)

The trial and acquittal of John Peter Zenger in the year 1735 is regarded as fundamental in establishing the freedom of the press in America. The essential point made in this famous libel trial is that a newspaper, magazine, or other publication may make damaging statements about a person with impunity without subjecting the writers or editors of such publications to prosecution, providing the statements made are, in fact, true.

This was not the case in colonial America in 1734. At that time Sir William S. Cosby was governor of the British Crown Colony of New York and John Peter Zenger, a German-born printer, was a reporter on the staff of the Weekly Gazette in that city. Cosby, a man of unsavory character and few, if any, scruples, had been governor of the island of Minorca in the Mediterranean. There he had illegally confiscated property and revenues for his own personal gain. In New York he embarked on a career of political corruption and diverting public funds to himself in a manner which William Marcy Tweed emulated in the same city more than a century later.

The situation came to a head when Rip Van Dam, Cosby's immediate predecessor as governor, refused to pay a portion of his wages to Cosby and Cosby took him to court, where Lewis Morris, chief justice of the New York Supreme Court, ruled against him. Infuriated, Cosby removed Morris from his post. Morris then ran for another office and was elected, despite Cosby's attempted trickery at the polls.

Zenger, in his Weekly Gazette, faithfully and truthfully attempted to report all details of Cosby's tactics but editor William Bradford refused to publish the report and Zenger was discharged.

Morris, Van Dam, James Alexander and William Smith then financed Zenger as an independent editor and publisher, and on November 1, 1733, Zenger brought out the first issue of the New York Weekly Journal. From the start Zenger's newspaper was definitely anti-Cosby, attacking and exposing him and his illegal, self-serving party. Four issues of the Journal were publicly burned at Cosby's order, and on November 17, 1734, Zenger, charged with seditious libel, was arrested. He was held in prison on excessive bail, but while he was incarcerated his wife, Anna, continued the publication of his newspaper.

Zenger's trial began on August 4, 1735. His attorneys, James Alexander and William Smith, had been disbarred on Cosby's orders, but friends enlisted the services of Andrew Hamilton, the outstanding member of the Philadelphia bar. The twelve-man jury hearing the case had nothing but contempt for Cosby and his overbearing ways, and Hamilton was far too prominent a man for the court to consider disbarment proceedings against him.

Zenger was charged with "printing and publishing" (but not writing) in the Journal statements which were "false, scandalous, malicious and seditious." James De Lancey was the presiding judge and an old friend of Cosby. The associate justice, Frederick Philipse, assisted him. Alexander and Smith, the disbarred lawyers, were in court to assist Hamilton, in addition to John Chambers. The latter had been originally designated by the court as defense counsel for Zenger but was overshadowed by Hamilton.

The trial proceeded, Zenger pleading "not guilty." Following the charges, as read by the attorney general, Hamilton stated to the court that Zenger did indeed publish the statements as alleged, and that, as the statements made were true, no libel existed.

The prosecution attempted to end the trial at this point, but Hamilton would not have it so, insisting that a truthful statement could not be a libelous one and stating further that he and his client stood ready to prove the truth of the statements involved. The court refused to permit Hamilton to prove the truth in order to justify a libel and ordered him to accept its ruling. Hamilton then turned to the jury, and in a dramatic speech appealed the matter to the "twelve good men and true."

Following the prosecution's closing arguments the jury retired to consider the evidence and render its verdict. After a ten-minute deliberation the findings of "not guilty" were returned.

Zenger, though never called as a witness, had, by his trial and acquittal, established that principle of freedom of the press which was to be incorporated in the Constitution of the United States some years later.

Suggested Readings

Alexander, James. A Brief Narrative of the Case and Trial of John Peter Zenger, Printer of the "New York Weekly Journal." Edited by Sidney N. Katz. Cambridge, Mass.: Harvard University Press, 1963.

Aymar, Brandt, and Edward Sagarin. "John Peter Zenger," in their A Pictorial History of the World's Great Trials. New York: Crown Publishers, 1967.

Brown, James Wright. "Life and Times of John Peter Zenger: A Statement of Facts Chronologically Arranged--As Gathered From Rutherford, Konkle, Cheslaw, Sheehan, Cooper, and Robb," Editor and Publisher, March 14, 21, 28: April 4 and 11, 1953.

Buranelli, Vincent, editor. The Trial of Peter Zenger. New York: New York University Press, 1957.

Emery, Edwin. The Press and America: An Interpretive History

of Journalism. Englewood Cliffs, N.J.: Prentice-Hall, 1962.
Galt, Tom. Peter Zenger: Fighter for Freedom. New York: Crowell, 1951.
Levy, Leonard W. Freedom of Speech and Press in Early American History: Legacy of Suppression. New York: Harper, 1963.
______. Freedom of the Press from Zenger to Jefferson. Indianapolis: Bobbs-Merrill, 1966.
Morris, Richard B. "The Case of the Palatine Printer," in his Fair Trial. New York: Knopf, 1952.
Mott, Frank L. American Journalism: A History, 1690-1960. New York: Macmillan, 1962.
Rutherford, Livingston, editor. John Peter Zenger: His Press, His Trial and a Bibliography of Zenger Imprints; also a Reprint of the First Edition of the Trial. New York: Dodd, Mead, 1904.
Schuyler, L. R. Liberty of the Press in the American Colonies Before the Revolutionary War. New York: Dodd, Mead, 1905.
The Story of John Peter Zenger. John Peter Zenger Memorial, Federal Hall Memorial. New York: 1953.

PONTIAC'S REBELLION (1763)

One of the outstanding American Indian warriors was Pontiac, Amerind chief of the Ottawas and leader of the Confederate tribes of the Ohio Valley and Lake Region. It was he who led his braves against the English in 1763 in what came to be called "Pontiac's Rebellion."

This chief, together with many fellow tribesmen, felt that they were being unjustly deprived of their ancient hunting lands by white settlers, and desired to reestablish Indian autonomy. Desiring to drive the English from their frontier holdings, Pontiac organized a confederacy embracing virtually all Indian tribes from the head of Lake Superior to the Gulf Coast. It was arranged that warriors from each tribe were, early in May, 1763, to attack simultaneously the English garrison in their immediate neighborhoods. Pontiac elected to lead the assault on the garrison at Detroit.

At this time there were fourteen English posts in the wilderness between the Pennsylvania frontier and Lake Superior, of which the three most important were located at Detroit, Fort Pitt and Mackinaw. The Indians succeeded in capturing all posts except those at Detroit, Ligonier, Niagara and Fort Pitt. Mackinaw fell to the Indian onslaught and the entire garrison was massacred. The attack at Detroit failed when an Indian girl alerted Major Gladwin, commander of the post, thus enabling him to prepare for the coming attack.

Pontiac and his braves then began a siege which lasted for five months. Eventually British reinforcements succeeded in entering Detroit and Pontiac's warriors began to desert. He hoped to receive help from the French but this hope was dispelled when it was learned that England and France had signed a peace treaty.

Realizing that his cause was lost, Pontiac raised the siege and on August 17, 1765, signed a formal treaty of peace. This he confirmed at Oswego in 1766.

Pontiac's Rebellion furnished the background for the Proclamation of 1763, made by George III of England, and a remonstrance of frontier grievances signed by Matthew Smith and James Gibson on February 13, 1764. The latter was submitted to John Penn, lieutenant-governor of Pennsylvania.

Suggested Readings

Alvord, Clarence W. The Mississippi River in British Politics. New York: Russell & Russell, 1916.

Carter, Clarence E. Great Britain and the Illinois Country, 1763-1774. Port Washington, N.Y.: Kennikat Press, 1970.

Coffin, Victor. The Province of Quebec and the Early American Revolution. Port Washington, N.Y.: Kennikat Press, 1974.

Commager, Henry Steele, editor. "Frontier Grievances," (Doc. No. 34) in his Documents of American History, 8th edition. New York: Appleton, 1968.

________. "The Proclamation of 1763," (Doc. No. 33) in his Documents of American History, 8th edition. New York: Appleton, 1968.

Parkman, Francis. The Conspiracy of Pontiac. New York: Collier Books, 1962.

Roland, Albert. Great Indian Chiefs. New York: Crowell, 1966.

Shepherd, William R. History of Proprietary Government in Pennsylvania. New York: AMS Press, 1896.

Turner, Frederick Jackson. The Frontier in American History. New York: Holt, Rinehart, 1947, 1962.

Wissler, Clark. The American Indian. Gloucester, Mass.: Peter Smith, 1922.

________. Indians of the United States. New York: Doubleday, 1940.

THE MASON-DIXON LINE (1763-1767)

William Penn was the son of Admiral Sir William Penn, a wealthy British naval officer. He inherited from his father a large financial claim against Charles II of England. As payment of the debt he petitioned the King for a grant of land in the New World, and in 1681 received the grant of an area which is now Pennsylvania. Here he established a Quaker colony.

George Calvert, first Baron Baltimore, was, in 1622, granted Newfoundland by James I of England but objected to it because of the poor climate. In 1632 he was granted the territory comprising what is now Maryland but died before the charter was issued. His son

Cecilius, second Baron Baltimore, received the charter from Charles I of England.

The boundary line between the two colonies had long been in dispute. In 1767 it was resolved, based on computations by Charles Mason, an English astronomer, and Jeremiah Dixon, an English surveyor. These two were employed by the proprietors of Pennsylvania and Maryland to survey the boundaries between the two colonies. The 312-mile line, started in 1763, was completed on October 18, 1767. The border line was marked with approximately 300 waist-high limestone obelisks with the arms of William Penn on one side and those of Baron Baltimore on the other. It stretches almost from the Atlantic Ocean to the Ohio River and constitutes the boundary between what are now, after several wars and compacts, the states of Pennsylvania, Delaware, Maryland and West Virginia.

Part of the boundary was re-surveyed and marked in 1849/50 and again surveyed in 1901/03 by a commission appointed by Pennsylvania and Maryland. Another survey, this by the National Geodetic Survey, a federal agency, is currently being made. This last survey has shown that Mason and Dixon implanted many of their obelisks as much as four feet off center and somehow worked a slight eastward bulge into the part of the boundary that drops down to separate Maryland from Delaware. Maryland got approximately fifteen feet of Delaware at the point of deepest curvature. These errors, comparatively minor, are attributed to the primitive surveying instruments used by the original surveyors.

The Mason-Dixon line, as it came to be called, was considered part of the boundary between the free and the slave states before the Civil War. The term was used during the debates over the Missouri Compromise. It meant, in this sense, not only the old disputed boundary line but also the line of the Ohio River from the Pennsylvania boundary to its mouth, the line of the east, north and west boundaries of Missouri and from that point westward to parallel 36° 30'. This was the line established by the Missouri Compromise except with regard to Missouri.

Suggested Readings

Andrews, Matthew P. History of Maryland: Province and State. New York: Doubleday, 1929.

Clark, Charles B. The Eastern Shore of Maryland and Virginia. New York: Lewis, 1950.

Dunaway, Wayland Fuller. A History of Pennsylvania. New York: Prentice-Hall, 1948.

Federal Writers Project (WPA). Maryland: A Guide to the Old Line State. New York: Oxford University Press, 1946.

______. Pennsylvania: A Guide to the Keystone State. New York: Oxford University Press, 1940.

Hull, William I. William Penn. Plainview, N.Y.: Books for Libraries, 1937.

Peare, Catherine O. William Penn: A Biography. New York: Lippincott, 1957.

Radoff, Morris L. The Old Line State: A History of Maryland.

Baltimore: Historical Records Association, 1956.
Stevens, Sylvester K. Pennsylvania: A Students' Guide to Localized History. New York: Teachers College Press, 1965.
_______. Pennsylvania History in Outline. Philadelphia: Pennsylvania Historical and Museum Commission, 1968.

THE STAMP ACT (1765)

The Stamp Act, also called the "British Stamp Act" and the "Colonial Stamp Act," was passed by the British Parliament without debate in 1765 for the purpose of raising revenue in the American colonies. By the terms of this Act, which became law on March 22, 1765, and went into effect on November 1 of that year, the colonists were required to purchase stamped paper brought from England for legal documents, diplomas, and certificates. They were also required, under the Act, to affix stamps to playing cards, brochures, calendars, pamphlets, almanacs, books, newspapers and certain other articles.

The Stamp Act aroused great opposition on the part of the colonists. When it went into effect and proclamations authorizing it were read publicly, the colonists lowered their flags to half-mast. They argued that they had not been represented in Parliament where the Act was passed, that this consisted of taxation without representation, and that such tax would result in the removal of much needed specie to England.

The Sons of Liberty, a patriotic organization whose membership included such American patriots as Samuel Adams and Paul Revere, opposed the Stamp Act and "led a campaign of physical violence" against the official stamp agents. Many agents were attacked by mobs of colonists and had their property destroyed. Colonial assemblies adopted resolutions of protest against the Act. Patrick Henry, addressing the Virginia House of Burgesses, introduced proposals against it. He concluded his speech with the famous lines, "Caesar had his Brutus; Charles the First his Cromwell; and George the Third--may profit by their example." To cries of "treason!" he replied, "If this be treason, make the most of it!"

Opposition culminated in the convening of the Stamp Act Congress at New York City in 1765, which was attended by delegates from nine of the thirteen colonies. The Congress was called to consider means of protesting against the Act and the taxes it authorized.

In 1766 the British realized that the expenses involved in collecting the tax were greater than the proceeds derived from it, and in that year it was repealed. However, Parliament enacted the Declaratory Act which affirmed the British government's right to pass acts legally binding on the colonists.

Suggested Readings

Andrews, Charles McLean. The Colonial Background of the American Revolution: Four Essays in American Colonial History. New Haven, Conn.: Yale University Press, 1924.
Bancroft, George. History of the United States of America, from the Discovery of the Continent. New York: Appleton, 1888.
Beer, George L. British Colonial Policy, 1754-1765. New York: Macmillan, 1907.
Bridenbaugh, Carl. Mitre and Sceptre: Transatlantic Faiths, Ideas, Personalities, and Politics. New York: Oxford University Press, 1962.
Chinard, Gilbert. Honest John Adams. Boston: Little, Brown, 1933.
Commager, Henry Steele, ed. "The Declaratory Act," (Doc. No. 41) in his Documents of American History, 8th edition. New York: Appleton, 1968.
________. "Instructions of the Town of Braintree, Massachusetts, on the Stamp Act," (Doc. No. 37) in his Documents of American History, 8th edition. New York: Appleton, 1968.
________. "Northampton County Resolutions on the Stamp Act," (Doc. No. 39) in his Documents of American History, 8th edition. New York: Appleton, 1968.
________. "Resolutions of the Stamp Act Congress," (Doc. No. 38) in his Documents of American History, 8th edition. New York: Appleton, 1968.
________. "The Stamp Act," (Doc. No. 35) in his Documents of American History, 8th edition. New York: Appleton, 1968.
________. "Virginia Stamp Act Resolutions," (Doc. No. 36) in his Documents of American History, 8th edition. New York: Appleton, 1968.
Gipson, Lawrence H. The Coming of the Revolution, 1763-1775. New York: Harper, 1954.
Howard, George E. Preliminaries of the Revolution, 1763-1775. New York: AMS Press, 1970.
Knollenberg, Bernhard. Origins of the American Revolution. New York: Macmillan, 1960.
Laprade, W. T. "The Stamp Act in British Politics," American Historical Review, Vol. XXV.
McIlwain, Charles H. The American Revolution: A Constitutional Interpretation. Ithaca, N.Y.: Cornell University Press, 1958.
Malone, Joseph J. Pine Tree and Politics: The Naval Stores and Forest Policy in Colonial New England. Seattle: University of Washington Press, 1964.
Morgan, Edmund S., and Helen M. Morgan. The Stamp Act Crisis. Chapel Hill, N.C.: University of North Carolina Press, 1953.
Tyler, M. C. Patrick Henry. New York: Gordon Press, 1965.
Van Tyne, Claude H. Causes of the War of Independence. Gloucester, Mass.: Peter Smith, 1970.

THE TOWNSHEND ACTS (1767)

The Stamp Act of 1765 was repealed by the British parliament in the following year. Comparative quiet then existed between the American colonies and the mother country until the spring of 1767. It was then that a series of laws known as the Townshend Acts was passed and these Acts precipitated bad feeling and additional trouble.

Parliament had cut the British land tax. Charles "Champagne Charlie" Townshend, chancellor of the British exchequer, indicated that, with the land tax reduced, he would tax the American colonists to make up the difference. The Revenue Act, one of the Townshend Acts, passed in 1767, levied import duties on tea, lead, paper, paints and glass shipped from Great Britain to America. The purpose of the Act was political, not economic. Townshend admitted that approximately 40,000 pounds annually would be realized from the application of the tax, whereas the slashing of the land tax deprived the British government of 400,000 pounds in yearly revenue. The monies collected under the Revenue Act were to be used to pay the salaries of British officials in the American colonies, thus making the officials independent of colonial legislatures and therefore better able to enforce British laws and orders.

The Americans regarded this as an attempt to weaken the authority of their legislatures as they had previously, by controlling the purse strings, been able to keep the British officials in line and responsive to the will of the colonists whom they served.

Another of the British acts which was extremely unpopular with the colonists was the Quartering Act of 1765. This required the proprietors of taverns and other public houses to lodge British troops in their facilities at the expense of provincial authorities when barracks were not available in the colonies. After 1765 American legislatures usually provided for the army's needs but maintained their constitutional integrity by avoiding a precise compliance with the letter of the law.

In New York the legislature was ordered suspended until it would agree to comply with the Act. This interference by Parliament in the affairs of the American colonists created widespread opposition.

Other Parliamentary Acts created anger and alarm in the colonies. Townshend reorganized the customs service in America to guarantee the collection of new taxes as well as achieve compliance with the older Navigation Acts. A special board, sitting in Boston, replaced the one which had controlled customs officers from Great Britain. The customs officers, who received personally a third of all fines collected by the Vice-Admiralty Courts, were extremely zealous in handling their assignments.

The activities of the customs officers were resented by the colonists. The customs officers appealed to Great Britain for protection and General Thomas Gage, British commander-in-chief in North America, was ordered to station troops in Boston where, on March 5, 1770, the so-called Boston Massacre occurred.

Colonial opposition to the Townshend Revenue Act was effective and powerful. British exports to America diminished in both volume and value because of non-importation agreements. British merchants complained and British political leaders, realizing the Act was foolish, repealed it in 1770 after Townshend had died and Lord North had become Prime Minister.

For the following three years the relations between the colonies and Great Britain were comparatively peaceful.

Suggested Readings

Alden, John R. General Gage in America. Baton Rouge, La.: Louisiana State University Press, 1948.

Brooke, John. The Chatham Administration, 1766-1768. New York: St. Martin's Press, 1956.

Butterfield, Herbert George. George III and the Historians. New York: Macmillan, 1957.

Commager, Henry Steele, ed. "The Declaratory Act," (Doc. No. 41) in his Documents of American History, 8th edition. New York: Appleton, 1968.

_______. "Quartering Act," (Doc. No. 42) in his Documents of American History, 8th edition. New York: Appleton, 1968.

_______. "The Townshend Revenue Act," (Doc. No. 43) in his Documents of American History, 8th edition. New York: Appleton, 1968.

Cramer, Kenyon C. The Causes of War. Chicago: Scott, Foresman, 1965.

Gipson, Lawrence H. The British Empire Before the American Revolution. New York: Knopf, 1965.

Jacobson, David L. John Dickinson and the Revolution in Pennsylvania, 1764-1766. Berkeley, Calif.: University of California Press, 1965.

Namier, Lewis B. England in the Age of the American Revolution. London: Macmillan, 1930.

_______. The Structure of Politics at the Accession of George III. London: Macmillan, 1929.

_______, and John Brooke. Charles Townshend. New York: St. Martin's Press, 1964.

Pares, Richard. King George III and the Politicians. Oxford: Oxford University Press, 1953.

Ubbelohde, Carl. The Vice-Admiralty Courts and the American Revolution. Chapel Hill, N.C.: University of North Carolina Press, 1960.

THE REGULATORS' REVOLT (1768-1771)

The inhabitants of the southern American colonies from Virginia to South Carolina and notably in North Carolina were largely non-slave-owning operators of small farms, "dissenting in religion and unrepresented in politics." They felt that they were being taken advantage of by the British government, represented in the year 1765 by William Tryon, governor of North Carolina.

In that year rebellion broke out in Granville and Mecklenburg Counties when a group of colonists protested the actions of the British authorities. In 1766 delegates from Orange County met on Sandy Creek and demanded that office holders representing the Crown should be held to more strict account than was the case. In 1768 a body of citizens called the "Regulators" was formed in Orange County. This organization refused to pay taxes imposed by the royal governor "unless and until the lawfulness of the demand was shown." It also interfered with the courts, the collection of rents, and in other ways.

In 1768 physical violence broke out. Two rioters who had been arrested and imprisoned were freed from jail by a mob of over 700 persons. Appeal for redress was made to Tryon, who showed no disposition to accede to the insurrectionists' demands.

In 1769 the inhabitants of Anson County, North Carolina petitioned the Assembly for relief from the injustices imposed by the governor. In this petition they charged that, among other things, "the poor inhabitants in general are much oppressed by reason of disproportion of taxes ... that no method is prescribed by law for the payment of taxes ... in produce (in lieu of a currency ...) to the people's great oppression ..." and "that lawyers, clerks and other pensioners, in place of being obsequious for the Country's use, are become a nuisance as the business of the people is often transacted without the least degree of fairness, the intention of the law evaded, exorbitant fees extorted, and the sufferers left to mourn under their oppressions."

The petitioners sought relief from what they considered injustices and suggested specific reforms, including the appointment of Benjamin Franklin "or some other known patriot to be appointed agent, to represent the unhappy state of this province to His Majesty, and to solicit the several Boards in England." Tryon showed no intention of modifying his gubernatorial policies and dissolved the Assembly which had come under the control of the Regulators.

In 1770 a new Assembly met and passed a riot act. Hermon Husbands, one of the leaders of the Regulators, was expelled from the Assembly and more violence followed. In the battle of Almance Creek nine Regulators were killed, many were wounded, and two were captured, tried, and executed. Husbands was tried and condemned to death but pardoned. The Regulators, however, were overwhelmingly defeated and their rebellion was at an end.

Suggested Readings

Andrews, Charles McLean. The Colonial Period in American History. New Haven, Conn.: Yale University Press, 1964.

Bassett, John Spencer. "The Regulators of North Carolina," American Historical Association Report, 1894.

Commager, Henry Steele, ed. "The Regulators of North Carolina," (Doc. No. 47) in his Documents of American History, 8th edi-

tion. New York: Appleton, 1968.
Dill, Alonzo T. Governor Tryon and His Palace. Chapel Hill, N.C.: University of North Carolina Press, 1955.
Gipson, Lawrence H. The Coming of the Revolution, 1763-1775. New York: Harper, 1954.
Haywood, M. DeL. Governor William Tryon. Chapel Hill, N.C.: University of North Carolina Press, 1955.
Heaps, Willard Allison. Riots, U.S.A., 1765-1970. New York: Seabury Press, 1970.
Lefler, Hugh T., and Albert R. Newsome. North Carolina: History, Geography, Government. New York: Harcourt, Brace, 1966.
Schaper, W. A. Sectionalism in South Carolina. New York: Da Capo Press, 1968.
Turner, Frederick Jackson. The Frontier in American History. New York: Holt, Rinehart, 1947, 1962.
_______. The Significance of Sections in American History. Gloucester, Mass.: Peter Smith, 1932.

THE BOSTON MASSACRE (1770)

The Boston Massacre involved the attack by British troops on a group of citizens of Boston, then the British colony of Massachusetts Bay.

Feeling between the colonists and the British authorities was distinctly unfriendly. Public resentment against the Townshend Acts and the British policy of quartering troops in the city led up to the incident.

On March 5, 1770, a mob of colonists, led by Crispus Attucks, an American mulatto or, perhaps, of mixed Indian and Negro blood, attacked the British soldiers. The soldiers retaliated by firing into the mob. Seven colonists were wounded and three, of whom Attucks was one, were killed instantly.

Subsequently the nine British soldiers and their commanding officer were charged with murder. At their trial they were defended by John Adams and Josiah Quincy. All were acquitted of the murder charge, although two were convicted of manslaughter and were given trifling sentences.

Suggested Readings

Alden, John R. General Gage in America. Baton Rouge, La.: Louisiana State University Press, 1948.
_______. The South in the Revolution. Baton Rouge, La.: Louisiana State University Press, 1957.
Barron, Thomas C. Trade and Empire: The British Customs Service in Colonial America. Cambridge, Mass.: Harvard University Press, 1967.

Beer, George L. British Colonial Policy, 1754-1765. New York: Macmillan, 1907.
Dickerson, Oliver M. The Navigation Acts and the American Revolution. Philadelphia: University of Pennsylvania Press, 1951.
Dickinson, Alice. Boston Massacre, March 5, 1770. New York: Watts, 1968.
Fleming, Thomas J. "The Boston Massacre," American Heritage Magazine, December, 1966.
Heaps, Willard Allison. Riots, U. S. A., 1765-1970. New York: Seabury Press, 1970.
Knollenberg, Bernhard. Origins of the American Revolution. New York: Macmillan, 1960.
Miller, John G. Sam Adams: Pioneer in Propaganda. Boston: Little, Brown, 1936.
Shy, John. Toward Lexington: The Role of the British Army in the Coming of the American Revolution. Princeton, N. J.: Princeton University Press, 1965.

THE BOSTON TEA PARTY (1773)

The "Boston Tea Party" was an action taken by a band of Boston citizens disguised as Indians on December 16, 1773, by which a shipment of tea belonging to the East India Company and on which the British parliament had placed a tax, was thrown into the harbor.

The American colonists had protested the principle of "taxation without representation" and felt that the tax imposed on a shipment of 342 chests of tea valued at 18,000 pounds was unjust and an example of the protested principle.

By Massachusetts law the three ships on which the tea was stored could not leave port without discharging their cargo. On the evening of December 16 a group of colonists boarded the ships as they lay at anchor and emptied the tea over the side. When the government of Boston refused to pay for the jettisoned tea, the British authorities closed the port.

In March, 1774, George III of England signed the Boston Port Act, the first of the "Intolerable Acts." This, scheduled to become law on June 1, was designed as punishment of the Boston colonists for the destruction of the tea the previous December. It provided for removing the seat of government of the Massachusetts Bay Colony from Boston to Salem and replacing Boston with Marblehead as the port of entry until financial restitution should be made and other specified conditions fulfilled.

This legislation was met with indignation on the part of the colonists. Boston was occupied by British troops and the harbor was blockaded. Other New England towns furnished food to the embattled colonists. The Act remained in effect until the outbreak of the American Revolution, and was one of the measures which culminated in the assembling of the first Continental Congress.

Suggested Readings

Bancroft, George. History of the United States of America, from the Discovery of the Continent. New York: Appleton, 1888.
Bassett, John Spencer. A Short History of the United States. New York: Macmillan, 1929.
Brant, Irving. James Madison: The Virginia Revolutionist. Indianapolis: Bobbs-Merrill, 1941.
Cary, John. Joseph Warren: Physician, Politician, Patriot. Urbana, Ill.: University of Illinois Press, 1961.
Channing, Edward. A History of the United States. New York: Macmillan, 1921.
Commager, Henry Steele, ed. "The Intolerable Acts," (Doc. No. 49) in his Documents of American History, 8th edition. New York: Appleton, 1968.
Dabney, Virginius, ed. The Patriots. New York: Atheneum, 1975.
Donoughue, Bernard. British Politics and the American Revolution: The Path to War, 1773-1775. New York: St. Martin's Press, 1964.
Hildreth, Richard. The History of the United States of America. Norwood, Pa.: Norwood Editions, 1877.
Howard, George E. Preliminaries of the Revolution, 1763-1775. New York: AMS Press, 1970.
Jensen, Arthur L. The Maritime Commerce of Colonial Philadelphia. Madison, Wis.: The State Historical Society of Wisconsin, 1963.
Labaree, Benjamin W. The Boston Tea Party. New York: Oxford University Press, 1964.
Mumby, Frank A. George III and the American Revolution. Millwood, N.Y.: Kraus Reprint, 1924.
Schlesinger, Arthur M., Sr. The Colonial Merchants and the American Revolution. New York: Columbia University Press, 1918.
Sutherland, Lucy S. The East India Company in Eighteenth-Century Politics. New York: Oxford University Press, 1952.
Trevelyan, Sir George Otto. The American Revolution. New York: David McKay, 1964.
Van Tyne, Claude H. Causes of the War of Independence. Gloucester, Mass.: Peter Smith, 1970.
Webb, Robert N. We Were There at the Boston Tea Party. New York: Grosset and Dunlap, 1956.
Wells, William V., ed. The Life and Public Services of Samuel Adams. Plainview, N.Y.: Books for Libraries, 1888.

PAUL REVERE'S RIDE (1775)

Paul Revere was an American patriot, silversmith, engraver, inventor, printer and soldier, and a part-time dentist, who made a set of false teeth for George Washington. He took part in the Boston Tea Party of 1773, designed and printed the first issue of Continental money, and designed and engraved the first official seal for the American colonies and the state seal for Massachusetts. He also furnished the armor plates for the frigate "Constitution."

In spite of these many accomplishments, Revere is best remembered for his famous ride from Charlestown to Lexington, Massachusetts, to warn Samuel Adams and John Hancock of the approach of 800 British troops and to warn the minutemen, as described in the poem by Henry Wadsworth Longfellow. The redcoats were on their way to confiscate the gunpowder stored in Concord by the American militia.

On the evening of April 18, 1775, Revere, by prearrangement with Joseph Warren, an American physician and Revolutionary War officer, waited at Charlestown until he saw the light of two lanterns blinking from the tower of the North Church in Boston, across the water. Then he and William Dawes, a fellow patriot, galloped off on horseback to warn the Americans that the British were coming by sea. At Lexington they stopped sufficiently long to warn Hancock and Adams in time to prevent their capture by the oncoming troops. Revere was detained at Concord by British scouts but Dawes was able to reach Lexington in time to warn the people.

Subsequently Revere was in command at Castle William and took part in the unsuccessful Penobscot expedition in the year 1789. After the Revolutionary War he returned to his trade of silversmith and much of his work, outstanding in its field, is extant today.

Suggested Readings

Alden, John R. _General Gage in America._ Baton Rouge, La.: Louisiana State University Press, 1948.

Bakeless, John. _Turncoats, Traitors and Heroes._ Philadelphia: Lippincott, 1959.

Bakeless, Katherine, and John Bakeless. "The Paul Revere Gang," in their _Spies of the Revolution._ Philadelphia: Lippincott, 1962.

Buranelli, Marguerite. _With Colors Flying: Highlights of the American Revolution._ New York: Crowell-Collier, 1969.

Commager, Henry Steele, and R. B. Morris, eds. _The Spirit of 'Seventy-six._ New York: Harper, 1967.

Curtis, Edward E. _The Organization of the British Army in the American Revolution._ New Haven, Conn.: Yale University Press, 1926.

Forbes, Esther. _America's Paul Revere._ Boston: Houghton Mifflin, 1946.

________. _Paul Revere & the World He Lived In._ Boston: Houghton Mifflin, 1942.

French, Allen. _The Day of Concord and Lexington: Nineteenth of April, 1775._ Boston: Little, Brown, 1925.

Green, Margaret. _Paul Revere: The Man Behind the Legend._ New York: Messner, 1964.

Lancaster, Bruce. From Lexington to Liberty: The Story of the American Revolution. New York: Doubleday, 1955.
Lawson, Robert. Mr. Revere and I. New York: Dell, 1974.
Longfellow, Henry Wadsworth. Paul Revere's Ride (poem). New York: Crowell, 1963.
Murdock, Harold. The Nineteenth of April, 1775: Concord and Lexington. Boston: Houghton Mifflin, 1923.
Nolan, Jeannette Covert. The Shot Heard 'Round the World. New York: Messner, 1963.
Russell, Francis. Lexington, Concord and Bunker Hill. New York: American Heritage Press, 1963.
Stevenson, Augusta. Paul Revere: Boy of Old Boston. Indianapolis: Bobbs-Merrill, 1946.
Sutton, Felix. We Were There at the Battle of Lexington and Concord. New York: Grosset and Dunlap, 1958.
Tourtellot, Arthur B. William Diamond's Drum: The Beginning of the War of the American Revolution. New York: Doubleday, 1959.
Ward, Christopher. The War of the Revolution. Edited by John R. Alden. New York: Macmillan, 1952.
Wells, William V., ed. The Life and Public Services of Samuel Adams. Plainview, N.Y.: Books for Libraries, 1888.

THE BATTLE OF BUNKER HILL (1775)

The Battle of Bunker Hill was the first large-scale engagement of the American Revolution. Although won by the British under General William Howe, the "heroic American defense action demonstrated that hastily organized militiamen could trade blow for blow with British regulars, thereby strengthening morale and the spirit of resistance throughout the rebelling colonies."

The Battle was fought on June 17, 1775 in Charlestown, now a part of Boston, Massachusetts. Boston Harbor was dominated by the 110-feet-high Bunker Hill and Breed's Hill, 75 feet high. The British, under General Thomas Gage, occupied Boston and the Americans hoped to dislodge them. On the night of June 16 some 1200 American troops, commanded by Colonel William Prescott, occupied and fortified Breed's Hill. On the morning of the 17th General Gage prepared to attack the American position. British warships had been brought within shelling range of Breed's Hill and about 3000 British troops under Howe were dispatched from Boston. The American forces had meanwhile been reinforced by 300 additional volunteers, including General Joseph Warren.

At 3:00 P.M. the British troops launched their initial attack. In this they were assisted by heavy shelling from the British warships. Colonel Prescott ordered his men not to shoot until "they could see the whites of their enemies' eyes." Following the Colonel's instructions, the Americans allowed the British to advance almost to the base of the earthworks before opening fire. The British suffered severe losses and retreated in confusion. General

Gage ordered a second charge, which was again repulsed. Unfortunately for the Americans, they had exhausted their ammunition and a third British assault made it necessary for them to withdraw.

General Warren was killed. American losses amounted to 300 wounded, 145 killed and 30 taken prisoner. The British, however, suffered over a thousand casualties. Charlestown was set afire by the British shells and burned to the ground. The British victory enabled them to retain their possession of Boston.

On June 17, 1825, the cornerstone of the Bunker Hill Monument commemorating the engagement was laid by Marquis de Lafayette, the French officer who served on the staff of General George Washington during the Revolution. The monument was dedicated on June 17, 1843, with Daniel Webster appearing as chief speaker.

Suggested Readings

Alden, John R. General Gage in America. Baton Rouge, La.: Louisiana State University Press, 1948.

Buranelli, Marguerite. With Colors Flying: Highlights of the American Revolution. New York: Crowell-Collier, 1969.

Commager, Henry Steele, and R. B. Morris, eds. The Spirit of 'Seventy-six. New York: Harper, 1967.

Curtis, Edward E. The Organization of the British Army in the American Revolution. New Haven, Conn.: Yale University Press, 1926.

French, Allen. The Day of Concord and Lexington: Nineteenth of April, 1775. Boston: Little, Brown, 1925.

Frothingham, Richard. History of the Siege of Boston and of the Battles of Lexington, Concord and Bunker Hill. New York: Da Capo Press, 1903.

Ketchum, Richard M. The Battle for Bunker Hill. Garden City, N.Y.: Doubleday, 1962.

Lancaster, Bruce. From Lexington to Liberty: The Story of the American Revolution. New York: Doubleday, 1955.

Leckie, Robert. Great American Battles. New York: Random House, 1968.

Murdock, Harold. The Nineteenth of April, 1775: Concord and Lexington. Boston: Houghton Mifflin, 1923.

Nolan, Jeannette Covert. The Shot Heard 'Round the World. New York: Messner, 1963.

Russell, Francis. Lexington, Concord and Bunker Hill. New York: American Heritage Press, 1963.

Sutton, Felix. We Were There at the Battle of Lexington and Concord. New York: Grosset and Dunlap, 1958.

Tourtellot, Arthur B. William Diamond's Drum: The Beginning of the War of the American Revolution. New York: Doubleday, 1959.

Ward, Christopher. The War of the Revolution. Edited by John R. Alden. New York: Macmillan, 1952.

THE DECLARATION OF INDEPENDENCE (1776)

One of the most important documents of American history is the Declaration of Independence, adopted by the Continental Congress on July 4, 1776. This document, directed to George III of England, though not naming him specifically, proclaimed the independence of the thirteen English colonies in America, and recounted the grievances of the colonists against the British crown. It marked the culmination of a political process which had begun as a protest against oppressive restrictions imposed by Great Britain on manufacturing, political liberty, and colonial trade. This had developed into a revolutionary struggle which led to the colonies declaring their independence, and which was followed by the Revolutionary War, from which the colonies emerged victoriously as a new independent nation.

On June 7, 1776, Richard Henry Lee, in the name of the Virginia delegates to the Continental Congress, introduced three resolutions. One of these stated that "the colonies are, and of right ought to be, free and independent states, that they are absolved from all allegiance to the British Crown, and that all political connection between them and the State of Great Britain is and ought to be totally dissolved." John Adams seconded Lee's motion, but action was deferred. A five-member committee, consisting of Adams, Benjamin Franklin, Thomas Jefferson, Robert R. Livingston and Roger Sherman, appointed on June 11, was preparing a declaration in line with Lee's resolution. The draft was prepared by Jefferson, with Adams and Franklin making a few minor changes before it was submitted to the Congress, which it was on June 28. On July 2 a resolution declaring independence was adopted, and on July 4, after a few additional small alterations incorporating Lee's resolution had been made, was engrossed, signed by John Hancock and issued as the Declaration of Independence. Copies were sent to the legislatures of the states.

The delegates of twelve of the colonies accepted the Declaration unanimously. The New York delegates did not vote, as they lacked authorization to do so. On July 9 the New York Provincial Congress voted to endorse the Declaration. It was engrossed on parchment and on August 2 was signed by the 53 members of the Continental Congress then present. The three absentees signed it at a later date.

Copies were widely distributed. When the national government was organized in 1789 the Declaration was assigned to the Department of State. It was deposited with the patent office in 1841 and returned to the Department of State in 1877. It had been placed on exhibition but, because of the deterioration of the parchment and fading of the ink, it was withdrawn from public display in 1894.

Thomas Jefferson's draft, along with other historical American documents, was put on display in 1947/48 on the Freedom Train which made a tour of the United States.

Suggested Readings

Bailyn, Bernard. The Ideological Origins of the American Revolution. Cambridge, Mass.: Harvard University Press, 1967.

Becker, Carl L. The Declaration of Independence. New York: Harcourt, Brace, 1941.

Boyd, Julian P. The Declaration of Independence. Princeton, N.J.: Princeton University Press, 1945.

Commager, Henry Steele, ed. "The Declaration of Independence," (Doc. No. 66) in his Documents of American History, 8th edition. New York: Appleton, 1968.

________. "Resolution for Independence," (Doc. No. 65), in his Documents of American History, 8th edition. New York: Appleton, 1968.

Dumbauld, Edward. The Declaration of Independence and What It Means Today. Norman, Okla.: University of Oklahoma Press, 1950.

Friedenwald, Herbert. The Declaration of Independence. New York: Da Capo Press, 1974.

Hawke, David. A Transaction of Freemen. New York: Scribner's, 1964.

Hazelton, John H. The Declaration of Independence: Its History. New York: Dodd, Mead, 1906.

Robbins, Caroline. The Eighteenth Century Commonwealthman. Cambridge, Mass.: Harvard University Press, 1959.

Rossiter, Clinton. Seedtime of the Republic. New York: Harcourt, Brace, 1953.

THE VOYAGE OF THE "TURTLE" (1776)

David Bushnell, an American inventor, with the assistance of his friend Dr. Benjamin Gale, designed and built the first submarine in the United States navy. This vessel, christened the "Turtle," was the first underwater craft in the world to be used for military purposes. Bushnell had interested Governor Jonathan Trumbull of Connecticut in the idea of using his submarine for military purposes and the governor had arranged a demonstration for General Israel Putnam of the Continental army. Putnam, impressed, had secured government financial aid for further development of submarines.

Compared to such sophisticated undersea vessels as today's atomic-powered "Seadragon" or "Skate," the "Turtle," built in the 1770's, was simple in the extreme. It was roughly spherical in shape with a diameter of approximately six feet. It was constructed of oak and the outer surface was covered with tar to render it watertight. Designed to be operated by one man, its air supply would last not more than thirty minutes once the entrance hatch was closed and clamped tight from the outside. The vessel was guided by the operator through a system of hand-operated propellers and rudders. Lead ballast attached to the keel held it in an upright position. The craft was submerged when a valve was opened, admitting water to a ballast tank and rose when the tank was emptied by means of a manually-operated pump.

A long screw on the outside of the "Turtle" had attached to it a bomb made of oak. This bomb, which was watertight, was equipped with a time fuse and carried a gunpowder charge. Bushnell's plan called for the operator to maneuver the vessel underwater to an enemy ship, attach the bomb to the ship's hull by means of the screw and disengage the submarine, which was to beat a hasty retreat.

On the evening of September 6, 1776, Sergeant Ezra Lee, an American volunteer, entered his undersea craft which was moored in New York Harbor. The vessel to be attacked and sunk was the British 64-gun flagship "Eagle." Sergeant Lee made his way to the enemy ship but was unable to attach his bomb to the hull because the screw could not penetrate the "Eagle's" copper sheathing. As time was growing short, Lee jettisoned the bomb and made his way back to shore. The bomb exploded but did no damage to either the "Eagle" or the submarine.

Lee made several other attempts to destroy British ships in New York harbor, but without success. In 1799 Robert Fulton, in France, constructed an improved submarine, the "Nautilus." This undersea craft used compressed air, thus increasing the length of time it could remain submerged.

Suggested Readings

Abbott, Henry L. The Beginning of Modern Submarine Warfare. Edited by Frank Anderson. Hamden, Conn.: Shoe String Press, 1966.

Allen, Gardner W. A Naval History of the American Revolution. New York: Russell & Russell, 1962.

Bishop, Farnham. The Story of the Submarine. New York: Appleton, 1943.

Cable, Frank T. The Birth and Development of the American Submarine. New York: Harper, 1924.

Coggins, Jack. Ships and Seamen of the American Revolution. Harrisburg, Pa.: Stackpole Books, 1969.

Jones, Virgil C. The Civil War at Sea. New York: Holt, Rinehart, 1963.

Parsons, W. Barclay. Robert Fulton and the Submarine. New York: Columbia University Press, 1922.

Preston, Anthony, David Lyon and John Batchelor. Navies of the American Revolution. Englewood Cliffs, N.J.: Prentice-Hall, 1975.

Stephens, Edward. Submarines: The Story of Underwater Craft from the Diving Bell of 300 B.C. to Nuclear-powered Ships. New York: Golden Press, 1962.

Thomas, Herbert. Doctors of Yale College, 1702-1815, and the Founding of the Medical Institution. Hamden, Conn.: Shoe String Press, 1960.

Verne, Jules. 20,000 Leagues Under the Sea (science fiction).

New York: Macmillan, 1962.
Zim, Herbert Spencer. Submarines: The Story of Undersea Boats. New York: Harcourt, Brace, 1942.

THE SPOONER TRAGEDY (1778)

"The feminine lead [in the Spooner Tragedy] was played by a beautiful, oversexed and discontented young matron." Bathsheba Spooner, the daughter of a Tory lawyer, judge and politician, had been married to Joshua Spooner in the year 1746 when she was barely eighteen. Her husband, a businessman of Brookfield, near Worcester, Massachusetts, was much older than she and by the time she was 32 she was thoroughly tired of her marriage and her husband.

Late in 1777 Ezra Ross, a young Revolutionary War soldier, was invalided home to Brookfield and Bathsheba Spooner became, first his nurse and then his mistress. She also became pregnant.

The two lovers discussed methods for killing Joshua Spooner but nothing was done to achieve this until Sergeant James Buchanan and Private Charles Brooks, British soldiers and escapees from an American prison camp, appeared on the scene. Bathsheba's husband was away on a business trip and the two soldiers, at the lady's invitation, stayed at her home.

Spooner returned from his trip and, seeing the British soldiers, ordered them to leave, although he did give them permission to spend the night in his barn. They left the Spooner home but remained in the neighborhood, staying at the residence of Mary Walker, a widow of Worcester.

After an abortive attempt to poison Spooner, his wife and the three men conceived a plan to kill him. This was carried out on an evening late in February, 1778. As the unsuspecting victim, returning from the local tavern, entered his gate, Brooks struck him over the head with a log. Then Brooks, Ross and Buchanan, after removing Spooner's watch and shoes, dropped him, apparently still alive, head first into the well.

In due course the murder was discovered and the body retrieved. Bathsheba and the three men were formally charged with killing the woman's husband. Evidence of their guilt was ample, tracks in the newly fallen snow showing exactly what had occurred. The well-curb was spattered with blood. Various articles, such as a pair of silver shoe buckles and a pair of breeches known to have belonged to Spooner, were found in the possession of the men.

All four prisoners pleaded not guilty. The case against them, however, though circumstantial, was conclusive. Bathsheba's attorney, Levi Lincoln, entered a plea of not guilty by reason of insanity, but the jury found her and the three men guilty of murder and Chief Justice William Cushing sentenced them to death by hanging.

Then Bathsheba Spooner announced that she was pregnant, thus invoking a law which forbade the execution of condemned female criminals who might "plead their belly." By so pleading, the sentence of a woman claiming pregnancy was postponed until after her child was born or time showed that she was not, in fact, pregnant. If she was lucky her sentence may have, in the meantime, been reversed or reduced. This law required that the woman must be examined by a jury "consisting of two men midwives and twelve matrons," which jury must be satisfied that she was "quick with child."

Bathsheba was so examined and the jury found her not to be with child as claimed. At her request a second examination was made by four midwives, one of whom was a woman. These four were assisted by two matrons. The midwives found her "now quick with child" but the matrons determined that she was "not now quick with child." This second jury, not consisting of the full complement of midwives and matrons as specifically required by law, had no legal standing and the verdict of the first jury held.

On the afternoon of July 2, 1778, the four murderers were hanged. Bathsheba Spooner had requested that, following her execution, her body be examined. In accordance with her request an examination of her corpse was made that evening by surgeons. A male fetus "of the growth of five months" was taken from her.

Suggested Readings

Abrahamsen, David, M.D. The Murdering Mind. New York: Harper & Row, 1973.

Bromberg, Walter. Mold of Murder: A Psychiatric Study of Homicide. New York: Grune & Stratton, 1961.

Bullock, Chandler. "The Bathsheba Spooner Murder Case," Worcester Historical Society Publications, New Series II, September, 1939.

Chandler, Peleg W. American Criminal Trials. Plainview, N.Y.: Books for Libraries, 1970.

Dempewolff, Richard. Famous Old New England Murders and Some That Are Infamous. Brattleboro, Vt.: Stephen Daye Press, 1942.

The Dying Declaration of James Buchanan, Ezra Ross and William Brooks, who Were Executed at Worcester on Thursday, July 2, 1778, for the Murder of Mr. Joshua Spooner. Worcester, Mass.: Isaiah Thomas, 1778.

Jesse, F. Tennyson. Murder and Its Motives. London: Harrap, 1952.

Lamb, Roger. Memoir of His Own Life. Dublin: 1811.

Lester, David, and Gene Lester. Crime of Passion: Murder and the Murderer. Chicago: Nelson Hall, 1975.

Morris, Richard B. "The Spooner Triangle Love Slaying," in his Fair Trial. New York: Knopf, 1952.

Sparrow, Gerald. Women Who Murder. New York: Abelard, 1970.

THE CONWAY CABAL AND THE TREASON OF CHARLES LEE (1778)

When George Washington was commanding the Continental army during the American Revolution certain fellow-officers attempted to undermine his authority and discredit him with Congress. One of these was the episode in American history known as the "Conway Cabal," which involved an abortive attempt to replace him with General Horatio Gates as commander-in-chief. It derived its name from General Thomas Conway, an Irish-born American revolutionary officer and soldier of fortune. A former colonel in the French army, Conway emigrated to the American colonies in 1777 and offered his services to the revolutionary army. They were accepted and he was given the rank of brigadier general.

Following his participation in the battles of Brandywine and Germantown he applied to Congress for a promotion. His application was disapproved on Washington's recommendation, the latter feeling that officers with more service than Conway were entitled to preference. Congress, however, overrode Washington's recommendation on December 14, 1777, and authorized Conway's promotion to the rank of major general.

Friends of General Gates had sought to put him in Washington's place. The Conway Cabal was instigated to accomplish this by discrediting Washington and comparing Gates' victories at Stillwell and Saratoga with Washington's defeats at Brandywine and Germantown.

A whispering campaign was started and Gates did nothing to discourage the gossip that he was a better and more deserving officer than was Washington. Word of this gossip reached Washington indirectly by way of a British general. This included the report of a remark erroneously attributed to Conway. Washington ended the affair with a frank letter addressed to Gates. Conway offered to resign his commission. General John Cadwalader challenged him to a duel on a charge that he was undermining Washington's authority. The duel was fought and Conway was severely wounded.

Subsequently Conway returned to France and joined the French army, in which he saw service in India. In 1787 he was made governor of the French possessions there and was later involved in the French Revolution in which he attempted to lead a royalist force against the insurgents.

Another anti-Washington officer was Charles Lee, an Anglo-American soldier who served in both the British and American armies. He was a member of General Edward Braddock's command in the Fort Duquesne expedition of 1755 and then settled in Berkeley County, Virginia. There he supported the American patriots in their struggle for freedom.

At the outbreak of the Revolutionary War he was appointed by Congress as the second ranking major general of the Continental army and late in 1776 joined General Washington. He was slow in carrying out Washington's orders and was extremely critical of him, notably in a letter he wrote to General Gates. Though implicated in the Conway Cabal, both he and General Thomas Mifflin repudiated their connection with it.

On December 13, 1776, Lee was captured by the British and it was assumed that he would be sent to England as a deserter. While a prisoner he submitted a secret plan to Sir William Howe, the British general. By this plan, if successful, the American forces could have been defeated.

In 1778 Washington arranged for Lee to be exchanged as a prisoner of war, and Lee returned to the American service deliberately to betray it. He was placed in command of a planned attack at Monmouth but instead of attacking, began a retreat. Fortunately for the Americans this maneuver was unsuccessful and the timely arrival of Generals Washington, Nathaniel Greene and Friedrich von Steuben and their forces saved the day.

Lee was tried by a military court martial and found guilty of disobedience, of misbehavior before the enemy, and of disrespect to the commander-in-chief. He continued to abuse Washington, fought a duel with John Laurens, Washington's aide-de-camp, and wrote an insolent letter to Congress.

In 1780 he was dismissed from the army and retired in disgrace to his Virginia estate. He died in 1782.

Suggested Readings

Alden, John Richard. General Charles Lee: Traitor or Patriot? Baton Rouge, La.: Louisiana State University Press, 1951.

Dupuy, Trevor Nevitt. The Military Life of George Washington. New York: Watts, 1969.

Freeman, Douglas Southall. George Washington. New York: Scribner's, 1948-1954.

Heitman, F. B. Historical Register of Officers of the Continental Army During the War of Revolution. Washington, D.C.: Government Printing Office, 1893.

Lee, Charles. The Lee Papers. New York: New York Historical Society, 1872-1875.

Moore, George H. The Treason of Charles Lee. Port Washington, N.Y.: Kennikat Press, 1970.

Patterson, Samuel W. Horatio Gates: Defender of American Liberties. New York: AMS Press, 1941.

Potter, John Mason. Plots Against the Presidents. New York: Astor, 1968.

Sparks, Jared. "The Life of Charles Lee, Major General in the Army of the Revolution," in his Library of American Biography. New York: 1864.

Vidal, Gore. Burr (historical fiction). New York: Random House, 1973.

Whiteley, Emily Stone. Washington and His Aides-de-Camp. New York: Macmillan, 1936.

Wright, Esmond. Washington and the American Revolution. London: English University Press, 1957.

THE BENEDICT ARNOLD DEFECTION (1780)

One of the tragedies of the American Revolutionary War was the defection of the American major general Benedict Arnold to the British in 1780.

Arnold, a merchant in Connecticut and a militia captain when the war broke out, was promoted to colonel and commissioned to raise troops for the American army. He joined Ethan Allen as a volunteer and assisted in the capture of Fort Ticonderoga on May 10, 1775. General George Washington appointed him to command an expedition against Quebec, which expedition was unsuccessful, and in which he was wounded.

The Continental Congress promoted Arnold to the rank of brigadier general and in 1776 he stopped a British thrust from Canada down Lake Champlain. In that same year he was defeated in a naval battle on the lake. After participating gallantly in the battle of Ridgefield, Connecticut, he was promoted to major general. He felt, however, that this last promotion was long overdue and that he had not received proper credit for repulsing General John Burgoyne's army of Britishers at Saratoga. He quarreled with his superior, General Horatio Gates, who relieved him of his command. In spite of this, and nominally without military authority, he took an active part in the second battle of Saratoga, in which he was severely wounded. Historians are generally agreed that Arnold and General Philip John Schuyler were actually responsible for the defense for which Gates received credit.

In 1778 Arnold was appointed commander at Philadelphia. Here he married, lived extravagantly, incurred heavy debts and was eventually court-martialled for alleged arbitrary exercise of his military authority and pro-Tory political favoritism. Found guilty, he was sentenced to be reprimanded by Washington, the commander-in-chief. Washington made the reprimand as considerate as possible and appointed Arnold commander at West Point.

In 1780 Arnold, who believed himself to have been victimized by jealous superiors and slighted by Congress, negotiated with Sir Henry Clinton, British commander-in-chief in North America, to surrender West Point to the latter. Details of the surrender were worked out with Major John André, Clinton's aide-de-camp. André, following a secret meeting with Arnold, set out for the British lines on horseback, the plans of West Point in his boots. Near Tarrytown, New York, he was captured, taken to Washington's headquarters at Tappan, New York, tried as a spy by a military court, and hanged October 2, 1780.

News of the capture of André, by a remarkable error, was sent directly to Arnold at West Point where Washington had just arrived. Arnold made a hasty departure and escaped to the British sloop "Vulture" which was at anchor in the Harlem River.

The British paid Arnold approximately $30,000 and commissioned him a brigadier general in their army. Early in 1781 he led British troops in an attack on Richmond, Virginia, and later in the war led an expedition that burned New London, Connecticut, and massacred the garrison at Fort Griswold.

Arnold's military service was not extended by the British following the Revolutionary War. He removed in 1783 to St. John, New Brunswick, where he became a merchant. Charged with dishonesty, he was forced to depart four years later. He returned to London, commanded a privateer in the war with France, but was not successful. He died in London in 1801 in disgrace and poverty.

Suggested Readings

Arnold, Isaac N. The Life of Benedict Arnold. Saint Clair Shores, Mich.: Scholarly Press, 1880.
Boylan, Brian. Benedict Arnold: The Dark Eagle. New York: Norton, 1973.
Bradford, Gamaliel. Damaged Souls. New York: Houghton Mifflin, 1923.
________. Wives. New York: Harper, 1925.
De Leeuw, Cateau. Benedict Arnold: Hero and Traitor. New York: Putnam, 1970.
Douglas, Jack. Benedict Arnold Slept Here. New York: Putnam, 1975.
Flexner, James T. The Traitor and the Spy: Benedict Arnold and John André. Boston: Little, Brown, 1975.
Freeman, Douglas Southall. George Washington. New York: Scribner's, 1948-1954.
Gocek, Matilda A. Benedict Arnold: A Readers' Guide and Bibliography. Monroe, N.Y.: Library Research Associates, 1973.
Lomask, Milton. Beauty and the Traitor: The Story of Mrs. Benedict Arnold. Philadelphia: Macrae Smith, 1967.
Nolan, Jeannette Covert. Benedict Arnold: Traitor to His Country. New York: Messner, 1956.
Roberts, Kenneth L. The March to Quebec (historical fiction). New York: Doubleday, 1940.
Sellers, Charles G. Benedict Arnold. New York: Hill & Wang, 1930.
Sherwin, Oscar. Benedict Arnold: Patriot or Traitor? New York: Gordon Press, 1960.
Van Doren, Carl. The Secret History of the American Revolution. New York: Augustus M. Kelley, 1941.
Wallace, Willard M. The Traitorous Hero. Plainview, N.Y.: Books for Libraries, 1954.

THE QUOCK WALKER CASE (1783)

The famous law case of Quock Walker vs. Nathaniel Jennison, tried in the Massachusetts Supreme Judicial Court in 1783, established the fact that the institution of slavery did not exist in that state, as specified in the first article of the Massachusetts Bill of Rights (1780).

During the American Revolutionary War the status of slavery in Massachusetts was uncertain. In 1770, a decade before the publication of the above-mentioned document, the Superior Court of Massachusetts had ruled in the case of James vs. Lechmere that the plaintiff, a Negro, "was entitled to his freedom under the laws of the province and the terms of the royal charter." Chief Justice Theophilus Parsons, in ruling on the case of Winchendon vs. Hatfield, said, "several Negroes born in this country [America] of imported slaves demanded their freedom of their masters by suit at law, and obtained it by judgment of court. The defense of the master was faintly made, for such was the temper of the times, that a restless, discontented slave was worth little; and when his freedom was obtained in a course of legal proceedings, the master was not holden for his future support if he became poor. But in the first action, involving the right of the master, which came before the Supreme Judicial Court after the establishment of the Constitution, the judges declared that by virtue of the first article of the Declaration of Rights, slavery in this state was no more."

The Quock Walker case was similar to the one ruled on by Justice Parsons. Nathaniel Jennison had been indicted for assault on the Negro Quock Walker. Jennison's defense was that his assault was justified, on the grounds that Walker was his slave. Chief Justice William Cushing, however, speaking for the Court, ruled that slavery had been abolished in Massachusetts by the first article of the Massachusetts Bill of Rights. Jennison was found guilty of assault.

The anti-slavery attitude of the residents of Massachusetts led to the Abolitionist Movement headed by such men as William Lloyd Garrison, who made Boston his headquarters and in 1831 began publication of his anti-slavery newspaper, Liberator. Slavery was finally abolished throughout America by President Abraham Lincoln's Emancipation Proclamation of January 1, 1863, and by the Thirteenth Amendment to the Constitution of the United States, dated December 18, 1865.

Suggested Readings

Adams, Charles Francis. Life of John Adams. Saint Clair Shores, Mich.: Scholarly Press, 1881.

Chinard, Gilbert. Honest John Adams. Boston: Little, Brown, 1933.

Commager, Henry Steele, ed. "The Constitution of the United States," (Doc. No. 87) in his Documents of American History, 8th edition. New York: Appleton, 1968.

_______. "The Emancipation Proclamation," (Doc. No. 222) in his Documents of American History, 8th edition. New York: Appleton, 1968.

_______. "Massachusetts Bill of Rights," (Doc. No. 70) in his Documents of American History, 8th edition. New York: Appleton, 1968.

_______. "The Quock Walker Case," (Doc. No. 71) in his Documents of American History, 8th edition. New York: Appleton,

1968.
Hurd, John C. The Law of Freedom and Bondage. Westport, Conn.: Negro Universities Press, 1858-1862.
Kurtz, Stephen G. The Presidency of John Adams. Philadelphia: University of Pennsylvania Press, 1957.
Locke, Mary S. Anti-Slavery in America. New York: Johnson Reprint, 1969.
Moore, George H. Notes on the History of Slavery in Massachusetts. Westport, Conn.: Negro Universities Press, 1968.
Nevins, Allan. American States During and After the Revolution, 1755-1798. Clifton, N.J.: Augustus M. Kelley, 1968.
Smith, Page. John Adams. Westport, Conn.: Greenwood Press, 1962.

SHAYS'S REBELLION (1786-1787)

The insurrection led by Daniel Shays, previously a captain in the Revolutionary army, occurred in Western Massachusetts in 1786-87. The reason for Shays's Rebellion, as it came to be called, lay in the fact that the residents of Massachusetts were suffering great economic distress, due primarily to the excessive taxes imposed on land.

Shays and his followers demanded a reduction of taxes, the abolition of the Court of Common Pleas and the immediate issue of large amounts of paper currency. Armed mobs prevented the sitting of courts at Concord, Worcester, Northampton and Great Barrington. In September, 1786, Shays, leading an "army" of 600 rioters, broke up a session of the supreme court at Springfield.

A force of militia, numbering 1200 men, routed Shays and his followers on January 25, 1787 when they attempted to seize the federal arsenal at Springfield. Shays was defeated again at Petersham on February 2 of the same year and fled to Vermont. He was condemned to death by the Massachusetts supreme court but was later pardoned.

Suggested Readings

Adams, James Truslow. New England in the Republic, 1776-1850. Saint Clair Shores, Mich.: Scholarly Press, 1926.
Commager, Henry Steele, ed. "Shays's Rebellion," (Doc. No. 81) in his Documents of American History, 8th edition. New York: Appleton, 1968.
Hart, Albert B., ed. The Commonwealth History of Massachusetts: Colony, Province and State. New York: Russell & Russell, 1967.
Heaps, Willard Allison. Riots, U.S.A., 1765-1970. New York: Seabury Press, 1970.
McLaughlin, Andrew C. Confederation and Constitution. New York:

Macmillan, 1962.
Miller, John C. Sam Adams: Pioneer in Propaganda. Boston: Little, Brown, 1936.
Minot, George R. History of the Insurrections in Massachusetts in the Year 1786 and the Rebellion Consequent Thereon. Plainview, N.Y.: Books for Libraries, 1970.
Taylor, Robert J. Western Massachusetts in the Revolution. Providence, R.I.: Brown University Press, 1954.
Warren, J. P. "The Confederation and the Shays Rebellion," American Historical Review, Vol. XI.
Wells, William V., ed. The Life and Public Services of Samuel Adams. Plainview, N.Y.: Books for Libraries, 1888.

CHISHOLM VS. GEORGIA (1793)

The Supreme Court decision rendered in the case of Alexander Chisholm vs. the state of Georgia in 1793 led to the Eleventh Amendment of the Constitution of the United States, ratified January 8, 1798. Historians consider it "probably the most important of the early cases which came before the Supreme Court" and state further that "in the decision of the Court can be found a foreshadowing of the nationalism enunciated by [Chief Justice John] Marshall a decade later."

This case was the result of certain citizens of the state of South Carolina, including Alexander Chisholm, executors of the estate of an English creditor, seeking to obtain compensation from the state of Georgia for property confiscated during the Revolutionary War.

The Constitution, in Article III, Section 2, stated that the judicial power of the United States should extend to "controversies between two or more states" and "between a state and citizens of another state." The question arose, under this provision, as to whether or not a citizen could bring legal action against a state in the federal courts. Alexander Hamilton, writing in the Federalist, stated that such suits could not be brought, and James Madison, in the debates in the Virginia Ratifying Convention also held this view. Said Madison, "It is not in the power of individuals to call any state into court. The only operation it can have, is that, if a state should wish to bring a suit against a citizen, it must be brought before the federal court."

In the case of Chisholm vs. Georgia, however, Justice James Wilson upheld Chisholm's right to sue the state of Georgia in the Supreme Court. In his opinion Wilson quoted from Article III, Section 2, of the Constitution.

The state of Georgia denied the validity of the judgment, refused to appear in court and "threatened to punish by death any official who should attempt to execute the decree of the Court." Other states also protested Wilson's ruling and subsequently the Eleventh Amendment to the Constitution was introduced and ratified. This reads, "The Judicial power of the United States shall not be construed to extend to any suit in law or equity, commenced or prosecuted against one of the United States by Citizens of another State, or by Citizens or Subjects of any Foreign State." This Amendment deprived the federal courts of jurisdiction in cases against one of the states by citizens of another state.

Suggested Readings

Boudin, Louis B. Government by Judiciary. New York: William Goodwin, 1932.

Brant, Irving. James Madison: The Virginia Revolutionist. Indianapolis: Bobbs-Merrill, 1941.

Carson, Hampton L. The History of the Supreme Court. New York: Burt Franklin, 1972.

Commager, Henry Steele, ed. "Chisholm vs. Georgia," (Doc. No. 95) in his Documents of American History, 8th edition. New York: Appleton, 1968.

_______. "The Constitution of the United States," (Doc. No. 87) in his Documents of American History, 8th edition. New York: Appleton, 1968.

Friedman, Leon, and Fred L. Israel, eds. The Justices of the U.S. Supreme Court, 1789-1869. New York: Bowker, 1969.

Gay, Sydney H. James Madison. Edited by John T. Morse, Jr. New York: AMS Press, 1898.

Hamilton, Alexander, James Madison and John Jay. The Federalist. Edited by Benjamin F. Wright. Cambridge, Mass.: Harvard University Press, 1961.

Hamilton, John C. The History of the Republic of the United States of America as Traced in the Writings of Alexander Hamilton. New York: 1857-1864.

Johnson, Gerald White. The Supreme Court. New York: Morrow, 1962.

Lodge, Henry Cabot. Alexander Hamilton. Edited by John T. Morse, Jr. New York: AMS Press, 1898.

McCloskey, Robert G. The American Supreme Court. Chicago: University of Chicago Press, 1960.

Phillips, Ulrich B. "Georgia and State Rights," American Historical Association Reports, 1901. Vol. II.

Rossiter, Clinton. Alexander Hamilton and the Constitution. New York: Harcourt, Brace, 1964.

Warren, Charles. The Supreme Court in United States History. Boston: Little, Brown, 1923.

THE WHISKEY REBELLION (1794)

The Whiskey Rebellion (or Insurrection) of 1794 is important to American history because it furnished the first real test of the federal government's law-enforcement power. The excise tax of March 3, 1791, which applied to whiskey, worked extreme hardship on the grain farmers of western Pennsylvania and Virginia who were accustomed to turn their corn into whiskey because this was the only form in which it could be transported to market economically. This tax was strongly opposed by the farmers, most of whom were also distillers, and who considered it discriminatory and unfair.

Organized resistance rapidly reached serious proportions, with some federal revenue agents being tarred and feathered by irate farmers. In September, 1792, President Washington issued a proclamation urging the citizens to obey the law, but this had little effect. In the spring of 1794 warrants for the arrest of several distillers who had failed to comply with the law were issued, and a federal officer was killed in one of the ensuing riots.

On August 7, 1794, Washington issued a second proclamation in which he ordered the insurgents to "disperse and retire peaceably to their respective abodes." He also requested the governors of Maryland, Pennsylvania, New Jersey and Virginia to mobilize the militia, and sent three commissioners to Parkinson's Ferry, Pennsylvania, to negotiate with the insurgents. These negotiations proved fruitless, and on October 14, 1794, Washington ordered the militia to proceed from Carlisle, Pennsylvania to the troubled area.

This brought results: the insurgents offered no resistance and David Bradford, their leader, fled.

Oaths of allegiance to the United States were exacted and those who refused to take such an oath were arrested. Two such were tried and convicted of treason, but were pardoned by Washington.

Suggested Readings

Bacheller, Irving, and Herbert S. Kates. Great Moments in the Life of Washington. New York: Grosset & Dunlap, 1932.

Baldwin, Leland B. Whiskey Rebels: The Story of a Frontier Uprising. Pittsburgh: University of Pittsburgh Press, 1939.

Brackenridge, Henry M. The History of the Western Insurrection in Western Pennsylvania Commonly Called the Whiskey Insurrection. New York: Arno Press, 1974.

Brunhouse, Robert L. The Counter-Revolution in Pennsylvania, 1776-1790. Harrisburg, Pa.: The Pennsylvania Historical and Museum Commission, 1942.

Commager, Henry Steele, ed. "Washington's Proclamation on the Whiskey Rebellion," (Doc. No. 97) in his Documents of American History, 8th edition. New York: Appleton, 1968.

Fee, Walter R. The Transition from Aristocracy to Democracy in New Jersey, 1789-1829. Somerset, N.J.: Somerset Press, 1933.

Ferguson, Russell J. Early Western Pennsylvania Politics. Pittsburgh: University of Pittsburgh Press, 1938.

Fox, Dixon R. The Decline of Aristocracy in the Politics of New York. New York: Columbia University Press, 1919.

Freeman, Douglas Southall. George Washington. New York:

Scribner's, 1948-1954.
Gilpatrick, Delbert H. Jeffersonian Democracy in North Carolina, 1789-1816. New York: Columbia University Press, 1931.
Heaps, Willard Allison. Riots, U.S.A., 1765-1970. New York: Seabury Press, 1970.
Link, Eugene P. Democratic-Republican Societies, 1790-1800. New York: Columbia University Press, 1942.
Scudder, Horace E. George Washington, An Historical Biography. Boston: Houghton Mifflin, 1924.

THE X.Y.Z. AFFAIR (1797-1798)

The X.Y.Z. Affair concerned the American statesmen Elbridge Gerry, John Marshall, and Charles Cotesworth Pinckney, who were sent to France by President John Adams in October, 1797, to negotiate with the French foreign minister Charles Maurice de Talleyrand-Périgord regarding political differences between their countries.

France was then at war with Great Britain, and the United States had declined to assist France, although such assistance had been stipulated in the Franco-American Treaty of 1778. This situation was complicated by the fact that the United States had concluded Jay's Treaty with Great Britain in 1794 and, under this treaty, had accepted the British viewpoint concerning the rights of neutrals and had forbidden French ships to use American ports. This, in turn, had resulted in the French attacking American vessels. It was in an attempt to reconcile these differences that Gerry, Marshall and Pinckney were sent to France.

Talleyrand, through secret agents, informed the American commission that no negotiations could be made unless the United States agreed to a personal gift of $250,000 to him and a "loan" of $10,000,000 to France. This proposal was declined and President John Adams, upon being advised of the situation, submitted the matter to Congress, which also refused to accept Talleyrand's conditions.

The dispatches of the American commission were made public in April, 1798. Although complete in other respects, the letters "X", "Y" and "Z" replaced the names of Talleyrand's representatives. The public exposure of France's attempted extortion caused general indignation. Congress prepared for war but war was not declared, although several clashes between American and French vessels did occur. In 1798 France repudiated agents "X", "Y" and "Z", and in 1799 President Adams, desiring to avoid armed conflict between the two countries, sent a new commission to the French capital. Before this commission arrived in Paris Napoleon I had assumed leadership of France and an agreement was made with him that put his country and the United States on a basis of friendship.

Suggested Readings

Adams, Charles Francis. Life of John Adams. East Saint Clair Shores, Mich.: Scholarly Press, 1881.
Adams, James Truslow. The Epic of America. Boston: Little, Brown, 1931.
Allen, Gay Wilson. Our Naval War With France. Boston: Houghton Mifflin, 1909.
Beveridge, Albert J. The Life of John Marshall. Boston: Houghton Mifflin, 1916-1919.
Bond, Beverly, Jr. The Monroe Mission to France, 1794-1796. Baltimore: Johns Hopkins University Press, 1907.
Chinard, Gilbert. Honest John Adams. Boston: Little, Brown, 1933.
Daur, Manning J. The Adams Federalists. Baltimore: Johns Hopkins University Press, 1953.
De Conde, Alexander. The Quasi-War: The Politics and Diplomacy of the Undeclared War With France, 1797-1801. New York: Scribner's, 1966.
Gilman, Daniel C. James Monroe. Boston: Houghton Mifflin, 1888.
Hazen, Charles D. Contemporary American Opinion of the French Revolution. Baltimore: Johns Hopkins University Press, 1897.
Hyneman, Charles S. The First American Neutrality. Urbana, Ill.: University of Illinois Press, 1934.
Kurtz, Stephen G. The Presidency of John Adams. Philadelphia: University of Pennsylvania Press, 1957.
Magruder, Allan B. John Marshall. Edited by John T. Morse, Jr. New York: AMS Press, 1898.
Smith, Page. John Adams. Westport, Conn.: Greenwood Press, 1962.

THE ALIEN AND SEDITION ACTS (1798)

The Alien and Sedition Acts consisted of four separate Acts passed by Congress in 1798. These were: 1) the Naturalization Act (June 18); 2) the Alien Act (June 25); 3) the Alien Enemies Act (July 6); and 4) the Sedition Act (July 14).

The Naturalization Act increased the residence requirement for naturalization from five to fourteen years. It was repealed in 1802. The Alien Act, which expired in 1800, gave the President the power to arrest and deport any alien considered dangerous. The Enemy Alien Act, which also expired in 1800, dealt with "subjects of foreign powers at war with the United States" and provided for their arrest and deportation. The Sedition Act was directed against a group of Federalist pamphleteers and editors of French and English extraction, including Thomas Cooper, Joseph Priestley, Benjamin F. Bache, Count de Volbey, James Callender and V. duPont. Expiring March 3, 1801, it declared as criminal offenses the printing or publishing of "false, scandalous, and malicious writing or writings against the government of the United States," the President, or Congress; the fostering of opposition against the lawful acts of Congress or of the President; or the assisting of a foreign power in plotting against the United States.

Enacted by a Federalist Congress and signed by President John Adams, himself a Federalist, the first three of these laws were aimed primarily at those who had been opposed to the Adams administration. They were also a response to the X. Y. Z. Affair which had precipitated a spirit of nationalism and an intense anti-foreign sentiment in America.

The Sedition Act was a Federalist attempt to check political opposition in general, but particularly by those Americans whose sympathies were for France. Under it about 25 Republican Party members were arrested and ten were convicted of violating its provisions.

These statutes, with their severe penalties for those who violated them, were the cause of a violent popular reaction. This hostility was reflected in a number of documents, such as the Kentucky Resolutions (November 16, 1798), drafted by Thomas Jefferson and sponsored in the Kentucky legislature by John Breckinridge, and the Virginia Resolutions (December 24, 1798), drafted by James Madison and introduced by John Taylor of Caroline.

The Kentucky and Virginia Resolutions drew forth replies from several states. They represented not so much a constitutional as a social philosophy and were drawn up with the primary purpose of presenting a democratic protest against what was considered a dangerous usurpation of power by the national government. They were also important as "the first statement in American political history of the principle of nullification" which later became a basic tenet of the doctrine of states' rights. There was grave doubt as to the constitutionality of the Sedition Act but this question never came before the courts.

Suggested Readings

Adams, Charles Francis. Life of John Adams. Saint Clair Shores, Mich.: Scholarly Press, 1881.

Anderson, F. M. "Contemporary Opinion of the Virginia and Kentucky Resolutions," American Historical Review, Vol. V.

Bassett, John Spencer. The Federalist System. New York: Cooper Square, 1905.

Bowers, Claude G. Jefferson and Hamilton. Boston: Houghton Mifflin, 1967.

Chinard, Gilbert. Honest John Adams. Boston: Little, Brown, 1933.

Commager, Henry Steele, ed. "The Alien and Sedition Acts," (Doc. No. 101) in his Documents of American History, 8th edition. New York: Appleton, 1968.

________. "The Constitution of the United States," (Doc. No. 87) in his Documents of American History, 8th edition. New York: Appleton, 1968.

________. "The Kentucky and Virginia Resolutions of 1798," (Doc.

No. 102) in his Documents of American History, 8th edition. New York: Appleton, 1968.

_______. "State Replies to the Virginia and Kentucky Resolutions," (Doc. No. 104) in his Documents of American History, 8th edition. New York: Appleton, 1968.

Daur, Manning J. The Adams Federalists. Baltimore: Johns Hopkins University Press, 1953.

Fay, Bernard. The Two Franklins, Fathers of American Democracy. Saint Clair Shores, Mich.: Scholarly Press, 1971.

Fleming, Thomas J. " 'A Scandalous, Malicious and Seditious Libel,' " American Heritage Magazine, December, 1967.

Harlow, Alvin F. "Martyr for a Free Press," American Heritage Magazine, October, 1955.

Hunt, Gaillard, and others. The Life of James Madison. New York: Russell & Russell, 1966.

Kurtz, Stephen G. The Presidency of John Adams. Philadelphia: University of Pennsylvania Press, 1957.

McLaughlin, Andrew G. The Courts, The Constitution and Parties. New York: Da Capo Press, 1972.

McMaster, John B. A History of the People of the United States, from the Revolution to the Civil War. New York: Appleton, 1938.

Miller, John C. Crisis in Freedom: The Alien and Sedition Acts. Boston: Little, Brown, 1951.

Schouler, James. History of the United States of America Under the Constitution. Millwood, N.Y.: Kraus Reprint, 1908-13.

Shaler, Nathaniel S. Kentucky: A Pioneer Commonwealth. New York: AMS Press, 1885.

Smith, James M. Freedom's Fetters: The Alien and Sedition Laws and American Civil Liberties. Ithaca, N.Y.: Cornell University Press, 1956.

Smith, Page. John Adams. Westport, Conn.: Greenwood Press, 1962.

Von Volst, Hermann E. Constitutional and Political History of the United States. New York: AMS Press, 1892.

Warfield, Ethelbert D. The Kentucky Resolutions of 1798. Plainview, N.Y.: Books for Libraries, 1969.

THE LOUISIANA PURCHASE (1803)

In 1803 the United States purchased from France lands which today include Arkansas, Iowa, Missouri, Minnesota west of the Mississippi River, the Dakotas, and parts of Kansas, Colorado, Montana, Wyoming and Louisiana. This was the largest area of territory ever acquired by the United States at one time, and included the city of New Orleans.

Louisiana had originally been owned by France, had passed to Spain following the Seven Years' War in 1763, and was returned to France by secret treaty in 1800. It was, by the terms of the treaty, to revert to Spain if France ever gave it up.

In 1802 relations between France and the United States became strained. Napoleon Bonaparte, Emperor of the French, sent an army to Santo Domingo to put down the Negro rebellion there and then to lay siege to New Orleans. He also withdrew the right of deposit to American merchants, which right permitted them to store goods at that city for transshipment.

Robert R. Livingston was then the U. S. minister to France. President Thomas Jefferson sent James Monroe to Paris to assist Livingston with negotiations to nullify the French actions.

Four alternatives existed. These were:

1. The purchase of Florida and New Orleans.
2. The purchase of New Orleans alone.
3. The purchase of land on the east bank of the Mississippi River on which an American port could be built.
4. The acquiring of perpetual rights of navigation and deposit.

Negotiations were conducted by Livingston and Monroe with the French foreign minister Charles Maurice de Talleyrand-Périgord, and were at first unsuccessful. Then, when the French army in Santo Domingo met with disaster and war with England seemed imminent, Napoleon reconsidered his position. Through Talleyrand, on April 11, 1803, he offered to sell the United States all of Louisiana. The American negotiators, although astonished at this proposal and acting beyond their authority, accepted. Early in May documents antedated to April 30 were signed and Louisiana was ceded to the United States. The purchase price, including interest payments, amounted to $27, 267, 622.

Suggested Readings

Adams, Henry. The History of the United States During the Administrations of Jefferson and Madison. New York: Scribner's, 1889-1891.

Barbé-Marbois, François. Histoire de la Louisiane et de la Cession. Paris: 1829. English translation: 1830.

Brooks, Philip C. "Spain's Farewell to Louisiana, 1803-1821," Mississippi Valley Historical Review, June, 1940.

Fletcher, Mildred S. "Louisiana as a Factor in French Democracy from 1763 to 1800," Mississippi Valley Historical Review, December, 1930.

Gayarré, Charles. History of Louisiana. New York: AMS Press, 1974.

Gilman, Daniel C. James Monroe. Boston: Houghton Mifflin, 1888.

Hirschfeld, Burt. Four Cents an Acre: The Story of the Louisiana Purchase. New York: Messner, 1965.

James, James A. "Louisiana as a Factor in American Diplomacy, 1795-1800," Mississippi Valley Historical Review, June, 1914.

Judson, Clara. Thomas Jefferson, Champion of the People. Chicago: Wilcox, 1952.

La Fargue, André. "The Louisiana Purchase: The French Viewpoint," Louisiana Historical Quarterly, January, 1940.

Lyon, E. Wilson. Louisiana in French Diplomacy. Norman, Okla.: University of Oklahoma Press, 1934.
________. The Man Who Sold Louisiana: The Career of François Barbé-Marbois. Norman, Okla.: University of Oklahoma Press, 1942.
Ogg, Frederic A. The Opening of the Mississippi. New York: Cooper Square, 1968.
Whitaker, Arthur P. The Mississippi Questions, 1795-1803. New York: Appleton, 1934.

MARBURY VS. MADISON (1803)

The Supreme Court case of Marbury vs. Madison was the first in which the Supreme Court held a law of Congress void. John Marshall, later Chief Justice of the court which heard the case, had served briefly as President John Adams' Secretary of State. The Federalists, under Adams, were about to relinquish power to Thomas Jefferson and the Republicans. Under the Judiciary Act of February 27, 1801, Adams appointed William Marbury a justice of the peace but, because of the negligence of Marshall, his commission, together with several others, though signed by Adams, was not delivered. After Jefferson became President he instructed James Madison, the new Secretary of State, to withhold it. Marbury brought suit for a writ of mandamus ordering Madison to deliver the commission to him.

Chief Justice Marshall, in ruling on the matter, discussed two questions: first, the ethical one of withholding the commissions and, second, the right of the Supreme Court to issue a mandamus.

On the first question he ruled that, in what is generally considered obiter dicta, Jefferson had no right to withhold a commission which had been signed by President Adams and bore the seal of the United States. He ruled on the second question that the law previously passed by Congress giving the Court the right to issue a mandamus was unconstitutional and therefore void.

It was not until 1857 when Chief Justice Roger B. Taney, in handing down the Supreme Court decision in the case of Dred Scott vs. Sandford, that the Court held another act of Congress null and void.

Suggested Readings

Beveridge, Albert J. The Life of John Marshall. Boston: Houghton Mifflin, 1916-1919.
Boudin, Louis B. Government by Judiciary. New York: William Goodwin, 1932.
Carson, Hampton L. The History of the Supreme Court. New York: Burt Franklin, 1972.
"The Case of the Missing Commissions," American Heritage Maga-

zine, June, 1963.
Commager, Henry Steele, ed. "Marbury v. Madison," (Doc. No. 109) in his Documents of American History, 8th edition. New York: Appleton, 1968.
Corwin, Edward S. Court over Constitution. Princeton, N.J.: Princeton University Press, 1938.
_______. The Doctrine of Judicial Review. Princeton, N.J.: Princeton University Press, 1914.
Cotton, Joseph P., ed. Constitutional Decisions of John Marshall. New York: Da Capo Press, 1967.
Garraty, John A., ed. Quarrels That Have Shaped the Constitution. New York: Harper, 1962.
Haines, Charles G. The American Doctrine of Judicial Supremacy. Los Angeles: University of California Press, 1932.
_______. The Role of the Supreme Court in American Government and Politics. Berkeley, Calif.: University of California Press, 1944.
Johnson, Gerald White. The Supreme Court. New York: Morrow, 1962.
Kelly, Alfred H., and Winifred A. Harbison. The American Constitution: Its Origins and Development. New York: Norton, 1970.
McCloskey, Robert G. The American Supreme Court. Chicago: University of Chicago Press, 1960.
McLaughlin, Andrew C. A Constitutional History of the United States. New York: Appleton, 1935.
_______. The Courts, The Constitution and Parties. New York: Da Capo Press, 1972.
Magruder, Allan B. John Marshall. Edited by John T. Morse, Jr. New York: AMS Press, 1898.
Thayer, James B. John Marshall. New York: Da Capo Press, 1974.
Warren, Charles. The Supreme Court in United States History. Boston: Little, Brown, 1923.

THE LEWIS AND CLARK EXPEDITION (1803-1806)

In 1803 President Thomas Jefferson appointed Captain Meriwether Lewis and Captain William Clark to head an expedition to explore the lands in North America acquired from France in that year by the transaction known as the Louisiana Purchase. Traveling westward from the Ohio River, the two leaders recruited volunteers from various military posts along the way. The party, consisting of 23 soldiers, a Negro slave, three interpreters and the two leaders, camped at the junction of the Mississippi and Missouri Rivers during the winter of 1803-04. On May 14 of the latter year, after recruiting an additional sixteen men in St. Louis, the expedition finally started. By October it had reached a point near the present site of Bismarck, North Dakota, where they spent the winter. In April, 1805, they resumed their journey and in June reached a point near present-day Great Falls, Montana.

After portaging around the falls the explorers ascended the Jefferson River, reaching the head of navigation on August 12, 1805. Here they abandoned river travel, obtaining horses and a guide. The guide, Sacajawea, was a Shoshone Indian squaw, known as the Bird Woman.

The party proceeded westward through the Rocky Mountains to the Clearwater River. Resuming river travel, the explorers then descended the Clearwater and Columbia Rivers, reaching the Pacific Ocean on November 15, 1805. Here they spent the winter.

The return journey commenced on March 23, 1806, the party reaching St. Louis on September 23 of that year.

Suggested Readings

Adams, Henry. The History of the United States During the Administrations of Jefferson and Madison. New York: Scribner's, 1889-1891.

Andrist, Ralph K. To the Pacific With Lewis and Clark. New York: American Heritage Press, 1967.

Bakeless, John. The Adventures of Lewis and Clark. Boston: Houghton Mifflin, 1962.

________. Lewis and Clark: Partners in Discovery. New York: Morrow, 1947.

Barclay, Isabel. Worlds Without End. Garden City, N.Y.: Doubleday, 1956.

Biddle, Nicholas. History of the Expedition Under the Command of Captains Lewis and Clark. Philadelphia: Lippincott, 1961.

Clark, William R. Explorers of the World. New York: Natural History Press, 1964.

Cutright, Paul R. Lewis and Clark: Pioneering Naturalists. Urbana, Ill.: University of Illinois Press, 1969.

Davis, Julia. No Other White Man. New York: Dutton, 1937.

De Voto, Bernard, ed. The Journals of Lewis and Clark. Boston: Houghton Mifflin, 1953.

Dillon, Richard. Meriwether Lewis: A Biography. New York: Coward-McCann, 1965.

Farnsworth, Frances Joyce. Winged Moccasins: The Story of Sacajawea. New York: Messner, 1954.

Frazier, Neta. Sacajawea: The Girl Nobody Knows. New York: David McKay, 1967.

Gass, Patrick. A Journal of the Voyages and Travels of a Corps of Discovery. Minneapolis: Ross and Haines, 1958.

Hays, Wilma P. The Meriwether Lewis Mystery. Philadelphia: Westminster Press, 1971.

Hirschfeld, Burt. Four Cents an Acre: The Story of the Louisiana Purchase. New York: Messner, 1965.

Lamb, Harold. New Found World: How North America Was Discovered and Explored. New York: Doubleday, 1955.

Laut, A. C. Pathfinders of the West. New York: Macmillan, 1904.

Mirsky, Jeannette. The Westward Crossings. New York: Knopf, 1946.

Munves, James. We Were There With Lewis and Clark. New York: Grosset and Dunlap, 1959.
Salisbury, Albert, and Jane Salisbury. Two Captains West. Seattle: Superior Publishing Co., 1950.
Seymour, Flora W. Sacajawea: Bird Girl. Indianapolis: Bobbs-Merrill, 1945.

THE AARON BURR CONSPIRACY (1806-1807)

Because of their political differences, Alexander Hamilton, Secretary of the Treasury, and Aaron Burr, Vice President of the United States, fought a duel on July 11, 1804, in which Hamilton was killed and Burr's political career blasted. An ambitious, unscrupulous lawyer and politician, Burr, who had almost achieved the Presidency but lost out to his enemy Thomas Jefferson, turned to the West where he schemed to found a state of his own. To achieve this he needed both men and money, which he lacked.

Historians are not in agreement on the nature of his plans. One contention is that he planned to separate Louisiana from the Union by force and set it up as an independent state. The other view is that he proposed to conduct a filibustering expedition against Vera Cruz and Mexico City. The real controversy is whether his conspiracy was aimed at Louisiana or Mexico. It is also quite possible that he had both schemes in mind--to secure Louisiana first and then operate against Vera Cruz.

One of Burr's associates was General James Wilkinson, an unscrupulous American soldier and adventurer. The latter, feeling that Burr's plans were doomed to failure, wrote to President Jefferson exposing Burr's intentions but concealing his own connection with the plan. Jefferson, on November 27, 1806, issued a proclamation warning citizens against the Burr conspiracy.

Burr's friends advised him that he was subject to arrest as the result of Jefferson's proclamation, and he fled to New Orleans with some 60 of his followers, later abandoning them and setting off for West Florida. He was arrested and sent to Richmond, Virginia, for trial.

The case aroused wide interest. Chief Justice John Marshall presided at the hearing, at which Burr was charged with treason. By the Constitution, treason is the levying of war against the government, or giving aid and comfort to the enemy, and two witnesses to the same overt act are necessary for conviction.

Marshall ruled that a man must be present when the overt act was committed in order to be guilty of treason within the meaning of the Constitution. Although Burr was generally known to have planned the whole movement, he was acquitted, as he was in Kentucky when his followers assembled on the Ohio River.

Both Marshall and Luther Martin, Burr's attorney, were enemies of President Jefferson. The latter was summoned to testify and to bring certain papers with him. This summons was disregarded, on the grounds that the president was not to be at the command of the federal courts. In refusing this subpoena Jefferson laid out the lines beyond which the court was not to go.

Suggested Readings

Abernethy, Thomas P. The Burr Conspiracy. New York: Oxford University Press, 1954.

Adams, Henry. The History of the United States During the Administrations of Jefferson and Madison. New York: Scribner's, 1889-1891.

Bassett, John Spencer. A Short History of the United States. New York: Macmillan, 1929.

Beveridge, Albert J. The Life of John Marshall. Boston: Houghton Mifflin, 1916-1919.

Bradford, Gamaliel. Damaged Souls. New York: Houghton Mifflin, 1923.

________. Wives. New York: Harper, 1925.

Commager, Henry Steele, ed. "Jefferson's Message on the Aaron Burr Conspiracy," (Doc. No. 110) in his Documents of American History, 8th edition. New York: Appleton, 1968.

Crouse, Anna, and Russel Crouse. Alexander Hamilton and Aaron Burr. New York: Random House, 1958.

Dos Passos, John. "The Conspiracy and Trial of Aaron Burr," American Heritage Magazine, February, 1966.

Jacobs, James R. Tarnished Warrior. Plainview, N.Y.: Books for Libraries, 1972.

Levy, Leonard W. Jefferson and Civil Liberties: The Darker Side. Cambridge, Mass.: Belknap Press, 1963.

McCaleb, Walter F. The Aaron Burr Conspiracy. New York: Wilson-Erickson, 1936.

Morris, Richard B. "The Trial of Aaron Burr," in his Fair Trial. New York: Knopf, 1952.

Nye, Russel B. A Baker's Dozen: Thirteen Unusual Americans. East Lansing, Mich.: Michigan State University Press, 1956.

Parton, James. Famous Americans of Recent Times. New York: Johnson Reprint, 1867.

________. The Life and Times of Aaron Burr. Boston: Houghton Mifflin, 1892.

Safford, William H., ed. The Blennerhassett Papers. New York: Arno Press, 1975.

________. The Life of Harman Blennerhassett. Plainview, N.Y.: Books for Libraries, 1972.

Schachner, Nathan. Aaron Burr: A Biography. New York: Stokes, 1937.

Todd, Charles Burr. Life of Colonel Aaron Burr, etc. (pamphlet). New York: 1879.

Vidal, Gore. Burr (historical fiction). New York: Random House, 1973.

Wandell, Samuel H. Aaron Burr in Literature. Port Washington, N.Y.: Kennikat Press, 1936.

________, and Meade Minnigerode. Aaron Burr. New York: Putnam, 1925.

Wilkinson, James. Memoirs of My Own Time. New York: AMS Press, 1816.
Wise, William. Aaron Burr. New York: Putnam, 1968.

THE "CHESAPEAKE"-"LEOPARD" AFFAIR (1807)

The War of 1812 between Great Britain and the United States stemmed from two basic causes: seizures of American ships by the British and the impressment of sailors from American ships, also by the British. France also seized American ships, but England had the stronger navy and her offenses were more numerous. This seizing was based on charges that American ships were in violation of British rules of war, the most noted of which was the Rule of War of 1756. This declared that "a trade not open in peace could not lawfully be opened in time of war." American shipowners circumvented this rule on technical grounds until the complaints of British shippers resulted in British courts invariably finding against Americans when their ships were taken.

Impressment involved British naval officers stopping American ships, boarding them and, without regard to neutrality laws, taking off sailors whom they declared to be British subjects. These sailors were forcibly enlisted in the British navy, replacing deserters and bringing short crews up to strength. Deserters, alleged and otherwise, were reinstated as British sailors when captured.

Several incidents of seizure and impressment occurred which President Thomas Jefferson and later President James Madison were willing to overlook, the first of these being the "Chesapeake"-"Leopard" Affair of 1807. In that year a British squadron was anchored in Lynnhaven Bay, near Norfolk, Virginia. The American ship "Chesapeake," Captain Barron commanding, preparing for a cruise to the Mediterranean, had reportedly shipped some British sailors, deserters from the British navy. The British Admiral Berkely had ordered that the "Chesapeake" be intercepted at sea and searched. Captain Barron, under instructions from President Jefferson to include no British deserters in his crew, believed that he had obeyed. However, one such deserter, shipping under an assumed name, had escaped his notice.

On June 22, 1807, the "Leopard" intercepted the "Chesapeake" at sea and signaled that she had dispatches. The "Leopard's" officer, on coming aboard, handed Barron an order from Admiral Berkely and demanded that any British deserters be handed over. Barron denied the presence of any such deserters aboard his vessel.

After the British officer returned to the "Leopard" that ship opened fire on the "Chesapeake" which, totally unprepared for combat, was quickly overcome. The British then boarded her a second time and removed three Americans who, having been impressed on a British ship, had deserted. They also removed the one British deserter who had enlisted under an assumed name.

The "Chesapeake" managed to reach Norfolk, where the news of the British action was the cause of an outburst of anger which raised anti-British sentiment in America to a high pitch. This affair, plus other aggressive acts on the part of the British, eventually led to the War of 1812.

Suggested Readings

Alderman, Clifford Lindsey. Wooden Ships and Iron Men. New York: Walker, 1964.

Babcock, Kendric C. The Rise of American Nationality, 1811-1819. Westport, Conn.: Greenwood Press, 1906.

Bassett, John Spencer. A Short History of the United States. New York: Macmillan, 1929.

Bemis, Samuel F., ed. American Secretaries of State and Their Diplomacy. New York: Cooper Square, 1927-1929.

Carr, Albert H. Z. The Coming of War. Garden City, N.Y.: Doubleday, 1960.

Forester, Cecil S. The Age of Fighting Sail. Garden City, N.Y.: Doubleday, 1956.

Goodman, W. H. "The Origins of the War of 1812," Mississippi Valley Historical Review, September, 1941.

Jameson, John F., ed. Privateers and Piracy in the Colonial Period. Clifton, N.J.: Augustus M. Kelley, 1923.

Mahan, Alfred T. Sea Power in Its Relations to the War of 1812. Westport, Conn.: Greenwood Press, 1969.

Stivers, R. E. Privateers and Volunteers. New York: Arco Press, 1975.

Updyke, Frank A. The Diplomacy of the War of 1812. Baltimore: Johns Hopkins University Press, 1915.

THE EMBARGO ACT (1807)

The Embargo Act of 1807, imposed by the United States against foreign commerce, was the direct result of the interference with American trade by France and England. The Act, recommended by President Thomas Jefferson and enacted by Congress on December 22, 1807, prohibited American vessels from sailing to foreign ports. It did permit coastwise commerce after the posting of heavy bond, and foreign vessels were allowed to sail from American ports provided they carried no cargo.

England and France were at war, and both countries had declared blockades. The United States was the chief source of foodstuffs for both belligerents and each tried to eliminate commerce between the other and America.

As a result of this situation, American shippers found their vessels searched and seized, their cargoes impounded, and many of their sailors impressed. However, neither England nor France could enforce its blockade and American shippers, because of the large profits to be made, defied both countries.

The British and French countered by licensing privateers to attack neutral ships, chiefly American, and about $60,000,000 in cargo and 1,600 American ships were lost to these privateers. America, in turn, counterattacked by passing the Embargo Act.

The economic effect of the Act on the American economy was disastrous. At the time of its passage, in spite of British and French depredations on American shipping, business was booming and foreign trade had steadily increased. American products were selling at high prices and additional ships were being built for the transportation of foreign shipments. Not only did the Act fail to induce France and England to cease interfering with American sea-borne commerce, but it quickly ended the era of prosperity for American farmers and businessmen. The dollar value of export trade sank to only a fraction of what it had been. Prices of agricultural products declined. Merchants were unable to dispose of their inventories of goods. Industries such as shipbuilding virtually ceased and unemployment was widespread. While a few home and manufacturing industries came into being opposition to the Act was widespread.

On March 1, 1809, the Embargo Act was replaced with the Non-Intercourse Act. This prohibited commerce with France and Great Britain.

Suggested Readings

Adams, Henry. The History of the United States During the Administrations of Jefferson and Madison. New York: Scribner's, 1889-1891.

_______. The Life of Albert Gallatin. Gloucester, Mass.: Peter Smith, 1879.

_______. The United States in 1800. Ithaca, N.Y.: Cornell University Press, 1969.

Adams, James Truslow. The Epic of America. Boston: Little, Brown, 1931.

Babcock, Kendric C. The Rise of American Nationality, 1811-1819. Westport, Conn.: Greenwood Press, 1906.

Beard, Charles A., and Mary R. Beard. A Basic History of the United States. New York: Doubleday, 1944.

Benton, Thomas Hart. Abridgement of the Debates of Congress, 1789-1856. New York: AMS Press, 1970.

Brant, Irving. James Madison: The Virginia Revolutionist. Indianapolis: Bobbs-Merrill, 1941.

Carey, Mathew. Olive Branch, or Faults on Both Sides, Federal and Democratic. Plainview, N.Y.: Books for Libraries, 1969.

Channing, Edward. The Jeffersonian System. New York: Cooper Square, 1968.

Commager, Henry Steele, ed. "Commercial Warfare," (Doc. No. 112) in his Documents of American History, 8th edition. New York: Appleton, 1968.

Dodd, William E. The Life of Nathaniel Macon, 1757-1837. New York: Burt Franklin, 1903.

Mahan, Alfred T. Sea Power in Its Relations to the War of 1812.

Westport, Conn.: Greenwood Press, 1969.
Malone, Dumas. Jefferson and the Ordeal of Liberty. Boston: Little, Brown, 1962.
Sears, Louis M. Jefferson and the Embargo. New York: Octagon Books, 1967.
Steinberg, Alfred. James Madison. New York: Putnam, 1965.
Stevens, John A. Albert Gallatin. Edited by John T. Morse, Jr. New York: AMS Press, 1898.
Warren, Charles. The Supreme Court in United States History. Boston: Little, Brown, 1923.
Wilkie, Katharine E., and Elizabeth R. Mosley. Father of the Constitution: James Madison. New York: Messner, 1963.

FLETCHER VS. PECK (1810)

The Supreme Court decision in the case of Fletcher vs. Peck was the first one in which the Court held a state law void under the Constitution.

In this famous and precedent-setting case the act of a state legislature was in question. In 1795 the Georgia legislature had passed an act for the sale of certain of the state's western lands. In 1796 a new legislature declared the original act null and void on the ground that it was passed fraudulently.

Peck claimed land under the annulled grant and brought suit in the federal courts. His contention was that the state of Georgia was in violation of the Constitutional clause which forbids a state to pass a law "impairing the obligation of a contract." Georgia's claim was that a land grant, made by a state in the disposal of a domain, was not a contract.

Two constitutional questions were raised by this case: the interpretation of the contract clause of the Constitution and the power of the Supreme Court to consider the legislative acts of a state.

The Court's decision, rendered by Chief Justice John Marshall, was that, first, a grant is a contract, and second, Georgia's attempt to repeal the grant was illegal.

A problem immediately arose: who was to execute the decision of the Supreme Court against a state? While the President would ordinarily be the one to do this, no power existed to compel him should he decline. The state of Georgia, realizing this, defied the Court and the difficulty was settled by compromise in 1814 when Congress paid sums of money to the claimants under the Georgia land grants.

Suggested Readings

Beveridge, Albert J. The Life of John Marshall. Boston: Houghton Mifflin, 1916-1919.
Carson, Hampton L. The History of the Supreme Court. New

York: Burt Franklin, 1972.
Commager, Henry Steele, ed. "The Constitution of the United States," (Doc. No. 87) in his Documents of American History, 8th edition. New York: Appleton, 1968.
_______. "Fletcher v. Peck," (Doc. No. 113) in his Documents of American History, 8th edition. New York: Appleton, 1968.
Corwin, Edward S. John Marshall and the Constitution. New Haven, Conn.: Yale University Press, 1919.
Cotton, Joseph P., ed. Constitutional Decisions of John Marshall. New York: Da Capo Press, 1967.
Haines, Charles G. The Role of the Supreme Court in American Government and Politics. Berkeley, Calif.: University of California Press, 1944.
Hunting, Warren B. The Obligation of Contracts Clause of the United States Constitution. Baltimore: Johns Hopkins University Press, 1919.
Johnson, Gerald White. The Supreme Court. New York: Morrow, 1962.
McCloskey, Robert G. The American Supreme Court. Chicago: University of Chicago Press, 1960.
McGrath, C. Peter. Yazoo: Law and Politics in the New Republic, the Case of Fletcher vs. Peck. Providence: Brown University Press, 1966.
Magruder, Allan B. John Marshall. Edited by John T. Morse, Jr. New York: AMS Press, 1898.
Thayer, James T. John Marshall. New York: Da Capo Press, 1974.
Warren, Charles. The Supreme Court in United States History. Boston: Little, Brown, 1923.
Wright, Benjamin F., Jr. The Contract Clause of the Constitution. Cambridge, Mass.: Harvard University Press, 1938.

THE FATE OF THEODOSIA BURR ALSTON (1812)

Following his 1807 treason trial at Richmond, Virginia, in which he was acquitted, Aaron Burr, former Vice President of the United States, went to Europe in a vain attempt to interest the English and French authorities in his privateering schemes. In 1812 he returned to America to practice law in New York City. Here he found himself ostracized and the victim of general public dislike, owing to the above-mentioned trial in which many people felt he should have been convicted. Further, many of Alexander Hamilton's politically powerful friends believed that Burr had committed an irreparable wrong in killing him in their duel at Weehawken, New Jersey, in 1804.

Burr's daughter Theodosia was born in 1783. As a young girl she was charming, beautiful, and a precocious student. Under her father's tutelage she mastered French, Latin and Greek and at the age of fourteen, following her mother's death, acted as hostess in her father's home.

She grew up to be a gracious woman, the proverbial apple of her father's eye. In 1801 she married Joseph Alston, a South Carolina plantation owner, and went to live with him in his Charleston mansion.

Over the years father and daughter conducted a voluminous correspondence. When Burr indicated his intention to return to America from his sojourn in Europe, his daughter, overjoyed, planned to travel to New York to visit him. She was in poor health, her only son Aaron Burr Alston having become ill and died at the age of twelve, leaving her prostrated with grief.

In December, 1812, Theodosia Burr Alston departed Charleston for New York aboard the "Patriot," a small schooner which had been engaged in privateering. She was accompanied by her maid and a physician whom her father had sent to attend his daughter on the voyage. Joseph Alston, now Governor of South Carolina and a general officer in the state militia, was unable to accompany her. Though small, the "Patriot" was well known for her sailing qualities. Her captain was a proficient seaman and it was estimated that the trip to New York should take "not more than five or six days." The schooner's guns had been dismounted and stowed below and she carried a cargo of rice from Alston's plantations. The War of 1812 was in progress and the captain of the "Patriot" carried a letter from Alston addressed to the commander of the British fleet, explaining the circumstances of the voyage and requesting a safe passage to New York.

The "Patriot" cleared Charleston on the afternoon of December 30. She crossed the bar the following morning and was never seen again. No member of her crew is known to have reached safety. No report of her fate was ever made.

It was later learned that, three days out from Charleston, a severe storm had occurred off Cape Hatteras. This storm was so intense that several British ships of the line, all much larger than the "Patriot," had been sent to the bottom. Historians are generally agreed that the most logical solution to the mystery of the disappearance of the "Patriot" is that she had foundered in the storm and that all hands were lost. However, other possibilities exist and a number of legends concerning the fate of Theodosia Burr Alston have appeared over the years.

One thought is that the schooner may have been taken by a British man-of-war because of her previous activities as a privateer. Her guns, as mentioned above, had been removed and she would therefore have been unable to defend herself against an armed British warship. This theory, however, does not explain the fate of those aboard the vessel and Theodosia Burr Alston was a person of such prominence that news of her capture by the British could not long have been concealed.

It was also suggested that the schooner may have been captured by pirates, of which a few still lurked along the east coast. If this were the case, it was surmised, the women aboard the "Patriot" could have been taken prisoner by the pirates or else made to walk the plank. If taken prisoner, it was possible that they could have been transported to the West Indies, there to become the wives or concubines of a pirate chief. No evidence to support this theory ever came forth.

For many years after 1813 reports appeared concerning "a beautiful and cultured woman" being seen aboard a ship manned by pirates or being detained in an island prison along with the captured crew of a pirate vessel. One rumor was that the missing woman had been seen in Europe "with a British naval officer who was showing her marked attentions." Various "reformed pirates" came forth with stories of various sorts which undertook to establish the facts of the ultimate fate of the missing woman, but none of these stories was ever substantiated.

Suggested Readings

Minnigerode, Meade. Lives and Times. Plainview, N.Y.: Books for Libraries, 1975.

Parton, James. Famous Americans of Recent Times. New York: Johnson Reprint, 1867.

_______. The Life and Times of Aaron Burr. Boston: Houghton Mifflin, 1892.

Pidgin, Charles Felton. Theodosia: The First Gentlewoman of Her Times. Boston: Houghton Mifflin, 1901.

Safford, William H., ed. The Blennerhassett Papers. New York: Arno Press, 1975.

_______. The Life of Harman Blennerhassett. Plainview, N.Y.: Books for Libraries, 1972.

Smith, Edward H. "Severed from the Race," in his Mysteries of the Missing. New York: Lincoln MacVeagh--The Dial Press, 1927.

Todd, Charles Burr. The Life of Colonel Aaron Burr, etc. (pamphlet). New York: 1879.

Vidal, Gore. Burr (historical fiction). New York: Random House, 1973.

Wandell, Samuel H., and Meade Minnigerode. Aaron Burr. New York: Putnam, 1925.

THE WRITING OF THE "STAR SPANGLED BANNER" (1814)

Francis Scott Key, an American lawyer, essayist and poet, is best remembered as the author of the "Star Spangled Banner." His essay, "The Power of Literature and Its Connection with Religion," which appeared in 1834, and "Poems," published posthumously in 1857, are all but forgotten today, but the "Star Spangled Banner" will always be remembered.

The War of 1812 was still being fought when the poem was written. Key had, under a flag of truce, boarded the British ship "Surprise" to negotiate the release of a friend who was being held prisoner. The British were bombarding Fort McHenry, the most important of the ring of American fortifications around Baltimore, and Key had been detained temporarily aboard the vessel during the attack.

The bombardment continued through the night of September 13-14, 1814. At daybreak the sight of the American flag still flying over the fort inspired Key to write the poem, the "Star Spangled Banner." He directed that the words be sung to the tune of "Anacreon in Heaven," a drinking song composed in England by John Stafford Smith sometime between 1770 and 1775. The song, with Key's words, was first sung in a Baltimore tavern.

On March 3, 1931, the words and music were declared by Congress to be the national anthem of the United States. Equally well-known is the patriotic hymn, "My Country, 'Tis of Thee," written by Samuel Francis Smith to the music of the British anthem, "God Save the King."

Suggested Readings

Georgiady, Nicholas P., and Louis Romano. Our National Anthem. Chicago: Follett, 1963.

Grove, Sir George. Dictionary of Music and Musicians. New York: Macmillan, 1954.

Key, Francis Scott. First Book Edition of the Star-Spangled Banner. New York: Watts, 1961.

________. Poems of the Late Francis S. Key, Esq. New York: 1857.

Key-Smith, F. S. Francis Scott Key. Washington, D.C.: Pratt, 1911.

Pickett, Mrs. L. C. "Poet of the Flag," in her Literary Hearthstones of Dixie. Philadelphia: Lippincott, 1912.

Quaife, Milo M. The History of the United States Flag. New York: Harper, 1961.

Sonneck, Oscar George Theodore. The Star Spangled Banner (song). New York: Da Capo Press, 1969.

Stevenson, Augusta. Francis Scott Key, Maryland Boy. Indianapolis: Bobbs-Merrill, 1946.

Swanson, Neil H., and A. S. Swanson. The Star-Spangled Banner. Philadelphia: Winston, 1958.

Weybright, Victor. Spangled Banner: The Story of Francis Scott Key. New York: Farrar & Rinehart, 1935.

THE HARTFORD CONVENTION (1814-1815)

The New England Federalists had been staunchly opposed to the War of 1812 between the United States and Great Britain. This opposition stemmed from its crippling effect on foreign commerce and on the fishing industry. The objects of the Hartford Convention were to devise means of safeguarding the privileges of the individual states against alleged encroachments of the federal government and to find ways of defending themselves against foreign nations.

The idea of a convention had been originally suggested by Senator Timothy Pickering of Massachusetts in 1812, but it was not until 1814 that the idea was revived. Governor Caleb Strong of Massachusetts, dissatisfied with the federal government's failure to pay the expenses of the state militia when he called it out against the British, felt that his state had been abandoned and would have to look out for its own interests. He called a meeting of the state legislature which, in turn, chose twelve delegates to a convention at Hartford, Connecticut. This was approved by the state of Connecticut, which appointed seven delegates, and also by the state of Rhode Island, which appointed four. Two came from New Hampshire and one from Vermont.

The meeting was conducted behind closed doors and lasted from December 15, 1814, to January 5, 1815. George Cabot, a former senator, was made president. The recommendations of the Convention, which took the form of seven amendments to the federal constitution, called for a number of drastic changes in the relationship between the individual states and the federal government. The ideas promulgated also upheld the opinion that a state should conduct her defense when invaded.

The legislatures of Massachusetts and Connecticut approved the proposed constitutional amendments and sent delegates to Washington, D.C. to urge that they be adopted. However, when disunion seemed inevitable, news of the signing of the Treaty of Ghent on December 24, 1814, ending the war was received and the whole movement collapsed.

Suggested Readings

Adams, Henry. The History of the United States During the Administrations of Jefferson and Madison. New York: Scribner's, 1889-1891.

______, ed. Documents Relating to New England Federalism, 1800-1815. New York: Burt Franklin, 1905.

Bassett, John Spencer. A Short History of the United States. New York: Macmillan, 1929.

Carey, Mathew. Olive Branch, or Faults on Both Sides, Federal and Democratic. Plainview, N.Y.: Books for Libraries, 1969.

Coles, Harry L. The War of 1812. Chicago: University of Chicago Press, 1965.

Commager, Henry Steele, ed. "Reports and Resolutions of the Hartford Convention," (Doc. No. 115) in his Documents of American History, 8th edition. New York: Appleton, 1968.

Dangerfield, George. The Era of Good Feelings. New York: Harcourt, Brace, 1952.

Dwight, Theodore. History of the Hartford Convention. New York: Da Capo Press, 1833.

Horsman, Reginald. The War of 1812. New York: Knopf, 1969.

Lodge, Henry Cabot, ed. The Life and Letters of George Cabot. New York: Da Capo Press, 1974.

Morison, Samuel Eliot. Harrison Grey Otis, Seventeen Sixty Five to Eighteen Forty Eight: The Urbane Federalist. Boston: Houghton Mifflin, 1969.

Perkins, Bradford. Castlereagh and Adams. Berkeley, Calif.:

University of California Press, 1964.
Smelser, Marshall. The Democratic Republic, 1801-1815. New York: Harper, 1968.

THE FIRST BATTLE OF NEW ORLEANS (1815)

The first of the two battles of New Orleans was fought on January 8, 1815, between some four thousand American troops commanded by General Andrew Jackson and more than ten thousand British troops under General Sir Edward Pakenham. Jackson's troops were reinforced by the French pirate and smuggler, Jean La Fitte, and his men. In return for his assistance, La Fitte and his fellow buccaneers were pardoned by President James Madison, but later they resumed their freebooting activities.

Jackson, learning that the British planned to attack and conquer the Gulf of Mexico region of the United States and that New Orleans was threatened by Pakenham's forces, hurried to that city, which was virtually undefended. On his way there he called out the militia from Tennessee, Kentucky, and Georgia.

The British, transported in a fleet of fifty vessels, had sailed from their base at Jamaica in the West Indies. They anchored in Lake Borgne, east of New Orleans, on December 10, 1814. Early on December 23 a division of the British army was landed some eight miles below the city on a narrower strip of land between the swamp and the river. A sharp battle followed and the British took refuge under the levee until reinforcements from their ships could reach them. This gave Jackson sufficient time to construct breastworks and in this he was aided by Pakenham's unwillingness to resume the fighting until all his reinforcements, including the artillery, were landed.

On January 8, 1815, Pakenham attempted to storm Jackson's defenses. The attempt to carry the American works was unsuccessful. Generals Pakenham and Gibbs were killed and General Keene was seriously wounded. British losses in this part of the army numbered 1,971 killed and wounded. American losses were only thirteen.

The British Colonel Thornton with 600 men crossed to the west bank of the Mississippi to attack some batteries there. He succeeded in this attempt and held the west bank. However, the Americans had repulsed the attack on the east bank and this so crippled the British that they abandoned the campaign and withdrew to their fleet. By the end of January they sailed for England.

The American victory at New Orleans saved the mouth of the Mississippi from conquest and made Andrew Jackson a national hero. The battle had no effect on the outcome of the War of 1812 as, unknown to those concerned, the War had been terminated by the signing of the Treaty of Ghent on December 24, 1814.

New Orleans was the scene of another battle which occurred in April, 1862, during the Civil War. This encounter between

Union and Confederate forces resulted in the capture of the city, which was then occupied by Union troops under General Benjamin Franklin Butler.

Suggested Readings

Adams, Henry. The History of the United States During the Administrations of Jefferson and Madison. New York: Scribner's, 1889-1891.
Andrist, Ralph K. Andrew Jackson: Soldier and Statesman. New York: American Heritage Press, 1963.
Bassett, John Spencer. The Life of Andrew Jackson. New York: Macmillan, 1925.
Brackenridge, Henry M. The History of the Late War Between the United States and Great Britain. Boston: Gregg Press, 1971.
Bradford, Gamaliel. Damaged Souls. New York: Houghton Mifflin, 1923.
Brooks, Charles B. The Siege of New Orleans. Seattle: University of Washington Press, 1961.
Brown, Wilburt S. The Amphibious Campaign for West Florida and Louisiana. University, Ala.: University of Alabama Press, 1969.
Chidsey, Donald Barr. The Battle of New Orleans. New York: Crown, 1961.
Coles, Harry L. The War of 1812. Chicago: University of Chicago Press, 1965.
De Grummond, Jane L. The Baratarians and the Battle of New Orleans. Baton Rouge: Louisiana State University Press, 1961.
Fortescue, John W. A History of the British Army, 1813-1815. London: Macmillan, 1920.
Gayarré, Charles. An Historical Sketch of Pierre and Jean Lafitte, the Famous Smugglers of Louisiana. Austin, Tex.: Jenkins Publishing Co., 1885.
Gleig, George R. The Campaigns of the British Army at Washington, Baltimore and New Orleans. Totowa, N.J.: Rowman & Littlefield, 1972.
Jacobs, James R., and Glenn Tucker. The War of 1812. New York: Hawthorn, 1969.
James, Marquis. Andrew Jackson: The Border Captain. Indianapolis: Bobbs-Merrill, 1933.
_______. Andrew Jackson: Portrait of a President. Indianapolis: Bobbs-Merrill, 1933.
Leckie, Robert. Great American Battles. New York: Random House, 1968.
Lossing, Benson J. Pictorial Field-Book of the War of 1812. Somersworth, N.H.: New Hampshire Publishing Co., 1973.
McConnell, Roland C. Negro Troops of Antebellum Louisiana. Baton Rouge: Louisiana State University Press, 1968.
Nolan, Jeannette Covert. Andrew Jackson. New York: Messner, 1949.

Parton, James. The Life of Andrew Jackson. New York: Johnson Reprint, 1860.
Reeder, Red. The Story of the War of 1812. New York: Duell, Sloan, 1960.
Remini, Robert V. Andrew Jackson. New York: Twayne, 1966.
Saxon, Lyle. Lafitte the Pirate. New Orleans: Robert L. Craeger, 1937.
Sumner, William Graham. Andrew Jackson. Boston: Houghton Mifflin, 1909.
Tallant, Robert. The Pirate Lafitte and the Battle of New Orleans. New York: Random House, 1951.
Vance, Marguerite. The Jacksons of Tennessee. New York: Dutton, 1953.

M'CULLOCH VS. MARYLAND (1819)

Two questions were decided by the Supreme Court decision in the case of M'Culloch vs. Maryland, which was decided in 1819. The first of these--"has Congress the right to incorporate a bank?"--was decided in the affirmative. The second was: "was a state tax on the bank constitutional?" Here Chief Justice John Marshall, delivering the unanimous opinion of the Court, stated that "the States have no power, by taxation or otherwise, to retard, impede, burden, or in any manner control, the operation of the constitutional laws enacted by Congress to carry into execution the powers vested in the general government."

This famous and important case arose when M'Culloch, cashier of the Baltimore branch of the Bank of the United States which was chartered by Congress, refused to pay a tax on its notes, which tax was imposed by the state of Maryland.

In rendering the Court's decision, Marshall construed the powers of the national government. He stated that the Court would declare a law passed by Congress of no effect if such law was unconstitutional, but the Court, not being a law-making body, would not attempt to annul a law of Congress made in the field proper to the activity of Congress.

The Court regarded the creation of a bank as something within the purview of Congress and as a thing useful "in the happy and prosperous government of the nation," and therefore constitutional.

So far as the state's right to tax the bank was concerned, the Court's decision was based on the thought that if a small tax could be imposed, so could a large one, making it possible for the state to tax the bank out of existence. Marshall's words in this connection, that "the power to tax involves the power to destroy," are still remembered.

Suggested Readings

Bassett, John Spencer. A Short History of the United States. New York: Macmillan, 1929.
Beveridge, Albert J. The Life of John Marshall. Boston: Houghton Mifflin, 1916-1919.
Boudin, Louis B. Government by Judiciary. New York: William Goodwin, 1932.
Carson, Hampton L. The History of the Supreme Court. New York: Burt Franklin, 1972.
Catterall, Ralph C. The Second Bank of the United States. Chicago: University of Chicago Press, 1903.
Channing, Edward. A History of the United States. New York: Macmillan, 1921.
Commager, Henry Steele, ed. "M'Culloch v. Maryland," (Doc. No. 118) in his Documents of American History, 8th edition. New York: Appleton, 1968.
Corwin, Edward S. John Marshall and the Constitution. New Haven, Conn.: Yale University Press, 1919.
Cotton, Joseph P., ed. Constitutional Decisions of John Marshall. New York: Da Capo Press, 1967.
Faulkner, Robert E. The Jurisprudence of John Marshall. Princeton, N.J.: Princeton University Press, 1968.
Govan, Thomas B. Nicholas Biddle: Nationalist and Public Banker. Chicago: University of Chicago Press, 1959.
Hammond, Bray. Banks and Politics in America, from the Revolution to the Civil War. Princeton, N.J.: Princeton University Press, 1957.
Johnson, Gerald White. The Supreme Court. New York: Morrow, 1962.
Loth, David G. Chief Justice John Marshall and the Growth of the Republic. New York: Norton, 1949.
McCloskey, Robert G. The American Supreme Court. Chicago: University of Chicago Press, 1960.
Magruder, Allan B. John Marshall. Edited by John T. Morse, Jr. New York: AMS Press, 1898.
Thayer, James B. John Marshall. New York: Da Capo Press, 1974.
Warren, Charles. The Supreme Court in United States History. Boston: Little, Brown, 1923.

THE DARTMOUTH COLLEGE CASE (1819)

The Supreme Court decision popularly known as the "Dartmouth College Case" (officially Trustees of Dartmouth College vs. Woodward) is important in American judicial history because of its far-reaching implications in constitutional law. In 1769 Dartmouth College had been established by a charter granted by the English King George III. This charter was altered in 1816 by the state of New Hampshire which, under the altered charter, organized a new board of trustees. President John Wheelock of the College, who had been removed in 1815, was reinstated and the name of the institution changed to Dartmouth University.

The old board declined to accept this change, arguing that it involved an impairment of contract and, as such, was unconstitutional. After unsuccessful litigation in the state courts the trustees brought the case before the United States Supreme Court. They were represented by the attorney and statesman Daniel Webster. The essential question to be decided was, "is a charter granted to a corporation inviolate by the legislature?" The court held that a charter is indeed a contract and, as such, "may not be recalled by the legislature provided the grantee observes the conditions on which it was granted."

This decision, handed down by Chief Justice John Marshall, reversed the ruling of the Supreme Court of New Hampshire. It became a precedent in all cases arising under acts of incorporation, spelling out the principle that no legislature may impair obligations imposed by a contract.

Suggested Readings

Beveridge, Albert J. *The Life of John Marshall.* Boston: Houghton Mifflin, 1916-1919.

Brubacher, John S., and Willis Rudy. *Higher Education in Transition.* New York: Harper, 1968.

Carson, Hampton L. *The History of the Supreme Court.* New York: Burt Franklin, 1972.

Commager, Henry Steele, ed. "Trustees of Dartmouth College v. Woodward," (Doc. No. 119) in his *Documents of American History*, 8th edition. New York: Appleton, 1968.

Corwin, Edward S. *John Marshall and the Constitution.* New Haven, Conn.: Yale University Press, 1919.

Cotton, Joseph P., ed. *Constitutional Decisions of John Marshall.* New York: Da Capo Press, 1967.

Curtis, George T. *Constitutional History of the United States.* New York: Da Capo Press, 1974.

Faulkner, Robert K. *The Jurisprudence of John Marshall.* Princeton, N.J.: Princeton University Press, 1968.

Fisher, Sydney George. *The True Daniel Webster.* Philadelphia: Lippincott, 1911.

Johnson, Gerald White. *The Supreme Court.* New York: Morrow, 1962.

Loth, David C. *Chief Justice John Marshall and the Growth of the Republic.* New York: Norton, 1949.

McCloskey, Robert G. *The American Supreme Court.* Chicago: University of Chicago Press, 1960.

Magruder, Allan B. *John Marshall.* Edited by John T. Morse, Jr. New York: AMS Press, 1898.

Rudolph, Frederick. *The American College and University.* New York: Knopf, 1962.

Shirley, John M. *Dartmouth College Causes and the Supreme Court of the United States.* New York: Da Capo Press, 1971.

Thayer, James B. John Marshall. New York: Da Capo Press, 1974.
Warren, Charles. The Supreme Court in United States History. Boston: Little, Brown, 1923.
Webster, Daniel. The Letters of Daniel Webster. East Saint Clair Shores, Mich.: Scholarly Press, 1902.

THE MISSOURI COMPROMISE (1820)

Missouri, acquired from France in 1803 as part of the Louisiana Purchase, became a territory in 1812 and petitioned for statehood in March, 1818. This was not acted upon until February, 1819. At this time the question arose as to whether the terms of the Louisiana Purchase, which guaranteed the inhabitants their liberty, property, and religion, "covered property in slaves, of which there were some two or three thousand in the Territory."

The Tallmadge Amendment, proposed on February 13, 1819, excluding slavery from the State, passed the House but was defeated in the Senate. For the rest of the year the slavery question was the largest issue in the country, with the number of slave and free states equally divided.

On January 26, 1820, the Taylor Amendment was passed in the House and the Thomas Amendment of February 17, 1820 was passed in the Senate. The Taylor Amendment prohibited slavery in Missouri and the Thomas Amendment provided to admit Missouri as a slave state but prohibited slavery north of 36° 30' north latitude in the rest of the Louisiana Purchase.

On January 3, 1820, the House had passed a bill to admit Maine to the Union. Earlier in the session Alabama had been admitted, so the admission of Maine would give the free states a majority. Maine's admission offered a way out of the difficulty. Maine was admitted, as was Missouri with the Thomas Amendment after the resolution of problems emanating from the constitution drawn up by the Missouri Convention. These were resolved by a conference committee.

The Missouri Compromise remained in effect until repealed by the Kansas-Nebraska Act of 1854.

Suggested Readings

Adams, Henry. John Randolph. Boston: Houghton Mifflin, 1893.
Ames, H. V. State Documents on Federal Relations. New York: Da Capo Press, 1900.
Bassett, John Spencer. A Short History of the United States. New York: Macmillan, 1929.
Brown, Everett S., ed. The Missouri Compromise and Presidential Politics, 1820-1825, from the Letters of William Plumer, Jr. St. Louis: Missouri Historical Society, 1926.
Commager, Henry Steele, ed. "The Missouri Compromise," (Doc.

No. 121) in his Documents of American History, 8th edition. New York: Appleton, 1968.
Dangerfield, George. The Era of Good Feelings. New York: Harcourt, Brace, 1952.
Dixon, Susan. History of the Missouri Compromise and Slavery in American Politics. New York: Johnson Reprint, 1903.
King, Charles R., ed. The Life and Correspondence of Rufus King. New York: Da Capo Press, 1971.
MacDonald, William. Select Documents Illustrative of the History of the United States, 1776-1861. New York: Burt Franklin, 1968.
McMaster, John B. A History of the People of the United States, from the Revolution to the Civil War. New York: Appleton, 1938.
Moore, Glover. The Missouri Controversy. Lexington: University of Kentucky Press, 1953.
Ray, P. Orman. The Repeal of the Missouri Compromise. Boston: J. S. Canner, 1965.
Rhodes, James F. History of the United States from the Compromise of 1850. New York: Macmillan, 1893-1906.
Shoemaker, Floyd C. Missouri's Struggle for Statehood, 1804-1821. New York: Russell & Russell, 1969.
Snydor, Charles S. The Development of Southern Sectionalism. Baton Rouge: Louisiana State University Press, 1948.
Trexler, Harrison A. Slavery in Missouri. Baltimore: Johns Hopkins University Press, 1914.
Turner, Frederick Jackson. The Rise of the New West. Gloucester, Mass.: Peter Smith, 1959.
Von Holst, Hermann E. Constitutional and Political History of the United States. New York: AMS Press, 1892.
Wiltse, Charles M. John C. Calhoun. Indianapolis: Bobbs-Merrill, 1941-1944.

COHENS VS. VIRGINIA (1821)

The famous 1821 Supreme Court case of Cohens vs. Virginia was precipitated by a comparatively insignificant incident. P. J. and M. J. Cohen were prosecuted for selling lottery tickets in the state of Virginia, in violation of a Virginia statute. Although the Cohens, as defendants, claimed through their attorneys the protection of an 1802 act of Congress establishing a lottery, they were found guilty and fined by the Virginia court.

The Cohens then sued out a writ of error to the Supreme Court of the United States under Section 25 of the Judiciary Act of 1789. This Act specifies, in part, that when "the highest court of law or equity of a state renders a judgment in which the validity of a treaty or statute of, or an authority exercised under, the United States" is questioned and the state court's decision is against the validity of the statute of the United States, such statute "may be re-examined, and reversed or affirmed to the Supreme

Court of the United States upon a writ of error...."

The case of Cohens vs. Virginia involved the constitutionality of Section 25 of the Judiciary Act and the interpretation of the 11th Amendment to the Constitution. This Amendment, quoted in full, reads, "The Judicial power of the United States shall not be construed to extend to any suit in law or equity, commenced or prosecuted against any one of the United States by Citizens of another State or by Citizens or Subjects of any Foreign State."

In delivering the unanimous decision of the Court, Chief Justice John Marshall ruled that the Judiciary Act of 1789 was not contrary to the Constitution. The Court ruled further that it did have appellate jurisdiction in the case because it involved a matter of constitutional interpretation.

Following the establishment of jurisdiction, the Court heard the case and decided it on its merits. Its finding was that the 1802 Act of Congress authorizing the establishment of a lottery was confined to the city of Washington, which gave the Cohens no authority to sell lottery tickets elsewhere. Consequently, the Quarterly Session Court of Norfolk, Virginia was within its legal rights in trying and convicting the defendants for violation of a Virginia law. The judgment of the Virginia court was thereby affirmed.

Suggested Readings

Beveridge, Albert J. The Life of John Marshall. Boston: Houghton Mifflin, 1916-1919.

Carson, Hampton L. The History of the Supreme Court. New York: Burt Franklin, 1972.

Commager, Henry Steele, ed. "Cohens v. Virginia," (Doc. No. 123) in his Documents of American History, 8th edition. New York: Appleton, 1968.

________. "The Constitution of the United States," (Doc. No. 87) in his Documents of American History, 8th edition. New York: Appleton, 1968.

________. "The Judiciary Act of 1789," (Doc. No. 91) in his Documents of American History, 8th edition. New York: Appleton, 1968.

Corwin, Edward S. John Marshall and the Constitution. New Haven, Conn.: Yale University Press, 1919.

Cotton, Joseph P., ed. Constitutional Decisions of John Marshall. New York: Da Capo Press, 1967.

Faulkner, Robert K. The Jurisprudence of John Marshall. Princeton, N.J.: Princeton University Press, 1968.

Friedman, Leon, and Fred L. Israel, eds. The Justices of the U.S. Supreme Court, 1789-1869. New York: Bowker, 1969.

Johnson, Gerald White. The Supreme Court. New York: Morrow, 1962.

Loth, David G. Chief Justice John Marshall and the Growth of the Republic. New York: Norton, 1949.

McCloskey, Robert G. The American Supreme Court. Chicago: University of Chicago Press, 1960.

Magruder, Allan B. John Marshall. Edited by John T. Morse, Jr. New York: AMS Press, 1898.
Pritchett, C. Herman. The American Constitutional System. New York: McGraw-Hill, 1971.
Thayer, James B. John Marshall. New York: Da Capo Press, 1974.
Warren, Charles. The Supreme Court in United States History. Boston: Little, Brown, 1923.

THE "GLOBE" MUTINY (1824)

In the year 1824 a 22-year-old megalomaniac from Nantucket, Massachusetts, provoked one of the most grisly mutinies in the history of the Pacific Ocean. Samuel B. Comstock had dreams of establishing a kingdom of which he would be the supreme ruler. He resolved to achieve this by murdering the officers of some ship, seizing it, sailing it to an island in the South Seas and then murdering the survivors, leaving him alone as king.

With this ghastly plan in mind, Comstock, at the age of eighteen, set about learning navigation, harpooning, steering and the other arts of the master sailor. After a voyage or two he and his brother George shipped aboard the "Globe," a whaler bound for the Pacific. Captain Thomas Worth commanded the vessel and serving under him were first mate William Beetle, second mate John Lumbard, and third mate Nathaniel Fisher.

Late in 1823 the "Globe" stopped at Honolulu to take on supplies. Six sailors jumped ship and were replaced by seven newcomers, recruited by Captain Worth from saloons and brothels. These seven were, literally, the dregs of humanity. Four of them, Silas Payne, John Oliver, William Humphries and Joseph Thomas, were to play important parts in the mutiny to come.

Thomas was flogged for insolence to the mate and captain. Following this Comstock decided on his course of action. He would recruit Oliver, Thomas and Payne, together with any other crew members who were willing, and seize the ship at Fanning Island. It was here that Comstock planned to establish his kingdom.

At this time, although the "Globe" was accompanied by another ship, the "Lyra," Comstock went ahead with his plan. The men who had been recruited at Honolulu had agreed to join him. He murdered the captain with an ax and Payne attempted to stab first mate Beetle as he lay in his bunk but bungled the job. Comstock then killed Beetle with his ax, following which he attacked mates Fisher and Lumbard. The bodies of the dead and dying officers were cast into the sea. No word of this reached the "Lyra," which shortly afterwards departed on another course.

William Humphries was suspected of planning the overthrow of the mutineers. He was tried by a kangaroo court, found guilty, and hanged. The remaining crew members proceeded to Mili Atoll, an island at the southern tip of the Marshall Islands.

Before long the crew of the "Globe" realized that Comstock intended to kill them all, ally himself with the natives of the island, and declare himself king. Trouble developed between Payne and Comstock and the former, assisted by Oliver and two henchmen, shot and killed Comstock on the beach of the atoll, where he was buried.

Following the death of Comstock, Payne and Oliver took command of the island, ordering George Comstock, Gilbert Smith and four others to return to the ship. Under Smith's leadership the six men slipped away from the island in the "Globe" and, incredibly enough, for none of them could navigate, sailed towards South America, 7,500 miles away. Four months later the ship and her crew, including the nutineer Joseph Thomas, arrived at Valparaiso and in due course, with an augmented crew, she returned to Edgartown, Massachusetts. Smith was tried for mutiny and acquitted for lack of positive proof.

Of the men remaining on the island, led by Payne and Oliver, eventually all but two were massacred by the natives. The survivors were Will Lay, a young sailor, and Cyrus Hussey, a cooper. These two became slaves of the natives, but were rescued by the U.S. schooner "Dolphin" in November, 1825.

Suggested Readings

Fuller, Edmund, ed. Mutiny! New York: Crown Publishers, 1953.

Hadfield, Robert L. Mutiny at Sea. New York: Dutton, 1938.

Hoyt, Edwin P. The Mutiny on the "Globe." New York: Random House, 1975.

Lay, William, and Cyrus Hussey. A Narrative of the Mutiny on Board the Ship "Globe" of Nantucket, in the Pacific Ocean, January, 1824.... New London, Conn.: 1828.

Michener, James A., and A. Grove Day. "Rascals in Paradise: The "Globe" Mutineers," in their Rascals in Paradise. New York: Random House, 1957.

Morison, Samuel Eliot. "Historical Notes on Giblert and Marshall Islands," American Neptune, Vol. 4, 1944.

Nordhoff, Charles, and James Norman Hall. Mutiny on the Bounty (fiction). Boston: Little, Brown, 1932.

Paulding, Lieut. Hiram. Journal of a Cruise of the United States Schooner "Dolphin" ... in Pursuit of the Mutineers on the Whale Ship "Globe".... New York: 1831.

Putnam, George C. Salem Vessels and Their Voyages. Salem, Mass.: Essex Institute, 1924-1930.

Stackpole, Edouard. The Sea-Hunters. Westport, Conn.: Greenwood Press, 1973.

Whipple, Addison Beecher Colvin. Yankee Whalers in the South Seas. Rutland, Vt.: Charles E. Tuttle, 1972.

GIBBONS VS. OGDEN (1824)

In 1807 Robert Fulton, an American painter, engineer and inventor, proved that the steamboat could be both mechanically and economically feasible when his "Clermont" steamed from New York City to Albany and back. Fulton's associate in this venture was Robert R. Livingston, an American lawyer and statesman.

The two partners were granted a monopoly on steam navigation by the New York legislature. This monopoly passed on to Aaron Ogden who, under it, operated a steamboat line between Elizabethtown, New Jersey, and New York City. A rival line was established by Thomas Gibbons and Ogden sought to restrain Gibbons from operating steamboats on the Hudson River in competition to him, on the grounds that he, Ogden, was legally conducting his steamship business under the terms of the Fulton-Livingston monopoly.

The case came before the United States Supreme Court in 1824. Chief Justice John Marshall, in rendering the Court's decision, established the principle of freedom for interstate commerce. He stated that "the word [commerce] used in the Constitution ... comprehends ... navigation ... and a power to regulate navigation ... within its meaning ... and a power to regulate navigation is ... expressly granted ... [by the Constitution.]"

Marshall went on to say that, under the Constitution, the power of Congress "must be exercised within the territorial jurisdiction of the several States ... and comprehends navigation within the limits of every State in the Union, so far as that navigation may be, in any manner, connected with 'commerce with foreign nations or among the several States....' "

He then held that, when a State enacts a law, the validity of which interferes with or is contrary to an act of Congress "passed in pursuance of the Constitution," the acts of the State "must yield to the law of Congress: and the decision sustaining the privilege they confer against a right given by a law of the Union must be erroneous." Thus Marshall reversed and annulled the decision of the Court of New York and declared the Fulton-Livingston monopoly of no effect.

Suggested Readings

Beveridge, Albert J. The Life of John Marshall. Boston: Houghton Mifflin, 1916-1919.

Carson, Hampton L. The History of the Supreme Court. New York: Burt Franklin, 1972.

Commager, Henry Steele, ed. "Gibbons v. Ogden," (Doc. No. 129) in his Documents of American History, 8th edition. New York: Appleton, 1968.

Corwin, Edward S. John Marshall and the Constitution. New Haven, Conn.: Yale University Press, 1919.

Cotton, Joseph P., ed. Constitutional Decisions of John Marshall. New York: Da Capo Press, 1967.

Faulkner, Robert K. The Jurisprudence of John Marshall. Princeton, N.J.: Princeton University Press, 1968.
Frankfurter, Felix. The Commerce Clause Under Marshall, Taney, and Waite. Chapel Hill, N.C.: University of North Carolina Press, 1937.
Friedman, Leon, and Fred L. Israel, eds. The Justices of the Supreme Court, 1789-1869. New York: Bowker, 1969.
Garraty, John A., ed. Quarrels That Have Shaped the Constitution. New York: Harper, 1962.
Haskins, George L. "Marshall and the Commerce Clause of the Constitution," in Jones, W. Melville. Chief Justice John Marshall: A Reappraisal. Ithaca, N.Y.: Cornell University Press, 1956.
Johnson, Gerald White. The Supreme Court. New York: Morrow, 1962.
Loth, David G. Chief Justice John Marshall and the Growth of the Republic. New York: Norton, 1949.
McCloskey, Robert G. The American Supreme Court. Chicago: University of Chicago Press, 1960.
Magruder, Allan B. John Marshall. Edited by John T. Morse, Jr. New York: AMS Press, 1898.
Pritchett, C. Herman. The American Constitutional System. New York: McGraw-Hill, 1971.
Thayer, James B. John Marshall. New York: Da Capo Press, 1974.
Warren, Charles. The Supreme Court in United States History. Boston: Little, Brown, 1923.

THE TARIFF OF ABOMINATIONS (1828)

The word "tariff" is defined by Webster as "a schedule, system or scheme of duties imposed by a government on goods imported or exported, or especially on imports."

Before the United States was established England imposed customs duties on colonial commerce and almost all colonial assemblies levied similar customs virtually from their inception. The colonial imposts were designed to raise revenue and also to protect colonial trade against foreign competition, discourage the use of liquor and "ostentatious apparel" and permit retaliation against discriminatory treatment abroad.

Subsequently the United States imposed various tariffs, the first being that of 1789. Provision for enacting tariff acts was made in the Constitution (Article I), and this first tariff was designed to raise revenue and also encourage the domestic manufacture of certain commodities, including earthenware, glass, and other products.

The first out-and-out protective tariff act was that of 1816. This was designed to foster the manufacture of a lengthy list of articles and was defended as a means of protecting industries already established rather than promoting new ones. The highest duties this Act provided were to remain in force for only three years, on the theory that by the end of that time American manufacturers would be able to hold their own against foreign competition. This, however, proved not to be the case and when other industries, not protected under the tariff, clamored for similar high rates on their commodities, a general upward revision of such rates was made in 1824. This was bitterly denounced by the agricultural South as it resulted in their having to pay higher prices for the manufactured articles they needed and could not produce themselves.

The woolen manufacturers of the North and certain other industrialists demanded still greater tariff protection against imports. These demands eventually led to the enactment of the tariff of 1828, known as the "tariff of abominations." This called for the highest rates up to that time and was extremely unpopular in the South. In 1832 a law was passed establishing rates approximately those of 1824.

In spite of this South Carolina was not appeased, and in 1828 and again in 1832 its legislature declared both acts null and void. A political crisis followed, sufficiently serious to threaten the disruption of the Union. President Andrew Jackson, in 1832, threatened to use force to compel South Carolina's submission. Henry Clay introduced the so-called Compromise Tariff of 1833 which provided for a gradual reduction of certain high custom duties until 1842. These reductions were made for a short time, but, coupled with the economic crisis of 1837, resulted in a serious reduction in national revenues. Tariff rates were increased once more and with the passage of the Tariff Act of 1842 the level of duties became approximately that of 1832.

Suggested Readings

Ames, H. V. State Documents on Federal Relations. New York: Da Capo Press, 1900.

Bassett, John Spencer. The Life of Andrew Jackson. New York: Macmillan, 1925.

______. A Short History of the United States. New York: Macmillan, 1929.

Boucher, Chauncey S. The Nullification Controversy in South Carolina. Chicago: University of Chicago Press, 1916.

Bowers, Claude G. Party Battles of the Jackson Period. New York: Octagon Books, 1965.

Brown, William G. The Lower South in American History. Saint Clair Shores, Mich.: Scholarly Press, 1970.

Calhoun, John C. "The South Carolina Exposition," in his Works of John C. Calhoun. Edited by Richard E. Cralle. New York: Russell & Russell, 1968.

Commager, Henry Steele, ed. "Jackson's Proclamation to the People of South Carolina," (Doc. No. 144) in his Documents of American History, 8th edition. New York: Appleton, 1968.

______. "South Carolina Ordinance of Nullification," (Doc. No. 143) in his Documents of American History, 8th edition. New

York: Appleton, 1968.
________. "The South Carolina Protest Against the Tariff of 1828," (Doc. No. 135) in his Documents of American History, 8th edition. New York: Appleton, 1968.
Lomask, Milton. John Quincy Adams, Son of the American Revolution. New York: Ariel Books, 1965.
McMaster, John B. A History of the People of the United States, from the Revolution to the Civil War. New York: Appleton, 1938.
Morse, John T. John Quincy Adams. New York: AMS Press, 1898.
Parton, James. The Life of Andrew Jackson. New York: Johnson Reprint, 1860.
Stanwood, Edward. American Tariff Controversies in the 19th Century. New York: Russell & Russell, 1967.
Taussig, Frank W. The Tariff History of the United States. New York: Putnam, 1923.
Turner, Frederick Jackson. The Rise of the New West. Gloucester, Mass.: Peter Smith, 1959.
Von Holst, Hermann E. John C. Calhoun. New York: AMS Press, 1899.

THE KITCHEN CABINET (1829-1832)

Andrew Jackson, seventh President of the United States, grew up in the frontier wilderness of the Carolinas. He received "little formal education or refinement, and had a most combative nature." However, he read law and in 1787, at the age of 20, was admitted to the bar. A successful attorney, politician and general officer, he was possessed of a strong will, a violent temper, and during his life fought a number of duels.

Jackson's reputation as a war hero and his previous experience in politics led to his election as President in 1828 and again in 1832. He took office for his first term on March 4, 1829. John C. Calhoun of South Carolina, former Vice President under John Quincy Adams, was elected Jackson's Vice President.

The cabinet was already announced. Martin Van Buren, a Jackson supporter, was Secretary of State. Other cabinet officers were Samuel D. Ingham, Secretary of the Treasury; John H. Eaton, Secretary of War; John Branch, Secretary of the Navy; William T. Barry, Postmaster General; and John M. Berrien, Attorney General. These men were nearly evenly divided between Jackson's own followers and the friends of Calhoun. They had been selected after much conference between the two factions, in some cases over Jackson's protests. Historians feel that this indicated that the Democratic Party "had come to be a definite organization, of which the President was only the leader."

It turned out that the cabinet was not to be a body of political advisers. Those members who had supported Calhoun did not enjoy the President's confidence as did Van Buren, Barry, and

Eaton. Those officials, along with W. B. Lewis, James Hamilton, Francis P. Blair, Andrew Jackson Donelson, Isaac Hill, Amos Kendall, and others constituted President Jackson's "Kitchen Cabinet." Kendall was reputedly the leader of the group. These men, operating behind the scenes, gave Jackson political advice and counsel, and he consulted them rather than the official cabinet on important matters of state.

The "Kitchen Cabinet" was pro-Jackson, pro-Van Buren and anti-Calhoun. In 1831 the official cabinet was reorganized, enabling Jackson to have a cabinet which included no Calhounite among its members. With this change Jackson began to consult his regular cabinet more freely, and the "Kitchen Cabinet" lost its importance and influence.

Suggested Readings

Andrist, Ralph K. Andrew Jackson, Soldier and Statesman. New York: American Heritage Press, 1963.
Bassett, John Spencer. The Life of Andrew Jackson. New York: Macmillan, 1925.
________. A Short History of the United States. New York: Macmillan, 1929.
Bowers, Claude G. Party Battles of the Jackson Period. New York: Octagon Books, 1965.
Burgess, John W. The Middle Period. Norwood, Pa.: Norwood Editions, 1897.
James, Marquis. Andrew Jackson: The Border Captain. Indianapolis: Bobbs-Merrill, 1933.
________. Andrew Jackson: Portrait of a President. Indianapolis: Bobbs-Merrill, 1937.
Kendall, Amos. The Autobiography of Amos Kendall. Gloucester, Mass.: Peter Smith, 1872.
MacDonald, William. Jacksonian Democracy, 1829-1837. Saint Clair Shores, Mich.: Scholarly Press, 1971.
McMaster, John B. A History of the People of the United States, from the Revolution to the Civil War. New York: Appleton, 1938.
Nolan, Jeannette Covert. Andrew Jackson. New York: Messner, 1949.
Parton, James. The Life of Andrew Jackson. New York: Johnson Reprint, 1860.
Remini, Robert V. Martin Van Buren and the Making of the Democratic Party. New York: Columbia University Press, 1959.
Schlesinger, Arthur M., Jr. The Age of Jackson. Boston: Little, Brown, 1945.
Shepard, Edward M. Martin Van Buren. Edited by John T. Morse, Jr. New York: AMS Press, 1899.
Sumner, William Graham. Andrew Jackson. Boston: Houghton Mifflin, 1909.
Van Deusen, Glyndon G. The Jacksonian Era. New York: Harper, 1959.
Vance, Marguerite. The Jacksons of Tennessee. New York:

Dutton, 1953.
Von Holst, Hermann E. John C. Calhoun. New York: AMS Press, 1899.
Wellman, Paul I. The House Divides. Garden City, N.Y.: Doubleday, 1966.
Wilson, Woodrow. Division and Reunion, 1829-1889. Gloucester, Mass.: Peter Smith, 1909.

THE WHITE MURDER (1830)

Captain Joseph White was, in the year 1830, a wealthy Salem, Massachusetts, shipowner, 82 years of age. On the morning of April 7 he was found murdered in his bed. He had been stabbed thirteen times and his skull had been fractured. The body was discovered by Benjamin, his servant.

Various persons were suspected of the murder, including Captain White's nephew, Stephen White; Captain Joseph Knapp, Jr., his great-nephew by marriage; and Knapp's brother Frank. Also suspected were the Crowninshield brothers, Richard and George, black-sheep members of a prominent Salem family. The latter two were arrested, on evidence which today is not known, and were incarcerated in the Salem jail.

For two weeks the Vigilance Committee which had been formed to investigate the murder and, if possible, bring the perpetrator or perpetrators to justice, made no progress. A letter demanding $350, "the refusal of which will ruin you" and signed "Charles Grant, Junior," posted from Belfast, Maine, was received by Joseph Knapp, Sr., the father of Captain Knapp. Knapp Senior showed this letter to his sons Phippen and Joseph, who suggested that it be referred to the Vigilance Committee. On the strength of this letter the Committee hired the attorney Rufus Choate and arrested "Charles Grant, Junior," in Belfast. "Grant" turned out to be a man named Palmer, an ex-convict who stated that Captain Joseph Knapp had offered him and Knapp's brother Frank $1,000 to kill Captain White. He had refused to participate in the crime but later, learning that a reward had been offered, wrote his letter in the hope that he might realize a financial gain from it.

The Knapp brothers were arrested and lodged in the Salem jail with the Crowninshields. The Reverend Colman, Joseph Knapp's pastor and a member of the Vigilance Committee, had become convinced that Joseph Knapp was guilty. He visited him in jail to secure a confession under a pledge of immunity for turning state's evidence. Knapp made an oral confession and told Colman where to find the club which had been concealed after being used in the murder. Colman, Dr. Barstow and Mr. Pettibone retrieved the club from a rat hole under the steps of a local church and then returned to the jail. Here the prisoner made a written confession. In this confession he stated that he had assisted in the murder and implicated his brother Frank and the Crowninshields as accessories.

Richard Crowninshield had actually done the killing, entering Captain Joseph White's house through a window which Joseph Knapp had left unbarred.

Richard Crowninshield, in jail, learned of Joseph Knapp's confession and, realizing that he would be convicted of murder, hanged himself in his cell, believing that if he, as a principal in the murder, were not convicted, his brother George and Frank Knapp could not be convicted as accessories. Joseph Knapp, having been granted immunity in return for his confession and agreement to turn state's evidence, would also go free.

The Vigilance Committee realized that with Richard Crowninshield dead and not available to be tried, it would be extremely difficult to secure a conviction of the other three defendants. Daniel Webster was engaged for a fee of $1,000 to assist the aging attorney general with the prosecution. Attorneys Dexter, Gardiner and Rantoul represented the accused. Frank Knapp, now charged as the principal, was to be tried first, with Joseph Knapp and George Crowninshield as accessories.

The trial commenced. Joseph Knapp, Jr., called to the stand, refused to testify. The Reverend Colman attempted to remember what Knapp had confessed to him but he and other witnesses scarcely survived cross-examination.

Daniel Webster made an eloquent closing speech for the prosecution but the trial ended with a hung jury. Another jury was impaneled and Frank Knapp was tried a second time. Here the witnesses were more sure of themselves than they had been on the occasion of the first trial and such original testimony as "My belief is" changed to "I have no doubt."

Frank Knapp was, at his second trial, found guilty. He was sentenced to hang and the sentence was carried out on September 28, 1830. Joseph Knapp was then tried, having forfeited his immunity by his refusal to testify in the trial of his brother Frank. He too was found guilty and was hanged on December 31, 1830.

George Crowninshield was able to show that he had spent the night of the murder with one Mary Bassett, a prostitute, and was acquitted. To the end of his life he maintained that Captain Joseph White had been murdered by his nephew Stephen, that he might inherit his uncle's money.

Suggested Readings

Abrahamsen, David, M.D. The Murdering Mind. New York: Harper & Row, 1973.

Carroll, Mary Tarver. Keep My Flag Flying, Daniel Webster. New York: Longmans, Green, 1945.

Curtis, Charles P. "The Young Devils and Daniel Webster," American Heritage Magazine, June, 1960.

Dempewolff, Richard. Famous Old New England Murders and Some That Are Infamous. Brattleboro, Vt.: Stephen Daye Press, 1942.

Fisher, Sydney George. The True Daniel Webster. Philadelphia: Lippincott, 1911.

Fuess, Claude M. Daniel Webster. Boston: Little, Brown, 1930.
Guttmacher, M. S. The Mind of the Murderer. Freeport, N.Y.: Books for Libraries, 1960.
Jesse, F. Tennyson. Murder and Its Motives. London: Harrap, 1952.
Lodge, Henry Cabot. Daniel Webster. Philadelphia: Richard West, 1973.
McDade, Thomas M., comp. The Annals of Murder. Norman, Okla.: University of Oklahoma Press, 1961.
Snyder, William L. Great Speeches by Great Lawyers. New York: Baker, Voorhis, 1881.
Trials of Captain Joseph J. Knapp, Jr. and George Crowninshield, Esq., for the Murder of Captain Joseph White of Salem. Boston: Charles Ellms, 1830.
Webster, Daniel. The Letters of Daniel Webster. East Saint Clair Shores, Mich.: Scholarly Press, 1902.

THE NAT TURNER INSURRECTION (1831)

In the year 1831 the American South was definitely proslavery--much more so than it had been three decades earlier. The activities of William Lloyd Garrison and other abolitionists were there received with scorn. Garrison, writing his newspaper the Liberator in Boston, said, "I shall contend for the immediate enfranchisement of our slave population.... I am in earnest--I will not equivocate--I will not retreat a single inch, and I will be heard!"

Other abolitionists made their headquarters in Oberlin, Ohio. In the South many bitter things were said about those who would "recklessly incite the slaves to murder their masters," and the "black terror" was feared by many southern communities.

In 1831 such an uprising, led by Nat Turner, a Negro slave, did occur in Southampton County, Virginia. Though only lasting two days, this insurrection resulted in revisions of the slave codes so severe that the Negroes could never again revolt.

Turner, born in 1800, was an intelligent man who, unlike most of his fellow slaves, could read and write. He was something of a religious fanatic and among the slaves enjoyed a powerful influence as a self-proclaimed prophet and preacher.

While many free Negroes lived in Southampton County, Turner grew to manhood a slave, the property of Benjamin Turner, a white farmer. His studies of the Bible, his intelligence and, for a Negro of the day, his "book learning," made him feel that he should lead his fellow slaves to freedom. Historians are generally agreed that it was this belief that he was divinely appointed to "lead the children to the Promised Land as did Moses in the days of old," rather than any influence of Garrison and the other abolitionists, which prompted him to organize and direct his insurrection in August, 1831.

In February and again in August of that year Turner received what he believed to be revelations from Jehova, directing him to lead his fellow blacks to freedom. These "revelations" were actually eclipses of the sun and atmospheric disturbances causing the sun to grow dim, but Turner interpreted them as the sign for which he had been waiting.

On Sunday, August 21, 1831, Turner met in the woods near Cabin Pond with six confederates: Jack, Hark, Will, Henry, Nelson and Sam. Here the group planned to "rise that night and kill all the white people." Fortunately for them, most of the whites of the neighborhood, as members of the militia, were away attending a camp meeting. The insurrection was to be sudden, swift and violent, and no whites were to be spared.

Around two o'clock in the morning of August 22 the slaves left the woods and proceeded to the farm of Joseph Travis. Here they killed Travis, his wife, Putnam Moore (who had become Turner's owner) and another white, hacking them to pieces with axes. Later that night they also killed Travis' infant child.

The insurrection proceeded, with the Negroes massacring whites at farm after farm. Turner himself found it almost impossible to kill anyone although, before the uprising had run its course, he did murder Margaret Whitehead, the daughter of a local farmer.

Other Negroes joined the original seven and the killings continued. Turner's "army" ultimately numbered fifty or sixty men, some volunteers, others conscripted, and included free blacks. Around noon the rebels reached the highway and Turner decided to march on the town of Jerusalem, Virginia, immediately, hoping to find guns and powder there. At Parker's farm he and his men encountered a party of armed whites. In the skirmish that followed the Negroes were scattered but not defeated. The militia had returned from their camp meeting and Governor John Floyd had sent military units to Southampton County. Federal troops were also dispatched from Fortress Monroe, Virginia. Volunteers were marching to the scene of the trouble. Eventually over 3,000 armed whites had arrived and others were mobilizing.

At Dr. Simon Blunt's farm the Turner forces were ambushed and defeated. Hark was killed and others were either killed or captured. Turner made his escape. In spite of an all-out manhunt he managed to remain at liberty until Sunday, October 30, when he was captured by Benjamin Phipps, a white, who discovered him hiding near Cabin Pond.

Imprisoned at the Southampton County jail, Turner was interviewed by Thomas Gray, an attorney who sought to learn the reason for the insurrection--something incomprehensible to the whites of Southampton who believed they treated their black property extremely well. Turner indicated that he had "acted on the instructions of Almighty God, whose prophet he was."

On November 5, Turner, represented by attorney William C. Parker, was tried in Jerusalem. He was found guilty of fomenting an insurrection and sentenced to hang. The sentence was carried out on November 11. In addition to Turner, 48 other

Negroes were tried on similar charges and eighteen of these were convicted and hanged. Ten others were given lesser sentences.

Suggested Readings

Apthecker, Herbert. American Negro Slave Revolts. New York: Columbia University Press, 1943.
Elkins, Stanley W. Slavery: A Problem in American Institutional and Intellectual Life. Chicago: University of Chicago Press, 1959.
Foner, Eric. Nat Turner. New York: Prentice-Hall, 1971.
Gray, Thomas R. The Confession of Nat Turner: The Leader of the Late Insurrection in Southampton, Va. Miami: Mnemosyne Publishers, 1969.
Heaps, Willard Allison. Riots, U.S.A., 1765-1970. New York: Seabury Press, 1970.
Nye, Russel B. A Baker's Dozen: Thirteen Unusual Americans. East Lansing, Mich.: Michigan State University Press, 1956.
Oates, Stephen B. "Children of Darkness," American Heritage Magazine, October, 1973.
Phillips, Ulrich B. American Negro Slavery. New York: Appleton, 1918.
Robert, Joseph C. The Road from Monticello: Study of the Virginia Slavery Debate of 1832. Durham, N.C.: Duke University Press, 1941.
Snydor, Charles S. The Development of Southern Sectionalism. Baton Rouge, La.: Louisiana State University Press, 1948.
Stampp, Kenneth M. The Peculiar Institution: Slavery in the Ante-bellum South. New York: Knopf, 1956.
Styron, William. The Confessions of Nat Turner. New York: Random House, 1967.
Thelwell, Mike, et al. William Styron's Nat Turner: Ten Black Writers Respond. Boston: Beacon Press, 1968.
Tragle, Henry I. The Southampton Slave Revolt. New York: Random House, 1973.

THE NULLIFICATION CONTROVERSY (1832)

Nullification in American history is the process by which a state suspends, or claims the right to suspend, the operation of a federal law which it considers unconstitutional, within its borders. It is based on the theory that the Union was created by the states which, as specified by the tenth amendment to the Constitution, declares that "the powers not delegated to the United States by the Constitution, nor prohibited by it to the States, are reserved to the States respectively, or to the people."

In the 1820's the southeastern section of the nation was suffering an economic depression. The soil of the older southern states was, from constant planting of the same crops, worn out.

This meant that this section could not compete with the new gulf states, and the planters of South Carolina felt that the tariffs, such as the 1828 "Tariff of Abominations," with its high protective duties, was out of line with reality. John C. Calhoun of South Carolina, as an advocate of states' rights, had, in 1828, anonymously published his "South Carolina Exposition and Protest" against it. For Calhoun, nullification was the means by which a minority in the nation, if it happened to be a majority in a single state, could either force the national majority to compromise with the minority by a revision of federal legislation or obtain a constitutional amendment which would permit the federal government to overrule the constitutional opinion of the nullifying state.

In November, 1832, a special state convention was held in South Carolina. The convention took the position that the delegated power of Congress to levy import duties was intended, not to encourage manufacturing at home, but to raise revenue. The tariff laws of 1828 and 1832 were declared unconstitutional and the attendees at the convention announced that these laws would become null and void in South Carolina after February 1, 1833. They stated further that succession "would be the result of federal coercion."

President Andrew Jackson, on December 10, 1832, issued a proclamation denouncing the doctrine of nullification. In January, 1833, a bill, known as the "Force Bill," was introduced in Congress authorizing the President to use force to collect import duties. Nullification he called an "impractical absurdity," and he stated that "disunion by armed force is treason." The Force Bill was passed in February, 1833.

While the Force Bill was being debated Henry Clay brought forward a compromise tariff bill calling for the gradual reduction of tariff duties over the next ten years. Calhoun, who had resigned the vice-presidency after the passage of the tariff of 1832 and had been immediately elected to the Senate, objected to the Force Bill. However, he and Clay worked to push the new compromise tariff through Congress and on March 3, 1833, the same day that the Force Bill was signed into law, Jackson also signed the compromise tariff of 1833.

After the compromise tariff was passed South Carolina repealed its Ordinance of Nullification, but it also declared the Force Bill null and void. President Jackson ignored this last face-saving gesture as, if the tariff duties were collected, the Force Bill became irrelevant.

Suggested Readings

Bancroft, Frederic. Calhoun and the South Carolina Nullification Movement. Baltimore: Johns Hopkins University Press, 1928.
Bassett, John Spencer. A Short History of the United States. New York: Macmillan, 1929.
Boucher, Chauncey S. The Nullification Controversy in South Carolina. Chicago: University of Chicago Press, 1916.
Capers, Gerald M. John C. Calhoun: Opportunist. Gainsville:

University of Florida Press, 1960.
Coit, Margaret I. John C. Calhoun. Boston: Houghton Mifflin, 1950.
Commager, Henry Steele, ed. "Jackson's Proclamation to the People of South Carolina," (Doc. No. 144) in his Documents of American History, 8th edition. New York: Appleton, 1968.
_______. "South Carolina Ordinance of Nullification," (Doc. No. 143) in his Documents of American History, 8th edition. New York: Appleton, 1968.
Freehling, William W. Prelude to Civil War. New York: Harper, 1965.
James, Marquis. Andrew Jackson: Portrait of a President. Indianapolis: Bobbs-Merrill, 1937.
Taussig, Frank W. The Tariff History of the United States. New York: Putnam, 1923.
Wiltse, Charles M. John C. Calhoun. Indianapolis: Bobbs-Merrill, 1941-1944.

THE BANK OF THE UNITED STATES CONTEST (1832-1833)

Andrew Jackson, seventh President of the United States, was an irascible general officer, war hero, duelist, man of action, lawyer and believer in popular rule. He abhorred anything smacking of class privilege and felt that the Second Bank of the United States, chartered by Congress in 1816, was a dangerous monopoly which should be destroyed.

The Bank had been chartered for twenty years, with a capital of $35,000,000, and was to act as a fiscal agent for the federal government. Although privately owned and managed, this bank, as chartered by Congress, had control over the nation's currency system.

Jackson considered the act of Congress which had created the bank as unconstitutional and also opposed such a bank on economic grounds. He felt that it failed to establish a "uniform and sound currency," and he favored a hard money policy in which paper money would be based on specie.

The bank's charter was due to expire in 1836. In his first annual message, delivered December 8, 1829, Jackson indicated that it was not too soon for both Congress and the people to consider the wisdom of a recharter. It was obvious that he wished to remove from the bank's hands its power and large profits.

In his second annual message, on December 6, 1830, the President unfolded a detailed plan for the type of bank he thought advisable, which was essentially one that would be attached to the treasury department and managed by public officials.

Here the matter rested, but only for the time being as the problem would have to be met, one way or another, in 1836. Nicholas Biddle, appointed director of the bank by President James Monroe in 1819 and elected its president in 1822, opposed Jackson's desire to do away with the bank and had maneuvered to keep it alive. In January, 1832, Biddle formally asked Congress for a renewed charter, although it was a foregone conclusion that if a bill granting such a charter was passed, Jackson would veto it.

Biddle and his associates lobbied for their charter and in July it passed both Senate and House. President Jackson immediately vetoed the bill, declaring the bank unconstitutional and a monopoly. In his veto message he appealed to the people's hostility toward monopolistic capitalist institutions. Attempts to pass the bill over Jackson's veto failed. The veto appealed to the people and Jackson was reelected President in 1832, defeating candidates Henry Clay, John Floyd and William Wirt.

Nicholas Biddle did not accept Jackson's reelection as the final verdict on the bank question. Closing up such a bank suddenly would require the calling in of loans and withdrawing bank money from circulation, to the serious detriment of commerce. Jackson, therefore, proceeded to break the power of the bank in 1833, withdrawing deposits and placing them in state banks over a period of time so that, by 1836, the state banks would be in a position to furnish the services previously performed by the Second Bank of the United States and thus lessen the probability of financial panic.

Various political maneuvers followed, involving such government officials as Roger B. Taney, William H. Duane, Henry Clay, Thomas Hart Benton, and others. The bill to renew the bank's charter was never passed. Nicholas Biddle secured a state charter and on March 1, 1836, the Second Bank of the United States became the Bank of the United States of Pennsylvania. Jackson was censured by Henry Clay, who carried his resolutions of censure through the Senate. Thomas Hart Benton of Missouri moved to expunge these resolutions and this was accomplished in January, 1837.

Suggested Readings

Andrist, Ralph K. Andrew Jackson, Soldier and Statesman. New York: American Heritage Press, 1963.

Bassett, John Spencer. The Life of Andrew Jackson. New York: Macmillan, 1925.

________. A Short History of the United States. New York: Macmillan, 1929.

Benson, Lee. The Concept of Jacksonian Democracy: New York as a Test Case. Princeton, N.J.: Princeton University Press, 1961.

Blau, Joseph L., ed. Social Theories of Jacksonian Democracy. New York: Liberal Arts Press, 1954.

Bowers, Claude G. Party Battles of the Jackson Period. New York: Octagon Books, 1965.

Catterall, Ralph C. The Second Bank of the United States. Chicago: University of Chicago Press, 1903.

Clark, M. S., and D. A. Hall. Legislative and Documentary History of the Bank of the United States. Clifton, N.J.: Augustus M. Kelley, 1967.

Commager, Henry Steele, ed. "Jackson's Veto of the Bank Bill,"

(Doc. No. 147) in his Documents of American History, 8th edition. New York: Appleton, 1968.
Dewey, David R. Financial History of the United States. Clifton, N.J.: Augustus M. Kelley, 1967.
_______. The Second United States Bank. National Monetary Commission Report, 1910.
Gallatin, Albert. Considerations on the Currency and Banking Systems of the United States. Westport, Conn.: Greenwood Press, 1968.
Govan, Thomas B. Nicholas Biddle: Nationalist and Public Banker. Chicago: University of Chicago Press, 1959.
Hammond, Bray. Banks and Politics in America, from the Revolution to the Civil War. Princeton, N.J.: Princeton University Press, 1957.
James, Marquis. Andrew Jackson: Portrait of a President. Indianapolis: Bobbs-Merrill, 1937.
MacDonald, William. Jacksonian Democracy, 1829-1837. Saint Clair Shores, Mich.: Scholarly Press, 1971.
McGrane, Reginald C. The Panic of 1837: Some Financial Problems of the Jacksonian Era. Chicago: University of Chicago Press, 1924.
Meyer, Marvin C. The Jacksonian Persuasion: Politics and Belief. New York: Vintage Books, 1960.
Nolan, Jeannette Covert. Andrew Jackson. New York: Messner, 1949.
Parton, James. The Life of Andrew Jackson. New York: Johnson Reprint, 1860.
Schlesinger, Arthur M., Jr. The Age of Jackson. Boston: Little, Brown, 1945.
Sumner, William Graham. Andrew Jackson. Boston: Houghton Mifflin, 1909.
Taylor, George R., ed. Jackson versus Biddle: The Struggle over the Second Bank of the United States. Lexington, Mass.: Heath, 1972.
Van Deusen, Glyndon G. The Jacksonian Era. New York: Harper, 1959.
Vance, Marguerite. The Jacksons of Tennessee. New York: Dutton, 1953.
Ward, John W. Andrew Jackson: Symbol for an Age. New York: Oxford University Press, 1962.
Wellman, Paul I. The House Divides. New York: Doubleday, 1966.
Wilson, Woodrow. Division and Reunion, 1829-1889. Gloucester, Mass.: Peter Smith, 1909.

THE CONNECTICUT BLACK LAW (1833)

Although the institution of slavery did not legally exist in the state of Connecticut prior to the Civil War, many people there felt that the Negro should be "kept in his place." The slave codes of most southern states forbade teaching slaves to read and write, and some states, such as Virginia, applied this prohibition to all Negroes, both slave and free. While these laws were generally disregarded, they were sometimes enforced, as in the case of Mrs. Douglas who, in 1853, was convicted of teaching slaves to read and write, contrary to the laws of Virginia. She served one month in jail.

In 1831 Miss Prudence Crandall, an American teacher and reformer, opened a private school for girls in Canterbury, Connecticut. In 1833 she admitted a Negro girl as a member of the student body, which action aroused the ire of her neighbors. Miss Crandall retaliated by opening her school to "young ladies and little misses of color," and enrolled fifteen or twenty Negro pupils. In this she was encouraged by the abolitionists but her neighbors attempted to destroy the school by boycott, abuse and personal insult. An obsolete vagrancy law was also imposed. Public meetings were held and petitions were circulated.

On May 24, 1833, the state of Connecticut passed its "Black Law." This prohibited anyone from establishing a school for the education of Negroes who were not inhabitants of that state, or from teaching in any such school.

Miss Crandall disregarded the law and, in spite of heavy opposition, continued to instruct her colored pupils. For this she was arrested, tried, and convicted. However, the Court of Errors reversed the conviction on a technicality in July, 1834.

Soon after this Miss Crandall abandoned her education project after her house was attacked and partially destroyed. This case intensified the bitter feeling between the abolitionists and the anti-abolitionists, which feeling was to endure long after the Civil War.

Suggested Readings

Commager, Henry Steele, ed. "Trial of Mrs. Douglas for Teaching Colored Children to Read," (Doc. No. 178) in his Documents of American History, 8th edition. New York: Appleton, 1968.

Frederickson, George M., comp William Lloyd Garrison. Englewood Cliffs, N.J.: Prentice-Hall, 1968.

Garrison, Wendell P., and Francis J. Garrison. William Lloyd Garrison, 1805-1879: The Story of His Life as Told by His Children. New York: Arno Press, 1969.

Grimke, Archibald H. William Lloyd Garrison, the Abolitionist. Westport, Conn.: Negro Universities Press, 1891.

James, Edward T., ed. "Prudence Crandall," in his Notable American Women. Cambridge, Mass.: The Belknap Press of Harvard University Press, 1971.

May, Samuel J. Some Recollections of Our Antislavery Conflict. New York: Arno Press, 1869.

Nye, Russell B. William Lloyd Garrison and the Humanitarian Reformers. Boston: Little, Brown, 1955.

Phillips, Ulrich B. American Negro Slavery. New York: Appleton,

1918.
Stanton, Elizabeth Cady, Susan Brownell Anthony and Matilda Joslyn Gage, comps. The History of Woman Suffrage. Cincinnati: Collectors Editions, 1971.
Van Dusen, Albert. Connecticut. New York: Random House, 1961.

THE AVERY TRIAL (1833)

On the morning of December 21, 1832, the dead body of Sarah Maria Cornell, a cotton mill loom tender, was found by John Durfee, a farmer of Tiverton, Rhode Island. The girl's body was hanging from a five-foot stake on Durfee's farm. A cord, one end of which was fastened to the stake near the top, was looped around the girl's neck, which was broken. Subsequent examination showed her to be four months pregnant, although she was unmarried.

Sarah Cornell had been a member of the Methodist church, of which the Reverend Mr. Bidwell was the local pastor. In her effects was a letter, undated and unmailed, submitting her resignation from the church Bible class. It was addressed to Bidwell. A scrap of paper, also found in her effects, read, "If I should be missing, enquire of the Rev. Mr. Avery in Bristol. He will know where I am gone. Dec. 20. S. M. Cornell."

The Reverend Ephraim K. Avery was the pastor of the Methodist church at Bristol, Rhode Island. Reverend Bidwell, his friend, rode to his home and advised him of the presumed suicide. The coroner's jury, following the inquest, found that Sarah Maria Cornell had committed suicide by hanging and was "influenced to commit said crime by the wicked conduct of a married man." Harvey Harnden, deputy sheriff for Fall River, Massachusetts, began to suspect the Reverend Avery of murder. A second inquest was held and a new jury returned a verdict accusing Avery of being the "principal or accessory" in the girl's death.

Avery was questioned by the authorities at Bristol, who attempted to close the matter by stating that no evidence against the pastor existed. Following the hearing the pastor left town. Harnden pursuaded Judge Randall to issue a warrant against Avery for suspicion of murder, together with a request for extradition in case he had found refuge in another state.

After a long chase through several states Harnden found Avery in Rindge, New Hampshire, arrested him and returned him to Rhode Island. His trial opened on May 6, 1833, Avery pleading "not guilty."

During the course of the trial the story of the Reverend Ephraim Avery and Sarah Maria Cornell came out. She had been a member of his congregation at Lowell, Massachusetts, where he had been assigned as pastor. A girl with spunk and the courage of her convictions, she had spoken out against certain churchmen, calling them "a pack of damned fools," for which she had been branded "a lewd woman and thief" and was read out of the church. Basically religious, she had entreated Avery to readmit her to the church, and he had refused to do so.

The girl saw Avery at various times, as was shown by several letters introduced at the trial. Some of these letters were exchanged between Sarah Cornell and Avery and one, written by Sarah to her sister Lucretia Rawson, stated that Avery had forced his attentions on her and caused her pregnancy. It was also shown that Sarah had said she would have her revenge on the Reverend Avery and the Methodist church, "though it cost me my life."

On Sunday, June 2, 1833, the jury returned a "not guilty" verdict and Avery declared, "I am free from the thrall." The verdict was an unpopular one. Avery's subsequent career as a Methodist minister was unsuccessful. He was hissed on several occasions when he attempted to preach. Eventually he resigned the ministry and turned to farming, dying a pauper in 1869.

Suggested Readings

Abrahamsen, David, M.D. The Murdering Mind. New York: Harper & Row, 1973.

Aristides, pseud. Strictures on the Case of Ephraim A. Avery, Originally Published in the "Republican Herald," Providence, R.I. With Corrections, Revisions, and Additions. Providence: William Simmons, Jr.-Herald Office, 1833.

Avery, Ephraim K. Explanation of the Circumstances Connected with the Death of Sarah Maria Cornell. Providence: William S. Clark, Printer, 1834.

The Correct, Full and Impartial Report of the Trial of Rev. Ephraim K. Avery Before the Supreme Judicial Court of the State of Rhode Island, at Newport, May 6, 1833, for the Murder of Sarah M. Cornell. Providence: Marshall & Brown, 1833.

Dempewolff, Richard. Famous Old New England Murders and Some That Are Infamous. Brattleboro, Vt.: Stephen Daye Press, 1942.

Drury, Luke. A Report of the Examination of Rev. Ephraim K. Avery, Charged with the Murder of Sarah Maria Cornell. Providence: 1833.

Howe, George. "The Minister and the Mill Girl," American Heritage Magazine, October, 1961.

McDade, Thomas M., comp. The Annals of Murder. Norman, Okla.: University of Oklahoma Press, 1961.

One Who Early Knew Her, pseud. Brief and Impartial Narrative of the Life of Sarah Maria Cornell, Who Was Found Dead (Suspended by the Neck and Suspected to Have Been Murdered) Near Fall River [Mass.] December 22, 1832. New York: G. Williams, 1833.

"Sibley's Harvard Graduates," Massachusetts Historical Society, 1956.

The Terrible Hay-Stack Murder. Philadelphia: Barclay & Co., 1876.

Williams, Catherine Read (Arnold). Fall River, An Authentic Narrative; By the Author of "Tales, National Revolutionary." Boston: Lilly, Wait & Co., 1834.

THE MOON HOAX (1835)

The first "permanent penny daily newspaper" was the *New York Sun*, founded by Benjamin H. Day, a printer, in 1833. Day hoped to create a paper which would appeal to the everyday working man who could not afford the other New York newspapers, which sold for six cents per copy and contributed little in the way of human interest material and much in the way of stodgy editorials and political comment.

The *Sun* prospered. In 1835 Day hired a young Anglo-American journalist named Richard Adams Locke as an editorial and feature-story writer. At that time Sir William Herschel, the eminent astronomer, was in Cape Town, South Africa, making certain celestial observations. Locke conceived the idea of publishing a series of news stories in the *Sun* describing the remarkable, though imaginary, life on the moon as observed by Herschel through his "giant telescope with its 24-inch lens, weighting nearly 15,000 pounds when polished and its estimated magnifying power of 42,000 times."

Day liked Locke's idea and told him to proceed with it. Locke's first article concerning the moon appeared on the second page of the *Sun* on Saturday, August 22, 1835. On Tuesday, August 25, a second story appeared, "quoting" the supplement to the *Edinburgh Journal of Science*, giving further details concerning the telescope and stating that the existence of life on the moon "had been affirmatively settled."

Further stories, all the result of Locke's vivid imagination, appeared in the *Sun* over the next several weeks. The public eagerly bought copies of the newspaper as fast as they came off the press, to learn of the moon's vegetable and animal life which survived in "an atmosphere very like that of the earth." Some animals "observed through the giant telescope" were "graceful and frisky"; others were such as to be "classed on earth as monsters." The most sensational of these creatures was one which had bat-like wings and walked upright. These came to be called "the man-bats of the moon."

Other New York newspapers picked up the stories and reprinted portions of them. Locke engaged Norris and Baker, lithographers, to supply illustrations to accompany his descriptions of the astronomical discoveries. Sales of the *Sun* increased as the reading public became more and more enthralled by each day's new discoveries.

At Cape Town Sir John Herschel was unaware of his "gigantic scientific discoveries" as represented in the *Sun* until Caleb Weeks, a circus proprietor who had gone to Africa to obtain some giraffes for his traveling show, handed him a copy of the newspaper. When he had read it he confessed to being overcome; "he could never hope to live up to the fame that had been heaped upon him."

The Moon Hoax could not go on forever. Locke, in advising a reporter friend of his on a rival newspaper not to reprint the *Sun* articles, let slip the fact that he had written the material himself and that it was based on imagination rather than fact. Next day the rival newspapers exposed the hoax for what it was.

Edgar Allan Poe was inspired by Locke's Moon Hoax to publish one of his own in the Sun. This was the "Balloon Hoax" which appeared on April 13, 1844. This was a journalistic account of an imaginary balloon voyage made by eight men. In Poe's story the balloonists traveled from England to Sullivan's Island, South Carolina, in three days.

Suggested Readings

Emery, Edwin. The Press and America: An Interpretative History of Journalism. Englewood Cliffs, N.J.: Prentice-Hall, 1962.

Lee, James Melvin. History of American Journalism. Garden City, N.Y.: Garden City Publishing Co., 1923.

Locke, George A. The Moon Hoax. Boston: Gregg Press, 1975.

MacDougall, Curtis D. Hoaxes. New York: Macmillan, 1940.

Mott, Frank L. American Journalism: A History, 1690-1960. New York: Macmillan, 1962.

Nevins, Allan. The "Evening Post": A Century of Journalism. New York: Russell & Russell, 1968.

Nicholson, Marjorie. Voyages to the Moon. New York: Macmillan, 1948.

O'Brien, Frank M. The Story of the "Sun." New York: Appleton, 1928.

Paul, Raymond. Who Murdered Mary Rogers? Englewood Cliffs, N.J.: Prentice-Hall, 1971.

Payne, George Henry. History of Journalism in the United States. New York: Appleton, 1920.

Poe, Edgar Allan. "The Balloon-Hoax" (science fiction), in his Complete Stories and Poems of Edgar Allan Poe. Garden City, N.Y.: Doubleday, 1966.

Smith, H. Allen. The Compleat Practical Joker. Garden City, N.Y.: Doubleday, 1959.

Weisberger, Bernard A. The American Newspaperman. Chicago: University of Chicago Press, 1961.

THE "ORDEAL" OF MARIA MONK (1835-1836)

Maria Monk was the nominal author of a sensational pre-Civil War anti-Catholic book which enjoyed an instant success and was reprinted in facsimile as recently as 1962.

The only daughter of William Monk, an army barracks orderly at St. John's, Quebec, Maria, as a child, experienced intermittent mental aberration. This was attributed by her widowed mother to a brain injury incurred in her youth. She quickly "became notorious for her perfervid imagination and her waywardness" and was confined briefly to a Catholic asylum for prostitutes near the Hôtel Dieu Hospital and Convent in Montreal. She was discovered to be pregnant and was dismissed from the Hospital in 1834.

About this time she met the Reverend William K. Hoyt (or Hoyte), a fanatical anti-Papist and head of the Canadian Benevolent Association, a nativist missionary society. She became Hoyt's mistress and went with him to New York City where, in 1835, her child was born.

Hoyt had conceived a plan to take advantage of Maria's vivid imagination. He enlisted the support of several American anti-Catholic agitators, including George Bourne, the Reverend W. C. Brownlee, Theodore Dwight, and the Reverend John L. Slocum. These men were all associated with the American Protestant Vindicator, a nativist paper. They proceeded to write Maria Monk's "autobiography." This, freely embroidering her already fantastic stories, appeared serially in the Vindicator in 1835 and in book form the following year. Titled Awful Disclosures of Maria Monk, it was an instant and outstanding success.

The book was designed to appeal to the most bigoted of readers. It described Maria's conversion to Catholicism, her acceptance as a nun by the authorities of the Hôtel Dieu Convent and the terrifying "truth" there revealed to her. She charged that priests and nuns lived together, despite their sacred vows, and that many babies born of such illicit unions were strangled and thrown into cellar holes. She described, in great detail, a secret tunnel leading from the Hôtel Dieu to the home of a priest, the murders of nuns "who refused to obey the lustful will of priests," and of her escape with her own child, the father of whom was a priest.

The publication of this "autobiography" touched off bitter controversy. Many Catholics declared it a hoax. William Leete Stone, a Protestant New York editor, made a careful investigation of the Hôtel Dieu and declared Maria Monk's account a complete fraud. Protestants and Catholics alike in Montreal indignantly repudiated the slanders against the Hospital, considered one of the finest in Canada. Maria's mother, in a sworn statement, branded her daughter's account completely false.

Hoyt and his associates replied with a flood of publications, and sales of Awful Disclosures increased. Another woman, said to have also escaped from the Hôtel Dieu, was described in Downfall of Babylon, an anti-Catholic paper and rival of the American Protestant Vindicator.

The Hoyt faction split in a wrangle over the division of royalties from the book. Maria transferred her allegiance from Hoyt to the Reverend Slocum, who published a sequel, Further Disclosures by Maria Monk.

In the fall of 1837 Maria Monk disappeared for a time and was found in Philadelphia with a new name and a new male companion. She gave birth to a second child in 1838.

In 1849, while an inmate of a New York brothel, she was arrested for picking the pocket of a customer. She died a short time later.

Awful Disclosures continued to sell well, over 300,000 copies being purchased prior to the Civil War. Popular editions continued to appear until well into the 20th century.

Suggested Readings

Billington, Ray Allen. "Maria Monk and Her Influence," Catholic Historical Review, October, 1936.
Brownlee, Rev. W. C. Popery. New York: 1836.
Desmond, Humphrey J. The Know Nothing Party. Washington, D. C.: Blair, 1904.
Documents Relating to the Ursuline Convent in Charleston. Boston: 1842.
Haynes, George H. "The Causes of Know Nothing Success in Massachusetts," American Historical Review Quarterly, 1897.
The Lady Superior, pseud. Answer to "Six Months in a Convent." Boston: 1835.
Monk, Maria. Awful Disclosures of Maria Monk, As Exhibited in a Narrative of Her Sufferings as a Novice, and Two Years as a Black Nun, in the Hôtel Dieu Nunnery at Montreal. Facsimile edition, edited by Ray E. Billington, 1962.
_______. Further Disclosures of Maria Monk. New York: Harper, 1837.
Stone, William Leete. Maria Monk and the Nunnery of the Hôtel Dieu. Boston, 1836.
Thompson, Ralph. "The Maria Monk Affair," Colophon, Part 17, 1934.
"The Truth About Maria Monk," Watson's Magazine, May, 1916.
Wright, Richardson. Forgotten Ladies. Philadelphia: Lippincott, 1928.

THE UNDERGROUND RAILROAD (1835-1860)

One of the important political issues of the 19th century in the United States was that of slavery. The agricultural economy of the South encouraged the use of slaves on cotton and tobacco plantations. The North, with its emphasis on manufacturing rather than agriculture, was not economically suited to the institution of involuntary bondage.

In the decades preceding the Civil War anti-slavery sentiment ran high, particularly in the North. Abolitionists, such as William Lloyd Garrison, considered slavery a blot on American civilization. The American Colonization Society, founded in 1816 and headed by such prominent persons as Bushrod Washington, a Supreme Court Justice, sought to promote the emancipation of Negro slaves by freeing them and sending them to Africa. By 1830 the Society had sent 1,162 Negroes to Liberia on the west coast of Africa, where most of them died of fever.

It became evident that the colonization idea was impractical. By 1835 cotton farming had increased in the South and the plantation owners wished to develop additional acreage, for which they needed slave labor.

About this time the "Underground Railroad" appeared. This organization, operated by abolitionists, was established to help slaves escape from the South and elude pursuit. "Stations" were established at regular intervals in the homes of participants called "agents," and "conductors" went South and escorted fugitives secretly from one "station" to another until at last safety was reached in a free state or in Canada, beyond the reach of the Fugitive Slave Law.

"Stockholders" financed the operation with contributions of money, food and clothing. Many Negroes in Canada made trips to the South as "conductors" to help their colored friends escape from slavery.

The persons connected with the "Underground Railroad," while of great probity in other matters, were convinced that the institution of slavery was morally wrong and saw no crime in freeing slaves from bondage. Such white Northern citizens as Levi Coffin, Thomas Garrett, Josiah Grinnell, Samuel J. May, Theodore Parker, Robert Purvis, and the Reverend Charles Torrey were actively engaged in the operation.

It is estimated that approximately 2,000 slaves each year escaped from their masters from 1830 to 1860.

Suggested Readings

Allee, Marjorie H. Susanna and Tristram (fiction). Boston: Houghton Mifflin, 1929.

Barnes, Gilbert H. The Anti-Slavery Impulse, 1830-1844. New York: Harcourt, Brace, 1933.

Chadwick, John W. Theodore Parker: Preacher and Reformer. East Saint Clair Shores, Mich.: Scholarly Press, 1971.

Commager, Henry Steele. Theodore Parker: Yankee Crusader. Gloucester, Mass.: Peter Smith, 1936.

Curtis, Anna L. Stories of the Underground Railroad. New York: Island Workshop, 1941.

Foner, Laura, and Eugene D. Genovese, eds. Slavery in the New World. New York: Pantheon Books, 1969.

Frederickson, George M., comp. William Lloyd Garrison. Englewood Cliffs, N.J.: Prentice-Hall, 1968.

Garrison, Wendell P., and Francis J. Garrison. William Lloyd Garrison, 1805-1879: The Story of His Life as Told by His Children. New York: Arno Press, 1969.

Genovese, Eugene D. The World the Slaveholders Made. New York: Random House, 1971.

Grimke, Archibald H. William Lloyd Garrison: The Abolitionist. Westport, Conn.: Negro Universities Press, 1891.

Henkle, Henrietta. Flight to Freedom: The Story of the Underground Railroad. New York: Crowell, 1958.

Nye, Russel B. William Lloyd Garrison and the Humanitarian Reformers. Boston: Little, Brown, 1955.

Rawick, George P. The American Slave: A Composite Autobiography. Westport, Conn.: Negro Universities Press, 1970.

Sears, Lorenzo. Wendell Phillips: Orator and Agitator. New York: Benjamin Blom, 1967.

Siebert, Wilbur H. The Underground Railroad from Slavery to

Freedom. New York: Macmillan, 1898.
Still, William. Underground Rail Road: A Record of Facts, Authentic Narratives, Letters, & c. Narrating the Hardships, Hair-breadth Escapes and Death Struggles of the Slaves in Their Efforts for Freedom. Chicago: Johnson, 1970. (First published in 1872).
Swift, Hildegarde H. Railroad to Freedom (fiction). New York: Harcourt, Brace, 1932.

THE BATTLE OF THE ALAMO (1836)

The Alamo, originally built in 1772 at San Antonio, Texas, as a Franciscan mission, subsequently used as a fort and now a state monument, was the scene of one of the more significant encounters of the Texas war of independence against Mexico.

Following the Louisiana Purchase from France in 1803, the American people considered Texas part of the Republic and resisted Spanish power. An American invasion in 1812 led by Augustus McGee and Bernardo Gutiérrez resulted in the capture of San Antonio and the defeat of a number of Spanish-Mexican forces before it, in turn, met defeat. Other expeditions led by James Long were equally unsuccessful.

In 1823 Stephen Austin made an agreement with Mexico, which had meanwhile secured its independence from Spain. This agreement followed an earlier grant made to Austin's father Moses in 1821, and gave the younger Austin authority to establish a colony in Texas. Other similar grants were made later.

In 1830 the Mexican government passed laws restricting American immigration to Texas, and in 1835 war broke out between the Mexican government and the American settlers, with Sam Houston commander-in-chief of the Texas armies. San Antonio was captured by the Americans in December, 1835.

On February 23, 1836, a Mexican army of approximately 4,000 men commanded by General Antonio López de Santa Anna arrived at San Antonio. The Texas garrison, under the command of Colonel William Barrett Travis and aided by Colonels David Crockett and James Bowie, consisted of some 155 Americans. The garrison was surrounded by Santa Anna's army. On March 1 the Texans were reinforced by an additional 32 men and withstood Santa Anna's assault until March 6. On this date the Mexicans were able to breach the walls of the Alamo. Hand-to-hand fighting followed, and the members of the garrison were killed, with the exception of three women, two children and a Negro slave. At the subsequent battle of San Jacinto the Texas army, under Houston, defeated the Mexicans and captured General Santa Anna.

Suggested Readings

Alter, Robert E. Two Sieges of the Alamo. New York: Putnam, 1965.
Beals, Carleton. Stephen F. Austin, Father of Texas. New York: McGraw-Hill, 1953.
Beals, Frank L. Davy Crockett. Chicago: Wheeler, 1941.
Blair, Walter. Davy Crockett: Frontier Hero. New York: Coward McCann, 1955.
Callcott, Wilfred H. Santa Anna: The Story of an Enigma Who Once Was Mexico. Hamden, Conn.: Shoe String Press, 1964.
Connor, Seymour Vaughn. North America Divided: The Mexican War, 1846-1848. New York: Oxford University Press, 1971.
Cousins, Margaret. We Were There at the Battle of the Alamo. New York: Grosset and Dunlap, 1958.
Crockett, Davy. The Adventures of Davy Crockett, Told Mostly by Himself. New York: Scribner's, 1934.
Downey, Fairfax. Texas and the War With Mexico. New York: American Heritage Press, 1961.
Holbrook, Stewart H. Davy Crockett. New York: Random House, 1955.
Jones, Oakah L. Santa Anna. New York: Twayne, 1968.
Moseley, Elizabeth R. Davy Crockett, Hero of the Wild Frontier. Champaign, Ill.: Garrard, 1967.
Reeder, Red. The Story of the Mexican War. New York: Meredith Press, 1967.
Warren, Robert Penn. Remember the Alamo! New York: Random House, 1958.
Werstein, Irving. The War With Mexico. New York: Norton, 1965.

THE "CAROLINE" AFFAIR (1837)

The "Caroline," an American steamboat owned by William Wells of Buffalo, N.Y. and commanded by Captain Gilman Appleby, left Buffalo on December 29, 1837, bound for the American port of Schlosser on the Niagara River. At that time radicals in Upper Canada, led by the Scottish-born William Lyon Mackenzie, in a struggle for reform, had resorted to armed insurrection, and the "Caroline" had been used to carry supplies to them on Navy Island.

Flying the American flag, the "Caroline" had passed Black Rock Harbor when it was fired upon by a party of insurgents on the Canadian shore. No one was injured.

Arriving at Navy Island, the vessel discharged freight and passengers and then proceeded to Schlosser. Here several Americans, unable to find lodgings in town, requested permission to spend the night aboard, which permission was granted. About midnight on December 29 the "Caroline" was boarded by "some seventy or eighty men, all of whom were armed." These men, members of the Canadian militia, seized the steamer by force of arms. She was cut loose from the dock, set afire and, carried by the river current, went over Niagara Falls. Of the 33 persons aboard, twelve died on the ship or in the cataract below the falls.

Great Britain, in 1840, claimed that the destruction of the "Caroline" was a legitimate act of war. The United States contended that the Canadians had entered its territory in time of peace, and demanded redress. In the same year a Canadian named Alexander McLeod was arrested in New York City. He had boasted of participating in the "Caroline" affair and was charged with murder. The British ministry threatened war if he were not released, but the trial was held nevertheless. McLeod's acquittal probably prevented serious trouble between Great Britain and the United States.

Suggested Readings

Bemis, Samuel F. *A Diplomatic History of the United States.* New York: Cooper Square, 1950.

Commager, Henry Steele, ed. "The 'Caroline' Affair," (Doc. No. 156) in his *Documents of American History*, 8th edition. New York: Appleton, 1968.

Corey, Albert H. *The Crisis of 1830-1842 in Canadian-American Relations.* New Haven, Conn.: Yale University Press, 1941.

Elliott, Charles W. *Winfield Scott: The Soldier and the Man.* New York: Macmillan, 1937.

Fuess, Claude M. *Daniel Webster.* Boston: Little, Brown, 1930.

Guillet, Edwin C. *The Lives and Times of the Patriots.* Toronto: University of Toronto Press, 1938.

Lindsey, Charles. *William Lyon Mackenzie.* Toronto: University of Toronto Press, 1909.

Lodge, Henry Cabot. *Daniel Webster.* Philadelphia: Richard West, 1973.

McMaster, John B. *A History of the People of the United States, from the Revolution to the Civil War.* New York: Appleton, 1938.

Remini, Robert V. *Martin Van Buren and the Making of the Democratic Party.* New York: Columbia University Press, 1959.

Smith, Arthur Douglas Howden. *Old Fuss and Feathers: The Life and Exploits of Winfield Scott.* Plainview, N.Y.: Books for Libraries, 1972.

Tiffany, O. E. "Relations of the United States to the Rebellion of 1837," in *Publications of the Buffalo Historical Society*, Vol. VIII.

THE BROOK FARM ASSOCIATION (1841-1847)

The idea of a community in which there is common ownership of land, the education of men and women alike, and religious toleration is by no means new. Thomas More in his *Utopia*, published in 1516, described an island on which these concepts prevailed. While More's island was imaginary, several cooperative communities were actually established, one of the most famous being the Brook Farm Association for Industry and Education, founded under the direction of George Ripley, an American editor, literary critic and social reformer in the year 1841.

Brook Farm was established at West Roxbury, Massachusetts, some nine miles south of Boston and occupied a farm of 200 acres. All members of the group were to contribute equally to the work of the community. Further, the members were to share equally in the proceeds from their work and in the social and educational opportunities the community offered. The work included farming, manufacturing, and teaching.

In addition to Ripley, members of the Brook Farm Association included Charles Anderson Dana, John Sullivan Dwight, Nathaniel Hawthorne, William Ellery Channing, Elizabeth Peabody, Ralph Waldo Emerson, and Margaret Fuller. Many prominent literary figures from Boston and Concord frequently visited the community.

In 1843 Albert Brisbane, an American author and advocate of the social theories of François Marie Charles Fourier, gained control of the Brook Farm Association and the name was changed to Brook Farm Phalanx. Under Brisbane's stewardship it was, for two years, the American headquarters of the Fourierst movement.

In 1846 a disastrous fire discouraged the members of the group and the project was terminated in 1847.

Suggested Readings

Bestor, Arthur. Backwoods Utopias. Philadelphia: University of Pennsylvania Press, 1950.

Codman, John T. Brook Farm. New York: AMS Press, 1971.

Commager, Henry Steele, ed. "The Constitution of the Brook Farm Association," (Doc. No. 162) in his Documents of American History, 8th edition. New York: Appleton, 1968.

Curtis, Edith R. Season in Utopia: The Story of Brook Farm. New York: Russell & Russell, 1971.

Egbert, Donald D., and Stow Persons, eds. Socialism and American Life. Princeton, N.J.: Princeton University Press, 1878.

Frothingham, Octavius B. George Ripley. New York: AMS Press, 1883.

_______. Transcendentalism in New England: A History. Philadelphia: University of Pennsylvania Press, 1972.

Hawthorne, Nathaniel. The Blithedale Romance (fiction). New York: Norton, 1958.

Hillquit, Morris. History of Socialism in the United States. New York: Russell & Russell, 1965.

More, Sir Thomas. Utopia of Sir Thomas More. Edited by H. B. Cotterill. New York: St. Martin's Press, 1908.

Negley, Glenn, and J. Max Patrick. The Quest for Utopia. Washington, D.C.: Consortium Press, 1972.

Nordhoff, Charles. Communistic Societies in the United States. New York: Schocken Books, 1965.
Noyes, John H. History of American Socialism. New York: AMS Press, 1972.
Ripley, George. Discourses on the Philosophy of Religion. New York: 1836.
Sams, Henry W., ed. The Autobiography of Brook Farm. Englewood Cliffs, N.J.: Prentice-Hall, 1958.
Swift, Lindsay. Brook Farm: Its Members, Scholars and Visitors. Gloucester, Mass.: Peter Smith, 1973.
Wallechinsky, David, and Irving Wallace. "Brook Farm," in their The People's Almanac. Garden City, N.Y.: Doubleday, 1975.

THE MARY CECILIA ROGERS MURDER (1841)

On the afternoon of Wednesday, July 28, 1841, the dead body of Mary Cecilia Rogers, a beautiful twenty-year-old former clerk in a New York tobacco shop owned by John Anderson, was found in the Hudson River off Hoboken, New Jersey. The discovery was made by Henry Mallin, a singer, and four friends who were out for a walk in what was then "a rustic playground of shaded paths, picnic groves, and pleasant taverns."

Late in the day Dr. Richard F. Cook performed a post mortem and determined that the girl had been strangled with a strip of lace torn from her underskirt and that the body had then been placed in the water where it was later found. He attributed the deed to "a gang of toughs."

In 1838 the girl had mysteriously disappeared from her New York home for three days. This disappearance was never satisfactorily explained. When she was again missing from her home on July 25, 1841, Daniel Payne, her former fiance, a dissolute cork cutter by trade, inserted an advertisement in the July 28 issue of the New York Sun offering a reward for information concerning her but not mentioning her by name.

Andrew Crommelin, an attorney and Mary's former suitor, identified her body. Archibald Padley, a friend of Crommelin, confirmed the identification. Later he was arrested and charged with the murder but was exonerated.

The New York Newspapers, particularly James Gordon Bennett's New York Herald and the New York Sun, edited by Moses Y. Beach, gave much sensational publicity to the case. Various rumors were published, investigated, and found to be groundless. Several newspapers invented their own theories of the crime.

Payne was a prime suspect, although it was later learned that Phoebe Rogers, Mary's mother, had induced her daughter to break her engagement with him. He was able to account for his movements on the Sunday on which the girl was said to have died, and so was exonerated "from even a shadow of suspicion."

Another rumor placed Mary in the company of "a dark, swarthy man" shortly before her death. Joseph Morse, an engraver, became implicated in the affair but was able to prove his innocence and the charges made against him were dropped.

In the summer of 1841 a divorcée named Frederica Loss was operating a tavern known as Nick Moore's House at Weehawken, New Jersey. She was assisted in her work by her three sons, Charles, Ossian, and Oscar Killenbarack. Charles and Ossian found, in a thicket near their mother's tavern, some remnants of clothing which were identified as having been worn by Mary Rogers at the time of her death. The boys took the clothing found in the thicket to their mother who, a week later, turned it over to Gilbert Merritt, a Hoboken justice of the peace. The mother stated that on Sunday, July 25, a young lady resembling Mary Rogers had visited her tavern in the company of "a young man of dark complexion." Charles and Oscar Killenbarack were charged with the murder but were later discharged.

On Thursday, October 7, 1841, Daniel Payne committed suicide by drinking a slow-acting poison near the thicket where the clothing had been found. He left a suicide note which read, "To the world--Here I am on the very spot. May God forgive me for a misspent life."

Late in October, 1842, Frederica Loss died from a gunshot wound. The weapon was accidentally discharged by one of her sons. The shot was not immediately fatal and she was often delirious. A story in the New York Tribune described a deathbed confession made by Mrs. Loss. This included a statement she had made to Merritt to the effect that Mary Rogers had had an abortion on Sunday, July 25, at Nick Moore's House. She stated further that the girl had died while under the hands of the physician and that her son Oscar was to take the corpse to the river and throw it in at the place where it was later found. Some items of her clothing were hidden in the thicket. It was then that the two older sons were arrested. This Tribune story, while sensational, was proven to be completely false, and the two Killenbarack brothers were released.

Another suspect in the Rogers case was Madam Restell, "the most notorious abortionist in New York." An editorial in the Police Gazette accused her of being implicated in the matter. This came to nothing.

No one was ever formally convicted of the murder of Mary Cecilia Rogers. Edgar Allan Poe, in 1842/43, published a three-part serial, "The Mystery of Marie Rogêt," in Showden's Ladies' Companion. Poe attempted, through his fictional detective C. Auguste Dupin, to solve the mystery by writing a parallel story in which the locale is Paris, not New York, and the murdered girl a French grisette rather than an American cigar clerk. In his story Poe points the finger of guilt at the unnamed "tall, swarthy man" seen with the victim on the day she disappeared. Raymond Paul, in his Who Murdered Mary Rogers?, charges Daniel Payne with the crime.

Suggested Readings

Allen, Hervey. Israfel: The Life and Times of Edgar Allan Poe. New York: Doran, 1934.
Byrnes, Thomas. Professional Criminals of America. New York: Cassel & Co., 1886.
Clemens, Will. "The Tragedy of Mary Rogers," Era Magazine, XIV, 1904.
Collins, Ted. "Mary Rogers," in his New York Murders. New York: Duell, Sloan, 1944.
Crouse, Russel. Murder Won't Out. New York: Doubleday, 1932.
Duke, Thomas S. Celebrated Criminal Cases of America. San Francisco: Barry, 1910.
Lane, Winthrop B. "The Mystery of Mary Rogers," Collier's, March 8, 1930.
Paul, Raymond. Who Murdered Mary Rogers? Englewood Cliffs, N.J.: Prentice-Hall, 1971.
Pearce, Charles E. Unsolved Murder Mysteries. London: Ballantyne, 1924.
Pearson, Edmund. Five Murders: With a Final Note on the Borden Case. New York: Crime Club, 1928.
________. Instigation of the Devil. New York: Scribner's, 1930.
________. "Mary Rogers and a Heroine of Fiction," Vanity Fair, July, 1929.
Poe, Edgar Allan. "The Mystery of Marie Rôget" (fiction), in his Complete Stories and Poems of Edgar Allan Poe. Garden City, N.Y.: Doubleday, 1966.
Radin, Edwin D. "The Mystery of Mary Rogers," Ellery Queen's Mystery Magazine, November, 1949.
Trumble, Alfred. Great Crimes and Criminals of America. New York: Harper, 1881.
Wagenknecht, Edward. Edgar Allan Poe: The Man Behind the Legend. New York: Oxford University Press, 1963.
Wallace, Irving. The Fabulous Originals. Millwood, N.Y.: Kraus Reprint, 1956.
Walsh, John. Poe the Detective: The Curious Circumstances Behind "The Mystery of Marie Rôget." New Brunswick, N.J.: Rutgers University Press, 1968.

THE "CREOLE" AFFAIR (1841)

In 1841 the brig "Creole" was conveying a shipment of Negro slaves from Hampton Roads, Virginia to New Orleans, Louisiana. After the vessel was at sea, the slaves revolted, murdered the captain, took possession of the "Creole" and sailed for the British port of Nassau, in the Bahamas. Having arrived there they, under British law, became free.

The American government made an effort to have the slaves returned. Daniel Webster, then Secretary of State in President John Tyler's cabinet, stated that because the slaves were chattels belonging to citizens of the United States, they must be given up. This view was opposed by the abolitionist and legislator Joshua Reed Giddings, a member of the House of Representatives. Giddings introduced into the House resolutions to the effect that the slaves violated no American laws when they resumed their natural rights to liberty. He was censured by the House for these resolutions and submitted his resignation, but was subsequently reelected and served until 1859.

Giddings' position established the principle that, while slavery existed in the United States, it so existed not by federal law but by laws enacted by the several states. Consequently, slavery did not exist in the province of the federal government.

The slaves who had escaped on the "Creole" never returned to the United States, and the dispute with Great Britain over the affair finally lapsed.

Suggested Readings

Chitwood, Oliver P. John Tyler: Champion of the Old South. New York: Russell & Russell, 1964.

Fisher, Sydney George. The True Daniel Webster. Philadelphia: Lippincott, 1911.

Fuess, Claude M. Daniel Webster. Boston: Little, Brown, 1930.

Hadfield, Robert L. Mutiny at Sea. New York: Dutton, 1938.

Lodge, Henry Cabot. Daniel Webster. Philadelphia: Richard West, 1973.

Morgan, Robert J. A Whig Embattled: The Presidency of John Tyler. Lincoln, Neb.: University of Nebraska Press, 1954.

Reeves, Jesse. American Diplomacy Under Tyler and Polk. Gloucester, Mass.: Peter Smith, 1907.

Stewart, James B. Joshua R. Giddings and the Tactics of Radical Politics. New York: University Press Book Service, 1970.

Tyler, Lyon G. Letters and Times of the Tylers. New York: Da Capo Press, 1970.

Webster, Daniel. The Letters of Daniel Webster. East Saint Clair Shores, Mich.: Scholarly Press, 1902.

DORR'S REBELLION (1842)

Thomas Wilson Dorr was an American lawyer and politician. He opposed the provisions of certain state constitutions that made the ownership of property a requirement for voters and office holders. By 1835 this requirement had been abandoned by a number of states, including Delaware, Mississippi, Georgia, and Tennessee. Many other constitutional reforms were made by these and other states, but Rhode Island still had in force the old colonial charter which limited the franchise to freeholders.

When, in 1774, Rhode Island renounced slavery, the triangular trade involving slaves, sugar and rum was no longer a source of income. The economic gap which this caused was soon filled by the rapid growth of manufacturing. A heavy influx of European immigrants provided an ample labor pool, but these people were not property holders and so were denied suffrage.

A contest to change the old system followed but the city property owners and country landholders were too strong for the laboring class. The demand for reform found a dedicated and active leader in Dorr. Mass meetings and parades were held in various Rhode Island cities and violence seemed imminent. The legislature eventually called a convention but it was to be administered by the property-holding voters and, as such, was unsatisfactory to the disfranchised party. The latter called a convention of its own which prepared a constitution and submitted it to the public. This, called the "People's Constitution," received 13,944 votes late in 1841.

Dorr's strong following caused apprehension in the ranks of the old party. They, in turn, prepared their own "Freemen's Constitution." This, when voted upon, received 8,013 affirmative votes and was declared lost. The "People's Constitution" provided for white manhood suffrage; the "Freeman's Constitution" imposed certain residence requirements for voters.

The Dorr party announced that their constitution, having received more votes than that of the property owners, was law and ordered an election for legislature and governor, a step which the existing legislature declared illegal. The election, however, was held and Dorr was elected governor, which position he maintained for two weeks. The rival governor also functioned, as a result of which Rhode Island had two governments.

Little violence occurred. On May 18, 1842, a pitched battle was averted when Dorr and a number of his followers marched on the arsenal with cannon, only to find that their pieces would not fire. The following summer he fortified himself in the northwest part of Rhode Island. The militia was sent against him and several of his followers were arrested but he escaped.

In 1844 Dorr was arrested, tried for treason and sentenced to life imprisonment at hard labor, but he did not serve the term. His followers had made an effective exhibition of strength and a liberal constitution was adopted. Dorr was released in 1845 under an act of general amnesty. According to historians, "to Dorr's efforts, right or wrong, the new constitution was chiefly due. The victory of democracy in Rhode Island wiped out the last considerable vestige of land privilege."

Suggested Readings

Bassett, John Spencer. A Short History of the United States. New York: Macmillan, 1929.

Brennan, Joseph. Social Conditions in Industrial Rhode Island. Washington, D.C.: Catholic University of America Press, 1940.

Carroll, Charles. Rhode Island: Three Centuries of Democracy. New York: Knopf, 1932.

Coleman, Peter J. The Transformation of Rhode Island, 1790-1860. Providence, R.I.: Brown University Press, 1963.
Heaps, Willard Allison. Riots, U.S.A., 1765-1970. New York: Seabury Press, 1970.
McMaster, John B. A History of the People of the United States, from the Revolution to the Civil War. New York: Appleton, 1938.
Morgan, Robert J. A Whig Embattled: The Presidency of John Tyler. Lincoln, Neb.: University of Nebraska Press, 1954.
Mowry, Arthur M. The Dorr War: or, The Constitutional Struggle in Rhode Island. Providence, R.I.: Preston & Rounds, 1901.
Porter, Kirk H. A History of Suffrage in the United States. New York: Greenwood Press, 1969.
Tanner, Earl C. Rhode Island: A Brief History. Providence, R.I.: Brown University Press, 1957.
Weeden, William B. Early Rhode Island: A Social History of the People. New York: Da Capo Press, 1909.
Williamson, Chilton. American Suffrage: From Property to Democracy, 1760-1860. Princeton, N.J.: Princeton University Press, 1960.

THE RALEIGH LETTER (1844)

The April 17, 1844 document known as the "Raleigh Letter" (from the place in which it was written) was sent by Henry Clay to the editors of the National Intelligencer and was instrumental in his losing the Presidency in the election of that year.

It seemed evident that Clay would be the Whig nominee for the Presidency and that the Democrats would nominate Martin Van Buren. The annexation of Texas was a prime political issue of the day and feeling ran high. Early in the year 1844 Van Buren visited Clay at Ashland and the two men agreed to eliminate the Texas question from the campaign. In accordance with this agreement both men on April 17 published letters opposing immediate annexation. The letters were similar in content, Clay saying, "I consider the annexation of Texas at this time, without the assent of Mexico, as a measure compromising the national character, involving us certainly in war with Mexico, probably with other foreign powers, dangerous to the integrity of the Union, inexpedient in the present financial condition of the country, and not called for by any general expression of public opinion."

This viewpoint pleased the North but was bitterly opposed in the South and cost Van Buren the nomination. In Clay's case it did not prevent the Whig party from nominating him, but did embarrass him to the extent that he sought to explain it away in a series of letters known as the "Alabama Letters."

In the election of 1844 James K. Polk, the Democratic nominee on a platform calling for the "re-annexation" of Texas, was elected over Clay and James G. Birney, the Liberty Party candidate. Texas was annexed to the United States by Joint

Resolution of Congress, was approved on March 1, 1845 as a southern slave-holding state, and was admitted to the Union on December 29 of that year. The Mexican War which followed originated in a dispute over boundaries.

Suggested Readings

Adams, Ephraim D. British Interests and Activities in Texas, 1838-1846. Gloucester, Mass.: Peter Smith, 1963.

Commager, Henry Steele, ed. "The Annexation of Texas," (Doc. No. 165) in his Documents of American History, 8th edition. New York: Appleton, 1968.

_______. "Clay's Raleigh Letter," (Doc. No. 164) in his Documents of American History, 8th edition. New York: Appleton, 1968.

_______. "Polk's Message on War With Mexico," (Doc. No. 168) in his Documents of American History, 8th edition. New York: Appleton, 1968.

_______. "Texas and Oregon," (Doc. No. 166) in his Documents of American History, 8th edition. New York: Appleton, 1968.

Garrison, George P. Texas: A Conquest of Civilizations. New York: AMS Press, 1903.

_______. Westward Expansion, 1841-1850. New York: Haskell House, 1906.

McCormac, Eugene I. James K. Polk: A Political Biography. New York: Burt Franklin, 1971.

Polk, James K. The Diary of James K. Polk During His Presidency, 1845 to 1849. Millwood, N.Y.: Kraus Reprint, 1910.

Reeves, Jesse. American Diplomacy Under Tyler and Polk. Gloucester, Mass.: Peter Smith, 1907.

Ripley, Roswell S. The War With Mexico. New York: Burt Franklin, 1849.

Schurz, Carl. Henry Clay. New York: Frederick Ungar, 1968.

Shepard, Edward M. Martin Van Buren. Edited by John T. Morse, Jr. New York: AMS Press, 1899.

Smith, Justin H. The Annexation of Texas. New York: AMS Press, 1972.

Smith, Theodore C. The Liberty and Free Soil Parties in the Northwest. New York: Arno Press, 1969.

THE MORMON MIGRATION TO UTAH (1846-1847)

The Mormon Church, a religious organization known officially as the "Church of Jesus Christ of Latter-Day Saints," was established by Joseph Smith, Jr., in 1830. His brother Hyrum, his friend Oliver Cowdery, and others formed the nucleus of the congregation which became the Mormon Church. This was formally established at Fayette, N.Y., on April 6, 1830, later moved to the Middle West, and finally migrated to what is now Salt Lake City, Utah.

Smith, by his own account, began to have visions in 1820, telling him that "the Church of Christ had been withdrawn from the earth and that God had chosen him to restore it." In 1823 he professed to have been shown some inscribed gold plates by Moroni, an angel. In 1827 he translated these inscriptions. His translation, dictated to his scribes from behind a curtain, was and is called The Book of Mormon. It was published in Palmyra, N.Y., in 1830, and is regarded by Mormon church members as comparable to Jewish and Christian scriptures.

The new religion spread very rapidly. In 1831 Smith moved to Kirtland, Ohio, and in 1832 was joined by Sidney Rigdon and Brigham Young. They constructed a temple but, because of popular opposition to their sect, moved to Independence, Missouri, and later to Commerce, Illinois, renamed Nauvoo.

In 1843 Smith claimed to have had a vision in which polygamy was authorized to members of his church. This caused a schism in the church and a number of dissident Mormons attacked this concept in the Nauvoo Expositor, a newspaper established to fight against this new concept. Smith promptly suppressed the paper, destroyed the press upon which it had been printed, and burned the printing office. For this he was arrested and tried, but the justice of the peace who heard the case was a loyal Mormon and dismissed the defendants. Hostility of non-church members caused Smith's arrest on a vague charge of treason. He was confined in jail at Carthage, Illinois, from which he was taken by a mob and shot on June 27, 1844.

Brigham Young, who had been converted to Mormonism by Hyrum Smith in 1832 and had become a church elder, assumed the leadership of the Mormon Church following Smith's death. A stern man and capable administrator, Young had assisted in the move from Independence to Nauvoo. Anti-Mormon sentiment was high in Illinois, and Young, as acting president of the church, decided to remove himself and his followers from that state. In 1846/47 he organized and supervised the migration of almost 5,000 Mormons "across the Great Plains and Rocky Mountains into the arid Great Basin." The caravan arrived in the Great Salt Lake Valley where, on July 24, 1847, Great Salt Lake City, later Salt Lake City, was founded. On December 5, 1847, Young was formally elected head of the Mormon Church.

By 1852 there were over 15,000 Mormons in Utah. The settlers developed farms, constructed irrigation systems, opened businesses of various sorts, established schools and set up a legislature. In August of that year Young, basing his pronouncement on a revelation said to have come to Joseph Smith in 1843, publicly endorsed the doctrine of polygamy. This led to President James Buchanan's appointment in 1857 of a new territorial governor to replace Young, who had been appointed to that post. In 1852 the issue of polygamy and certain other doctrines introduced by Young had resulted in the formation of the Reorganized Church of Jesus Christ of Latter Day Saints. This organization now has over 170,000 members and is headquartered in Independence, Missouri. In 1871 Young was indicted on a charge of polygamy but was not convicted. The original church, which abandoned polygamy in 1890, thus confirming a presidential proclamation of the same year, is located in Salt Lake City, today one of the important cities of the United States. It has a membership of over 1,900,000.

Suggested Readings

Anderson, Maybelle H., ed. Appleton Milo Harmon Goes West. Berkeley, Calif.: Gillick Press, 1946.

Arrington, Leonard J. Great Basin Kingdom: An Economic History of the Latter-Day Saints. Cambridge, Mass.: Harvard University Press, 1959.

Brooks, Juanita, ed. On the Mormon Frontier: The Diary of Hosea Stout. Salt Lake City: University of Utah Press, 1964.

Burt, Olive. Brigham Young. New York: Messner, 1956.

Clayton, William. William Clayton's Journal. Salt Lake City: Deseret News, 1921.

Hafen, Le Roy R., and Ann W. Hafen. Handcarts to Zion: The Story of a Unique Western Migration. Glendale, Calif.: Arthur H. Clark, 1960.

________. The Utah Expedition. Glendale, Calif.: Arthur H. Clark, 1958.

Hinckley, Gordon B. What of the Mormons? Salt Lake City: Church of Jesus Christ of Latter-Day Saints, 1954.

Kjelgaard, James A. The Coming of the Mormons. New York: Random House, 1953.

Oaks, Dallin H., and Marvin S. Hill. Carthage Conspiracy. Champaign, Ill.: University of Illinois Press, 1975.

Riegel, O. W. Crown of Glory: The Life of James J. Strang, Moses of the Mormons. New Haven, Conn.: Yale University Press, 1935.

Smith, Joseph. The Book of Mormon. Palmyra, N.Y.: 1830.

________. Doctrine and Covenants of the Church of Jesus Christ of Latter-Day Saints. Edited by Orson Pratt. Westport, Conn.: Greenwood Press, 1880.

________. The History of the Church. Salt Lake City: Deseret Books, 1897.

Stegner, Wallace. Gathering of Zion: The Story of the Mormon Trail. New York: McGraw-Hill, 1964.

Tyler, Daniel. A Concise History of the Mormon Batallion in the Mexican War. Chicago: Rio Grande Press, 1964.

Young, Ann E. Wife No. Nineteen: The Story of a Life in Bondage, Being a Complete Exposé of Mormonism and Revealing the Sorrows, Sacrifices, and Suffering of Women in Polygamy. New York: Arno Press, 1875.

THE ORDEAL OF THE DONNER PARTY (1846-1847)

In July, 1846, a covered-wagon train headed by George Donner left Missouri for California, traveling by a new route suggested to them by L. W. Hastings.

By the end of October the party had reached Truckee Lake, now known as Donner Lake, on the eastern slope of the Sierra Nevada mountains, above the present city of Reno. Shortly after their arrival a heavy snowfall made it impossible for the immigrants to cross the pass leading to the Sacramento Valley in California. It was equally impossible for them to reverse their direction and return to safety at a lower level.

The party spent the winter in the Sierras, suffering incredible hardships. Some died of exposure and starvation; others resorted to cannibalism. Of the original 87 immigrants who started the trek, five had died before reaching the mountain camps, 34 either at the camps or on the mountains while trying to cross, and one just after reaching the Sacramento Valley. Thus, only 47 survived. Historians consider this disaster as the most spectacular and tragic in the history of western migration.

Suggested Readings

Altrocci, Julia Cooley. Snow-Covered Wagons (poetry). New York: Macmillan, 1936.
Clark, Thomas D. Frontier America: The Story of the Westward Movement. New York: Scribner's, 1969.
Coulter, Bruce N. Wagons Across the Mountains. New York: Dodd, Mead, 1957.
Croy, Homer. Wheels West. New York: Hastings House, 1955.
Houghton, Eliza P. Donner. The Expedition of the Donner Party and Its Tragic Fate. Chicago: McClung, 1911.
McGlashan, Charles Fayette. The History of the Donner Party. Ann Arbor, Mich.: University of Michigan Press, 1966.
Mirsky, Jeannette. The Westward Crossings. New York: Knopf, 1946.
Murphy, Virginia Reed. "Mountain Ordeal," Century Illustrated Monthly Magazine, July, 1891. Reprinted in Kartman, Ben, and Leonard Brown, eds. Disaster! New York: Pellegrini & Cudahy, 1948.
Pigney, Joseph. For Fear We Shall Perish: The Story of the Donner Party Disaster. New York: Dutton, 1961.
Powers, Alfred. True Adventures on Westward Trails. Boston: Little, Brown, 1954.
Stewart, George Rippey. Ordeal By Hunger: The Story of the Donner Party. Boston: Houghton Mifflin, 1960.
Stookey, Walter M. Fatal Decision: The Tragic Story of the Donner Party. Salt Lake City: Deseret Book Co., 1950.

THE ONEIDA COMMUNITY (1848)

The Oneida Community, a communistic society established by John Humphrey Noyes at Oneida, New York, in 1848, was the offspring of a similar society founded by him at Putney, Vermont, in 1836.

Noyes was an American Congregational minister, born in Brattleboro, Vermont and educated at Dartmouth College and the theological school of Yale College. In 1834 he was deprived of his license to preach for professing a second conversion. Two years later he, with a few followers, established his Putney Community. He promulgated the doctrine that Christ had returned to earth before the end of the Apostolic Age and that his work of saving from sin was complete. Therefore, he reasoned, no one is bound by any moral code. His views were expressed in the Perfectionist, a paper published by him, and he and his followers called themselves the Perfectionists.

In 1846 Noyes' tenets of complex marriage and free love aroused public condemnation of his Putney Community and he was forced to disband it. However, in 1848 he and those members who remained faithful to him set up a similar organization, the Oneida Community. Like the original group, this was a cooperative venture. Members renounced personal property and binding personal relationships, including marriage. All properties were held in common.

At Oneida the group established a number of manufacturing enterprises which, for the most part, were commercially successful. Between 1848 and 1880 the operating capital increased from $67,000 to $600,000.

Although the Oneida Community, unlike most similar social organizations, was financially successful, the outside antagonism experienced at Putney was again manifest. In 1880 Noyes was forced to move again, this time to Canada. He left Oneida in order to escape prosecution for adultery, the majority of his followers having abandoned his tenets the previous year. The communal property system was replaced by a joint stock company, incorporated in 1881, known as Oneida Community, Limited. This was designed to carry on the various manufacturing activities of the old cooperative organization. Manufactured products included mouse and other traps, canned fruits and vegetables, and silk. These products were gradually abandoned and today the company is known as a producer of plated and sterling silverware.

Suggested Readings

Bestor, Arthur. Backwoods Utopias. Philadelphia: University of Pennsylvania Press, 1950.

Cross, Whitney R. Burned-Over District. New York: Harper, 1950.

Egbert, Donald D., and Stow Persons, eds. Socialism and American Life. Princeton, N.J.: Princeton University Press, 1878.

More, Sir Thomas. Utopia of Sir Thomas More. Edited by H. B. Cotterill. New York: St. Martin's Press, 1908.

Negley, Glenn, and J. Max Patrick. The Quest for Utopia. Washington, D.C.: Consortium Press, 1972.

Nordhoff, Charles. Communistic Societies in the United States. New York: Schocken Books, 1965.

Noyes, John H. The Berean. New York: Arno Press, 1969.

________. Bible Communism. New York: AMS Press, 1972.
________. History of American Socialisms. New York: AMS Press, 1972.
________. Home Talks. New York: AMS Press, 1875.
________. Male Continence. New York: AMS Press, 1972.
________. Scientific Propagation. New York: AMS Press, 1873.
Noyes, Pierrepont B. My Father's House: An Oneida Boyhood. Gloucester, Mass.: Peter Smith, 1937.
Nye, Russel B. A Baker's Dozen: Thirteen Unusual Americans. East Lansing, Mich.: Michigan State University Press, 1956.
Parker, Robert A. A Yankee Saint: John Humphrey Noyes and the Oneida Community. New York: Porcupine Press, 1972.
Tyler, Alice. Freedom's Ferment. New York: Harper, 1965.
Wallechinsky, David, and Irving Wallace. "Oneida," in their The People's Almanac. Garden City, N.Y.: Doubleday, 1975.

THE SENECA FALLS WOMEN'S RIGHTS CONVENTION (1848)

Today's "Women's Liberation" movement is not something that sprang up overnight. The concept of campaigning for equality between the sexes in America goes back to the early 19th century when such suffragettes as Frances Wright and Ernestine Rose championed the cause of women's rights. From 1830 on Frances Wright "scandalized" contemporary America by appearing on the lecture platform to attack religion and the existing system of education as well as to defend the principle of equal rights for women. She also advocated a system of marriage "based on moral obligation only," and the abolition of slavery.

The anti-slavery crusade was the immediate origin of the women's rights movement of the mid 19th century. In 1840 the World Anti-Slavery Convention, held in London, excluded a group of American women delegates. These women felt that the cause of emancipation affected them as it did slaves.

On July 19 and 20, 1848, Elizabeth Cady Stanton organized the first women's rights convention at Seneca Falls, New York. Associated with her were Susan B. Anthony, Matilda Joslyn Gage, Lucretia Coffin Mott, and others. They published the "Seneca Falls Declaration of Sentiments and Resolutions" on July 19. This document in many ways resembles the Declaration of Independence of 1776. Both documents commence with the words, "When, in the course of human events it becomes necessary...." The Seneca Falls Declaration states further that "we hold these truths to be self-evident, that all men and women are created equal," and closely follows the format of the earlier Declaration, including the listing of the "injustices" which men had imposed on women over the centuries.

Part 2 of this Declaration lists a number of resolutions, promulgating the creed that women, being in no way inferior to men, should have the rights and privileges accorded men by society. Particularly specified were woman's "sacred right to the elective franchise" and the belief "that the same amount of virtue, delicacy, and refinement of behavior that is required of woman in the social state, should also be required of man, and the same transgressions should be visited with equal severity on both man and woman."

Susan B. Anthony organized the National Woman Suffrage Association in 1869 and was president of that organization from 1892 to 1900. Elizabeth Cady Stanton was its first president, holding that office from 1869 to 1890.

Suggested Readings

Anthony, Katharine. Margaret Fuller. Folcroft, Pa.: Folcroft Library Editions, 1920.

________. Susan B. Anthony: Her Personal History and Her Era. New York: Doubleday, 1954.

Blatch, Harriot S. Mobilizing Woman Power. New York: Jerome S. Ozer, 1974.

Catt, Carrie Chapman, and Nettie R. Shuler. Woman Suffrage and Politics: The Inner Story of the Suffrage Movement. New York: Scribner's, 1923.

Commager, Henry Steele, ed. "Woman's Rights," (Doc. No. 172) in his Documents of American History, 8th edition. New York: Appleton, 1968.

Coolidge, Olivia. Women's Rights: The Suffrage Movement in America, 1848-1920. New York: Dutton, 1966.

Flexner, Eleanor. A Century of Struggle: The Women's Rights Movement in the United States. Cambridge, Mass.: Harvard University Press, 1959.

Hallowell, Anna D. James and Lucretia Mott. Westport, Conn.: Hyperion Press, 1975.

Hecker, Eugene A. A Short History of Women's Rights. Westport, Conn.: Greenwood Press, 1971.

Kraditor, Aileen S. The Ideas of the Woman Suffrage Movement, 1890-1920. New York: Columbia University Press, 1965.

Lutz, Alma. Susan B. Anthony: Rebel, Crusader, Humanitarian. Boston: Beacon Press, 1959.

Sinclair, Andrew. The Better Half. New York: Harper, 1965.

Stanton, Elizabeth Cady. Eighty Years and More: Reminiscences, 1815-1897. Cincinnati: Collectors Editions, 1970.

________, Susan Brownell Anthony and Matilda Joslyn Gage, comps. The History of Woman Suffrage. Cincinnati: Collectors Editions, 1971.

Stanton, Theodore, and Harriot S. Blatch. Elizabeth Cady Stanton as Revealed in Her Letters. New York: Arno Press, 1921.

Williamson, Chilton. American Suffrage: From Property to Democracy, 1760-1860. Princeton, N.J.: Princeton University Press, 1960.

THE ASTOR PLACE RIOT (1849)

Two of the great tragic actors of the 19th century were William C. Macready and Edwin Forrest. Macready's acting was characterized by what was considered "restraint and good taste" and he was extremely popular in his native England. He considered the American Forrest's acting style as "exaggeration and extravagance."

At first Macready's critical opinion of Forrest's acting was expressed only in his diary, but eventually Forrest heard of it and the two became bitter rivals. In 1849 Macready was performing in America. On Monday, May 7 of that year, he appeared at the Astor Place Opera House in New York City, starring in "Macbeth." Forrest had opened at the Broadway Theater on April 23 and was playing to packed houses. He announced a production of "Macbeth" for May 7, in direct competition to Macready's production.

Macready's audience included a large contingent of Bowery B'hoys, rowdies who had come especially to heckle the English tragedian. The first two scenes of the play went smoothly. Then the rowdies began to stamp their feet, yell, groan and hiss, and pelt the actors and actresses with eggs "of doubtful purity," assorted vegetables and, ultimately, chairs. The curtain was rung down halfway through the third act.

Forrest's performance at the Broadway Theater was also constantly interrupted, but "only to permit him to acknowledge the bursts of applause from an enraptured audience." The police did nothing to quell the disturbance at either theater.

Macready announced his intention to return to England immediately, but was induced by several leading New Yorkers to change his mind and continue with the play. On Thursday, May 10, the American Committee, an organization of rabble-rousers led by Edward Zane Carroll Judson, which advocated "America for Americans," attended Macready's performance en masse. Mayor Caleb S. Woodhull of New York City met with Chief of Police G. W. Matsell, Sheriff J. J. V. Westervelt, Recorder Tallmadge, General Sandford and the two theater managers, Colonel William Niblo and James H. Hackett. The mayor's suggestion to cancel further performances by Macready was refused; Matsell was ordered to post a full police detail and Sandford to "hold a sufficient military force in readiness."

When the doors of the Astor Place Opera House opened, some 1,800 theater-goers swarmed in. Others, unable to enter the crowded building, milled about outside. Two hundred policemen were distributed throughout the audience and others remained on guard near the entrance. General Sandford's soldiers had been mustered at their downtown drillroom.

The curtain rose and the play commenced. The pattern of Monday night was repeated. Boos, hisses and groans accompanied the presentation. At the end of the first act three rowdies were arrested and dragged struggling from the audience. They were locked in a vacant room in the basement. One of the Bowery B'hoys broke an upper window, stuck his head out and shouted to the crowd in the street that three of his cohorts had been arrested. The crowd immediately assailed the building with rocks, paving stones, and bricks.

The police outside, badly outnumbered, were unable to quell the riot and a messenger was sent to bring General Sandford and his troops to the scene. On the stage the actors were attempting to continue their performance.

The soldiers arrived about nine o'clock as Act III of "Macbeth" began. Infantry and cavalry slowly forced the crowd back but they were repulsed by the stones and rocks hurled by the rioters, who assumed that the soldiers would not use their guns. Finally, after warning the mob that the troops would use their firearms if order was not immediately restored, Sheriff Westervelt ordered the troops to fire over the heads of the rioters. This proved ineffectual; the mob renewed its rock throwing and a second order to fire was given. The instructions, "Don't hit above the legs!," was not heard by all the soldiers and a dozen or more rioters were dropped while the others fell back. A slaughtering third volley broke the spirit of the mob and it retreated. At Macready's insistence the play had been completed and the audience was escorted from the theater by the police.

The Astor Place Riot resulted in 31 deaths and 48 persons known injured. Others, not wishing to be listed with the rioters, had departed to have their wounds treated privately. The theater itself was a shambles.

At the coroner's inquest it was decided that "the circumstances existing at the time justified the authorities in giving the order to fire."

Suggested Readings

Alger, William B. The Life of Edwin Forrest. New York: Benjamin Blom, 1877.

Barrett, Lawrence. Edwin Forrest. Saint Clair Shores, Mich.: Scholarly Press, 1881.

Eaton, Walter Pritchard. "Edwin Forrest," Atlantic Monthly, August, 1938.

Heaps, Willard Allison. Riots, U.S.A., 1765-1970. New York: Seabury Press, 1970.

Hughes, Glenn. A History of the American Theater, 1700-1950. New York: Samuel French, 1951.

Macready, William C. The Diaries of William Charles Macready. Edited by William C. Toynbee. New York: Benjamin Blom, 1912.

________. The Journal of William Charles Macready, 1832-1851. Carbondale, Ill.: Southern Illinois University Press, 1970.

Moody, Richard. Edwin Forrest: First Star of the American Stage. New York: Knopf, 1960.

Moses, Montrose J. The Fabulous Forrest. New York: Benjamin Blom, 1972.

O'Dell, George C. Annals of the New York Stage. New York: AMS Press, 1970.

Quinn, Arthur H. A History of the American Drama from the Beginning to the Civil War. New York: Irvington, 1943.
Rees, James ("Colley Cibber," pseud.). The Life of Edwin Forrest. Philadelphia: Lippincott, 1874.
Seilhamer, George O. History of the American Theater. New York: Benjamin Blom, 1968.

THE DISAPPEARANCE OF DR. GEORGE PARKMAN (1849)

Professor John White Webster, an American chemist and member of the faculty of Harvard College, was tried and found guilty of the murder of Dr. George Parkman, a physician, real estate speculator and money lender. He was hanged at the Leverett Street jail, Boston, on August 30, 1850.

Both men were graduates of Harvard College and had been acquaintances for more than forty years. Webster was in debt to Parkman in the amount of $2,432. The latter was a ruthless man and harsh in his business dealings with his debtors. Webster, an imprudent man, had mortgaged his valuable mineral collection to Parkman as security for the debt. Later he attempted to sell this collection to Robert Gould Shaw, Parkman's brother-in-law. Shaw informed Parkman of this, which caused Parkman to threaten and insult the easygoing Webster.

On November 23, 1849, the situation reached a climax. Parkman disappeared some time after five o'clock that afternoon and was never seen again. Shortly before his disappearance Webster had paid him some money, thereby discharging a portion of his debt.

Ephraim Littlefield, the janitor at Harvard Medical School, discovered a portion of a human body in a privy next to Webster's laboratory at the school. Webster was arrested and charged with the murder of Parkman. Subsequently a piece of human jaw was found in the ashes of Webster's furnace and another part of a body was found in a tea chest in a corner of his private room. This portion had a knife in it.

Chief Justice Lemuel Shaw presided at Webster's trial, along with associate justices Samuel S. Wilde, Charles A. Dewey and Theron Metcalf. The trial was characterized by circumstantial evidence. It was never conclusively shown that Parkman had actually been murdered, or that the human relics were portions of his dismembered body. Nevertheless, Webster was found guilty of murder and hanged.

Suggested Readings

Bemis, George. Report of the Case of John W. Webster. Boston: Little, Brown, 1850.
Carlson, Eric T., M.D. "The Unfortunate Dr. Parkman," American Journal of Psychiatry, December, 1966.

Chase, Frederic Hathaway. Lemuel Shaw, Chief Justice. Boston: Houghton Mifflin, 1918.
Dempewolff, Richard. Famous Old New England Murders and Some That Are Infamous. Brattleboro, Vt.: Stephen Daye Press, 1942.
French, John A. The Trial of John W. Webster for the Murder of Dr. George Parkman in the Medical College. Boston: Boston Herald Steam Press, 1850.
Holmes, Oliver Wendell. The Benefactors of the Medical School of Harvard University: With a Biographical Sketch of the Late Dr. George Parkman. Boston: Ticknor, Reeds & Fields, 1850.
______. George Parkman, Benefactor of the Harvard Medical School. Privately printed, 1850.
Irving, H. B. "Professor Webster," in his A Book of Remarkable Criminals. New York: Doran, 1918.
Kunstler, William H. "Murder in Payment: The Commonwealth of Massachusetts versus John White Webster," in his First Degree. New York: Oceana Publications, 1960.
Lawson, John D., ed. American State Trials, 1569-1920. Wilmington, Del.: Scholarly Resources, 1972.
Littlefield, Ephraim, pseud. "How I Found Dr. Parkman," Harvard Alumni Bulletin, November 10, 1949.
McDade, Thomas M. "The Parkman Case," American Book Collector, May, 1959.
Morris, Richard B. "Grand Guignol at Harvard Medical School," in his Fair Trial. New York: Knopf, 1952.
Pearson, Edmund. "America's Classic Murder, or The Disappearance of Dr. Parkman," in his Masterpieces of Murder. Boston: Little, Brown, 1963.
Sullivan, Robert. The Disappearance of Dr. Parkman. Boston: Little, Brown, 1971.
Symons, Julian. A Pictorial History of Crime. New York: Crown Publishers, 1966.
Thomson, Helen. Murder at Harvard. Boston: Houghton Mifflin, 1971.

THE CALIFORNIA GOLD RUSH (1849)

Although gold was discovered by James Wilson Marshall in a mill race at Coloma, California, on the American River on January 24, 1848, the man usually thought of in connection with its discovery and the subsequent influx of gold seekers the following year is Captain John Augustus Sutter, a Swiss-born Western pioneer and adventurer.

In 1839 Sutter had settled in the Sacramento Valley of California, then a Mexican province. He became a Mexican citizen and official and received a grant of 40,000 acres from the Mexican government. He established a colony which he called New Helvetia, now Sacramento. Here he constructed a fort, a mill, and other buildings.

In 1846 California was acquired by the United States. Following the annexation many settlers came to New Helvetia. In order to supply these settlers with the timber they needed, Sutter engaged Marshall to build a new sawmill in a valley in the Sierra foothills. On that historic day in January Marshall noticed shining flakes in the mill's tail race. These, when examined by Marshall and Sutter, were determined to be placer gold. It was this discovery which set off the California Gold Rush of 1849.

Sutter desired to keep the find a secret until he could conclude a treaty with the local Indian chiefs giving him and Marshall exclusive mining privileges. News of the discovery, however, was spread by workmen at the mill and nuggets from Coloma began to circulate at Sutter's Fort and in Yerba Buena (San Francisco).

At first Marshall's find caused little interest. It was mentioned casually in the Californian and the California Star, two small local newspapers. Then, in June, with such men as Sam Brannan shouting the news, gold fever manifested itself, and in a week "the Northern coast was virtually depopulated." By the end of June three-quarters of the houses in San Francisco were deserted. Ships in the harbor were abandoned by their crews, many of them never to sail again. The garrison at the presidio had deserted en masse and trade was at a virtual standstill except for such items as could conceivably prove useful at the diggings. Many rich strikes were made and by the middle of 1848 the total yield of gold was estimated at $600,000.

Word of Marshall's find spread to the Atlantic Coast and "the entire nation was gripped with a mining fever so virulent that ordinary concerns were forgotten." Tens of thousands prepared to travel to the gold fields. Three routes were available: by sea around Cape Horn; by sea to Panama, Mexico or Nicaragua, thence overland to the Pacific Ocean and from there by sailing vessel North to California; or by one of the overland trails.

By the spring of 1849 a heavy traffic was moving over all three routes. More than 20,000 persons had left various East Coast and Gulf ports by mid-April, and by June thousands more were struggling to reach the "land of gold." Other thousands essayed the overland route. In one three-week period more than 18,000 persons crossed the Missouri River and headed West. Before snow blocked the mountain passes of the Rockies and the Sierras 36,000 persons entered California by the overland routes and an equal number disembarked from ships which had arrived in San Francisco Bay.

Immigrants from virtually every country in Europe, from China, Australia and South American countries found their way to San Francisco and from there to the mines in the Sierras beyond Sacramento. The sudden arrival of so many thousands of men, often with their families, completely disrupted all normal life. When Marshall made his momentous discovery in January, 1848, there were approximately 12,000 white residents in California.

Four years later the population had increased to over a quarter of a million.

For a period a boom-town atmosphere prevailed and violence, political corruption and lawlessness ran riot. A vigilance committee was formed to deal with the situation in 1851 and was reorganized in 1856. Eventually the placer gold was exhausted and hard rock mining, requiring expensive machinery, replaced the individual miners who panned gold by hand.

John Augustus Sutter did not profit by Marshall's discovery. Gold-seekers settled on his land as squatters and stole his cattle and sheep. Many of his workmen deserted to become miners. Disputes regarding the ownership of land sprang up between Sutter and the squatters. Eventually Sutter's land claims came before the United States Supreme Court, which found the titles to most of his lands invalid. In 1852 he was forced into bankruptcy and, while California granted him a pension of $250 a month for his services in settling the state, his petitions to the United States Congress for recompense were never acknowledged. He died in poverty in 1880.

Suggested Readings

Andrist, Ralph K. The California Gold Rush. New York: American Heritage Press, 1961.

Caughey, John W. Gold Is the Cornerstone. Berkeley, Calif.: University of California Press, 1948.

________, ed. Rushing for Gold. Berkeley, Calif.: University of California Press, 1949.

Coy, Owen C. The Great Trek. Los Angeles: Powell Publishing, 1931.

Geiger, Vincent, and Wakeman Bryarly. Trail to California: Overland Journey. Revised by David M. Potter. New Haven, Conn.: Yale University Press, 1962.

Greever, William S. The Bonanza West: The Story of the Western Mining Rushes. Norman, Okla.: University of Oklahoma Press, 1963.

Holt, Stephen, pseud. We Were There With the California Forty-niners. New York: Grosset and Dunlap, 1956.

Jackson, Joseph Henry. Anybody's Gold: The Story of California's Mining Towns. New York: Appleton, 1941.

Lewis, Oscar. California Heritage. New York: Crowell, 1949.

Morgan, Dale L. The Overland Diary of James A. Pritchard. Denver: F. A. Rosenstock, 1959.

Paul, Rodman W. California Gold: The Beginning of Mining in the Far West. Cambridge, Mass.: Harvard University Press, 1947.

________. Mining Frontiers of the Far West. New York: Holt, Rinehart, 1963.

Wellman, Paul I. Gold in California. Boston: Houghton Mifflin, 1958.

THE COMPROMISE OF 1850 (1850)

Following the Mexican War and the acquisition of new territories by the United States as a result of that war, the question of slavery was inescapably precipitated into American national politics. In August, 1846, David Wilmot, an American lawyer and representative from Pennsylvania, introduced his "Wilmot Proviso," which provided that slavery should be prohibited in any territories thus acquired from Mexico. This Proviso was adopted by the House but defeated in the Senate.

In 1848 Oregon was organized as a territory but efforts to similarly organize California and New Mexico were unsuccessful. In 1849 California adopted a constitution which prohibited slavery within its boundaries. Slavery was a vital political issue of the day, and the United States was experiencing an "ominous and growing movement for disunion." It was recognized by the more cool-headed conservatives that compromise between the slave states of the South and the free states of the North was vitally necessary.

This situation precipitated one of the greatest debates in Congressional history, involving the foremost statesmen of the day, including Henry Clay, Daniel Webster, John C. Calhoun, Jefferson Davis, Salmon P. Chase, William H. Seward and others.

On January 29, 1850, Clay introduced a series of compromise resolutions. These, submitted to a committee of which Clay was the chairman, were reported as separate bills, one an "Omnibus Bill" dealing with the organization of the territories, another prohibiting trading in slaves in the District of Columbia, and a third providing an improved fugitive slave law.

John C. Calhoun, close to death, made his appearance and had his speech denouncing Northern aggressions read by a friend. Daniel Webster made his Seventh of March speech supporting the compromise measures on slavery proposed by Clay.

The "Omnibus Bill" was finally passed in separate acts. California was admitted to the Union as a free state and the slave trade was prohibited in the national capital. The new fugitive slave law, which gave the responsibility for its enforcement to the federal courts, was also adopted. This combination of resolutions came to be known as the "Compromise of 1850."

Suggested Readings

Bassett, John Spencer. A Short History of the United States. New York: Macmillan, 1929.

Benton, Thomas Hart. Abridgement of the Debates of Congress, 1789-1856. New York: AMS Press, 1970.

Channing, Edward. A History of the United States. New York: Macmillan, 1921.

Commager, Henry Steele, ed. "The Compromise of 1850," (Doc. No. 174) in his Documents of American History, 8th edition. New York: Appleton, 1968.

Corder, Eric. Prelude to Civil War: Kansas-Missouri, 1854-61.

New York: Crowell, 1970.
Craven, Avery O. The Growth of Southern Nationalism, 1848-1861. Baton Rouge, La.: Louisiana State University Press, 1953.
Current, Richard N. Daniel Webster and the Rise of National Conservation. Boston: Little, Brown, 1955.
Davis, Jefferson. The Rise and Fall of the Confederate Government. Cranbury, N.J.: A. S. Barnes, 1881.
Dodd, William E. The Life of Jefferson Davis. New York: Russell & Russell, 1966.
Fuess, Claude M. Daniel Webster. Boston: Little, Brown, 1930.
Garrison, George P. Westward Expansion, 1841-1850. New York: Haskell House, 1906.
Greeley, Horace. The American Conflict. Westport, Conn.: Negro Universities Press, 1864-66.
Hamilton, Holman. Prologue to Conflict: The Crisis and Compromise of 1850. Lexington: University of Kentucky Press, 1964.
_______. Zachary Taylor: Soldier in the White House. Indianapolis: Bobbs-Merrill, 1941-1951.
Hart, Albert B. Salmon Portland Chase. New York: Haskell House, 1969.
Hoyt, Edwin P. Zachary Taylor. Chicago: Reilly & Lee, 1966.
Johnson, Alexander, and James P. Woodburn. American Orations. Plainview, N.Y.: Books for Libraries, 1896.
Johnson, Allen. Stephen A. Douglas. New York: Da Capo Press, 1970.
Lodge, Henry Cabot. Daniel Webster. Philadelphia: Richard West, 1973.
McMaster, John B. A History of the People of the United States, from the Revolution to the Civil War. New York: Appleton, 1938.
Nichols, Roy Franklin. Franklin Pierce: Young Hickory of the Granite Hills. Philadelphia: University of Pennsylvania Press, 1931.
Rayback, Robert J. Millard Fillmore: Biography of a President. East Aurora, Ill.: Henry Stewart, 1972.
Rhodes, James F. History of the United States. Port Washington, N.Y.: Kennikat Press, 1920.
Schurz, Carl. Henry Clay. New York: Frederick Ungar, 1968.
Von Holst, Hermann E. Constitutional and Political History of the United States. New York: AMS Press, 1892.
_______. John C. Calhoun. New York: AMS Press, 1899.
Wiltse, Charles M. John C. Calhoun. Indianapolis: Bobbs-Merrill, 1941-1944.

THE SAN FRANCISCO VIGILANCE COMMITTEES (1851 and 1856)

Following James Wilson Marshall's discovery of gold in California in 1848, the city of San Francisco, to that time little more than a Mexican settlement, experienced an explosive influx of gold hunters. In addition to those who hoped to become wealthy through prospecting, many unsavory characters poured into the city. A large number of the latter were "ruffians from the frontier towns of Australia and escaped convicts and ticket-of-leave men from the British penal settlements at Sydney and Tasmania, then called Van Diemen's Land."

By the end of 1849 so many of these arrivals from Australia had settled in the city by the Golden Gate that the district where they congregated became known as Sydney-Town and they came to be called "Sydney Ducks" or "Sydney Coves." Sydney-Town was a district of cheap dance halls, saloons, brothels and commercialized vice. Law and order were virtually nonexistent. John W. Geary, First Alcalde, attempted to keep order but was unable to do so, the lawless element being protected by unscrupulous, bribe-taking politicians and city officials. Murder, robbery, arson, burglary, assault and theft went unchecked.

This state of affairs brought about the organization of the Vigilance Committees. After a series of fires deliberately set by the Sydney Ducks, approximately 200 prominent citizens of San Francisco gathered in early June, 1851, in a building owned by Sam Brannan, following which meeting the first Vigilance Committee was formed. This was to be a law-enforcement body and its object was to maintain "the peace and good order of society, and the preservation of the lives and property of the citizens of San Francisco."

On the afternoon of June 10, 1851, an Australian convict named John Jenkins, having been apprehended while stealing a small safe, was captured, tried by the Committee, and publicly hanged in Portsmouth Square. The members of the Committee further enforced law and order by warning individual Sydney Ducks to leave the city at once or suffer the consequences of disobedience. A thirty-man sub-committee was appointed to board all incoming vessels and examine all suspicious persons, denying entrance to those who could not furnish "evidence of honesty and good character."

Lesser criminals left the city but the ringleaders remained behind, believing that the trouble would blow over and that they would continue to enjoy the protection of their friends the politicians.

Things came to a head in July, 1851, when a notorious criminal known as English Jim was apprehended attempting to rob a vessel moored in San Francisco Bay. He was captured by the ship's crew and turned over to the Vigilance Committee. He was found guilty of innumerable crimes and hanged on Market Street Wharf. Samuel Whittaker and Robert Mackenzie, two other Sydney Ducks, were arrested, tried and found guilty of arson, robbery and burglary. They were sentenced to be hanged. The sentence was carried out in spite of interference on the part of Governor John MacDougal.

This double execution was the last official act of the First Vigilance Committee. The inhabitants of Sydney-Town left hurriedly and no further need of the Committee, which was never formally dissolved, was felt until 1856, when it was once more activated by William J. Coleman. This was necessitated by the return of many criminals to the city and by the corrupt political machine which, dominated by David C. Broderick and others, protected them from the consequences of their crimes.

The murder of General W. H. Richardson by the gambler Charles Cora and of editor James King of William by his rival journalist James P. Casey precipitated the reorganization of the Committee. Cora and Casey were summarily tried and hanged. Two other murderers were similarly executed: Joseph Heatherington and a man named Brace.

These last hangings, in effect, completed the work of the Second Vigilance Committee. Four men had been hanged, 26 had been banished, and several hundred undesirables had been frightened away. The Committee then disbanded, late in 1856.

Suggested Readings

Asbury, Herbert. The Barbary Coast. New York: Knopf, 1933.

Bancroft, Hubert Howe. The History of California. San Rafael, Calif.: Bancroft Press, 1888.

Beebe, Lucius, and Charles Clegg. San Francisco's Golden Era: A Picture Story of San Francisco Before the Fire. Berkeley, Calif.: Howell-North Books, 1960.

Bryce, James. The American Commonwealth. New York: Macmillan, 1914.

Commager, Henry Steele, ed. "Constitution of the Committee of Vigilantes of San Francisco," (Doc. No. 184) in his Documents of American History, 8th edition. New York: Appleton, 1968.

Grey, William. Pioneer Times in San Francisco. San Francisco: 1881.

Lewis, Oscar. California Heritage. New York: Crowell, 1949.

_______. San Francisco: Mission to Metropolis. Berkeley, Calif.: Howell-North Books, 1966.

Royce, Josiah. California, from the Conquest in 1846 to the Second Vigilance Committee in San Francisco, 1856. New York: AMS Press, 1972.

Smith, Frank Meriwether, ed. San Francisco Vigilance Committee of 1856. San Francisco: Barry, 1883.

Soule, Frank, John Gihon, M.D. and James Nisbet. The Annals of San Francisco. New York: 1855.

Williams, Mary F. The History of the San Francisco Committee of Vigilance of 1851. New York: Da Capo Press, 1969.

THE KNOW-NOTHINGS (1852-1861)

In 1815 American public opinion was, in general, favorable towards immigration, but by 1830 the tide had swung the other way and demands were made for discrimination against those who came from foreign lands to settle in America. The Irish caused special alarm as they were more numerous than newcomers from other countries. As they were generally Catholics, it was feared that a large Irish vote would weaken the American doctrine of strict separation of church and state.

Organizations such as the Native American Association, the Order of the Sons of America, and the Order of the Star Spangled Banner appeared in various American cities. The objects of these groups were essentially to prevent foreign-born citizens from holding public office and to place restraints on foreign influences and ideas.

These various organizations culminated, in 1852, in the National American Party, a secret political organization popularly called the "Know-Nothings," due to the standard answer, "I don't know," given by its members when asked any question concerning the organization. The Party's membership was largely Whigs, as the Irish Catholics tended to become Democrats.

As a political organization the Know-Nothings sought electoral offices, and at first it was quite successful. In 1854, it carried both the Delaware and Massachusetts elections and polled a quarter of the total vote in New York State, as well as two-fifths of that of Pennsylvania. In the following year it elected governors and legislatures in Connecticut, New Hampshire, and Rhode Island, as well as New York. It was also generally successful in the South and West.

This "silent machine, without canvassers or other outward evidence of activity, but sweeping so much before it, struck terror to the old party leaders." Following the Spring election of 1855, it abandoned secrecy and boasted that it had over a million members, all enrolled as voters. This emergence from secrecy disclosed the fact that the Know-Nothing organization was chiefly the old, disintegrating Whig party under another name, which blasted all hope of building up from it a great Union party. In the 1856 election Millard Fillmore, its candidate, carried only one state, Maryland. By 1861 it had no representation in Congress and shortly thereafter vanished completely from the American political scene.

Suggested Readings

Beals, Carleton. Brass Knuckle Crusade. New York: Hastings House, 1960.

Billington, Ray Allen. The Protestant Crusade, 1800-1860. Gloucester, Mass.: Peter Smith, 1938.

Binkley, Wilfred E. American Political Parties: Their Natural History. New York: Knopf, 1945.

Desmond, Humphrey J. The Know Nothing Party. Washington, D.C.: Blair, 1904.

Haynes, George H. "The Causes of Know Nothing Success in Massachusetts," American Historical Review Quarterly, 1897.

McMaster, John B. With the Fathers. New York: Benjamin Blom, 1896.

Mandelbaum, Seymour J. The Social Setting of Intolerance. Chicago: Scott, Foresman, 1964.

Overdyke, W. Darrell. Knownothing Party in the South. Baton

Rouge, La.: Louisiana State University Press, 1950.
Pearson, Henry G. "Preliminaries of Civil War," in Hart, Albert B., ed. Commonwealth History of Massachusetts. New York: Russell & Russell, 1930.
Rayback, Robert J. Millard Fillmore: Biography of a President. East Aurora, Ill.: Henry Stewart, 1972.
Rhodes, James F. History of the United States. Port Washington, N.Y.: Kennikat Press, 1920.
Scisco, Louis D. Political Nativism in New York State. New York: AMS Press, 1968.
Soule, Leon G. The Know Nothing Party in New Orleans: A Reappraisal. Baton Rouge, La.: Louisiana State University Press, 1961.

THE GADSDEN PURCHASE (1853)

The Treaty of Guadalupe Hidalgo between the United States and Mexico, signed on February 2, 1848, ended the war fought by the two countries. It was negotiated by Nicholas Philip Trist, an American lawyer and diplomat who had been sent to Mexico as a special agent to negotiate peace. His blundering in handling the negotiations resulted in his recall. This he disregarded. He signed the treaty, which was later accepted by both countries.

The treaty defined the border between the United States and Mexico only vaguely, and the former wished to retain possession of the Masilla Valley, which was claimed by Mexico. Other provisions of the Treaty pertaining to the United States' responsibility for restraining marauding Indians had not been enforced and Mexico claimed heavy damages.

This situation was resolved in 1853 when the United States purchased from Mexico, for $10,000,000, a tract of land lying in what are now the states of Arizona and New Mexico. This tract, containing 45,535 square miles, is bounded on the East by the Rio Grande River, on the North by the Gila River, on the West by the Colorado River, and adjoins the Mexican border. The purchase, known as the "Gadsden Purchase," was named for James Gadsden, an American soldier, railroad promoter, and diplomat who negotiated the treaty which included the sale. Under the terms of this treaty the United States was able to abrogate the article dealing with Indians and cancel Mexico's claims for damages.

The treaty was ratified, with a few insignificant changes, by the United States Senate in 1854. It met with great opposition in Mexico and one of the Mexican negotiators, General Antonio López de Santa Anna, was declared a traitor and banished the following year.

Suggested Readings

Callahan, James M. American Foreign Policy in Mexican Relations. New York: Cooper Square, 1932.
Callcott, Wilfred H. Santa Anna: The Story of an Enigma Who Once Was Mexico. Hamden, Conn.: Shoe String Press, 1964.
Commager, Henry Steele, ed. "Treaty of Guadalupe Hidalgo," (Doc. No. 171) in his Documents of American History, 8th edition. New York: Appleton, 1968.
Dodd, William E. The Life of Jefferson Davis. New York: Russell & Russell, 1966.
Garber, Paul N. The Gadsden Treaty. Gloucester, Mass.: Peter Smith, 1959.
Jones, Oakah L. Santa Anna. New York: Twayne, 1968.
Lockwood, Frank C. "The Gadsden Purchase," Arizona Quarterly, Vol. 2, 1946.
Nevins, Allan. Ordeal of the Union. New York: Scribner's, 1947.
Nichols, Roy Franklin. Franklin Pierce: Young Hickory of the Granite Hills. Philadelphia: University of Pennsylvania Press, 1931.
Rayback, Robert J. Millard Fillmore: Biography of a President. New York: Harry Stewart, 1972.
Rippy, James Fred. The United States and Mexico. New York: Crofts, 1931.
Strode, Hudson. Jefferson Davis: American Patriot, 1808-1861. New York: Harcourt, Brace, 1955.

THE KANSAS-NEBRASKA ACT (1854)

With the westward expansion of the United States during the first half of the nineteenth century the question of slavery assumed increasing importance. Opinion was greatly divided between the pro-slavery South and the anti-slavery North. As states and territories were added to the Union, increasing agitation as to whether they should be admitted as slave or free territories arose.

This was the case when the inhabitants of the Nebraska country west of Missouri and Iowa found that slavery men had defeated an attempt to make it a territory because, under the Missouri Compromise, it would be free.

In December, 1853, an Iowa senator introduced a bill to create Nebraska Territory. This bill, sponsored by Stephen A. Douglas, the Whig senator from Illinois, came to be known as the Kansas-Nebraska Act (or Bill). It embodied a number of controversial provisions which intensified the feeling between the slave and the free states and did much to bring about the Civil War. It provided for the repeal of the Missouri Compromise and the creation of Kansas and Nebraska as new territories, the inhabitants of which were to decide for themselves the question of slavery or freedom. The Bill was enacted by Congress in 1854.

Douglas became an anathema to a large part of the North and a group of northern Democrats, led by Salmon P. Chase, issued "An Appeal of the Independent Democrats." This document, signed by six prominent abolitionists, was published in newspapers throughout the country. It called for the organization of an anti-slavery party and helped lead to the organization of the new Republican Party in July, 1854.

The passage of the Kansas-Nebraska Act (or Bill) with its provision for "popular sovereignty" or, as it was nicknamed, "squatter sovereignty," precipitated a violent struggle for the control of Kansas, which territory came to be known as "Bleeding Kansas." Other results of the Act included the drawing of national attention to Abraham Lincoln who, in the Lincoln-Douglas debates, attacked the theory of popular sovereignty. John Brown, the abolitionist, carried out several forays against the proslavery elements.

Suggested Readings

Beveridge, Albert J. Abraham Lincoln, 1809-1858. Boston: Houghton Mifflin, 1928.

Billington, Ray Allen. The Far Western Frontier. New York: Harper, 1956.

________. The Westward Movement in the United States. New York: Van Nostrand, 1959.

Blackmar, Frank W. The Life of Charles Robinson. Plainview, N.Y.: Books for Libraries, 1970.

Capers, Gerald M. Stephen A. Douglas, Defender of the Union. Boston: Little, Brown, 1959.

Commager, Henry Steele, ed. "Appeal to the Independent Democrats," (Doc. No. 179) in his Documents of American History, 8th edition. New York: Appleton, 1968.

________. "The Kansas-Nebraska Act," (Doc. No. 180) in his Documents of American History, 8th edition. New York: Appleton, 1968.

________. "The Lincoln-Douglas Debates," (Doc. No. 187) in his Documents of American History, 8th edition. New York: Appleton, 1968.

Craven, Avery O. The Coming of the Civil War. Chicago: University of Chicago Press, 1966.

________. The Growth of Southern Nationalism, 1848-1861. Baton Rouge, La.: Louisiana State University Press, 1953.

Dick, Everett. The Sod-House Frontier, 1854-1890. Lincoln, Neb.: Johnsen, 1954.

Donald, David. Charles Sumner and the Coming of the Civil War. New York: Knopf, 1960.

Hart, Albert B. Salmon Portland Chase. New York: Haskell House, 1969.

Hodder, F. H. "Genesis of the Kansas-Nebraska Act," Wisconsin Historical Society Proceedings, 1912.

Iger, Eve Marie. John Brown: His Soul Goes Marching On. New York: Scott, Foresman, 1969.

Johannsen, Robert W. Stephen A. Douglas. New York: Oxford University Press, 1973.
Johnson, Allen. Stephen A. Douglas. New York: Da Capo Press, 1970.
Malin, James C. The Nebraska Question. Gloucester, Mass.: Peter Smith, 1972.
Milton, George F. The Eve of Conflict: Stephen A. Douglas and the Needless War. Boston: Houghton Mifflin, 1934.
Nevins, Allan. Ordeal of the Union. New York: Scribner's, 1947.
Nichols, Roy Franklin. "The Kansas-Nebraska Act: A Century of Historiography," Mississippi Valley Historical Review, September, 1956.
Rhodes, James F. History of the United States. Port Washington, N.Y.: Kennikat Press, 1920.
Robinson, Charles. The Kansas Conflict. Plainview, N.Y.: Books for Libraries, 1892.
Sandburg, Carl. Abraham Lincoln: The Prairie Years and the War Years. New York: Harcourt, Brace, 1954.
Villard, Oswald Garrison. John Brown, Eighteen Hundred to Eighteen Fifty Nine: A Biography Fifty Years After. Boston: Houghton Mifflin, 1910.

THE "CRIME AGAINST KANSAS" SPEECH (1856)

Probably the most controversial question which led to the Civil War was that of slavery. As new territories or states were admitted to the Union, bitter controversies as to whether they should be "free" or "slave" broke out.

One of the geographical areas involved in the bitter debate on the subject was the Kansas Territory. Kansas was admitted to the Union as a free state in 1861; but while it was a territory, as determined by the Kansas-Nebraska Act of 1854, clashes between partisans of slavery and abolition were both frequent and violent.

In early 1856 the Kansas situation was being considered by Congress. That territory's population had grown tremendously and two contending governing bodies, one pro-slavery and the other anti-slavery, had been set up. Both were tainted with illegality and historians generally agree that they should have been overthrown and new elections held. Impassioned speeches were given by many Southerners on one side and by Stephen A. Douglas, Jacob Collamer, John P. Hale and Charles Sumner on the other.

Sumner's speech was vitriolic and bitter. Into his anti-slavery tirade, delivered before Congress on May 19, 1856, and known as the "Crime Against Kansas" Speech, he "put as much denunciation as his soul could utter." He called his speech "the most thorough philippic ever uttered in a legislative body." Into it he brought caustic personal attacks on several fellow senators, including Douglas and Andrew Pickens Butler of South Carolina.

Douglas, in rebuttal, gave an equally biting speech. Senator Lewis Cass, breaking the painful silence which followed Sumner's tirade, said, "I have listened with equal regret and surprise to the speech of the honorable senator from Massachusetts. Such a speech, the most un-American and unpatriotic that ever grated on the ears of the members of this high body, I hope never to hear again, here or elsewhere."

On May 22, three days after Sumner's discourse, Butler's nephew, Preston Smith Brooks, strode into the senate chamber armed with a gold-headed cane. He attacked Sumner savagely, breaking the cane and finishing the chastisement with the butt. Sumner never completely recovered from the beating.

Many Southerners approved Brooks's action. Some sent him, as gifts, canes to replace the one broken over the head of Sumner. Many Northerners, however, regarded his act with horror and campaigned to arouse the average man against the South.

This incident did nothing whatever for Kansas. Congress could not agree on a plan and the territory continued to be the scene of bitter strife between the abolitionists and the pro-slavery faction.

Suggested Readings

Bancroft, George. History of the United States of America, from the Discovery of the Continent. New York: Appleton, 1888.

Bassett, John Spencer. A Short History of the United States. New York: Macmillan, 1929.

Beard, Charles A., and Mary R. Beard. A Basic History of the United States. New York: Doubleday, 1944.

Corder, Eric. Prelude to Civil War: Kansas, Missouri, 1854-61. New York: Crowell, 1970.

Dixon, Susan. History of the Missouri Compromise and Slavery in American Politics. New York: Johnson Reprint, 1903.

Johnson, Alexander, and James A. Woodburn. American Orations. Plainview, N.Y.: Books for Libraries, 1896.

Johnson, Allen. Stephen A. Douglas. New York: Da Capo Press, 1970.

Ray, P. Orman. The Repeal of the Missouri Compromise. Boston: J. S. Canner, 1965.

Rhodes, James F. History of the United States from the Compromise of 1850. New York: Macmillan, 1893-1906.

Robinson, Charles. The Kansas Conflict. Plainview, N.Y.: Books for Libraries, 1892.

Schouler, James. History of the United States of America Under the Constitution. New York: Kraus Reprint, 1908-13.

Sumner, Charles. Charles Sumner: His Collected Works. Westport, Conn.: Negro Universities Press, 1969.

Von Holst, Hermann E. Constitutional and Political History of the United States. New York: AMS Press, 1892.

THE DRED SCOTT DECISION (1857)

The Dred Scott Decision, probably the most famous in the history of the United States Supreme Court, involved the determination of the constitutionality of the Missouri Compromise and of the legal right of a Negro to become a United States citizen.

Dred Scott was a Negro slave, the property of Dr. John Emerson. In 1834 he was taken by his owner from Missouri, a slave state, to the free state of Illinois, and from there to Fort Snelling, then in the Wisconsin Territory where, by the Missouri Compromise, enacted by Congress in 1820, slavery was expressly forbidden. While there, Scott married a female Negro slave and fathered a child, born in free territory.

In 1838 Scott, his wife and child were removed to Missouri. In 1846 he brought suit to obtain his freedom, on the ground that he had, by being taken to free territory, become free. The Supreme Court of Missouri, trying the case in 1852, ruled that, by his being returned to the slave state of Missouri, the status of slavery reattached to him and, as a slave, he was not a citizen of the United States.

The case came, in 1854, before the Federal Circuit Court, which took jurisdiction but found against Scott. The case was then, in 1855, brought before the United States Supreme Court on appeal, where, after lengthy debate, it was decided in 1857. Chief Justice Roger Brooke Taney delivered the majority opinion, which covered three essential points: 1) whether Scott was a citizen of the state of Missouri, so as to give the federal courts jurisdiction; 2) whether he had been set free by his sojourn in the free state of Illinois or in the free territory of Wisconsin; and 3) whether the Missouri Compromise was constitutional.

The findings of the Court on these points were: 1) that Scott was "not a citizen of Missouri, in the sense in which that word is used in the Constitution; and that the Circuit Court of the United States for that reason had no jurisdiction in the case and could give no judgment in it"; 2) "that neither Dred Scott himself, nor any member of his family were made free by being carried into [free] territory"; and 3) "the act of Congress which prohibited a citizen from holding or owning [slave] property . . . in [free territory] is not warranted by the Constitution and is therefore void."

Suggested Readings

Carson, Hampton L. The History of the Supreme Court. New York: Burt Franklin, 1972.

Catton, Bruce. "Black Pawn on a Field of Peril," American Heritage Magazine, December, 1963.

Commager, Henry Steele, ed. "Dred Scott v. Sandford," (Doc. No. 185) in his Documents of American History, 8th edition. New York: Appleton, 1968.

Corwin, Edward S. "The Dred Scott Decision in the Light of

Contemporary Legal Doctrines," American History Review, 1911.
Curtis, George T. Constitutional History of the United States. New York: Da Capo Press, 1974.
Eaton, Clement. The Mind of the Old South. Baton Rouge, La.: Louisiana State University Press, 1964.
Elkins, Stanley M. Slavery: A Problem in American Institutional and Intellectual Life. Chicago: University of Chicago Press, 1959.
Franklin, John Hope. The Militant South. Cambridge, Mass.: Harvard University Press, 1956.
Friedman, Leon, and Fred L. Israel, eds. The Justices of the Supreme Court, 1789-1869. New York: Bowker, 1969.
Hodder, F. H. "Some Phases of the Dred Scott Case," Mississippi Valley Historical Review, Vol. XVI.
Hopkins, Vincent J. Dred Scott's Case. New York: Fordham University Press, 1951.
Hurd, John C. The Law of Freedom and of Bondage. Westport, Conn.: Negro Universities Press, 1858-1862.
Isely, Jeter Allen. Horace Greeley and the Republican Party. Princeton, N.J.: Princeton University Press, 1947.
Johnson, Gerald White. The Supreme Court. New York: Morrow, 1962.
Latham, Frank B. The Dred Scott Decision, March 6, 1857. New York: Watts, 1968.
Liston, R. A. Slavery in America, the History of Slavery. New York: McGraw-Hill, 1970.
McCloskey, Robert G. The American Supreme Court. Chicago: University of Chicago Press, 1960.
Nye, Russel B. Fettered Freedom: Civil Liberties and the Slavery Controversy. East Lansing: Michigan State University Press, 1949.
Quaife, Milo M. The Doctrine of Non-Intervention With Slavery in the Territories. Chicago: N. C. Chamberlin, 1910.
Rozwenc, Edwin Charles. The Causes of the American Civil War. Boston: Heath, 1961.
______. Slavery as a Cause of the Civil War. Boston: Heath, 1949.
Smith, Charles B. Roger B. Taney: Jacksonian Jurist. Chapel Hill, N.C.: University of North Carolina Press, 1936.
Smith, William E. The Francis Preston Blair Family in Politics. New York: Macmillan, 1933.
Swisher, Carl B. Roger B. Taney. New York: Macmillan, 1935.
Tyler, Samuel, ed. Memoir of Roger Brooke Taney: Chief Justice of the Supreme Court of the U.S. New York: Da Capo Press, 1970.
Warren, Charles. The Supreme Court in United States History. Boston: Little, Brown, 1923.
Weisenburger, Francis P. The Life of John McLean: A Politician on the United States Supreme Court. Columbus: Ohio State University Press, 1937.

THE ALMANAC ACQUITTAL (1858)

One of the incidents in the life of Abraham Lincoln which has become a legend is the court trial of 1858 in which he defended William "Duff" Armstrong on a charge of murder. In this case he produced a copy of Jayne's Almanac and showed by it that the evidence given by Charles Allen, a prosecution witness, should be regarded with suspicion. This maneuver, plus a moving closing argument to the jury, resulted in an acquittal for Lincoln's client.

In the early 1830's when Lincoln was preparing for the bar he was befriended by Nancy Armstrong and her husband Jack, whom he had known as a boy in Clary's Grove, Illinois. In November, 1857, Mrs. Armstrong visited Lincoln, then a successful lawyer and politician, and asked him to defend her son Duff. Lincoln agreed to do so.

Duff Armstrong had attended a revival meeting at the Salt Creek Campground near his home. He had been drinking and on the evening of August 29, 1857 had been attacked by James P. "Pres" Metzker, a local farmer and bully much larger than himself. A fight ensued which was stopped by the onlookers.

Later in the evening Metzker got into another fight, this one with James H. Norris. The next morning Metzker was seen entering a local bar. His right eye was swollen shut. He drank a glass of whisky, mounted his horse and rode off. Duff Armstrong never saw him again.

On September 1 Metzker died. At the post-mortem it was revealed that he had received two head wounds. One was alleged to have been given him by Norris, who supposedly struck him with a piece of wagon harness. The other wound was attributed to a blow struck by Duff Armstrong with a slung shot. As it was judged that either blow could have been fatal, both Norris and Armstrong were arrested and charged with murder in the first degree.

The two men were tried separately. Norris's record was anything but good. He had previously killed a man named Thornburg but cleared himself, pleading self-defense. Norris was tried for the murder of Metzker, found guilty of manslaughter and sentenced to eight years in prison.

Duff Armstrong was tried at Beardstown, Illinois. Judge Harriott was on the bench. Hugh Fullerton, the state's attorney, prosecuted the case and Abraham Lincoln and William Walker conducted the defense.

The prosecution's star witness was Charles Allen, who testified that he had seen Armstrong strike Metzker with a slung shot about eleven o'clock at night on August 29, 1857. His testimony was that he had observed the act from about forty feet away by the light of the full moon.

Other evidence was produced by both defense and prosecution. Medical testimony was given showing that Metzker could have died as the result of a blow inflicted by Norris or else by falls from his horse as he rode away from the camp meeting. Dr. Charles A. Parker, a defense witness, testified to this.

In due course the defense rested. Walker led off the summation for the defense and then Lincoln took the floor. He showed, using a copy of Jayne's Almanac, that on the night in question the moon had been, not full, as stated by Allen, but that it had barely passed the first quarter. This was confirmed by another almanac consulted by the prosecution. Duff Armstrong was found not guilty.

Years later, when questioned concerning the matter, he stated: "There was no moon that night. If there was, it was hidden by the clouds. But it was light enough for anybody to see the fight. The fight took place in one of the bars, and each bar had two or three candles in it."

Suggested Readings

Fehrenbacher, Don E. Prelude to Greatness: Lincoln in the 1850's. Palo Alto, Calif.: Stanford University Press, 1962.
Gridley, J. N. "Lincoln's Defense of Duff Armstrong," Illinois State Historical Society Journal, III, 24-44.
Herndon, William H., and Jesse W. Weik. Abraham Lincoln: The True Story of a Great Life. New York: A. & C. Boni, 1892.
King, James L. "Lincoln's Skill as a Lawyer," North American Review, CLXVI, 1898.
"Lincoln Lore," Lincoln National Life Foundation Bulletin, August 6, 1935.
Morris, Richard B. "Armstrong's Acquittal by Almanac," in his Fair Trial. New York: Knopf, 1952.
Sandburg, Carl. Abraham Lincoln: The Prairie Years and the War Years. New York: Harcourt, Brace, 1954.
Tarbell, Ida M. The Life of Abraham Lincoln. New York: Doubleday, 1900.
Townsend, William H. Lincoln the Litigant. Boston: Houghton Mifflin, 1925.
________. "Lincoln's Defense of Duff Armstrong," American Bar Association Journal, XI, 1925.
Tracy, Gilbert A., ed. Uncollected Letters of Lincoln. Boston: Houghton Mifflin, 1917.
Woldman, Albert A. Lawyer Lincoln. Boston: Houghton Mifflin, 1936.

THE LINCOLN-DOUGLAS DEBATES (1858)

On June 17, 1858, at the close of the Republican State Convention at Springfield, Illinois, Abraham Lincoln gave his famous "House Divided" speech in which he announced that the war on slavery was uncompromising. "A house divided against itself cannot stand," he said. "I believe this government cannot endure permanently half slave and half free.... I do not expect the house to fall; but I do expect it will cease to be divided."

Stephen A. Douglas, the Democratic politician and opponent of the Republican Lincoln, was unpopular in the South but hoped to restore himself to favor before the national election of 1860 when he hoped to be voted into the Presidency of the United States. In the meantime his term in the United States Senate expired in 1859 and the Democrats, at the 1858 convention, nominated him to succeed himself. Lincoln had been named as his opponent.

On July 24, 1858, Lincoln challenged Douglas to a series of debates. The challenge was accepted and seven such debates were arranged, to be held in the Illinois towns of Ottawa, Freeport, Jonesboro, Charleston, Galesburg, Quincy and Alton from August 21 to October 15.

Of these seven debates, the second, held at Freeport, is considered the most significant. It was then that Lincoln forced Douglas to declare his so-called Freeport Doctrine which made Douglas unacceptable to the pro-slavery South. Lincoln asked Douglas, "Can the people of a United States territory, in any lawful way, against the wish of any citizen of the United States, exclude slavery from its limits prior to the formation of a state constitution?" Douglas, in his reply, said very emphatically that "In my opinion the people of a territory can, by lawful means, exclude slavery from their limits prior to the formation of a state constitution." Slavery, he stated, could not exist in a territory without local police regulations to protect it, and these could be made only by the local legislature which would oppose slavery if the people who elected the legislators were opposed to it. "Hence, no matter what the decision of the Supreme Court may be on that abstract question, still the right of the people to make a slave territory or a free territory is perfect and complete under the Nebraska Bill."

Douglas was elected to the senatorship over Lincoln but the debates brought Lincoln to national attention, paving the way to his candidacy in 1860, when he was elected President over Douglas, John C. Breckinridge, and John Bell.

Suggested Readings

Angle, Paul M. Created Equal? The Complete Lincoln-Douglas Debates of 1858. Chicago: University of Chicago Press, 1958.

Baringer, William E. Lincoln's Rise to Power. Boston: Little, Brown, 1937.

Bassett, John Spencer. A Short History of the United States. New York: Macmillan, 1929.

Beveridge, Albert J. Abraham Lincoln, 1809-1858. Boston: Houghton Mifflin, 1928.

Chubb, Jerome M., and Howard W. Allen, eds. Electoral Change and Stability in American Political History. New York: Free Press, 1971.

Cole, Arthur C. The Era of the Civil War. Plainview, N.Y.: Books for Libraries, 1972.

Commager, Henry Steele, ed. "The Lincoln-Douglas Debates," (Doc. No. 187) in his Documents of American History, 8th edition. New York: Appleton, 1968.

________. "Lincoln's House Divided Speech," (Doc. No. 186) in his Documents of American History, 8th edition. New York: Appleton, 1968.
Congressional Quarterly Service. Presidential Candidates from 1788 to 1964, Including Third Parties, 1832-1964, and Popular Electoral Vote: Historical Review. Washington, D. C. : Government Printing Office, 1964.
Current, Richard N. The Lincoln Nobody Knows. New York: Hill & Wang, 1958.
Fehrenbacher, Don E. Prelude to Greatness: Lincoln in the 1850's. Palo Alto, Calif. : Stanford University Press, 1962.
Jaffa, Harry V. Crisis of the House Divided: An Interpretation of the Issues in the Lincoln-Douglas Debates. Garden City, N. Y. : Doubleday, 1959.
________, and Robert W. Johannsen. In the Name of the People. Columbus: Ohio State University Press, 1959.
Johannsen, Robert W. Stephen A. Douglas. New York: Oxford University Press, 1973.
________, ed. The Letters of Stephen Arnold Douglas. Urbana, Ill. : University of Illinois Press, 1961.
________, ed. The Lincoln-Douglas Debates of 1858. New York: Oxford University Press, 1965.
Johnson, Alexander, and James A. Woodburn. American Orations. Plainview, N. Y. : Books for Libraries, 1896.
Johnson, Allen. Stephen A. Douglas. New York: Da Capo Press, 1970.
Macy, Jesse. Political Parties in the United States, 1846-1861. New York: Arno Press, 1973.
Morse, John T. Abraham Lincoln. Philadelphia: Richard West, 1973.
Nicolay, John G. A Short Life of Abraham Lincoln. New York: Century, 1902.
Nolan, Jeannette Covert. The Little Giant: The Story of Stephen A. Douglas and Abraham Lincoln. New York: Messner, 1942.
Rhodes, James F. History of the United States from the Compromise of 1850. New York: Macmillan, 1893-1906.
Roseboom, Eugene H. A History of Presidential Elections, from Washington to Richard M. Nixon. New York: Macmillan, 1970.
Sandburg, Carl. Abraham Lincoln: The Prairie Years and the War Years. New York: Harcourt, Brace, 1954.
Tarbell, Ida M. The Life of Abraham Lincoln. New York: Doubleday, 1900.
Thomas, Benjamin P. Abraham Lincoln: A Biography. New York: Knopf, 1952.

THE ADAMS EXPRESS ROBBERIES (1858-1859)

The successful solving of the Adams Express Robberies brought Allan Pinkerton, the Anglo-American detective, into national prominence. This case first came to Pinkerton's attention in mid-1858 when he received a letter from E. S. Stanford, vice president of the Adams Express Company, asking for help. An express pouch had arrived at Montgomery, Alabama, by messenger from Atlanta, Georgia, on April 26. This pouch, when opened by Nathan Maroney, the Adams Express agent at Montgomery, was supposed to contain, among other packages, two sent in error to Atlanta. One of these was a bundle of currency totalling $4,750; the other, containing $10,000, was missing. Chase, the messenger who had brought the locked pouch from Atlanta, swore that he had delivered it just as it had been handed to him and that it had remained in the locked safe of the express car all the way to Montgomery.

Stanford elected to ignore Pinkerton's reply to his letter and his investigation concerning the missing currency led nowhere. Maroney was suspected of stealing the money but nothing could be proved. In January, 1859, Maroney resigned his position with the Adams Express Company, and the resignation was accepted. He remained on the job until his successor could be appointed but before this could be done a second theft occurred. Four bundles containing currency totalling $40,000 were found to be missing from a pouch being shipped from Montgomery to Atlanta. Maroney prepared and locked the pouch in the presence of the messenger Chase, who was again to accompany the shipment to its destination. The loss was discovered when Chase delivered the pouch to the agent at Atlanta. It was then that Stanford telegraphed Pinkerton in Chicago asking again for assistance.

Allan Pinkerton responded by sending an operative named Porter to Montgomery. Porter put Maroney under observation. The latter was arrested, then freed on bail. It was generally felt that the Adams Express Company's case against him was exceedingly weak. Pinkerton sent Roch, another operative, to assist Porter.

In New York Pinkerton learned from the Express Company officials something of Maroney's background and also something of that of Belle, his supposed wife. Both had led gay, unconventional lives and associated with people of dubious reputation. Pinkerton traveled to Montgomery where, after surreptitiously observing the woman, he assigned Green, another operative, to follow her movements. "Mrs. Maroney" and her daughter by a previous marriage began an extended trip to Charleston and thence to New York, Green following them. Nathan Maroney also decided to take a "business trip" and told Porter, who was now friendly with him, of it. On April 5 Maroney left for Atlanta by train. He was shadowed by Roch, the Pinkerton detective. Maroney visited several Southern towns and eventually arrived at Natchez, Mississippi. Here he took possession of a trunk which had been shipped to him by express. He took the trunk to Montgomery, where it was stored in a hotel luggage room.

"Belle Maroney" and her daughter were traced to the home of her brother-in-law John Cox of Jenkintown, Pennsylvania. Pinkerton arranged for Mrs. Kate Warne, one of his "female operatives," to meet Belle, calling herself "Madame Imbert, the wife of Jules Imbert, the notorious forger serving a prison sentence." In May Nathan Maroney, trailed by Roch, went from Montgomery to Philadelphia. There he and Belle were married, confirming Pinkerton's suspicion that their previous relationship had not involved what he termed "the holy bond of matrimony."

Nathan Maroney was arrested in New York City and, unable to obtain bail, was incarcerated in jail. Here he met a fellow prisoner calling himself John R. White. White and his "nephew" Shanks, who visited him daily, were Pinkerton agents. Shanks mailed letters for Maroney after opening and reading them himself.

Mrs. Maroney made a trip to Montgomery and removed a sum of money from the trunk her husband had stored in the hotel there. She returned to Jenkintown where, on the advice of "Madame Imbert," with whom she had become friendly, she buried it in the cellar of her brother-in-law's house.

In New York the prisoners Maroney and White plotted to frame Chase, the Adams Express messenger. Maroney admitted to White that he had stolen the money from the pouches. He obtained the first $10,000, he said, because of a parcel which was misdirected. The additional $40,000 had come into his possession by dropping the parcels, not into the pouch, but past its mouth and down behind the counter. Chase recipted for four parcels of cash he never had in his possession.

White, released from the New York jail, traveled to Jenkintown where, posing as a friend of Maroney, he persuaded Belle Maroney to disclose the hiding place of the money. He and Mrs. Kate Warne dug up $39,515 from the cellar hiding place.

In October, 1859, Nathan Maroney was extradited. His trial was held at Montgomery. John R. White, the man to whom he had admitted his guilt and confessed in full, appeared in court as a witness for the prosecution. Maroney, having first pled "not guilty," changed his plea to "guilty." He was convicted and sentenced to ten years at hard labor. His wife, though guilty as an accessory after the fact, was not prosecuted.

Suggested Readings

Griffiths, Arthur. Mysteries of Police and Crime, 1898.

Horan, James D. The Pinkertons: The Detective Dynasty That Made History. New York: Crown Publishers, 1967.

________, and Howard Swiggett. The Pinkerton Story. New York: Putnam, 1951.

Lavine, Sigmund A. Allan Pinkerton: America's First Private Eye. New York: Dodd, Mead, 1963.

Orrmont, Arthur. Master Detective: Allan Pinkerton. New York: Messner, 1965.

Pinkerton, Allan. Criminal Reminiscences and Detective Sketches. Freeport, N.Y.: Books for Libraries, 1970. (Originally published 1878.)

________. Mississippi Outlaws and the Detectives. Don Pedro and the Detectives. Poisoner and the Detectives. New York: G. W. Carleton & Co., 1879.

________. Professional Thieves and the Detective. New York: G. W. Carleton & Co., 1881.

________. Strikers, Communists, Tramps and Detectives. New York: G. W. Carleton & Co., 1878.
Rowan, Richard Wilmer. The Pinkertons, a Detective Dynasty. Boston: Little, Brown, 1931.

THE JOHN BROWN RAID (1859)

John Brown was a fanatical American abolitionist. Early in life he acquired a fervent hatred of slavery, a hatred which was to dominate his life and career. He despised those who wished to do away with the institution by constitutional means, saying, "Without the shedding of blood there is no remission of sin."

In 1834 he initiated, in Pennsylvania, a project to educate young Negroes. In 1855 he moved to Kansas where, two years later, he and his sons killed five Missouri pro-slavery advocates. This was done at Pottawattomie, Kansas, in revenge for the murder of abolitionists.

In 1858 Brown was in New York, planning further activities. His idea was to "collect a band of devoted armed followers, seize and fortify a position in the mountains of Virginia or Maryland, and from it make raids into the farming communities to liberate slaves."

After a sojourn in Kansas, where he conducted a raid into Missouri, rescued eleven slaves and escaped with them to Canada, Brown returned to New England, solicited funds for his mountain refuge, and leased a farm to be used as a base of operations.

On October 16, 1859, together with eighteen followers including several of his sons, Brown seized the United States arsenal at Harpers Ferry, Virginia. He captured thirty or more citizens, cut the telegraph wires, and for 24 hours held his position.

At dawn on October 18 Brown and his followers were defeated by a detachment of marines commanded by Colonel Robert E. Lee and assisted by Lieutenant J. E. B. Stuart. Brown and four of his men were taken prisoner, seven escaped and ten, of whom two were Brown's sons, were killed. Brown was arrested, tried for various crimes including treason and murder, and was found guilty and sentenced to hang. At the conclusion of his trial he made a moving speech in which he attempted to explain his viewpoint and give the reasons for his actions. His sentence was carried out on December 2, 1859, at Charlestown, Virginia, now West Virginia.

Suggested Readings

Abels, Jules. Man on Fire: John Brown and the Cause of Liberty. New York: Macmillan, 1971.
Ansley, Delight. The Sword and the Spirit; a Life of John Brown. New York: Crowell, 1955.
Bassett, John Spencer. A Short History of the United States. New

York: Macmillan, 1929.
Benét, Stephen Vincent. John Brown's Body (narrative poem). New York: Doubleday, 1928.
Bradford, Gamaliel. Damaged Souls. New York: Houghton Mifflin, 1923.
Commager, Henry Steele, ed. "John Brown's Last Speech," (Doc. No. 189) in his Documents of American History, 8th edition. New York: Appleton, 1968.
Fleming, Thomas J. "The Trial of John Brown," American Heritage Magazine, August, 1967.
Furnas, Joseph Chamberlain. The Road to Harpers Ferry. New York: William Sloan Associates, 1959.
Hinton, Richard J. John Brown and His Men. New York: Funk & Wagnalls, 1894.
Iger, Eve Marie. John Brown: His Soul Goes Marching On. New York: Scott, Foresman, 1969.
Malin, James C. John Brown and the Legend of Fifty-Six. Philadelphia: Lippincott, 1942.
Morris, Richard B. "The Treason Trial of John Brown," in his Fair Trial. New York: Knopf, 1952.
Nevins, Allan. Ordeal of the Union. New York: Scribner's, 1947.
Nolan, Jeannette Covert. John Brown. New York: Messner, 1950.
Oates, Stephen B. To Purge This Land with Blood: A Biography of John Brown. New York: Harper, 1970.
Ruchanes, Louis, ed. John Brown: The Making of a Revolutionary. Originally published as A John Brown Reader. New York: Grosset and Dunlap, 1969.
Sanborn, Franklin B., ed. The Life and Letters of John Brown, Liberator of Kansas and Martyr of Virginia. Westport, Conn.: Negro Universities Press, 1971.
Villard, Oswald Garrison. John Brown, Eighteen Hundred to Eighteen Fifty Nine: A Biography Fifty Years After. Boston: Houghton Mifflin, 1910.
Warren, Robert Penn. John Brown: The Making of a Martyr. New York: Payson & Clark, 1929.

THE DANIEL E. SICKLES TRIAL (1859)

Daniel E. Sickles was an American lawyer, politician, Union Civil War Major General, rake, diplomat, and scion of Washington society. His wife, Teresa Bagioli Sickles, had an adulterous affair with the widower Philip Barton Key, son of Francis Scott Key, who wrote "The Star Spangled Banner." Philip Key was at that time district attorney of Washington, D. C.

On February 27, 1859, Sickles shot Key in Lafayette Square, a block away from the White House, shouting, "You villain, you have defiled my bed and you must die!" He had been informed by an anonymous letter that Key had been his wife's lover.

The dying Key was carried to the nearby Washington Club, where he expired. Sickles surrendered to United States Attorney

General Jeremiah Sullivan Black. He was taken to the city prison where, declining bail, he was incarcerated, first in a cell and then in the warden's quarters. Here he was permitted visitors and had the company of Dandy, his pet greyhound.

As a wealthy aristocrat of great influence and having many powerful friends, Sickles was able to command the best legal talent available for his defense. His attorneys included Edwin M. Stanton, later Lincoln's Secretary of War, as well as James Topham Brady, John Graham, Philip Phillips, and Messrs. Chilton, Magruder, and Ratcliffe. The prosecution was handled by Robert Ould, a plodding, methodical assistant district attorney who was aided by J. M. Carlisle, a Washington lawyer.

The battery of defense lawyers was faced with certain problems, one of which was their client's reputation as "a person of notorious profligacy of life." Further, his reputation as a brawler and as one who disdained paying his debts had him frequently in court. These characteristics did not make believable the picture of the man as a defender of marital fidelity.

The trial got under way. District Attorney Ould, to the relief of Sickles' attorneys, concentrated his argument on the homicide rather than on the defendant's character. He produced a stream of witnesses to the shooting. These testified that Sickles had fired three times at the unarmed Key, missing with the first shot but fatally wounding him with the others.

Ould rested the case for the prosecution and then the battery of defense lawyers swung into action. John Graham, in a two-day burst of oratory, depicted his client as a cuckolded husband rightfully avenging the seduction of his wife. He contended that Sickles was so upset on learning of his wife's unfaithfulness that he was rendered temporarily legally insane.

Teresa Bagioli Sickles, at her husband's insistence, had written a full confession. This, inscribed the day before Sickles shot and killed Key, was introduced at the trial and identified as being in Mrs. Sickles' handwriting by Bridget Duffy, her maid. Prosecution and defense attorneys wrangled as to whether or not this confession was admissible as evidence and Judge Thomas H. Crawford ruled that it might not be admitted because to do so would "destroy before the law the 'confidential identity' of the husband and wife."

Stanton and Ould engaged in a shouting match in which the former accused the prosecutor of being motivated by a "thirst for blood." Ould denied any such "thirst" and declared that he would continue to call Sickles' deed "murder" regardless of any other designation applied to it by the defense.

The trial went on. Legal technicalities were introduced by both sides. The question as to whether Key was armed when killed was debated at length. A witness named Albert A. Megaffey gave evidence indicating that Key might have been carrying a pistol when killed, but Judge Crawford declared Megaffey's testimony inadmissable. Evidence depicting Sickles as a philandering husband was introduced. Technicalities were used to bar the introduction of much evidence. This legal maneuvering went on for twenty days, with Stanton summing up for the defense in a long, powerful speech.

In his instructions to the jury Judge Crawford sided with the defense, saying, "If the jury have any doubt as to the case, either in reference to the homicide or the question of sanity, Mr. Sickles should be acquitted." The jury, after due deliberation, found the accused innocent "by reason of temporary mental aberration." The verdict was an extremely popular one.

Later that year Sickles took Teresa back as his wife. However, she never appeared in society with him again and died eight years later. Sickles went on to an outstanding military and political career. After Teresa's death he married a second time. This wife left him, after bearing him two children, because he was habitually unfaithful.

Suggested Readings

Fleming, Thomas J. "A Husband's Revenge," American Heritage Magazine, April, 1967.

Flower, Frank A. Edwin McMasters Stanton: The Autocrat of Rebellion. New York: AMS Press, 1905.

Fontaine, Felix G. The Washington Tragedy: Trial of Daniel E. Sickles. New York: 1859.

Graham, John. Opening Speech to the Jury, Sickles Trial. New York: 1859.

Lawson, John D., ed. American State Trials, 1569-1920. Wilmington, Del.: Scholarly Resources, 1972.

Morris, Richard B. "The Fate of the Flagrant Adulterer," in his Fair Trial. New York: Knopf, 1952.

Pinchon, Edgcumb. Dan Sickles. Garden City, N.Y.: Doubleday, 1945.

Pratt, Fletcher. Stanton: Lincoln's Secretary of War. Westport, Conn.: Greenwood Press, 1953.

Schwanberg, W. A. Sickles the Incredible. New York: Scribner's, 1956.

Seitz, Don C. The Dreadful Decade, Detailing Some Phases in the History of the United States from Reconstruction to Resumption, 1869-1879. Westport, Conn.: Greenwood Press, 1968.

Snyder, William L. Great Speeches by Great Lawyers. New York: Baker, Voorhis, 1881.

Thomas, Benjamin P., and Harold M. Hyman. Edwin M. Stanton: The Life and Times of Lincoln's Secretary of War. New York: Knopf, 1962.

THE CRITTENDEN COMPROMISE (1860)

By the end of the year 1860 it was clear that certain southern states planned to secede from the Union. President James Buchanan hoped to halt the worsening schism over slavery and revealed his views in his 1857 inaugural address and in the formation of his cabinet. The growing crisis of disunion was magnified when South

Carolina moved toward secession. Throughout his administration Buchanan endeavored to hold the Union together and sought to achieve a compromise which would be acceptable to both the North and the South.

Buchanan's hope that such a compromise could be achieved was reflected in Congress, which created a senate committee to report a plan of compromise. A number of resolutions were referred to the committee, the most notable being the "Crittenden Compromise," proposed by Senator John Jordan Crittenden of Kentucky in December, 1860. This Compromise was preceded by the election of Abraham Lincoln to the Presidency and by a declaration by Buchanan that "the federal government did not have the right to coerce a seceding state into submission."

The "Crittenden Compromise" was designed primarily to stipulate that in those United States territories north of the Missouri Compromise line (36° 30') slavery should be abolished, but that south of that line it should be protected. Crittenden proposed further that slavery might not be abolished in any state where it existed without that state's consent and that owners of fugitive slaves in states where slavery was established should be financially compensated by the federal government when such slaves had escaped with outside assistance.

Civil war was avoided during the Buchanan administration and Buchanan left the White House on March 1, 1861, with a feeling of relief. The "Crittenden Compromise" had not been either accepted or rejected at the time when Lincoln assumed the Presidency. The Republicans opposed the proposal on the grounds that their principles forbade anything which would admit slavery into another territory, and the senators from the cotton states opposed it as a matter of form. On December 28, 1860, the committee which was considering it reported that it could not come to an agreement.

The proposal never came to a vote in the senate. South Carolina had seceded before it was rejected by the House of Representatives in January, 1861, and the Civil War was formally declared in April of that year.

Suggested Readings

Auchampaugh, Philip G. James Buchanan and His Cabinet on the Eve of Secession. Lancaster, Pa.: J. S. Canner, 1965.

Bassett, John Spencer. A Short History of the United States. New York: Macmillan, 1929.

Buchanan, James. Mr. Buchanan's Administration on the Eve of the Rebellion. Plainview, N.Y.: Books for Libraries, 1865.

_______. The Works of James Buchanan. Edited by John B. Moore. Philadelphia: Lippincott, 1908-1911.

Coleman, Chapman. The Life of John J. Crittenden. New York: Da Capo Press, 1972.

Curtis, George T. The Life of James Buchanan, Fifteenth President of the United States. Plainview, N.Y.: Books for Libraries, 1883.

Keith, C. A. The Life of John J. Crittenden of Kentucky. New York: Scribner's 1926.
Klein, Philip S. President James Buchanan: A Biography. University Park, Pa.: Pennsylvania State University Press, 1962.
Nichols, Roy Franklin. The Disruption of American Democracy. New York: Free Press, 1967.
Nicolay, John G. A Short Life of Abraham Lincoln. New York: Century, 1902.
Sandburg, Carl. Abraham Lincoln: The Prairie Years and the War Years. New York: Harcourt, Brace, 1954.
Tarbell, Ida M. The Life of Abraham Lincoln. New York: Doubleday, 1900.
Thomas, Benjamin P. Abraham Lincoln: A Biography. New York: Knopf, 1952.

THE PONY EXPRESS (1860-1861)

On April 3, 1860, the American mail service known as the "Pony Express" was inaugurated by William Hepburn Russell. It was operated under the direction of the Central Overland California and Pike's Peak Express Company. It carried mail on horseback between St. Joseph, Missouri and Sacramento, California, and thence by steamer to San Francisco.

The schedule was, for that time, quite fast, eight days being allowed for each one-way trip. Stations were approximately 25 miles apart and each pony express rider covered 75 miles per day, changing horses at intervals. Before the Express was discontinued there were 100 stations, approximately 80 riders and between 400 and 500 horses.

The life of a pony express rider was extremely hazardous owing to frequent attacks by hostile Indians. One of the more prominent riders was Colonel William Frederick Cody, better known as "Buffalo Bill," afterwards an Indian scout, buffalo hunter, actor and proprietor of Buffalo Bill's Wild West Show.

The Pony Express carried mail until it was discontinued in October, 1861, on completion of the Pacific Telegraph Company.

Suggested Readings

Adams, Samuel Hopkins. Pony Express. New York: Random House, 1950.
Beals, Frank L. Buffalo Bill. Chicago: Wheeler, 1943.
Bloss, Roy S. Pony Express: The Great Gamble. Berkeley, Calif.: Howell-North Books, 1959.
Bradley, Glenn D. The Story of the Pony Express. Chicago: McClurg, 1913.
Carter, Kate B. Utah and the Pony Express. Salt Lake City: Utah Printing Co., 1960.
Chapman, Arthur. The Pony Express. New York: Putnam, 1932.

Cody, William Frederick. An Autobiography of Buffalo Bill (Colonel F. W. Cody). New York: Rinehart, 1920.
Davidson, Mary Richmond. Buffalo Bill, Wild West Showman. Champaign, Ill.: Garrard, 1962.
Driggs, Howard R. The Pony Express Goes Through. New York: Stokes, 1935.
Editors of Time-Life Books, with text by David Nevin. The Expressmen. New York: Time-Life Books, 1974.
Hafen, Le Roy R. The Overland Mail. Glendale, Calif.: Arthur H. Clark, 1926.
Hagen, Olaf. "The Pony Express Starts from St. Joseph," Missouri Historical Review, Vol. XLIII.
Jackson, W. Turrentine. "A New Look at Wells Fargo, Stagecoaches and the Pony Express," California Historical Society Quarterly, December, 1966.
Jensen, Lee. The Pony Express. New York: Grosset & Dunlap, 1955.
Lewis, Oscar. California Heritage. New York: Crowell, 1949.
Nathan, Mel C., and W. S. Boggs. The Pony Express. New York: The Collectors Club, 1962.
Regli, Adolph Casper. The Real Book About Buffalo Bill. Garden City, N.Y.: Garden City Books, 1952.
Scheele, Carl H. A Short History of the Mail Service. Washington, D.C.: Smithsonian Institution Press, 1970.
Settle, Raymond W., and Mary L. Settle. Saddles and Spurs: The Pony Express Saga. Harrisburg, Pa.: The Stackpole Co., 1955.
Smith, Waddell F. The Story of the Pony Express. New York: Hesperian House, 1960.
Steele, William O. We Were There with the Pony Express. New York: Grosset & Dunlap, 1956.
Stevenson, Augusta. Buffalo Bill, Boy of the Plains. Indianapolis: Bobbs-Merrill, 1948.
Thompson, Robert Luther. Wiring a Continent. Princeton, N.J.: Princeton University Press, 1947.
Visscher, William L. A Thrilling and Truthful History of the Pony Express. Chicago: Rand McNally, 1908.

THE BALTIMORE PLOT (1861)

In January, 1861, the detective Allan Pinkerton was hired by Samuel M. Felton, president of the Philadelphia, Wilmington, and Baltimore Railroad, to investigate rumors that secessionist plotters in Maryland were intending to destroy railroad property. Felton believed that their object was to cut off the government at Washington from the Northern states.

In the course of his investigation Pinkerton and his agents discovered a plot to assassinate President-elect Abraham Lincoln as he passed through Baltimore on his way to Washington, D.C. from Springfield, Illinois. The leader of the plot was hot-headed Captain Fernandina, a former barber.

In Perryman, Maryland, Timothy Webster, a Pinkerton operative, succeeded in joining a troop of volunteer cavalry. It was he who first learned of the assassination plot while attending a meeting of the superior officers of the troop, a select group to which he had been admitted. Harry Davies, another Pinkerton detective, using the alias Joseph Howard, made the acquaintance, in Baltimore, of a young secessionist named Hill. Hill, a member of the Palmetto Guards, another volunteer military company, sponsored the detective for membership in this organization. Here Davies met Captain Fernandina and was able to introduce him to Allan Pinkerton, whom he identified as 'Major E. J. Allen, just up from Georgia--an earnest worker in the cause." Fernandina discussed his plans with Davies, Pinkerton, and other members of his group. That night Pinkerton learned from Webster that the date for Lincoln's assassination had been set.

Webster had also learned that the conspirators planned to hold a meeting in Baltimore at which the President-elect's assassin would be selected. Hill and Davies attended the meeting where, by lot, eight men were designated. Only the leaders of the plan were aware of this, and each of the eight men who drew a red ballot believed that he was the one selected to carry out the deed. Both Davies and Hill drew white ballots and so had not been selected to kill Lincoln.

On February 11, 1861, Lincoln left Springfield, Illinois, for Washington. On the train with him were John G. Nicolay, his private secretary, Judge David Davis, several officers who were later to become Civil War generals, and Norman B. Judd, his campaign manager. The latter was well known to Allan Pinkerton, who had written him concerning the planned assassination.

On February 21 the Lincoln party arrived at Philadelphia from New York, as did Pinkerton. There the latter engaged a room at the St. Louis Hotel under the name of J. H. Hutchinson, and here Judd visited him at his request. Felton was with Pinkerton and these two convinced Judd that Lincoln's life was in danger and must be protected. The three men drove to the Continental Hotel where Lincoln was staying and told him of the assassination plot.

Lincoln found it hard to believe that he was so hated that his life could be in jeopardy, but was eventually convinced that such was the case. He agreed to consider the proposal of Pinkerton and the others--to slip through Baltimore secretly and go by rail direct to Washington.

Next morning Lincoln made a scheduled speech in Philadelphia but before his special train could leave for Harrisburg he received dispatches from Washington warning him of the same plot already exposed by Pinkerton. At Harrisburg Lincoln agreed to Pinkerton's plan. He had been scheduled to spend the night of February 22 at the mansion of Governor Andrew G. Curtin of Pennsylvania following a dinner given in his honor. At the conclusion of the meal he excused himself and, escorted by Curtin, walked to the street where he was joined by Judd and the other members of the Presidential party. They entered a waiting carriage and were conveyed to a special train consisting of a fast locomotive and one passenger coach waiting for immediate departure for Washington.

Lincoln, Ward Lamont, a member of his party, and two railroad officials boarded the train. Judd and the other party members stayed behind to conform to the general belief that Lincoln had remained in Harrisburg. The engine and coach left immediately, headed for Philadelphia. Pinkerton, taking no chances, had arranged for the telegraph wires between Harrisburg and Baltimore to be cut, in case an ally of the conspirators should learn of Lincoln's sudden departure and attempt to advise Captain Fernandina of it.

Lincoln's train raced through the night to West Philadelphia, stopping only once so that the locomotive could take on water. At West Philadelphia the Lincoln party transferred to the Washington train which was being held for special orders. Guarded by Pinkerton, George Bangs and Mrs. Kate Warne, the latter two Pinkerton detectives, Lincoln was installed in the sleeping section at the rear of the last car of the train.

At 10:55 the train started. It reached Baltimore at 3:30 A.M. Here there was a two-hour wait until Lincoln's sleeping car could be coupled to a connecting train from the west which was behind schedule.

The run from Baltimore to Washington was uneventful. The train bearing Lincoln and his guards arrived at 6:00 in the morning of February 23, 1861. The Baltimore Plot had been foiled.

Suggested Readings

Agar, Herbert. Abraham Lincoln. New York: Macmillan, 1952.

Horan, James D., and Howard Swiggert. The Pinkerton Story. New York: Putnam, 1951.

Lavine, Sigmund A. Allan Pinkerton: America's First Private Eye. New York: Dodd, Mead, 1963.

Miers, Earl Schenck. Abraham Lincoln in Peace and War. New York: Harper, 1964.

Nicolay, John G. A Short Life of Abraham Lincoln. New York: Century, 1902.

Orrmont, Arthur. Master Detective: Allan Pinkerton. New York: Messner, 1965.

Pinkerton, Allan. Criminal Reminiscences and Detective Sketches. Freeport, N.Y.: Books for Libraries, 1970. (Originally published 1878.)

_______. History and Evidence of the Passage of Abraham Lincoln from Harrisburg, Pa. to Washington, D.C., on the 22nd and 23rd of February, 1861. Chicago: 1868.

Potter, John Mason. Plots Against the Presidents. New York: Astor, 1968.

Rowan, Richard Wilmer. The Pinkertons, a Detective Dynasty. Boston: Little, Brown, 1931.

Sandburg, Carl. Abraham Lincoln: The Prairie Years and the War Years. New York: Harcourt, Brace, 1954.

Shaw, Albert. Abraham Lincoln. New York: Review of Reviews Corp., 1929.

Tarbell, Ida M. The Life of Abraham Lincoln. New York:

Doubleday, 1900.
Thomas, Benjamin P. Abraham Lincoln: A Biography. New York: Knopf, 1952.

THE FALL OF FORT SUMTER (1861)

The causes of the Civil War were many and complex but the basic reason for the struggle was that since its founding the United States had developed into two separate sections, each with interests opposing the other. Manufacturing and commercial activities were found chiefly in the North, while the South, with its large plantations, was largely agricultural.

This situation led to clashes between the North and the South on two primary issues: the question of the extension of slavery into the territory of the United States not yet admitted into the Union as states, and the question of states' rights. These two issues precipitated increasingly bitter quarrels and ultimately led to the Civil War. The immediate cause of the War was the decision of eleven Southern states to withdraw from the Union and establish a Southern Confederacy.

South Carolina, the first state to secede (December 20, 1860), declared that "the Union now subsisting between South Carolina and other states under the name of the United States of America is hereby dissolved." This was tantamount to a declaration of war. By June of 1861 ten other southern states had followed South Carolina's lead and withdrawn from the Union. Jefferson Davis was elected president of the newly-formed Confederate States of America.

There were then eight federal forts in the seceding states. Six of these were garrisons and were easily taken over by the secessionists. The other two were Fort Pickens at Pensacola, Florida, and Fort Sumter at Charleston, South Carolina, the latter being manned by 84 soldiers commanded by Major Robert Anderson. Anderson had, on December 26, 1860, transferred his troops from Fort Moultrie. Governor Francis W. Pickens of South Carolina then seized Fort Moultrie and the fortifications in Charleston Harbor and demanded that Fort Sumter surrender. The demand was refused by Major Anderson.

On January 9, 1861, an effort to land supplies and reinforcements at Fort Sumter was made by the American merchant vessel "Star of the West," but she was fired upon by the secessionists and forced to withdraw after coming within a mile and a half of the fort. Fort Pickens, meanwhile, with a garrison of 48 men, remained in federal hands, where it stayed all through the Civil War.

President Abraham Lincoln, on March 5, 1861, was shown a letter from Major Anderson stating that he was running low on provisions and needed a force of at least 20,000 men to maintain his post. On April 8 Lincoln advised Governor Pickens that an attempt to send provisions only to Fort Sumter would be made. However, General Pierre Beauregard, acting on orders from Jefferson

Davis, demanded that Anderson surrender. This Anderson refused to do, but stipulated that if he did not receive provisions or "controlling instructions from the federal government by noon of April 15, he would abandon the fort on that day." This reply was unsatisfactory to the Confederates, and at 4:30 on the morning of April 12, 1861, the first shot of the bombardment, which lasted for 34 hours, was fired. This put an end to all further negotiations and marked the beginning of the Civil War.

Lincoln had sent a relieving fleet to Charleston. This arrived on April 12 but was unable to enter the harbor. On April 13 Major Anderson agreed to surrender and on the following day he and his garrison marched out of the fort with full honors of war. There were no casualties during the bombardment but the fort was badly wrecked. Before embarking on the relief ships with his troops Anderson fired a fifty-gun salute to the flag and he and his men were greeted by loud cheers from the Confederates.

Fort Sumter was taken over and repaired by the Confederates, who held it until February 17, 1865, when General William Tecumseh Sherman's army approached Charleston. On April 14 of that year, Major Anderson, then promoted to general, raised the same flag that he had been forced to lower four years earlier. The fort is now a national monument, having been so designated in 1948.

Suggested Readings

Barnes, Eric W. The War Between the States. New York: Whittlesey House, 1959.

Bassett, John Spencer. A Short History of the United States. New York: Macmillan, 1929.

Bradford, Gamaliel. Confederate Portraits. Plainview, N.Y.: Books for Libraries, 1917.

Commager, Henry Steele, ed. "South Carolina Declaration of Causes of Secession," (Doc. No. 199) in his Documents of American History, 8th edition. New York: Appleton, 1968.

_______. "South Carolina Ordinance of Secession," (Doc. No. 198) in his Documents of American History, 8th edition. New York: Appleton, 1968.

Freeman, Douglas Southall. Lee's Lieutenants. New York: Scribner's, 1942.

_______. R. E. Lee, A Biography. New York: Scribner's, 1934-1935.

Grant, Bruce. American Forts, Yesterday and Today. New York: Dutton, 1965.

Jordan, Robert Paul. The Civil War. Washington, D.C.: National Geographic Society, 1969.

Levenson, Dorothy. The First Book of the Confederacy. New York: Watts, 1968.

McDowell, David. Robert E. Lee. New York: Random House, 1953.

McGiffin, Lee. Swords, Stars and Bars. New York: Dutton, 1953.

McMeekin, Isabella McLennan. Robert E. Lee, Knight of the South. New York: Dodd, Mead, 1950.
Miers, Earl Schenck. The How and Why Wonder Book of the Civil War. New York: Grosset & Dunlap, 1961.

THE "TRENT" AFFAIR (1861)

During the Civil War one of the desires of the Confederacy was to be recognized by foreign nations, particularly England, as a belligerent rather than as a group of insurgents. Confederate agents in London asked for "recognition of independence" by the British. This recognition was not granted, although Queen Victoria, on May 13, 1861, issued a proclamation of neutrality in which both the Confederate and Federal governments were given the rights of belligerency within British jurisdiction.

Confederate diplomats John Slidell and James Murray Mason had left for Europe, the former to represent his government in Paris and the latter to act as Confederate representative in London. They had managed to escape through the Federal blockade to Havana, a neutral port, and there embarked on the British mail steamer "Trent," bound for Southampton.

On November 8, 1861, the American ship "San Jacinto," Captain Charles Wilkes commanding, encountered the "Trent" on the high seas and removed Slidell and Mason by force.

News of this episode caused great joy in the American North. Captain Wilkes was the hero of the day, and Congress and the secretary of the navy extended him their thanks. President Lincoln, Postmaster General Montgomery Blair and Secretary of State William Henry Seward, however, regretted the occurrence. It was obvious to them that Great Britain would demand a disclaimer and that war with that country was a distinct possibility.

The British government prepared an offensive demand for the surrender of the two Confederate diplomats. Prince Albert, Victoria's consort, saw the dispatch and suggested a more diplomatic approach, which was used. In January, 1862, Mason and Slidell were released and departed for Europe. However, no apology was made by the Federal government. Seward, in a long reply to Great Britain's note, stated the American position: "Had Wilkes seized the 'Trent' and sent her before an admiralty court he would have been within his right. As it was, he had exercised the right of search, something the American government had never opposed."

Suggested Readings

Adams, Charles Francis, Jr. Charles Francis Adams, 1807-1886, by His Son. Palo Alto, Calif.: Stanford University Press, 1961.
Adams, Ephraim D. Great Britain and the American Civil War.

New York: Russell & Russell, 1957.
Bancroft, Frederic. The Life of William H. Seward. Gloucester, Mass.: Peter Smith, 1966.
Bassett, John Spencer. A Short History of the United States. New York: Macmillan, 1929.
Beard, Charles A., and Mary R. Beard. A Basic History of the United States. New York: Doubleday, 1944.
Duberman, Martin B. Charles Francis Adams, 1807-1886. Palo Alto, Calif.: Stanford University Press, 1961.
Franklin, Walter. Famous American Ships. New York: Simon & Schuster, 1958.
Sandburg, Carl. Abraham Lincoln: The Prairie Years and the War Years. New York: Harcourt, Brace, 1954.
Tarbell, Ida M. The Life of Abraham Lincoln. New York: Doubleday, 1900.
Van Deusen, Glyndon G. William Henry Seward. New York: Oxford University Press, 1967.
Woodward, W. E. A New American History. New York: Farrar & Rinehart, 1936.

THE SOUTHERN BLOCKADE (1861-1865)

By definition, a blockade is "a naval operation conducted by a country at war, with the object of closing to foreign commerce the ports of an enemy country and thereby aiding in the military defeat of that country by denying it access to supplies from without." Several such blockades have been employed by belligerent nations including one declared by France against England in 1805 and a counter-blockade declared by England against all continental Europe. Another blockade was declared by the Allies in World War I when they attempted to isolate the Central Powers in the Baltic Sea, and still another was used by the United States and Great Britain in World War II.

Probably the most effective military blockade of a long coast was that maintained by the Federal navy against the South during the American Civil War. This blockade was declared on April 19, 1861. Its object was to prevent the export of southern cotton to England and also to prevent the importing of manufactured goods to southern ports from that country. Immediately upon the proclamation of the blockade, a dozen ships were sent to the most important southern harbors. By the end of the Civil War three hundred ships, including tugs, merchant ships purchased for blockade duty, and newly constructed vessels were on the line.

The blockade was divided into four squadrons: the North Atlantic, South Atlantic, West Gulf and East Gulf. These squadrons covered the coast from Fortress Monroe, Virginia, to the Rio Grande River.

Blockade runners often attempted to elude the guarding ships and transport cargo to and from southern ports. Large profits were realized from successful runnings and fast-sailing ships, manned by foreign nationals who, if captured, could not legally be held as prisoners of war. Some blockade runners were successful, others were captured, and in some cases the crew of the runner ran her ashore and set her afire in order to prevent her falling into the hands of the enemy.

In 1862 the South endeavored to break the blockade with the use of ironclads. The first of these, originally the merchantman "Merrimac," rebuilt and rechristened the "Virginia," engaged in combat with the Federal ironclad "Monitor" at Newport News, Virginia. The battle ended in a stalemate and the blockade continued in existence until 1865 when the war ended.

Suggested Readings

Bradlee, Frances B. C. *Blockade Running During the Civil War.* Salem, Mass.: Essex Institute, 1925.

Cochran, Hamilton. *Blockade Runners of the Confederacy.* Indianapolis: Bobbs-Merrill, 1958.

Horner, Dave. "The Blockade Runners," in his *Shipwrecks, Skin Divers, and Sunken Gold.* New York: Dodd, Mead, 1965.

Jones, Virgil C. *The Civil War at Sea.* New York: Holt, Rinehart, 1963.

Mahan, Alfred T. *Admiral Farragut.* New York: Haskell House, 1969.

Porter, Admiral David D. *The Naval History of the Civil War.* London: Sampson, Low, 1887.

Rhodes, James F. *History of the United States from the Compromise of 1850.* New York: Macmillan, 1893-1906.

Scharf, J. Thomas. *History of the Confederate States Navy.* New York: Rogers & Sherwood, 1887.

Semmes, Raphael. *The Confederate Raider "Alabama"; Selections from Memoirs of Service Afloat During the War Between the States.* Edited by Philip V. Stern. Gloucester, Mass.: Peter Smith, 1887.

Soley, James Russell. *The Blockade and the Cruisers.* New York: Scribner's, 1883.

Sprunt, James. "Running of the Blockade," *Southern Historical Papers,* Vol. 24.

Taylor, Thomas E. *Running the Blockade.* London: J. Murray, 1923.

THE BATTLE OF THE IRONCLADS (1862)

The first naval battle in history between ironclad vessels was the Civil War encounter between the "Merrimac" and the "Monitor" on March 9, 1862. The "Merrimac" was originally a Confederate wooden war vessel which had been sunk and abandoned in the Elizabeth River by Federal troops early in 1861. It had then been raised by the Confederates, rebuilt as an ironclad steamer and renamed the "Virginia." It is, however, best remembered by its original name.

The Federal "Monitor" was also an ironclad, designed and constructed by John Ericcson, a Swedish-American engineer and inventor. This unique vessel, propelled by steam and using a screw propeller, had guns mounted in a revolving turret, inaugurating a new era in naval engineering. The freeboard was low and the general appearance of the "Monitor" was such that it was called "the Yankee cheesebox on a raft."

On March 8, 1862, the "Merrimac," under the command of Captain Franklin Buchanan, together with several small escorts, entered Hampton Roads from Norfolk and attacked and destroyed the wooden Federal vessels "Cumberland" and "Congress." On March 9 the "Monitor," captained by John L. Worden, engaged the "Merrimac," which she was able to outmaneuver. The latter vessel was commanded by Catesby ap Roger Jones, Buchanan having been wounded the previous day.

The encounter between the two ironclads, which lasted for several hours, was indecisive. The "Monitor" suffered a direct hit which drove splinters of metal into Worden's eyes, forcing him to relinquish command to Lieut. Samuel Dana Green, his executive officer.

Following the battle the "Merrimac" withdrew up the Elizabeth River. It was destroyed by the Confederates in May of 1862 when they evacuated the Norfolk navy yard. The "Monitor" sank in a gale off Cape Hatteras on December 31 of the same year. After 1880 every battleship employed the principle of the revolving turret first demonstrated in battle by the "Monitor."

Suggested Readings

Baxter, James P. Introduction to the Ironclad Warship. Cambridge, Mass.: Harvard University Press, 1933.

Beard, A. E. S. "The Man Who Saved the Union Navy in 1862," in her Our Foreign-born Citizens. New York: Crowell, 1958.

Donovan, Frank R., and Bruce Catton. Ironclads of the Civil War. New York: American Heritage Press, 1964.

Durkin, Joseph T. Stephen R. Mallory, Confederate Navy Chief. Chapel Hill, N.C.: University of North Carolina Press, 1954.

Gosnell, H. Allen. Guns on the Western Waters. Baton Rouge, La.: Louisiana State University Press, 1949.

Jones, Virgil C. The Civil War at Sea. New York: Holt, Rinehart, 1963.

Knox, Dudley W. A History of the United States Navy. New York: Harper, 1948.

Latham, Jean Lee. Man of the Monitor: The Story of John Ericcson. New York: Harper, 1962.

Nevins, Allan. Ordeal of the Union. New York: Scribner's, 1947.

Nichols, Roy Franklin, ed. Battles and Leaders of the Civil War. New York: Century, 1887-1888.

Porter, Admiral David D. The Naval History of the Civil War. London: Sampson, Low, 1887.

Pratt, Fletcher. The Monitor and the Merrimac. New York: Random House, 1951.
Rhodes, James F. The History of the Civil War. Edited by E. B. Long. New York: Frederic Ungar, 1961.
_______. History of the United States from the Compromise of 1850. New York: Macmillan, 1893-1906.
Scharf, J. Thomas. History of the Confederate States Navy. New York: Rogers & Sherwood, 1887.
Still, William. Iron Afloat: The Story of the Confederate Ironclads. Nashville: Vanderbilt University Press, 1971.
West, Richard S., Jr. Mister Lincoln's Navy. New York: Longmans, Green, 1957.

THE HOMESTEAD ACT (1862)

The Homestead Act of 1862 was designed to enable American citizens without capital to acquire homesteads. Such an Act had passed both houses of Congress in 1859 but was vetoed by President James Buchanan. With the election of the Republican Abraham Lincoln the following year, homestead legislation was assured.

The Act passed on May 20, 1862, provided that after January 1, 1863, any American citizen, either 21 years of age or the head of a family, could acquire a parcel of federal public land not exceeding 160 acres. The homesteader acquired title to such land after five years, provided he settled upon it and cultivated it. Lands in the thirteen original states and in Kentucky, Maine, Tennessee, Texas, Vermont and West Virginia were excluded from the provisions of the Act. Land so acquired by a homesteader could not be levied against by creditors for the satisfaction of debts contracted prior to the issuance of the land grant.

This Act and subsequent others, which though modifying it to some degree did not change its basic concepts, provided an incentive for the settlement of the western parts of the United States.

Millions of acres were taken by bona fide settlers under the Homestead Act. However, there were those who evaded the law. Squatters had occupied some tracts without bothering to comply with the required legal formalities. Cattlemen fenced in huge areas for their own use and maintained their illegal positions by force of arms. By using various underhanded means, companies and individuals obtained possession of desirable tracts and held them for speculative profit. Some railroads had been granted enormous parcels of land along their rights of way under specified conditions, and it was found that in many instances these conditions had not been met.

In the late 1800's President Grover Cleveland forced the return of several million acres which had been illegally obtained by unscrupulous individuals and business organizations. The supply of land was not inexhaustible and, in his last annual message, Cleveland urged that what remained should be "husbanded with great care." Unfortunately, no effective steps in this direction were taken for many years.

Suggested Readings

Carstensen, Vernon, ed. The Public Lands: Studies in the History of the Public Domain. Madison, Wis.: University of Wisconsin Press, 1968.

Commager, Henry Steele, ed. "Homestead Act," (Doc. No. 214) in his Documents of American History, 8th edition. New York: Appleton, 1968.

Gates, Paul W. Fifty Million Acres: Conflicts Over Kansas Land Policy, 1854-1890. New York: Atherton Press, 1966.

Hibbard, Benjamin H. A History of the Public Land Policies. Madison, Wis.: University of Wisconsin Press, 1965.

Hill, Robert T. The Public Domain and Democracy. New York: AMS Press, 1968.

Nevins, Allan. Grover Cleveland: A Study in Courage. New York: Dodd, Mead, 1932.

Ottoson, Howard W., ed. Land Use Policy and Problems in the United States. Lincoln, Neb.: University of Nebraska Press, 1963.

Robbins, Roy M. Our Landed Heritage: The Public Domain. Princeton, N.J.: Princeton University Press, 1962.

Shannon, Fred A. The Farmer's Last Frontier: Agriculture, 1860-1897. New York: Farrar & Rinehart, 1945.

Stephenson, George M. The Political History of the Public Lands from 1840 to 1862: From Pre-Emption to Homestead. Boston: Richard C. Gadger, 1917.

Zahler, Helen S. Eastern Workingmen and National Land Policy, 1829-1862. New York: Columbia University Press, 1941.

THE PRAYER OF TWENTY MILLIONS (1862)

The "Prayer of Twenty Millions" was an open letter addressed to President Abraham Lincoln. It was written by Horace Greeley, editor and publisher of the New York Tribune, and appeared in that newspaper on August 20, 1862. Lincoln replied in a letter to Greeley, dated August 22. This reply was also published by newspaper.

Greeley's sentiments were definitely anti-slavery and his views on the subject were reflected by the editorials he wrote in his newspaper. At the time he published his open letter to Lincoln he was not aware that the President was conducting secret discussions with his cabinet concerning the freeing of Negro slaves, which was finally achieved by the Emancipation Proclamation of January 1, 1863. While Greeley's views were shared by a large and powerful anti-slavery element in the North, he certainly did not speak for "twenty millions."

The "Prayer of Twenty Millions" chided the President for being "unduly influenced by ... certain fossil politicians hailing from the Border Slave States," and for annulling [General John C.] "Frémont's Proclamation and [General David] Hunter's Order Favoring Emancipation." It also took exception to Lincoln's enforcing General Henry W. Halleck's order "forbidding fugitives from slavery to rebels to come within his lines," and to his failure to execute the provisions of the second Confiscation Act touching slavery. Greeley demanded that Lincoln definitely commit himself to emancipation.

In his published reply Lincoln stated his policy so far as the slavery question was concerned. He said, "my paramount object in this struggle is to save the Union, and is not either to save or destroy slavery. If I could save the Union without freeing any slave, I would do it; and if I could save it by freeing all the slaves, I would do it; and if I could save it by freeing some and leaving others alone, I would also do that."

Lincoln's letter was widely read and had a great influence on public opinion. Greeley subsequently caused great embarrassment to both himself and the administration by "dallying with peace overtures from the South." Following the Civil War he signed a bail bond for Jefferson Davis, President of the Confederacy, who had been arrested and held without trial at Fortress Monroe, Virginia.

Suggested Readings

Bassett, John Spencer. A Short History of the United States. New York: Macmillan, 1929.

Commager, Henry Steele, ed. "The Emancipation Proclamation," (Doc. No. 222) in his Documents of American History, 8th edition. New York: Appleton, 1968.

________. "Lincoln and Greeley," (Doc. No. 219) in his Documents of American History, 8th edition. New York: Appleton, 1968.

________. "Opposition to the Emancipation Proclamation," (Doc. No. 223) in his Documents of American History, 8th edition. New York: Appleton, 1968.

Greeley, Horace. The American Conflict. Westport, Conn.: Negro Universities Press, 1864-66.

________. Recollections of a Busy Life. New York: Arno Press, 1970.

Isley, Jeter Allen. Horace Greeley and the Republican Party. Princeton, N.J.: Princeton University Press, 1947.

Parton, James. The Life of Horace Greeley. New York: Arno Press, 1972.

Sandburg, Carl. Abraham Lincoln: The Prairie Years and the War Years. New York: Harcourt, Brace, 1954.

Seitz, Don C. Horace Greeley, Founder of the New York Tribune. New York: AMS Press, 1926.

Tarbell, Ida M. The Life of Abraham Lincoln. New York: Doubleday, 1900.

Van Deusen, Glyndon G. Horace Greeley, Nineteenth Century Crusader. New York: Hill & Wang, 1964.

THE NAST CARTOONS (1862-1894)

Today political cartoons caricaturing public figures and candidates for office are to be found in virtually every newspaper and many magazines. These pictures, now taken for granted, were introduced in metropolitan American newspapers in the middle 1800's. One of the pioneer political cartoonists was the German-American Thomas Nast, whose work appeared in publications both in America and abroad. Nast was educated in the New York City public schools and at the school of the National Academy of Design. After serving as a draftsman on the staffs of Frank Leslie's Illustrated Newspaper and Harper's Weekly he went to England as an artist for the New York Illustrated News and while there did work for the London News.

In 1862 he returned to the United States and rejoined Harper's Weekly as staff cartoonist. Here he remained during the Civil War and the Reconstruction period, creating many cartoons burlesquing political events and ludicrously exaggerating the personal characteristics of those individuals involved in them. His cartoons advocating service in the Union army were so effective that President Lincoln referred to him as "our best recruiting sergeant."

Nast is remembered chiefly for his cartoons published from 1869 through 1872 attacking the corrupt Tweed Ring headed by the political boss and grafter William Marcy Tweed. He used the device of the conversation balloon which is still found in modern comic strips. He also created the political symbols of the tiger for Tammany Hall, the elephant for the Republican Party and the donkey for the Democratic Party. Horace Greeley, editor of the New York Tribune and a Presidential candidate in the election of 1872, was defeated for the office by Ulysses S. Grant. Nast's cartoons in Harper's Weekly held Greeley up to merciless ridicule and contributed substantially to his losing the election. Again, in the Presidential election of 1884 the Republican James G. Blaine was defeated by the Democrat Grover Cleveland in what has been called "the dirtiest campaign in American history." Nast's Harper's Weekly cartoons caricatured Blaine as "a magnetic candidate too heavy for the party elephant to carry."

In 1887 Nast left Harper's Weekly and in 1894 joined the staff of the Pall Mall Gazette in London. In 1902 he was appointed United States consul at Guayaquil, Ecuador, where he died of yellow fever the same year.

Suggested Readings

Bryce, James. The American Commonwealth. New York: Macmillan, 1914.
Lynch, Denis T. "Boss" Tweed: The Story of a Grim Generation. New York: Boni & Liveright, 1927.
Mandelbaum, Seymour J. Boss Tweed's New York. New York: Wiley, 1965.
Nevins, Allan. Grover Cleveland: A Study in Courage. New York: Dodd, Mead, 1932.
________, and Frank Weitenkampf. A Century of Political Cartoons. New York: Octagon Books, 1975.
Paine, Albert Bigelow. Thomas Nast: His Period and His Pictures. Gloucester, Mass.: Peter Smith, 1904.
Seitz, Don C. The Dreadful Decade, Detailing Some Phases in the History of the United States from Reconstruction to Resumption, 1869-1879. Westport, Conn.: Greenwood Press, 1968.
________. Horace Greeley, Founder of the New York Tribune. New York: AMS Press, 1926.
Shaw, Albert. Abraham Lincoln. New York: Review of Reviews Corp., 1929.
Thomas, Harrison C. The Return of the Democratic Party to Power in 1884. New York: AMS Press, 1919.
Werner, M. R. Tammany Hall. Garden City, N.Y.: Doubleday, 1928.

THE EMANCIPATION PROCLAMATION (1863)

The Emancipation Proclamation was issued by President Abraham Lincoln on January 1, 1863. It declared "all persons held as slaves within any State or designated part of a State the people whereof shall then be in rebellion against the United States shall be then, thenceforward and forever free...." The states affected were listed in the Proclamation and exempted slaves in parts of the Southern states then held by Union armies.

Following the outbreak of the Civil War in 1861, large numbers of Negro slaves surrendered themselves to Union military authorities, such as General Benjamin F. Butler, commander of Fortress Monroe, Virginia, and volunteered to fight for their freedom.

On March 13, 1862, the Federal government instructed all Union army officers not to return fugitive slaves to their owners despite the Fugitive Slave Laws. On April 10 Congress authorized the compensation of slave owners who freed their slaves, and on April 16 all slaves within the District of Columbia were freed, their owners being paid $300 per slave.

On January 22, 1862, Lincoln read a preliminary draft of an emancipation proclamation to his cabinet. Secretary of State William H. Seward suggested that the proclamation be withheld until a military victory had been won. The battle of Antietam on September 17, 1862, in which the Federal troops were victorious, gave

Lincoln his chance, and on September 22 he read a second draft of the proclamation to his cabinet. This, with modifications, was issued as a preliminary proclamation, and on January 1, 1863 the formal and definite final version was made. This conferred liberty on approximately 3,120,000 Negro slaves.

The Emancipation Proclamation had great influence on antislavery sentiment in France and England. The governments of these countries were generally sympathetic to the southern cause and the Proclamation made it impossible for them to interfere on behalf of the Confederacy. It also greatly assisted the Republican party in becoming a major political organization.

The Thirteenth Amendment to the Constitution, dated December 18, 1865, made it clear that the institution of slavery no longer existed in the United States.

Suggested Readings

Adams, Ephraim D. Great Britain and the American Civil War. New York: Russell & Russell, 1957.

Commager, Henry Steele, ed. "The Constitution of the United States," (Doc. No. 87) in his Documents of American History, 8th edition. New York: Appleton, 1968.

________. "The Emancipation Proclamation," (Doc. No. 222) in his Documents of American History, 8th edition. New York: Appleton, 1968.

________. "Opposition to the Emancipation Proclamation," (Doc. No. 223) in his Documents of American History, 8th edition. New York: Appleton, 1968.

Donald, David. Lincoln Reconsidered. New York: Vintage Books, 1956.

Franklin, John Hope. The Emancipation Proclamation. Garden City, N.Y.: Doubleday, 1963.

Jordan, Donaldson, and Edwin J. Pratt. Europe and the American Civil War. New York: Octagon Books, 1969.

McPherson, James M. The Struggle for Equality. Princeton, N.J.: Princeton University Press, 1964.

Nevins, Allan. Ordeal of the Union. New York: Scribner's, 1947.

Nicolay, John G. A Short Life of Abraham Lincoln. New York: Century, 1902.

Quarls, Benjamin. Lincoln and the Negro. New York: Oxford University Press, 1962.

Randall, James G. Constitutional Problems Under Lincoln. Gloucester, Mass.: Peter Smith, 1950.

________. Lincoln the President. New York: Dodd, Mead, 1945.

Rhodes, James F. The History of the Civil War. Edited by E. B. Long. New York: Frederic Ungar, 1961.

Sandburg, Carl. Abraham Lincoln: The Prairie Years and the War Years. New York: Harcourt, Brace, 1954.

Tarbell, Ida M. The Life of Abraham Lincoln. New York: Doubleday, 1900.

Williams, T. Harry. Lincoln and the Radicals. Madison, Wis.: University of Wisconsin Press, 1941.

Wright, John S. Lincoln and the Politics of Slavery. Reno: University of Nevada Press, 1970.

THE DEATH OF STONEWALL JACKSON (1863)

Thomas Jonathan ("Stonewall") Jackson is regarded by military historians as one of the ablest of Confederate general officers. Born in Clarksburg, Virginia (now West Virginia) in 1824, he graduated from the U.S. Military Academy in 1846. Following participation in the Mexican War he became, in 1851, an instructor at Virginia Military Institute, resigning his commission.

He entered the Confederate army as a colonel in 1861 when the Civil War broke out, and was shortly promoted to brigadier general. His valiant participation in the first Battle of Bull Run earned him the nickname of "Stonewall." In subsequent battles he distinguished himself both as a strategist and as a fearless soldier.

Jackson's military career came to an end during the battle of Chancellorsville, fought May 2, 3 and 4, 1863, about ten miles west by south of Fredericksburg, Virginia. General Joseph Hooker commanded a Union army of about 130,000 men; the Confederate forces, under General Robert E. Lee, numbered approximately 60,000. Jackson, under Lee, commanded a corps of about 26,000 men.

On May 2 Lee was ready to attack Hooker's forces in spite of their numerical superiority. Jackson is said to have suggested the plan of battle which followed and which resulted in victory for the Confederate forces--their last great victory in the Civil War.

On Lee's instructions, Jackson made a fifteen-mile detour and at five o'clock in the afternoon attacked Hooker's right flank, thoroughly overcoming General Oliver Otis Howard's IX Corps, which composed it. Darkness closed in and it was thought that Jackson would renew the battle the following day. At twilight he rode out past his own sentinels to reconnoiter in the enemy's rear. Half an hour later a group of cavalrymen galloped back toward the sentinels who, thinking they were being attacked by Federal troops, fired several shots.

One of the horsemen cried, "Boys, don't fire again; you have hit General Jackson!" "Stonewall" was carried through his own lines to a hospital where he lived for ten days before expiring on May 10.

Surgeons, in an attempt to save Jackson's life, amputated his left arm. General Lee dispatched a note to him, saying, "You are better off than I am, for while you have lost your left, I have lost my right arm!"

Suggested Readings

Barnes, Eric W. The War Between the States. New York: Whittlesey House, 1959.

Bradford, Gamaliel. Confederate Portraits. Plainview, N.Y.: Books for Libraries, 1917.
Channing, Edward. A History of the United States. New York: Macmillan, 1921.
Daniels, Jonathan. Stonewall Jackson. New York: Random House, 1959.
Davis, Burke. They Called Him Stonewall: A Life of Lt. General T. J. Jackson, C.S.A. New York: Rinehart, 1954.
Freeman, Douglas Southall. Lee's Lieutenants. New York: Scribner's, 1942.
_______. R. E. Lee: A Biography. New York: Scribner's, 1934-1935.
Greeley, Horace. The American Conflict. Westport, Conn.: Negro Universities Press, 1864-66.
Henderson, G. F. Stonewall Jackson and the American Civil War. Gloucester, Mass.: Peter Smith, 1960.
Jordan, Daniel Paul. The Civil War. Washington, D.C.: National Geographic Society, 1969.
McDowell, David. Robert E. Lee. New York: Random House, 1953.
McGiffin, Lee. Swords, Stars and Bars. New York: Dutton, 1953.
McMeekin, Isabella McLennan. Robert E. Lee, Knight of the South. New York: Dodd, Mead, 1950.
Maurice, Frederick. Robert E. Lee, the Soldier. Folcroft, Pa.: Folcroft Library Editions, 1928.
Rhodes, James F. The History of the Civil War. Edited by E. B. Long. New York: Frederic Ungar, 1961.
Schouler, James. History of the United States of America Under the Constitution. Millwood, N.Y.: Kraus Reprint, 1908-13.
Tate, Allen. Stonewall Jackson: The Good Soldier. Ann Arbor, Mich.: University of Michigan Press, 1957.

THE BATTLE OF GETTYSBURG AND PICKETT'S CHARGE (1863)

The Battle of Gettysburg has been declared by historians to be the turning point of the American Civil War. Fought at Gettysburg, Pennsylvania, July 1, 2 and 3, 1863, it involved more than 75,000 Confederate troops commanded by General Robert E. Lee opposing an army of almost 82,000 Federal troops under the leadership of General George Gordon Meade. The encounter resulted in victory for the Federal army, and the dramatic but unsuccessful charge of 15,000 Confederate troops under the command of General George Edward Pickett on July 3 marked the virtual end of the battle and of the Confederate hope for ultimate victory.

Lee had decided to invade the North for three reasons. First, he wished to transfer the fighting to Federal territory. Second, he hoped to counteract the Confederate loss at Vicksburg, Mississippi, with a decisive victory and the capture of major northern cities. Third, he felt that a Confederate victory would tend to influence the North to make peace on the basis of southern independence.

He moved into Pennsylvania and concentrated his army at Gettysburg. This army consisted of three corps, commanded respectively by Generals Richard Stoddert Ewell, Ambrose Powell Hill, and James Longstreet. On July 1 Hill's forces encountered General John Buford's Federal cavalry division, supported by General John Fulton Reynolds' infantry. This fighting resulted in the Federal troops being forced to retire to Culp's Hill and Cemetery Ridge, southeast of Gettysburg, after the arrival of Ewell's Confederate reinforcements from the north.

On July 2 General Meade assembled his forces in a line stretching westward between Culp's Hill and Cemetery Ridge, then south to Round Top and Little Round Top. The Confederates deployed in a long, thin, concave line. Lee ordered an attack. Longstreet, on one flank, was not able to advance until late afternoon. The Federal troops held Cemetery Ridge and Little Round Top but were driven back from Peach Orchard and Devil's Den, both advanced positions. Ewell, in the center of the Confederate line, was able to win part of Culp's Hill.

That night (July 2) Meade held a council of war at which the decision not to retreat was made. Early on the morning of July 3 Culp's Hill was stormed and recaptured from the Confederates. It was then that the Confederate Major General George Edward Pickett made his famous charge on Cemetery Ridge and the Union center, commanded by General Winfield Scott Hancock. On orders from his superior, General Longstreet, Pickett's division, supported by Hill's corps and General J. E. B. Stuart's cavalry, some 15,000 troops in all, charged Cemetery Ridge. In spite of withering Federal fire the Confederate column met and carried the Union infantry back beyond their own cannon where a new line met and held it. After a brief hand-to-hand encounter Hancock's troops repulsed the Confederates, who were killed, captured, or driven back.

Lee's army was badly shattered and he prepared a defense against the Federal countercharge which he thought would follow. Meade's plans, however, were defensive and no countercharge ensued. On July 5, the Confederates withdrew south to Virginia.

The Battle of Gettysburg was the last major Confederate invasion of the north. Less than two years later Lee surrendered to General Ulysses S. Grant at Appomattox Court House, Virginia.

Suggested Readings

Barnes, Eric W. The War Between the States. New York: Whittlesey House, 1959.

Bassett, John Spencer. A Short History of the United States. New York: Macmillan, 1929.

Bradford, Gamaliel. Confederate Portraits. Plainview, N. Y.: Books for Libraries, 1917.

Carter, Hodding. Robert E. Lee and the Road of Honor. New York: Random House, 1955.

Catton, Bruce. The Battle of Gettysburg. New York: American

Heritage Press, 1963.
_______. Gettysburg: The Final Fury. Garden City, N.Y.: Doubleday, 1974.
_______. Glory Road: The Bloody Road from Fredericksburg to Gettysburg. Garden City, N.Y.: Doubleday, 1956.
Dowdey, Clifford. Death of a Nation. New York: Knopf, 1958.
Frassanito, William A. Gettysburg, A Journey in Time. New York: Scribner's, 1975.
Freeman, Douglas Southall. Lee's Lieutenants. New York: Scribner's, 1942.
_______. R. E. Lee: A Biography. New York: Scribner's, 1934-1935.
Jordan, Robert Paul. The Civil War. Washington, D.C.: National Geographic Society, 1969.
Kantor, MacKinlay. Gettysburg. New York: Random House, 1952.
Leckie, Robert. Great American Battles. New York: Random House, 1968.
_______. Wars of America. New York: Harper, 1968.
McDowell, David. Robert E. Lee. New York: Random House, 1953.
McMeekin, Isabella McLennan. Robert E. Lee, Knight of the South. New York: Dodd, Mead, 1950.
Rhodes, James F. The History of the Civil War. Edited by E. B. Long. New York: Frederic Ungar, 1961.
Shaara, Michael. The Killer Angels (fiction). New York: David McKay, 1974.
Tucker, Glenn. High Tide at Gettysburg. Indianapolis: Bobbs-Merrill, 1958.
Williams, Kenneth P. Lincoln Finds a General. New York: Macmillan, 1949-1959.
Williams, T. Harry. Lincoln and His Generals. New York: Knopf, 1952.

THE DRAFT RIOTS (1863)

When the American Civil War commenced in 1861, the ranks of the Union armies were filled by volunteer enlistees. After two years of fighting the original patriotic ardor cooled and volunteering for military service virtually ended. Congress, realizing that conscription was necessary, passed the Enrollment Act on March 3, 1863. This law, which ordered a draft on practically all able-bodied men between the ages of 20 and 45 years, was extremely unpopular. It permitted those drafted to avoid military service by furnishing a substitute or making a payment of $300. The latter provision, called "the rich man's exemption," caused widespread dissatisfaction among the poorer citizens of New York City and elsewhere. The Act itself was attacked by the Democrats as unconstitutional.

The Enrollment Act became effective on Saturday, July 11, 1863. No disturbances occurred on this date, but on Monday,

July 13, an unruly crowd quickly mobilized. The draft headquarters in New York City were attacked and set ablaze, and the rioters prevented the fire fighters from extinguishing the fire. Flames shortly spread to the entire block.

The New York police, aided by a small detachment of U.S. Marines, attempted to quell the disturbance but had scant success. The draft protesters, joined by habitual thieves and rowdies, proceeded on a program of looting and murdering, particularly of Negroes. On the afternoon of July 13 the Colored Orphan Asylum, a charitable institution, was sacked and burned. The mob virtually ruled the city from Union Square to Central Park.

The rioting, which had subsided somewhat late Monday night, was resumed the following day. The police, aided by Federal troops, tried in vain to disperse the mobs. More looting took place, more Negroes were murdered, and several neighborhoods were burned. Rioting continued until July 15, when military detachments from West Point and Pennsylvania arrived. On that day temporary suspension of the draft was announced and by Thursday, July 16, law and order had been restored.

It was estimated that, during the three-day life of the Draft Riots, more than a thousand persons had been killed or wounded, over 50 large buildings had been destroyed, and property damage totaling somewhere between a million and a half and two million dollars was inflicted.

Subsequent investigations showed that the allotments of the Democratic enrollment districts in New York were excessive. This was corrected and the draft was resumed and proceeded quietly, although it remained unpopular. Out of 470,942 persons whose names were drawn in the two drafts of 1864, 94,636 failed to report for duty.

Suggested Readings

Anti-Negro Riots in the North, 1863. Introduction by James M. McPherson. New York: Arno Press, 1969.

Archer, Jules. Riot: A History of Mob Action in the United States. New York: Hawthorn Books, 1974.

Barnes, David M. The Draft Riots in New York. New York: Baker & Godwin, 1863.

Bassett, John Spencer. A Short History of the United States. New York: Macmillan, 1929.

Cook, Adrian. Armies of the Streets: The New York Draft Riots of 1863. Lexington, Ky.: University of Kentucky Press, 1974.

Dickinson, Anna. What Answer? Boston: Tichnor & Fields, 1868.

Fry, James Barnett. New York and the Conscription of 1863. New York: Putnam, 1885.

Headley, J. T. Great Riots of New York, 1712 to 1873. New York: E. B. Treat Co., 1873.

Heaps, Willard Allison. Riots, U.S.A., 1765-1970. New York: Seabury Press, 1970.

Hoehling, Adolph A. "The Draft Riots, 1863," in his Disaster: Major American Catastrophes. New York: Hawthorn Books,

1973.
McCague, James. Second Rebellion: The Story of the New York Draft Riots of 1863. New York: Dial Press, 1968.
Mitchell, Edward B. Memoirs of an Editor. New York: Scribner's, 1924.
Mitchell, J. B., comp. Race Riots in Black and White. Englewood Cliffs, N.J.: Prentice-Hall, 1970.
Morris, John V. "Five Zouaves and Draft Riots," in his Fires and Firefighters. Boston: Little, Brown, 1953.
Official Records, War of the Rebellion. Washington, D.C.: Government Printing Office, 1890-1891.
Stoddard, William O. The Volcano Under the City. New York: Fords, Howard & Hulbert, 1887.
Werstein, Irving. Draft Riots, July, 1863. New York: Messner, 1971.

THE GETTYSBURG ADDRESS (1863)

On November 19, 1863, President Abraham Lincoln spoke at the ceremonies dedicating the battlefield at Gettysburg, Pennsylvania as a national cemetery. His two-minute speech, known as the "Gettysburg Address," is recognized as a "classic model of the noblest kind of oratory" and "as one of the most moving expressions of the democratic spirit ever uttered." Many school children have memorized Lincoln's words, which begin, "Fourscore and seven years ago our fathers brought forth on this continent a new nation, conceived in liberty, and dedicated to the proposition that all men are created equal."

At the dedication ceremony the principal speaker was Edward Everett, an American Unitarian clergyman, orator and statesman. He delivered a two-hour address. Everett's speech was printed on the first page of newspapers all over the country and, by contrast, Lincoln's short address was considered so insignificant that most papers carried it on the inside pages.

On November 20 Everett wrote to Lincoln congratulating him on his brief remarks. He said, "I should be glad if I could flatter myself that I came as near to the central idea of the occasion in two hours as you did in two minutes." In his acknowledgment Lincoln said, "In our respective parts yesterday, you could not have been excused to make a short address, nor I a long one."

Although now regarded as one of the great orations of American history, the true worth of Lincoln's short speech did not come to be realized until several years after it was delivered. The Chicago Times, reporting the dedication of the cemetery in 1863, said editorially, "The cheek of every American must tinge with shame as he reads the silly, flat and dishwatery utterances of the man who has to be pointed out to intelligent foreigners as the President of the United States."

Although there is considerable variation in phraseology of the various versions of the Gettysburg Address, it is regarded as an outstanding example of American oratory.

Suggested Readings

Adler, M. J., and William Gorman. American Testament. New York: Praeger, 1975.
Barton, William E. Lincoln at Gettysburg. Gloucester, Mass.: Peter Smith, 1920.
Basler, Roy P. The Lincoln Legend: A Study in Changing Conceptions. Boston: Houghton Mifflin, 1935.
Carr, Carl E. Lincoln at Gettysburg. Chicago: McClurg, 1908.
Colver, Anne. Abraham Lincoln: For the People. Champaign, Ill.: Garrard Press, 1960.
Commager, Henry Steele, ed. "The Gettysburg Address," (Doc. No. 228) in his Documents of American History, 8th edition. New York: Appleton, 1968.
Frothingham, Paul R. Edward Everett, Orator and Statesman. Port Washington, N.Y.: Kennikat Press, 1925.
Lincoln, Abraham. Abraham Lincoln: His Speeches and Writings. Cleveland: World, 1946.
_______. Abraham Lincoln's Gettysburg Address, illustrated by Jack E. Levin. Radnor, Pa.: Chilton, 1965.
_______. Famous Speeches of Abraham Lincoln. Mount Vernon, N.Y.: Peter Pauper Press, 1935.
Nevins, Allan, ed. Lincoln and the Gettysburg Address. Urbana, Ill.: University of Illinois Press, 1964.
Plowden, David. Lincoln and His America, 1809-1865, With the Words of Abraham Lincoln. New York: Viking Press, 1970.
Sandburg, Carl. Abraham Lincoln: The Prairie Years and the War Years. New York: Harcourt, Brace, 1954.
Shaw, Albert. Abraham Lincoln. New York: Review of Reviews Corp., 1929.
Tarbell, Ida M. The Life of Abraham Lincoln. New York: Doubleday, 1900.

THE ANDERSONVILLE PRISONERS (1863-1865)

In November, 1863, the Confederate States of America established a military stockade near Andersonville, Georgia, some sixty miles southwest of Macon. This establishment, known as Andersonville Prison, was designed to be used for the confinement of captured Union enlisted men during the Civil War. It was supervised by Captain Henry Wirz, a Swiss-born Confederate officer, who was referred to by the prisoners as "Death on a Pale Horse."

The first prisoners arrived at Andersonville in February, 1864. Between that date and the cessation of hostilities when General Robert E. Lee surrendered to General Ulysses S. Grant on April 9, 1865, 49,485 Union soldiers had been incarcerated there. As many as 30,000 prisoners were held in confinement at one time, of which more than 13,700 died. Following the war, 12,789 graves were identified and marked, 925 of these being those of unknown soldiers.

The Andersonville stockade consisted of 27 acres of open ground bisected by a stream one foot deep and five feet wide, bordered by swampland. The prison area was enclosed by pine log walls, 15 to 20 feet in height. The prisoners were provided with no shelter of any kind and were exposed to the elements night and day. The food was inadequate and poorly cooked. What water was available was impure, and sanitation was minimal. Crowded together under these conditions it is not surprising that over a quarter of the inmates became victims of dysentery and scurvy.

In 1864 two Confederate medical officers were appointed to investigate conditions at Andersonville. Following their inspection they recommended that the majority of the prisoners be transferred to other locations, and some were conveyed to establishments at Florence, South Carolina, and Millen, Georgia.

Following the surrender of the Confederates, Captain Wirz was tried for murder by a special U.S. military court, the president of which was Major General Lew Wallace, later to achieve fame as the author of the novels Ben Hur, The Fair God, and The Prince of India. Wirz was found guilty and sentenced to hang. The sentence was carried out on November 10, 1865.

Suggested Readings

Atwater, Dorence. A List of the Union Soldiers Buried at Andersonville. New York: The Tribune Association, 1890.

Braun, H. A. Andersonville: An Object Lesson on Protection. Milwaukee: C. D. Fahsel, 1892.

Faller, L. W. Dear Folks at Home. Carlisle, Pa.: Cumberland County Historical Society, 1963.

Futch, Ovid L. The History of Andersonville Prison. Gainesville, Fla.: University of Florida Press, 1968.

Goss, Warren. A Soldier's Story of His Captivity at Andersonville, Belle Island, and Other Rebel Prisons. Boston: Lee & Shepard, 1866.

Griffit, R. C., et al. Report of the Unveiling and Dedication of Indiana Monument at Andersonville, Georgia (National Cemetery), Thursday, November 26, 1908. Indianapolis: Beerford, 1909.

Kantor, McKinlay. Andersonville (fiction). New York: Crowell, 1971.

Kellogg, R. H. Life and Death in Rebel Prisons. Hartford, Conn.: Stebbins, 1865.

Levitt, Saul. The Andersonville Trial (play). New York: Random House, 1960.

Long, Lessel. Twelve Months in Andersonville. Huntington, Ind.: Butler, 1886.

McElroy, John. Andersonville: A Story of Rebel Military Prisons. Toledo, Ohio: D. R. Locke, 1879.

________. This Was Andersonville. New York: McDowell, Obolensky, 1957.

Ransom, John L. Andersonville Diary. Auburn, N.Y.: Published by the author, 1881.

Spencer, Ambrose. A Narrative of Andersonville. New York:

Harper, 1866.
Stevenson, R. R. The Southern Side. Baltimore: Turnbull Brothers, 1876.
Wallace, Lew. An Autobiography. New York: Somerset, 1972.

SHERMAN'S MARCH TO THE SEA (1864)

General William Tecumseh Sherman became commander of the Union force in Chattanooga, Tennessee, following General Ulysses S. Grant's promotion to lieutenant general on February 29, 1864 and appointment as commander of all the Union troops in the field.

Early in May, 1864, Sherman appeared before Dalton, Georgia, 120 miles north of Atlanta. Opposed by Confederate forces commanded by General Joseph E. Johnston, he drove south, outmaneuvering Johnston by a series of flanking movements. By the end of June the Confederates, entrenched at Kenesaw Mountain, 25 miles from Atlanta, successfully withstood an attack by Sherman's troops, following which Sherman made another flanking movement and on July 9 reached the north bank of the Chattahoochee River, six miles from Atlanta.

Johnston was severely criticized for what the southerners considered a poor military showing, and on July 17 he was replaced by General John B. Hood. Sherman crossed the Chattahoochee and within eleven days Hood had fought and lost three battles--Peach Tree Creek, Atlanta, and Ezra Church. The city of Atlanta, however, was not taken until September 2, on which day Hood evacuated and Sherman occupied it. Sherman sent a jubilant message north: "Atlanta is ours and fairly won!"

Hood fought unsuccessful engagements against troops commanded by Generals George H. Thomas and John M. Schofield, and on January 23, 1865, was relieved of command at his own request.

On November 15, 1864, Sherman began his "march to the sea." Before starting he burned machine shops, cut telegraph lines and otherwise damaged Atlanta, as well as destroying the railroad to Chattanooga. His army, numbering approximately 60,000 men, set out for Savannah, Georgia, 360 miles distant. The men marched along parallel roads covering an area sixty miles wide, had provisions for 25 days, and were ordered to "forage liberally." During the march the Union soldiers laid to waste the country through which they passed. Sherman, in his official report, stated, "I estimate damage done to the state of Georgia and its military resources at one hundred million dollars, at least twenty million of which has inured to our advantage and the remainder is simply waste and destruction." It was in this connection that Sherman made the observation that "war is cruel and you cannot refine it," later corrupted to "War is Hell."

Sherman's troops reached Savannah on December 21, 1864, having encountered no serious opposition en route. General William

J. Hardee, the Confederate commander who was holding the town with a force of 15,000 men, rather than risk a siege, withdrew on December 20. On December 22 President Abraham Lincoln received a telegram from Sherman which read, "I beg to present you as a Christmas gift the city of Savannah."

Suggested Readings

Ambrose, Stephen E. Halleck: Lincoln's Chief of Staff. Baton Rouge, La.: Louisiana State University Press, 1962.

Barnes, Eric W. The War Between the States. New York: Whittlesey House, 1959.

Barrett, John G. Sherman's March Through the Carolinas. Chapel Hill, N.C.: University of North Carolina Press, 1956.

Bassett, John Spencer. A Short History of the United States. New York: Macmillan, 1929.

Catton, Bruce. Never Call Retreat. Garden City, N.Y.: Doubleday, 1964.

Freeman, Douglas Southall. Lee's Lieutenants. New York: Scribner's, 1942.

_______. R. E. Lee: A Biography. New York: Scribner's, 1934-1935.

Hood, John B. Advance and Retreat. Millwood, N.Y.: Kraus Reprint, 1968.

Horn, Stanley F. The Decisive Battle of Nashville. Baton Rouge, La.: Louisiana State University Press, 1956.

Johnston, Joseph E. Narrative of Military Operations Directed During the Late War Between the States. Millwood, N.Y.: Kraus Reprint, 1959.

Jordan, Robert Paul. The Civil War. Washington, D.C.: National Geographic Society, 1969.

Lawrence, Alexander A. A Present for Mr. Lincoln. Macon, Ga.: Ardivan Press, 1961.

Leckie, Robert. Great American Battles. New York: Random House, 1968.

Lewis, Lloyd. Sherman, Fighting Prophet. New York: Harcourt, Brace, 1932.

Liddell Hart, B. L. Sherman: Soldier, Realist, American. New York: Dodd, Mead, 1929.

Luvaas, Jay. The Military Legacy of the Civil War. Chicago: University of Chicago Press, 1959.

McDowell, David. Robert E. Lee. New York: Random House, 1953.

McMeekin, Isabella McLennan. Robert E. Lee, Knight of the South. New York: Dodd, Mead, 1950.

Rhodes, James F. The History of the Civil War. Edited by E. B. Long. New York: Frederic Ungar, 1961.

Sandburg, Carl. Abraham Lincoln: The Prairie Years and the War Years. New York: Harcourt, Brace, 1954.

Sherman, William T. The Memoirs of General William T. Sherman by Himself. Westport, Conn.: Greenwood Press, 1972.

Tarbell, Ida M. The Life of Abraham Lincoln. New York:

Doubleday, 1900.
Thorndike, Rachel S., ed. The Sherman Letters. New York: Da Capo Press, 1969.
Wiley, Bell I. The Life of Billy Yank and the Life of Johnny Reb. Indianapolis: Bobbs-Merrill, 1952.
Williams, T. Harry. Lincoln and His Generals. New York: Knopf, 1952.

THE ESCAPE FROM LIBBY PRISON (1864)

Libby Prison in Richmond, Virginia had, prior to the Civil War, been a warehouse operated by A. Libby & Son, chandlers and grocers. It was also known by the Federal army officers who were incarcerated there as "Hotel de Libby."

The prison was crowded, food was scanty, and sanitation virtually nonexistent. It was situated between Cary and Canal Streets, was three stories high on Cary Street and, with a cellar under the center of the building, four stories high on Canal Street. The prisoners were held in large community rooms on the two upper floors rather than in individual cells. The middle room on the first floor, known as the "kitchen," to which the prisoners had access, was used for cooking purposes and had three fireplaces in its east partition wall. The cooking was done by the prisoners on stoves located before the fireplaces. The doors and windows of the warehouse-turned-prison were barred. Lack of sanitation and great overcrowding caused the death of a great number of Union prisoners held there.

In 1863 there were between fifteen and twenty thousand Federal prisoners held in various places of confinement in Richmond. A plan for a general escape failed when an informer advised the Confederate authorities of it. Following this fiasco the inmates of Libby Prison resolved to engineer their own escape and include in the plan only men who could be trusted. A group of fifteen prisoners, including Colonel Thomas E. Rose, Major A. J. Hamilton and Lieutenant James M. Wells, decided to tunnel their way to freedom. The tunnel was to extend eastward under a short cross-street reaching from Canal Street to Cary Street and terminate under a carriage shed opposite the prison.

In order to succeed in their plan, the prisoners needed access to the building's cellar. This they achieved by removing bricks in one of the fireplaces behind a stove and then breaching the wall into the cellar. Under the guidance of Colonel Rose this was successfully accomplished, the work being done secretly at night. Prisoner sentries were posted, ready to give the alarm in case of discovery, and the bricks were replaced before daylight.

Having reached the cellar, an opening was made through the stone foundation wall. Using case knives and clam shells for tools, a tunnel, sixteen inches in diameter, eight or nine feet below the surface of the ground and approximately sixty feet long, was dug. The dirt removed from the tunnel was hidden under a pile of straw which had been stored in the rear end of the basement and which was not in use at that time.

Eventually the air in the tunnel became so foul that the diggers experienced extreme difficulty in breathing, and candles, needed for light, would not burn. While only the select few who had dug the tunnel were aware of the intended escape, it was planned, once the tunnel was completed, to let as many prisoners into the secret as could escape on a single night, with others hopefully escaping at a later date by the same route.

On February 9, 1864, about two hundred men assembled in the "kitchen" to make their break for freedom. One by one they crawled through the tunnel to the carriage house exit. Of the men who escaped that night, only 48 got entirely away.

In 1889 the "Hotel de Libby" was moved to Chicago, there to serve as a museum.

Suggested Readings

Allan, P. B. M. Prison-Breakers: A Book of Escapes from Captivity. Ann Arbor, Mich.: Gryphon Books, 1971.

Cavada, Lieut. Col. F. F. Libby Life: Experience of a Prisoner of War in Richmond, Va. Philadelphia: Lippincott, 1865.

Davenport, Basil. Great Escapes. New York: Sloan, 1952.

Faller, L. W. Dear Folks at Home. Carlisle, Pa.: Cumberland County Historical Society, 1963.

Fellowes-Gordon, Ian. The World's Greatest Escapes. New York: Taplinger, 1966.

Goss, Warren. A Soldier's Story of His Captivity at Andersonville, Belle Island, and Other Rebel Prisons. Boston: Lee & Shepard, 1866.

Hoffer, Frank William. The Jails of Virginia. New York: Appleton, 1933.

Hopkins, Tighe. The Way Out of Libby (pamphlet). Fort Wayne, In.: Fort Wayne Public Library of Allen County, 1955.

Kellogg, R. H. Life and Death in Rebel Prisons. Hartford, Conn.: Stebbins, 1865.

Putnam, George Haven. Prisoner of War in Virginia, 1864-65. New York: Putnam, 1912.

Scoggin, Mary Clara, comp. Escapes and Rescues. New York: Knopf, 1970.

Stevenson, R. R. The Southern Side. Baltimore: Turnbull Brothers, 1876.

Wells, James M. "Colonel Rose's Tunnel," in French, Joseph Lewis, ed. Thrilling Escapes. New York: Dodd, Mead, 1923.

Williams, Eric, comp. The Book of Famous Escapes. New York: Norton, 1954.

________. The Will to be Free: Great Escape Stories. Camden, N.J.: Nelson, 1970.

Yeats-Brown, F. Escape: A Book of Escapes of All Kinds. New York: Macmillan, 1933.

THE CREDIT MOBILIER OF AMERICA (1864-1868)

The Crédit Mobilier of America was a business organization established by Oakes Ames, an American congressman and financier, and others, to enrich themselves through contracts to construct various portions of the Union Pacific Railroad.

The federal government had chartered the railroad in 1862 and 1864, to connect the Missouri River and the Pacific Ocean. It was well financed. The original capital was $100,000,000, and in addition the federal government had made a 20-million-acre land grant along the right of way and had also guaranteed federal loans totaling approximately $60,000,000.

Ames and his associates, all owners of large amounts of Union Pacific stock, realized that the total cost of constructing the railroad would be approximately half the total amount received from the federal government, but not being satisfied with this huge potential profit they conceived the construction company device by which they could realize even greater gains. They purchased the Pennsylvania Fiscal Agency, a contract and loan company and obtained a charter for it as a construction company from the Pennsylvania state legislature, renaming it the Crédit Mobilier of America.

By this arrangement Ames and the Union Pacific stockholders involved with him in the scheme through stock ownership controlled both the railroad and the Crédit Mobilier. This enabled them to contract with the latter to build the railroad at inflated construction costs, paying for such construction with Union Pacific bonds and stock.

In the years 1864-68 Crédit Mobilier made profits estimated at $44,000,000, paying dividends of better than 500% per year. Union Pacific, on the other hand, was left with huge unpaid loans and virtually no assets other than its roadbed and certain equipment. The stock of Crédit Mobilier skyrocketed in price.

Eventually the suspicions of the public were aroused. The stockholders quarreled among themselves over the division of the booty. Ames allegedly bribed certain federal judges and congressmen with shares of Crédit Mobilier stock.

The Presidential election of 1872 brought charges of corruption in connection with the affair against Vice President Schuyler Colfax, Vice Presidential candidate Henry Wilson, and Speaker of the House James G. Blaine. Investigations by congressional committees resulted in the censure of Ames and James Brooks, the latter a Union Pacific director. Certain judges resigned or were impeached, and a number of political careers were ruined.

Suggested Readings

Adams, Charles Francis, and Henry Adams. Chapters of the Erie and Other Essays. Gloucester, Mass.: Peter Smith, 1886.

Commager, Henry Steele, ed. "The Pacific Railway Act," (Doc. No. 215) in his Documents of American History, 8th edition. New York: Appleton, 1968.

Coolidge, Louis A. Ulysses S. Grant. Edited by John T. Morse, Jr. New York: AMS Press, 1922.
Crawford, Jay B. The Crédit Mobilier of America. New York: AMS Press, 1880.
Haney, Lewis B. Congressional History of Railways. Clifton, N.J.: Augustus M. Kelley, 1908.
Hesseltine, William B. Ulysses S. Grant, Politician. New York: Dodd, Mead, 1935.
Josephson, Matthew. The Robber Barons. New York: Harcourt, Brace, 1962.
Lawson, Thomas W. Frenzied Finance. New York: Somerset, 1972.
Lingley, Charles Ramsdell, and Allen Richard Foley. Since the Civil War. New York: Appleton, 1935.
Meyer, Balthasar H. Railway Legislation in the United States. New York: Arno Press, 1973.
Muzzey, David Saville. James G. Blaine: A Political Idol of Other Days. New York: Dodd, Mead, 1934.
Riegel, Robert E. The Story of Western Railroads. Gloucester, Mass.: Peter Smith, 1926.
Seitz, Don C. The Dreadful Decade, Detailing Some Phases in the History of the United States from Reconstruction to Resumption. 1869-1879. Westport, Conn.: Greenwood Press, 1968.
Smith, Theodore C. The Life and Letters of James Abram Garfield. Hamden, Conn.: Shoe String Press, 1968.
White, Henry K. History of the Union Pacific Railway. Clifton, N.J.: Augustus M. Kelley, 1968.

THE ASSASSINATION OF PRESIDENT LINCOLN (1865)

The Civil War ended on April 9, 1865 with the surrender of Confederate General Robert E. Lee to Union General Ulysses S. Grant at Appomattox Court House, Virginia. Five days later, on the evening of April 14, President Abraham Lincoln was shot and fatally wounded by the actor John Wilkes Booth while attending a performance of "Our American Cousin" at Ford's Theater in Washington, D.C.

Lincoln had accepted an invitation from the management of the theater to view the play. In preparation for his visit two boxes were made into one and a rocking chair was provided for his use, with six straight-legged chairs being provided for extra guests, plus two stuffed chairs and a sofa. Decorations included an engraved picture of George Washington and several flags.

Lincoln had invited at least twelve people to attend the theater with him that night, including General and Mrs. Grant, Mr. and Mrs. Edwin M. Stanton, Schuyler Colfax, Richard Oglesby and Richard Yates. However, none of these people was able to accept. When the curtain rose on the first act of the play, the occupants of the Presidential box were Lincoln, his wife Mary, Major Henry Reed Rathbone and Rathbone's fiancee Clara Harris. Although the major was supposedly there to watch out for Lincoln's safety, he sat on the sofa in the front of the box, nowhere near the President, who occupied the rocking chair which had been placed in the left-hand corner.

On the morning of April 14 John Wilkes Booth learned of Lincoln's planned attendance at the theater that night. Booth was a fanatical sympathizer with the South and believed that, if Lincoln were killed, the southern cause might still triumph, despite Lee's surrender. He had conceived several plots to kidnap or assassinate the President and had spoken freely in Washington of his intentions. Associated with him were David Herold, George Atzerodt and Lewis Paine, all of whom had met with him in a Washington boarding house operated by Mary Surratt. A multiple massacre had been planned, in which Secretary of State William H. Seward and Vice President Andrew Johnson were to be killed, as well as President Lincoln.

Booth resolved to carry out the planned assassinations. He proceeded to the theater and made a hole in the wall for a bar to jam the door in the corridor leading to the box the Presidential party would occupy. He also bored a peephole in the door of the box itself.

On the evening of April 14 Lincoln and his party arrived at the theater and the performance commenced. John Parker, the special policeman assigned to sit outside the door to the President's box, had gone downstairs to watch the play from one of the dress circle seats. Booth, armed with a single-shot derringer, entered the box, shot Lincoln in the back of the head, and leaped over the rail to the stage. Brandishing a dagger, he shouted, "sic semper tyrannis!" When he jumped his spur caught in a flag draped in front of the box and Booth, off balance, broke his left ankle. He hobbled to the wings, mounted a horse which he had rented and which was being held for him at the stage entrance, and galloped away. Lincoln had been fatally shot and Major Rathbone, attempting to deprive Booth of his dagger, had received a severe slash above the elbow. After firing his derringer, Booth dropped it in the box, from which it was later recovered.

The dying President was carried across the street to a boarding house operated by a man named Petersen. Here he was laid across a bed. Physicians were summoned but were unable to save the dying man. Lincoln, without regaining consciousness, expired at twenty-two minutes past seven the following morning.

Booth fled to the South. He stopped at the farm of Dr. Samuel A. Mudd, who set his broken ankle. After twelve days he and David Herold, who had joined him, were surrounded by a troop of soldiers in a barn on Garrett's Farm near Bowling Green, Virginia. The barn was set afire. Herold surrendered but Booth, crying out, "I'll shoot it out with the whole damned detachment!," was either killed by Sergeant Boston Corbett or else committed suicide.

Secretary of War Edwin M. Stanton took command of the prosecution. A military court tried the conspirators. Atzerodt, Herold, Paine and Mrs. Surratt were sentenced to hang and Dr. Mudd was sentenced to imprisonment at Fort Jefferson in the Dry Tortugas.

Suggested Readings

Bishop, Jim. The Day Lincoln Was Shot. New York: Harper, 1955.
Brogran, George S. The Great American Myth. New York: Carrick and Evans, 1940.
Catton, Bruce. A Stillness at Appomattox. Garden City, N.Y.: Doubleday, 1953.
Dewitt, David M. The Assassination of Abraham Lincoln and Its Expiation. New York: Macmillan, 1909.
_______. The Judicial Murder of Mary E. Surratt. Saint Clair Shores, Mich.: Scholarly Press, 1970.
Eisenschiml, Otto. Why Was Lincoln Murdered? Boston: Little, Brown, 1937.
Hoyt, Edwin P. Andrew Johnson. Chicago: Reilly & Lee, 1965.
Kunhardt, Dorothy Meserve, and Philip B. Kunhardt, Jr. "Assassination!," American Heritage Magazine, April, 1965.
Potter, John Mason. Plots Against the Presidents. New York: Astor, 1968.
Roscoe, Theodore. The Lincoln Assassination, April 14, 1865: Investigation of a President's Murder Uncovers a Web of Conspiracy. New York: Franklin Watts, 1971.
Sandburg, Carl. Abraham Lincoln: The Prairie Years and the War Years. New York: Harcourt, Brace, 1954.
Tarbell, Ida M. The Life of Abraham Lincoln. New York: Doubleday, 1900.
Thomas, Benjamin P. Abraham Lincoln: A Biography. New York: Knopf, 1952.
_______, and Harold M. Hyman. Edwin M. Stanton: The Life and Times of Lincoln's Secretary of War. New York: Knopf, 1962.
Van Deusen, Glyndon G. William Henry Seward. New York: Oxford University Press, 1967.
Weichmann, Louis J. A True History of the Assassination of Abraham Lincoln and of the Conspiracy of 1865. Edited by Floyd E. Risvold. New York: Knopf, 1975.
Williams, T. Harry. Lincoln and His Generals. New York: Knopf, 1952.

THE KU KLUX KLAN (1865-1976)

The Ku Klux Klan originated in the American South, following the Civil War. The southern Negro had been emancipated and, at the close of reconstruction, was legally on an equal footing with the whites, a fact that many white southern males found galling.

A secret society, the Ku Klux Klan, or Invisible Empire, originated at Pulaski, Tennessee, in 1865, shortly after the close of the War. Its objectives were essentially to preserve white supremacy and discourage Negroes from participating in politics. The objectives were pursued by the Klan members who, hooded and wearing white sheets, enforced their orders by night raids on previously warned Negroes, by whippings and by other dire punishments. General Nathan Bedford Forrest was the first Grand Wizard of the order. It was formally disbanded in 1869.

The Klan was resurrected in 1915 by Colonel William Joseph Simmons, a native of Atlanta, Georgia. Its objective was, like its predecessor, "to keep the nigger in his place," but was also extended to include Catholics, Jews, and members of certain minority groups. During the first five years of its existence it made little headway, but in 1920 Edward Y. Clarke of the Southern Publicity Association was engaged by Simmons to put the Klan on a paying basis.

Clarke was an excellent salesman, organizer and fund-raiser. Under his stewardship the Klan grew. Memberships, sold by "Kleagles" on a commission basis, cost ten dollars, of which the salesman kept four. Clarke's official title was "Imperial Kleagle," and Simmons was the Klan's "Imperial Wizard." "King Kleagles" headed "realms" into which the country was divided, and the "realms," in turn, contained "domains," each headed by a "Grand Goblin." The "King Kleagles" and "Grand Goblins" retained a percentage of each new member's entrance fee, with the balance going to the Imperial Treasury at Atlanta.

Outrages attributed to night-riding Klan members, including whippings, tarrings and featherings, plus a disclosure of Klan activities in the _New York World_, caused a Congressional investigation in 1921. As a result of this investigation Clarke was expelled from the society and Simmons was replaced by Hiram Wesley Evans as "Imperial Wizard."

In spite of this the Klan continued to expand, particularly in the states of Arkansas, California, Connecticut, Indiana, Ohio, Oklahoma, Oregon, and Texas. The outrages, night riding and burning of fiery crosses on hilltops continued. Hoodlums and criminals took advantage of the Klan's existence; often their depredations were attributed to the Klan, against which the local law enforcement agents hesitated to move.

In 1924 it was estimated that the Klan had 6,000,000 members. That year, at the Democratic national convention in New York, a resolution denouncing the Klan was introduced and, after a bitter controversy, was defeated. Subsequently a number of states enacted legislation designed to restrict Klan activities and at this time several of its officials were imprisoned for embezzlement.

In the 1930's, during the great depression, the Klan again became active on a wide scale, directing its activities primarily against the southern trade union organizers and threatening Negroes who exercised their right to vote.

In 1940 the Klan joined with the German-American Bund in holding its 1940 rally at Camp Nordland, New Jersey. During World War II it found it advisable to curtail its activities, but it was revived again after the war was over in 1945. In 1947 the state of Georgia revoked its charter, but other societies with similar objectives were organized to replace it. The largest of these, the Association of Georgia Klans, was placed on the list of subversive organizations issued by the United States Attorney General in 1949.

In 1954 and again in 1965 the Klan came into national prominence during the civil rights disturbances in Alabama. In the latter year four members were arrested in Birmingham for the March 25 murder of Mrs. Viola Liuzzo, a white civil rights worker. President Lyndon Baines Johnson, in a television address, declared the Klan members to be "the enemies of justice who for decades have used the rope and the gun and the tar and the feathers to terrorize their neighbors." He stated that the government would declare war on them as a "hooded society of bigots." Following Johnson's denunciation the Committee on Un-American Activities voted unanimously to conduct a full investigation into the Klan's operations.

Collie Le Roy Wilkins, Jr., a Klan member, was tried for the murder of Mrs. Liuzzo in Haynesville, Alabama. The jury could not agree on his guilt or innocence and the court, declaring a mistrial, released him.

Suggested Readings

Allen, Frederick Lewis. Only Yesterday. New York: Harper, 1931.

Andrist, Ralph K., editor in charge. The American Heritage History of the 20's and 30's. New York: American Heritage, 1970.

Brown, William C. The Lower South in American History. Saint Clair Shores, Mich.: Scholarly Press, 1970.

Chalmers, David M. Hooded Americanism: The First Century of the Ku Klux Klan, 1865-1965. Garden City, N.Y.: Doubleday, 1965.

Commager, Henry Steele, ed. "Act to Enforce the Fourteenth Amendment," (Doc. No. 273) in his Documents of American History, 8th edition. New York: Appleton, 1968.

_______. "The Ku Klux Klan, Organization and Principles," (Doc. No. 271) in his Documents of American History, 8th edition. New York: Appleton, 1968.

Coughlin, Robert. "Konklave in Kokomo," in Leighton, Isabel, ed. The Aspirin Age, 1919-1941. New York: Simon & Schuster, 1949.

Dixon, Thomas. The Clansman (fiction). New York: Gordon Press, 1973.

_______. The Fall of a Nation: Sequel to The Birth of a Nation. New York: Arno Press, 1975.

Jackson, Kenneth T. The Ku Klux Klan in the City, 1915-1930. New York: Oxford University Press, 1967.

Lester, J. C., and D. L. Wilson. The Ku Klux Klan. Nashville: Wheeler, Osborn & Duckworth, 1884.

Lowe, David. Ku Klux Klan: The Invisible Empire. New York: Norton, 1967.

Mecklin, John. The Ku Klux Klan: A Study of the American Mind. New York: Russell & Russell, 1963.

Randel, William P. The Ku Klux Klan: A Century of Infamy. Philadelphia: Lippincott, 1965.
Rice, Arnold S. The Ku Klux Klan in Politics. New York: Haskell House, 1972.
Whitehead, Donald F. Attack on Terror: The FBI Against the Ku Klux Klan in Mississippi. New York: Funk & Wagnalls, 1970.

THE MILLIGAN CASE (1866)

The Milligan Case of 1866 involved a decision by the United States Supreme Court condemning military tribunals in areas where civil courts were open. Lambdin P. Milligan was a long-time citizen of the state of Indiana where "the federal authority was always unopposed and its courts always open to hear criminal accusations and redress grievances." He was accused of "conspiring against the government, affording aid and comfort to rebels, and inciting the people to insurrection." Although Milligan had never served in any branch of the armed forces, he was, "while at his home, arrested by the military power of the United States, imprisoned ... tried, convicted and sentenced to be hanged by a military commission organized under the direction of the military commander of the military district of Indiana." The question before the court, as stated by Justice David Davis, was, "Had this tribunal the legal power to try and punish this man?"

In the Milligan Case the Court held that the military tribunal had no such legal power, and recognized Milligan's petition for a writ of habeas corpus. Certain basic points involving the meaning of the Constitution were made. On March 3, 1863, Congress had authorized the President to suspend the writ of habeas corpus and, under this authority, President Abraham Lincoln had suspended the writ in cases "where officers held persons for offenses against the military or naval service." The Court maintained that, under the Constitution, civil courts took precedence over military courts in cases concerning "every one accused of crime who is not attached to the army or navy, or militia in actual service when such civil courts are available." This, however, was not the case "when war exists in a community and the courts and civil authorities are overthrown." For example, during the Civil War the seceding state of Virginia had been invaded by Union forces and the civil courts were closed. "The national authority was overturned and the courts driven out." Martial law prevailed there, for the invasion was "actual and present." This, however, was not the case in Indiana. "On her soil there was no hostile foot; if once invaded, that invasion was at an end, and with it all pretext for martial law [which cannot] arise from a threatened invasion."

The decision of the military tribunal was overruled. Some Supreme Court justices, including Chief Justice Salmon P. Chase, delivered an opinion "differing from the Court in several important points, but concurred in the judgment in the case."

Suggested Readings

Belden, Thomas G., and Marva R. Belden. So Fell the Angels. Boston: Little, Brown, 1956.
Carson, Hampton L. The History of the Supreme Court. New York: Burt Franklin, 1972.
Commager, Henry Steele, ed. "Ex Parte Merryman," (Doc. No. 209) in his Documents of American History, 8th edition. New York: Appleton, 1968.
______. "Ex Parte Milligan," (Doc. No. 256) in his Documents of American History, 8th edition. New York: Appleton, 1968.
Donald, David, ed. Inside Lincoln's Cabinet: The Civil War Diaries of Salmon P. Chase. Millwood, N.Y.: Kraus Reprint, 1954.
Dunning, William A. Essays on the Civil War and Reconstruction. Plainview, N.Y.: Books for Libraries, 1897.
______. Reconstruction: Political and Economic, 1865-1877. New York: Harper, 1907.
Hart, Albert B. Salmon Portland Chase. New York: Haskell House, 1969.
Johnson, Gerald White. The Supreme Court. New York: Morrow, 1962.
Klaus, Samuel, ed. The Milligan Case. New York: Knopf, 1929.
Lesh, U. S. A Knight of the Golden Circle (pamphlet). Fort Wayne, Ind.: Public Library of Fort Wayne and Allen County, 1956.
McCloskey, Robert C. The American Supreme Court. Chicago: University of Chicago Press, 1960.
Morrison, Olin Dee. Indiana at Civil War Time. Athens, Ohio: Author, 1961.
Sievers, Harry J. The Trial of Lambdin P. Milligan (pamphlet). Fort Wayne, Ind.: Public Library of Fort Wayne and Allen County, 1964.
Warden, Robert V. Account of the Private Life and Public Services of Salmon Portland Chase. Cincinnati: World, 1974.
Warren, Charles. The Supreme Court in United States History. Boston: Little, Brown, 1923.
Whiting, William. War Powers Under the Constitution of the United States. New York: Da Capo Press, 1970.

THE NATIONAL GRANGE OF THE PATRONS OF HUSBANDRY (1867)

Shortly after the Civil War farmers in the United States found their economic position extremely insecure. They were particularly concerned with the declining prices of farm products, with their increasing indebtedness to banks and merchants, and particularly with the discriminatory freight rates charged them by railroads. They were also concerned with the fact that the railroads were preempting public lands on a huge scale.

Railroads were administered by entrepreneurs such as "Commodore" Cornelius Vanderbilt, Jay Gould, James Fisk, Jr., Daniel Drew and, later, James J. Hill and Edward H. Harriman. These railroad magnates considered these miles of transportation lines their personal property rather than utilities designed to serve the public. They were, as one author put it, "men of business sagacity and foresight, but their ethical outlook was restricted and their sense of public responsibility not well developed."

In 1867 Oliver Hudson Kelly, a government clerk in Washington, D.C., together with some associates, founded the National Grange of the Patrons of Husbandry, better known as the "Grange." Its objectives were "to advance the social, economic and political interests of the farmers of the United States." Later it worked toward correcting the transportation abuses of the railroads and arousing cooperation among farmers in other ways. It established cooperative stores, factories, and purchasing agencies. These commercial enterprises were largely unsuccessful, due to mismanagement and intense competition from private businesses. However, the Grange was reasonably successful in its social activities and in the area of railroad reforms.

A number of political parties stemming from the Grange were organized in the early 1870's. These included the Anti-Monopoly Party and the Reform Party, which were successful in electing several state officers and three United States senators.

Several lawsuits dealing with railroad practices and known collectively as the "Granger Cases" were tried. Two of these, both decided by the Supreme Court, were Munn vs. Illinois and Peik vs. the Chicago and Northwestern Railway Company. These cases and others, as well as the passing of certain state laws, contributed to the passage of legislation regulating railroad rates and practices, and ultimately to the Interstate Commerce Act of 1887.

Although a number of "Granger Laws" were later repealed or drastically modified by such decisions as the one made in the 1886 Supreme Court case of Wabash, St. Louis and Pacific Railroad Company vs. Illinois, others were upheld by the Court and served as the basis for subsequent legislation dealing with railroad and public utility regulation.

The Grangers urged other types of legislation, including anti-trust laws and measures establishing postal savings and parcel post.

Suggested Readings

Adams, Charles Francis, and Henry Adams. Chapters of the Erie and Other Essays. Gloucester, Mass.: Peter Smith, 1886.

Buck, Solon J. The Agrarian Crusade: A Chronicle of the Farmer in Politics. New Haven, Conn.: Yale University Press, 1920.

_______. The Granger Movement. Cambridge, Mass.: Harvard University Press, 1913.

Commager, Henry Steele, ed. "The Granger Movement," (Doc.

No. 287) in his Documents of American History, 8th edition. New York: Appleton, 1968.

_______. "The Interstate Commerce Act," (Doc. No. 318) in his Documents of American History, 8th edition. New York: Appleton, 1968.

_______. "Munn v. Illinois," (Doc. No. 294) in his Documents of American History, 8th edition. New York: Appleton, 1968.

_______. "Wabash, St. Louis and Pacific Railroad Company v. Illinois," (Doc. No. 314) in his Documents of American History, 8th edition. New York: Appleton, 1968.

Commons, John R., et al., eds. Documentary History of American Industrial Society. New York: Russell & Russell, 1958.

Dewey, David R. National Problems, 1885-1897. Westport, Conn.: Greenwood Press, 1968.

Gardner, Charles M. The Grange, Friend of the Farmer. Washington, D.C.: The National Grange, 1949.

Kelley, Oliver H. Origin and Progress of the Order of the Patrons of Husbandry in the United States. Philadelphia: J. A. Wagenseller, 1875.

Meyer, Balthasar H. Railway Legislation in the United States. New York: Arno Press, 1973.

Riegel, Robert E. The Story of Western Railroads. Gloucester, Mass.: Peter Smith, 1926.

Ripley, William Z. Railroads: Rates and Regulation. New York: Arno Press, 1973.

Saloutos, Theodore. Farmer Merchants in the South, 1865-1933. Berkeley, Calif.: University of California Press, 1960.

Shannon, Fred A. The Farmer's Last Frontier: Agriculture, 1860-1897. New York: Farrar & Rinehart, 1945.

Taylor, Carl C. The Farmer's Movement, 1620-1920. New York: American Book Co., 1953.

Warren, Charles. The Supreme Court in United States History. Boston: Little, Brown, 1923.

THE PURCHASE OF ALASKA (1867)

Alaska, the northernmost state of the United States, occupies the northwest extremity of North America and includes two geologically related archipelagoes and a number of off-lying islands.

Old maps indicate that the coast of present-day Alaska was known to explorers at least as early as 1579. Vitus Jonas Bering, a Danish navigator employed by Russia, explored the area in 1741 and discovered a number of the islands. Russian fur traders visited the region during the next half century and in 1783 a Russian settlement was established on Kodiac Island. Captain James Cook, the British navigator and explorer, surveyed much of the coast and other British explorers also made discoveries. In 1825 the Russian and British governments concluded an agreement on the boundaries of Russian America.

A charter which had been granted the Russian-American Company lapsed in December, 1861. A Russian imperial governor administered Russian America until 1867, when Russia offered to sell the area to the United States. Negotiations were handled by Secretary of State William H. Seward and on March 30, 1867, a treaty providing for the purchase of the territory for $7,200,000 was drawn up. Seward was enthusiastic concerning the purchase, which was lobbied through Congress by Senator Robert Walker of Mississippi, "to the accompaniment of rumors of bribery."

The purchase was criticized by many American politicians and journalists, who referred to it as "Seward's folly" and "Seward's icebox," "Icebergia," and "Wallrussia." Little was known about Alaska and as the United States had just emerged from an expensive Civil War, the cost seemed prohibitive. However, Russia had appeared to be "well-disposed toward the United States during the War" and the inclination to acquire territory was strong. On April 9, 1867, the Senate ratified the treaty and the United States came into possession of some 600,000 square miles which turned out to be rich in coal, petroleum, fish, furs, timber and minerals.

On October 18, 1867, ceremonies transferring Alaska, the territory's new name, to the United States were held at Sitka.

Suggested Readings

Bancroft, Frederic. The Life of William H. Seward. Gloucester, Mass.: Peter Smith, 1966.

Callahan, James M. American Foreign Policy in Canadian Relations. New York: Cooper Square, 1937.

Chevigny, Hector. Russian America: The Great Alaskan Venture, 1741-1867. New York: Viking Press, 1965.

Commager, Henry Steele, ed. "The Purchase of Alaska," (Doc. No. 268) in his Documents of American History, 8th edition. New York: Appleton, 1968.

Dodd, William E. Robert J. Walker, Imperialist. Gloucester, Mass.: Peter Smith, 1914.

Farrar, Victor J. The Annexation of Russian America to the United States. Washington, D.C.: W. F. Roberts Co., 1937.

Sherwood, Morgan B., ed. Alaska and Its History. Seattle: University of Washington Press, 1967.

Shields, Archie W. The Purchase of Alaska. College, Alaska: University of Alaska Press, 1967.

Thomas, Benjamin P. Russo-American Relations, 1815-1867. Baltimore: Johns Hopkins University Press, 1930.

Van Deusen, Glyndon G. William Henry Seward. New York: Oxford University Press, 1967.

THE IMPEACHMENT TRIAL OF PRESIDENT ANDREW JOHNSON (1868)

Andrew Johnson, the seventeenth President of the United States, was the only President to be the subject of impeachment proceedings. The Constitution of the United States provides that the Chief Executive may be impeached for "Treason, Bribery, or other high Crimes and Misdemeanors." Johnson was tried for alleged "high Crimes and Misdemeanors" before the United States Senate in 1868.

A southern Democrat, Johnson had been the Republican Abraham Lincoln's running-mate in the election of 1860 and when Lincoln was assassinated by John Wilkes Booth in 1865 Johnson succeeded him as President. The Republican Congress did not agree with Johnson's lenient policies regarding the South following the Civil War, and anti-Johnson factions developed. An early attempt to impeach him, led by James M. Ashley of Ohio and Benjamin F. Butler, the former Civil War general officer, had been voted down in the House.

Edwin M. Stanton, Lincoln's Secretary of War, had been retained in the cabinet by Johnson when he assumed the Presidency. Republican Stanton was anti-Johnson and had acted as an informant concerning cabinet matters. On August 5, 1867, Johnson requested his resignation. This was contrary to the Tenure of Office Act of March 2, 1867, which prohibited the President from removing from office officials appointed by and with the advice of the Senate. Stanton took refuge behind the Act but left office. The Senate refused to concur with Johnson's action and Stanton returned to his duties. The President who, believing the Act unconstitutional, had vetoed it only to see it passed by the Senate over his veto, dismissed Stanton a second time. On February 24, 1868, the House of Representatives impeached Johnson for "high Crimes and Misdemeanors."

Chief Justice Salmon P. Chase presided at the hearing, which was managed by seven Radical Republicans, including Thaddeus Stevens and Benjamin F. Butler. Johnson was defended by former Attorney General Henry Stanberry, Benjamin R. Curtis, a former Supreme Court justice, and William M. Evarts, a prominent New York attorney. Eleven charges, centered upon four accusations, were made, the primary one concerning the allegedly illegal removal of Secretary Stanton from office.

The trial lasted from early March to late May and it became apparent that the House had little on which to base an impeachment and that the proceedings were motivated largely by hatred of the President. On May 16, 1868, the Senate voted on the eleventh charge, which Johnson's enemies felt the most likely to pass. Fifty-four members participated. The result was 35 votes for conviction and nineteen for acquittal. Thus, lacking one vote for the necessary 2/3 majority, the deciding "not guilty" vote, cast by Senator Edmund G. Ross of Kansas, saved Johnson from conviction.

Votes were taken on only two other articles, with identical results, and the trial was ended.

Suggested Readings

Aymar, Brandt, and Edward Sagarin. "The Impeachment Trial of Andrew Johnson," in their A Pictorial History of the World's Great Trials. New York: Crown Publishers, 1967.

Beale, Howard K. The Critical Year: A Study of Andrew Johnson and Reconstruction. New York: Frederic Ungar, 1958.

Binkley, Wilfred E. The Powers of the President. New York: Russell & Russell, 1973.

Bowers, Claude G. The Tragic Era: The Revolution After Lincoln. Boston: Houghton Mifflin, 1929.

Brodie, Fawn M. Thaddeus Stevens: Scourge of the South. New York: Norton, 1959.

Clemenceau, Georges. American Reconstruction 1865-70 and the Impeachment of President Johnson. New York: Lincoln MacVeagh and Dial Press, 1928.

Commager, Henry Steele, ed. "The Impeachment of President Johnson," (Doc. No. 269) in his Documents of American History, 8th edition. New York: Appleton, 1968.

________. "The Tenure of Office Act," (Doc. No. 262) in his Documents of American History, 8th edition. New York: Appleton, 1968.

________. "Veto of the Tenure of Office Act," (Doc. No. 263) in his Documents of American History, 8th edition. New York: Appleton, 1968.

Corwin, Edward S. The President: Office and Powers. New York: New York University Press, 1957.

DeWitt, David M. The Impeachment and Trial of Andrew Johnson, Seventeenth President of the United States: A History. New York: Macmillan, 1903.

Dunning, William A. Essays on the Civil War and Reconstruction. Plainview, N.Y.: Books for Libraries, 1897.

________. Reconstruction: Political and Economic, 1865-1877. New York: Harper, 1907.

Foster, G. Allen. Impeached: The President Who Almost Lost His Job. New York: Criterion Books, 1964.

Franklin, John Hope. Reconstruction: After the Civil War. Chicago: University of Chicago Press, 1961.

Hart, Albert B. Salmon Portland Chase. New York: Haskell House, 1969.

Hoyt, Edwin P. Andrew Johnson. Chicago: Reilly & Lee, 1965.

Jellison, Charles A. Fessenden of Maine: Civil War Senator. Syracuse: Syracuse University Press, 1962.

Johnson, Andrew. The Trial of Andrew Johnson on Impeachment. New York: Da Capo Press, 1970.

Lingley, Charles Ramsdell, and Allen Richard Foley. Since the Civil War. New York: Appleton, 1935.

Lomask, Milton. Andrew Johnson: President on Trial. New York: Farrar & Cudahy, 1960.

McKitrick, Eric L. Andrew Johnson and Reconstruction. Chicago: University of Chicago Press, 1960.
Milton, George F. The Age of Hate: Andrew Johnson and the Radicals. New York: Coward-McCann, 1930.
Pratt, Fletcher. Stanton: Lincoln's Secretary of War. Westport, Conn.: Greenwood Press, 1953.
Ross, Edmund G. History of the Impeachment of Andrew Johnson, President of the United States, by the House of Representatives and His Trial by the Senate for High Crimes and Misdemeanors in Office, 1868. Santa Fe, New Mexico: New Mexican Printing Company, 1896.
Simon, John Y., ed. The Personal Memoirs of Julian Dent Grant. New York: Simon & Schuster, 1975.
Stampp, Kenneth M. The Era of Reconstruction. New York: Knopf, 1965.
Stryker, Lloyd Paul. Andrew Johnson: A Study in Courage. New York: Macmillan, 1929.
Thomas, Benjamin P., and Harold M. Hyman. Edwin M. Stanton: The Life and Times of Lincoln's Secretary of War. New York: Knopf, 1962.
Trefousse, Hans Louis. Ben Butler: The South Called Him Beast! New York: Twayne Publishers, 1957.
________. Benjamin Franklin Wade: Radical Republican from Ohio. New York: Twayne Publishers, 1963.
Trial of Andrew Johnson, President of the United States, Before the Senate of the United States, on Impeachment by the House of Representatives for High Crimes and Misdemeanors. Published by order of the Senate. 3 vols. Washington, D.C.: Government Printing Office, 1868.
Winston, Robert W. Andrew Johnson: Plebeian and Patriot. New York: AMS Press, 1970.

THE CARDIFF GIANT HOAX (1869)

In 1869 George Hull, a cigar maker of Binghamton, New York, perpetrated a gigantic and profitable hoax, known as the "Cardiff Giant."

Hull, apparently a man who combined irascibility, a sense of humor, and a flair for profitable practical jokes, had become irritated by a Methodist clergyman who was continually quoting the line from Genesis, "There were giants on the earth in those days." He decided to do something to relieve his ire. Secretly purchasing a huge block of gypsum in Iowa, he had it transferred to Chicago and employed Edward Salle, a stonecutter, to carve the figure of a ten-foot giant from the stone.

When completed this statue weighed 2,990 pounds and, having been artificially aged with ink, sulphuric acid, and sand, appeared to be the body of a gigantic fossilized man. This Hull had crated and, billed as "machinery," shipped to the farm of William Newell, his cousin, near Cardiff, New York. Here he buried it and, on October 15, 1869, had it dug up, announcing the "discovery" as an important anthropological find.

Scientists and experts examined the Cardiff Giant, as it was called. A prominent paleontologist and a chemist, both professors at Yale University, declared it to be a genuine fossil. The director of the New York State Museum pronounced it an ancient statue, calling it "the most remarkable object yet brought to light in this country," a judgment in which Oliver Wendell Holmes, who inspected the Giant, concurred. In the meantime Newell was exhibiting the "find," charging as much as one dollar admission.

The president of Cornell University was not convinced of the authenticity of the alleged fossil. He engaged a famous sculptor to examine it. The sculptor, detecting marks of a chisel, declared the Giant to be a fake, man-made from a block of gypsum.

The ensuing controversy came to the attention of Phineas T. Barnum, the famous showman. He, ever seeking new exhibits, offered Newell $60,000 for a three-month lease, which offer was refused. Barnum then hired Professor Carl C. F. Otto, a sculptor, to make a similar Giant for him. When it was completed Barnum displayed it to paying audiences in a tent in Brooklyn, claiming it to be the original Cardiff Giant, "taller than Goliath whom David slew."

Newell sold his Giant to a syndicate which sought a court injunction to restrain Barnum from claiming his Giant to be the original article. Barnum's claim, upheld by the court, was that he was actually exhibiting a hoax of a hoax and was committing no crime in so doing.

Eventually the truth came out. Newspaper reporters ran down the facts of the making of the Cardiff Giant in Chicago and its subsequent burial and "discovery" on Newell's farm. Hull confessed to the fraud, which had cost him approximately $2,200 and from which he realized a profit of $35,000.

Suggested Readings

Adams, James Truslow, ed. Album of American History. New York: Scribner's, 1944-1949.

Asbury, Herbert. All Around the Town. New York: Knopf, 1929.

Barnum, Phineas Taylor. Here Comes Barnum: P. T. Barnum's Own Story Collected from His Books and Introduced by Helen Ferris. New York: Harcourt, Brace, 1932.

________. The Life of P. T. Barnum. Edited by George S. Bryan. New York: Knopf, 1927.

________. Struggles and Triumphs: or, The Life of P. T. Barnum, Written by Himself. Edited by George S. Bryan. New York: Knopf, 1927.

Beuton, Joel. The Life of Honorable Phineas T. Barnum. Philadelphia: Edgewood, 1891.

Bradford, Gamaliel. Damaged Souls. New York: Houghton Mifflin, 1923.

Bryan, Joseph. The World's Greatest Showman. New York: Random House, 1956.

Drummond, A. M. and Robert E. Card. The Cardiff Giant (historical "show"). Two parts. Interpolated traditional American songs. First produced at the Willard Straight Theater, Cornell University, Ithaca, N.Y., May 20, 1939.
Groh, Lynn. P. T. Barnum, King of the Circus. Champaign, Ill.: Garrard, 1966.
MacDougall, Curtis D. Hoaxes. New York: Macmillan, 1940.
O'Brien, Frank M. The Story of the Sun. New York: Appleton, 1928.
Root, Harvey W. The Unknown Barnum. New York: Harper, 1927.
Smith, H. Allen. The Compleat Practical Joker. Garden City, N.Y.: Doubleday, 1959.
Stowe, Harriet Beecher. Lives and Deeds of Our Self-made Men. Hartford: Worthington-Dustin, 1872.
Wallace, Irving. The Fabulous Showman. New York: Knopf, 1959.
Wells, Helen Frances. Barnum, Showman of America. New York: David McKay, 1957.
Werner, M. R. Barnum. New York: Harcourt, Brace, 1923.

THE GOLD CONSPIRACY (1869)

The Gold Conspiracy, also known as "Black Friday," occurred on Friday, September 24, 1869, and was the direct result of the efforts of the robber barons Jay Gould and James Fisk, Jr., to control the gold market by obtaining through purchase the gold reserve held by New York banks. Had this attempt at a corner been successful, short sellers (those who had sold gold without owning it, hoping to repurchase it later at a lower price) would have lost heavily. This would result because the only source from which they could purchase would be the holders of the corner, Gould and Fisk who would, of course, charge extremely high prices.

As trading proceeded in the Gold Room, the price of gold rose from 137 to 162, and prices on the other exchanges and commodity markets fluctuated wildly.

Historians are generally agreed that the corner would have been successful had not the U.S. treasury, on instructions from treasury secretary George Sewall Boutwell, released its $4,000,000 of gold reserves for trading.

Gould and Fisk were prepared for the possibility that the treasury would put its gold on the market. When they realized that this had happened, they sold gold short at the record-breaking high prices it had attained, repudiated many of their purchases of the metal, and realized a profit of approximately $11,000,000. Many other speculators, however, were bankrupted.

Suggested Readings

Adams, Charles Francis, and Henry Adams. Chapters of the Erie and Other Essays. Gloucester, Mass.: Peter Smith, 1886.
Bowers, Claude G. The Tragic Era: The Revolution After Lincoln. Boston: Houghton Mifflin, 1929.
Fuess, Claude M. Carl Schurz, Reformer. New York: Dodd, Mead, 1932.
Fuller, Robert H. Jubilee Jim: The Life of Colonel James Fisk, Jr. New York: Macmillan, 1928.
Hesseltine, William B. Ulysses S. Grant, Politician. New York: Dodd, Mead, 1935.
Josephson, Matthew. The Politicos, 1865-1896. New York: Harcourt, Brace, 1938.
________. The Robber Barons. New York: Harcourt, Brace, 1962.
Lawson, Thomas W. Frenzied Finance. New York: Somerset, 1972.
Lynch, Denis T. "Boss" Tweed: The Story of a Grim Generation. New York: Boni & Liveright, 1927.
Mandelbaum, Seymour J. Boss Tweed's New York. New York: Wiley, 1965.
Mayer, George H. The Republican Party, 1854-1966. New York: Oxford University Press, 1967.
Muzzey, David Saville. James G. Blaine: A Political Idol of Other Days. New York: Dodd, Mead, 1934.
Nevins, Allan. Hamilton Fish: The Inner History of the Grant Administration. New York: Dodd, Mead, 1936.
Seitz, Don C. The Dreadful Decade, Detailing Some Phases in the History of the United States from Reconstruction to Resumption, 1869-1879. Westport, Conn.: Greenwood Press, 1968.
________. Horace Greeley, Founder of the New York Tribune. New York: AMS Press, 1926.
Simon, John Y., ed. The Personal Memoirs of Julian Dent Grant. New York: Simon and Schuster, 1975.
Van Deusen, Glyndon B. Horace Greeley, Nineteenth Century Crusader. New York: Hill & Wang, 1964.
Warshow, Robert Irving. Jay Gould: The Story of a Fortune. New York: Greenberg, 1928.
White, Bouck. The Book of Daniel Drew. New York: Arno Press, 1973.
White, Leonard D. The Republican Era, 1869-1901. New York: Macmillan, 1958.

THE TAYLOR-SUTTON FEUD (1869-1877)

One of the famous feuds of Texas history was that which raged between the Suttons and the Taylors during the latter part of the 19th century. The cause of the feud was a difference of opinion regarding the ownership of unbranded cattle which roamed the Texas prairies and were claimed by both factions.

The trouble started following the Civil War when cattlemen, returning from military service to their untended ranches, found that their herds had mingled on the open range and the resulting calves were mavericks.

Two of the largest cattle-raising families in De Witt County, Texas, were the Taylors and the Suttons. Both had proceeded to apply their own brands to the mavericks and this led to the inevitable dispute which soon became open warfare.

Joe Tumlinson, a deputy United States marshal, attempted to put an end to the branding evils and range quarrels by organizing a band of men called "Tumlinson's Regulators." The Regulators were to keep order in the region, but violence broke out. One of the friends of the Taylors was shot and killed by a Regulator, and Jack Helmes, United States marshal, was shot and wounded by John Wesley Hardin, an outlaw. Then sympathetic friends of Billy, Buck and Jim Taylor volunteered their services in case the Taylors ever needed them, and James and William Sutton received similar offers of assistance from their associates.

Judge Henry Clay Pleasants, who officiated at the district court, resolved to eliminate the rapidly growing feud from his jurisdiction. One evening in the spring of 1869 Judge Pleasants and his family attended a public social gathering in the town of Clinton, then the governmental seat of De Witt County. Buck Taylor and his cousin Dick Chisholm attended the festivities. Both men were shot dead before they could dismount from their horses. It was thought that Bill Sutton and his friend Doc White might have been the assassins. In retaliation a rider for the Sutton outfit was shot.

Following these shootings the Taylors caused the arrest of Coots Tuggle, Mason Arnold, and Scrap Taylor on a charge of cattle theft. The three were taken from Sheriff W. J. Weissegar's posse by a group of armed men and hanged.

By this time each of the warring families had gathered several hundred followers. Judge Pleasants was advised that two miniature armies, each representing one of the feuding families, were riding to Clinton, the Suttons to prefer cattle stealing charges against the Taylors and the Taylors to prevent them from so doing. Pleasants took up a position in the town square and ordered both factions to leave town immediately--which, unbelievably, they did.

The two armies headed for the town of Cuero, Texas. Here the predicted shooting did not occur, although the two factions remained there for two days. Then reinforcements for the Suttons arrived and the Taylors agreed to peaceful arbitration.

The Taylor-Sutton feud flared up again when Gabe and Bill Slaughter were shot by Jim and Billy Taylor, who were avenging the deaths of Buck and Scrap Taylor. Billy was tried and convicted of murder but his sentence--ten years in prison--was reversed by the Court of Appeals. Jim Taylor and two of his followers, Winchester Smith and a man named Hendricks, were shot dead in a cotton field by members of the Sutton family.

Other killings followed. On September 19, 1876, Dr. Philip Brazell and one of his three sons died at the hands of seven assassins who apparently mistook him for another man with the same surname. This senseless double killing resulted in Judge Pleasants contacting the governor at Austin, who sent to Clinton a detachment of Texas Rangers led by Lieutenant Lee Hall.

Pleasants called the grand jury in connection with the Brazell murders. The jury promptly indicted Dave Augustine, William Cox, Charles H. Heissig, James Hester, William D. Meador, Jake Ryan and Joe Sitterlie. The Rangers served the warrants and the seven were arrested. Judge Pleasants officiated at their trial and members of Lieutenant Hall's Rangers were on hand to keep order. Defendants Augustine and Hester were acquitted. The prosecution of Heissig was dismissed. On a change of venue to Bexar County, the other four men were tried. Cox, Ryan and Sitterlie were found guilty of murder in the first degree, but the Appeals Court reversed the judgment on technical grounds. Meador, who was to be tried separately from the others, died before his trial could be held. The other three made bond and the case never came to trial again.

Suggested Readings

Cutler, James Elbert. Lynch-Law: An Investigation into the History of Lynching in the United States. New York: Negro University Press, 1969.

Douglas, C. L. Famous Texas Feuds. Dallas: The Turner Co., 1936.

Fehrenbach, T. R. Lone Star: A History of Texas and the Texans. New York: Macmillan, 1968.

Gard, Wayne. Frontier Justice. Norman, Okla.: University of Oklahoma Press, 1949.

Raper, Arthur F. The Tragedy of Lynching. Chapel Hill, N.C.: University of North Carolina Press, 1933.

Robson, William A. Civilization and the Growth of Law. New York: Macmillan, 1935.

Seagle, William. The Quest for Law. New York: Knopf, 1941.

Shay, Frank. Judge Lynch: His First Hundred Years. New York: I. Washburn, 1938.

Sonnichsen, C. L. I'll Die Before I'll Run. New York: Harper, 1951.

_______. Ten Texas Feuds. Albuquerque: University of New Mexico Press, 1957.

Zane, John Maxey. The Story of Law. Garden City, N.Y.: Garden City Publishing Co., 1927.

THE SUTRO TUNNEL (1869-1878)

On June 11, 1859, Henry Tompkins Paige Comstock, known as "Old Pancake," staked out a claim to a silver deposit in Western Nevada. Unaware that the Comstock Lode, as it came to be called, was to prove the greatest silver bonanza ever discovered by man, yielding some $340,000,000 from the time of its discovery until 1890, he sold his rights for $11,000.

The Comstock Lode is a compound fissure vein located on the eastern slope of Mount Davidson in Nevada, about twenty miles east of the California line. As soon as its potential was realized a number of mining companies were organized to exploit it. One of the most active of the mining magnates was William Chapman Ralston, cashier and later president of the Bank of California, headquartered in San Francisco. By a series of adroit financial maneuvers Ralston succeeded in gaining control of several of the mines on the Lode. An ambitious man, he took his profits from the mines he controlled and used them to establish industries in San Francisco. Eventually he had overexpanded to the point where he needed a steady flow of silver and gold bullion from the Comstock to keep his various business enterprises operating.

Mine shafts sunk into the Lode struck water and the mines became flooded. Pumps were unable to remove the water fast enough to permit further mining below the water level. It was then that Adolph Sutro, a Prussian-American mining engineer, conceived the idea of constructing a drainage tunnel, four miles in length, westward from the Carson River to a point beneath the Comstock Lode.

Sutro discussed his idea with Ralston and his business associates and with other mine owners. He contracted with these people to construct the tunnel, Sutro to arrange for financing and construction and the mine owners to pay him a two-dollar royalty for each ton of ore removed from the mines once the tunnel was completed. Ralston endorsed the project and on March 1, 1865, made a personal appeal for cooperation from the other mining companies on the Lode. In 1866 Ralston gave Sutro further assistance by furnishing him with letters of introduction to bankers in New York and London.

Then Ralston had a change of heart. William Sharon, Ralston's representative in Virginia City, Nevada, on the Lode, had conducted underground explorations and struck a bonanza in the Kentuck mine. By installing more powerful pumps it was possible to control the flow of water and continue mining operations. Congress had confirmed Sutro's title to 1,280 acres of land at the mouth of the projected tunnel and had confirmed the two dollar per ton royalty agreement between Sutro and the mine owners. Following Sharon's strike Ralston decided that the two-dollar royalty was an exorbitant price to pay when he could operate his mines without the assistance of the drainage tunnel.

From then on Ralston and his partners did everything in their power to prevent the construction of the Sutro Tunnel. As the Bank of California, dominated by Ralston, was an extremely powerful institution, it carried tremendous influence on businessmen and politicians who might have otherwise assisted in the tunnel project.

In spite of overwhelming odds against him, Adolph Sutro commenced construction of the tunnel on October 19, 1869. Aside from his financial difficulties he encountered others of a mechanical nature. Progress was slow as he was working largely through hard rock which necessitated blasting. Ventilation presented a terrific problem which became worse as the tunnel progressed into the mountain. The heat became unbearable and often the workmen encountered springs of boiling hot water which became torrents. Oozing clay swelled from pick holes, displacing rail tracks and support timbers. Cave-ins occurred and some workmen were killed.

Finally, on July 8, 1878, Sutro's workmen broke through to the shaft of the Savage mine at the 1,640-foot level.

Even though Sutro had succeeded in reaching the heart of the Comstock Lode his troubles were not over. The mine owners, unwilling to pay the agreed royalty, protested that they did not need the tunnel, would not use it and would not pay. It was not until a pump broke down and the Hale and Norcross, Chollar-Potosi and Savage mines became flooded with scalding water that the deluge was turned into the tunnel. Other mines struck water and their owners realized they must capitulate or face ruin. One by one the mine owners gave in, signed new contracts and prepared to engage in deep mining, free from the threat of flooding.

Sutro agreed to a new tonnage rate for the use of his tunnel: a two-dollar tax on all ore assaying forty dollars or over to the ton, and one dollar per ton for all ore assaying a lesser amount.

Two years later Sutro sold out his interest in the Sutro Tunnel Company and retired to San Francisco where he became a prominent businessman, real estate developer, and philanthropist. He served as the city's mayor from 1894 to 1896.

Suggested Readings

Angel, Myron, ed. The History of Nevada. New York: Arno Press, 1973.

Aron, Joseph. History of a Great Work and of an Honest Man. Paris: 1892.

Bancroft, Hubert Howe. The History of Nevada, Colorado and Wyoming. San Rafael, Calif.: Bancroft Press, 1888.

DeQuille, Dan, pseud. History of the Big Bonanza. New York: Knopf, 1947.

Emrich, Duncan, ed. Comstock Bonanza. New York: Vanguard Press, 1950.

Jackson, Joseph Henry. Anybody's Gold: The Story of California's Mining Towns. New York: Appleton, 1941.

James, George Wharton. Heroes of California. Boston: Little, Brown, 1910.

Lewis, Oscar. California Heritage. New York: Crowell, 1949.

________. The Silver Kings. New York: Knopf, 1947.

Lord, Eliot. Comstock Mining and Miners. Berkeley, Calif.: North-Howell Books, 1959.

Lyman, George D. Ralston's Ring. New York: Scribner's, 1937.

________. The Saga of the Comstock Lode. New York: Scribner's, 1934.

Mack, Effie Mono. Nevada. Glendale, Calif.: Arthur H. Clark,

1930.
Sutro, Adolph. Drain Tunnel for the Great Comstock Lode. San Francisco: 1865.
______. Mineral Resources of the United States. Baltimore: 1868.
Tilton, Cecil G. William Chapman Ralston, Courageous Builder. Boston: Christopher Publishing House, 1935.
Young, Bob, and Jan Young. Forged in Silver. New York: Messner, 1968.

THE MULLIGAN LETTERS (1869-1884)

James Gillespie Blaine, Republican candidate for the Presidency in 1884, had a checkered political career. In April, 1869, when he was Speaker of the House of Representatives, a bill favorable to the Arkansas land grant Little Rock and Fort Smith Railroad was before that House. Blaine was requested to use his influence to secure the passage of the bill and, with the assistance of a fellow member of the House, General John A. Logan, did so and the bill was passed. Blaine then, at the request of Warren Fisher, Jr., a Boston businessman, agreed to "participate in the affairs of the railroad" and, in writing, consented. Subsequently he sold substantial amounts of Little Rock Railroad stock to his friends in Maine, for which he was paid a "handsome commission."

Between 1869 and 1872 much correspondence was exchanged between Blaine and Fisher. When their relations ended in the latter year, Blaine understood that all their correspondence had been mutually surrendered.

In 1876 Blaine was a candidate for the Republican Presidential nomination. He and certain other politicians were rumored to have participated in questionable financial transactions dealing with railroads. Blaine denied complicity but did not discuss the matter fully, not wishing his Maine constituents to learn of the "handsome commissions" he had received for selling Little Rock Railroad stock.

In May, 1876, a congressional investigation into the matter of congressmen selling their political influence got under way, and Blaine was one of the witnesses. Fisher and James Mulligan, the latter a confidential clerk to Fisher and previously employed in a similar capacity by one of Mrs. Blaine's brothers, were also called upon to testify. Mulligan, in his testimony, indicated that he intended to produce a packet of letters which had passed between Fisher and Blaine. That evening Blaine visited Mulligan and persuaded him to lend him the letters. Blaine then, upon advice of counsel, declined to return the letters to Mulligan.

Shortly thereafter Blaine, realizing that to remain silent on the matter would be tantamount to admitting guilt, read selections from the Mulligan Letters to the House of Representatives and made an impassioned speech defending himself. The chairman of the investigation committee reluctantly admitted that "a dispatch which Blaine declared would exonerate him" had been suppressed and, with Blaine apparently cleared, the matter was temporarily dropped. Later he was elected to the Senate and was no longer under the jurisdiction of the House Committee.

The matter of the Mulligan Letters was revived in 1884 when Blaine received the Republican nomination for the Presidency. On September 15 Fisher and Mulligan made public additional letters, previously undisclosed. These were extremely damaging to Blaine, one in particular being drawn up by him which he requested Fisher to copy and publish as his own. This letter was marked "confidential" and "burn this letter." Fisher neither copied nor burned it.

The Democratic party, having nominated Grover Cleveland as its Presidential choice, made much of these new disclosures. The campaign of 1884 was characterized by much mud-slinging and vicious attacks on the characters of the candidates. Cleveland was elected by a narrow margin and became the 22nd President, later winning re-election in 1892 as the 24th President.

Suggested Readings

Cleveland, Grover. Presidential Problems. Plainview, N.Y.: Books for Libraries, 1975.
Ford, Henry J. The Cleveland Era. New Haven, Conn.: Yale University Press, 1919.
Lingley, Charles Ramsdell, and Allen Richard Foley. Since the Civil War. New York: Appleton, 1935.
Merrill, Horace S. Bourbon Leader: Grover Cleveland and the Democratic Party. Boston: Little, Brown, 1957.
Muzzey, David Saville. James G. Blaine: A Political Idol of Other Days. New York: Dodd, Mead, 1934.
Nevins, Allan. Grover Cleveland: A Study in Courage. New York: Dodd, Mead, 1932.
Paxson, Frederic L. Recent History of the United States. New York: Cooper Square, 1921.
Rhodes, James F. History of the United States from the Compromise of 1850. New York: Macmillan, 1893-1906.
Russell, Charles Edward. Blaine of Maine: His Life and Times. New York: Cosmopolitan Book Corp., 1931.
Stanwood, Edward. James Gillespie Blaine. Edited by John T. Morse, Jr. New York: AMS Press, 1905.
Thomas, Harrison C. The Return of the Democratic Party to Power in 1884. New York: AMS Press, 1919.

THE NOBLE ORDER OF THE KNIGHTS OF LABOR (1869-1917)

With the rise of the great industrial firms following the Civil War, many workers felt that they were being unjustly exploited by their employers. The workers realized that, as individuals, they were powerless to negotiate with management and that it would be necessary to establish unions which could represent them in dealings with their employers.

Several short-lived trade unions were established, but the first one which advocated the inclusion of all workers in the country regardless of their particular trades was the Noble Order of the Knights of Labor. This organization was established in Philadelphia in 1869 by Uriah Smith Stephens, a garment worker. A number of his fellow-employees assisted in the organization of the Order. It was a secret society, known only as ***** and using an elaborate mystic ritual. Workers in all trades were eligible for membership, although liquor dealers, lawyers, physicians and politicians were excluded.

The Order grew slowly at first, but with the economic depression of 1873, which caused much suffering among the working classes, many working men joined in the hope of improving their situations. In 1878 the first general assembly met at Reading, Pennsylvania, and adopted a constitution. This specified that the purposes of the Order, among others, were "to secure to the toilers a proper share of the wealth they create," to reserve public lands "for the actual settler--not another acre for railroads or speculators," and to prohibit "the employment of children in workshops, mines, and factories before attaining their fourteenth year." Other provisions in the constitution were "to secure for both sexes equal pay for equal work" and "the reduction of the hours of labor to eight per day," as well as "the substitution of arbitration for strikes whenever and wherever employers and employees are willing to meet on equitable grounds."

In 1881 the secret and fraternal nature of the Order was eliminated. It then began to function as a labor union, adopting a policy of militant action against employers. It participated in strikes against coal mine owners and railroad operators and became the leading labor organization in the United States.

For about five years after the 1878 assembly the Order used the strike on several occasions. However, the national leadership urged less violent tactics and in 1883 Terence Vincent Powderly, who advocated the boycott rather than the strike, was elected president. Under his direction the Order won a number of improvements in working conditions and increased its membership to more than 700,000.

The membership began to decline in 1886 for various reasons. These included the members' dissatisfaction with Powderly's opposition to a one-day general strike proposed as a means of winning the eight-hour day and his denunciation of the eight anarchists convicted of complicity in the Haymarket Square Riot of 1886, whose conviction was considered by many to be unjust.

The failure of a railroad strike aided further in the decline of the Order, as did the secession of a large number of craft unions which, in December, 1886, participated in the organization of the American Federation of Labor.

In 1894 a railroad strike, opposed by the Federation and which resulted in the total defeat of the strikers, resulted in the virtual collapse of the Order. It was formally dissolved in 1917.

Suggested Readings

Beard, Mary R. A Short History of the American Labor Movement. Westport, Conn.: Greenwood Press, 1968.
Browne, Henry J. The Catholic Church and the Knights of Labor. Washington, D.C.: Catholic University of America Press, 1949.
Commager, Henry Steele, ed. "Preamble to the Constitution of the Knights of Labor," (Doc. No. 298) in his Documents of American History, 8th edition. New York: Appleton, 1968.
Commons, John R., et al. The History of Labor in the United States. Clifton, N.J.: Augustus M. Kelley, 1918.
Dulles, Foster R. Labor in America: A History. Northbrook, Ill.: AHM Publishing Co., 1949.
Gompers, Samuel. Seventy Years of Life and Labour. Clifton, N.J.: Augustus M. Kelley, 1966.
Harvey, Rowland H. Samuel Gompers, Champion of the Toiling Masses. New York: Octagon Books, 1973.
Lingley, Charles Ramsdell, and Allen Richard Foley. Since the Civil War. New York: Appleton, 1935.
Powderly, Terence V. Thirty Years of Labor, 1859-1889. Columbus: Ohio State University Press, 1889.
Ware, Norman. The Labor Movement in the United States, 1860-1895. Gloucester, Mass.: Peter Smith, 1959.
Wright, Carroll B. The Industrial Evolution of the United States. New York: Russell & Russell, 1967.

THE "ROBERT E. LEE"-"NATCHEZ" STEAMBOAT RACE (1870)

In 1807 Robert Fulton demonstrated the practicality of the steamboat when his "Clermont" made the 150-mile trip from New York City to Albany in 32 hours. Fulton's vessel traveled on the Hudson River, and by 1820 other steamboats were plying various navigable American rivers, one of the most important of which was the Mississippi.

As time went by the boats were improved and became virtual floating palaces, carrying passengers and cargo far up the Mississippi and its tributaries. Rivalry between the various boats and their captains ran high and in July, 1870, one of the most famous steamboat races ever held on the river was that between the "Robert E. Lee" and the "Natchez." The "Lee" was commanded by Captain John W. Cannon and the "Natchez" by Captain Thomas P. Leathers. The race was run between the cities of New Orleans, Louisiana and St. Louis, Missouri, 1,200 miles upstream.

The race resulted from a challenge made by Leathers. At first Cannon did not wish to test the speed of his steamer against that of the "Natchez," but he was badgered into agreeing to the encounter. The two boats were considered the fastest on the Mississippi and the victor in such an encounter would benefit in terms of prestige and increased passenger and freight business.

The race began at five o'clock on the evening of June 30, 1870. The "Lee" backed away from the New Orleans wharf shortly before the "Natchez," and, swinging around, started up river, her 38-foot paddle wheels churning. The "Natchez" followed and the race was on. Betting on the outcome was heavy.

On the "Lee" a hot water pipe in the hold came apart. William Perkins, chief engineer, John Berry, his assistant, and an apprentice engineer named John Weist were able to repair the damage. The vessel continued on its way without slowing down and did not lose its lead.

At midnight engineer Perkins was advised that a leak had developed in one of the eight boilers of the "Lee." Weist, crawling under the leaking boiler, was able to diagnose the trouble and correct it. The "Lee" stayed ahead of the "Natchez."

Both steamers loaded fuel at the town of Natchez, Mississippi, which they reached on the forenoon of July 1. The loading was done from barges which were quickly lashed to each steamer. The fuel was placed aboard as the steamers proceeded upstream and the barges were then cast off, to be towed to their wharves by waiting tugs. The "Natchez" was then ten minutes behind the "Lee."

At Vicksburg, Mississippi, 380 miles above New Orleans, the "Natchez" was only eight minutes behind her rival.

The "Natchez" developed mechanical trouble a few miles above the junction of the Mississippi and Yazoo Rivers when a cold water pump ceased to function. The steamer was headed to shore and tied to a tree. Andy Pauley, chief engineer, repaired the damage and the race was resumed 34 minutes later. Farther upriver the "Natchez" ran aground but was able to back off and continue her pursuit of the "Lee."

At 10:00 A.M. on July 2 the "Natchez" took on fuel at Napoleon, Arkansas. It was then fifty minutes behind the other steamer. The "Lee" took on coal at Memphis, Tennessee, 815 miles north of New Orleans, at 11:10 P.M. The "Natchez," arriving at Memphis shortly after midnight, also took on coal and proceeded on her way. The "Lee" was now leading her rival by one hour and three minutes.

Above Memphis the Mississippi River is characterized by shallow waters, narrow winding channels and a maze of small islands. Again the "Natchez" ran aground but was not damaged and was able to back off and continue her journey upstream.

At Hickman, Kentucky, the "Lee" led the "Natchez" by twenty miles. At Cairo, Illinois, at the junction of the Mississippi and Ohio Rivers, both steamers took on fuel from waiting barges. Again, this time off the Illinois shore, the "Natchez" ran aground, and again she was able to free herself and resume her upriver journey.

The race was made more difficult by the fog which descended on the river above Cairo. Both boats passed safely by Thebes, Illinois, above which the fog continued and, because of shallow water, sand bars, narrow channels and hidden rocks, the problems of navigation multiplied. At Shepherd's Landing, Missouri, Captain Leathers decided to stop, tie the "Natchez" to a tree by the shore, and wait until morning. A citizen of the town told him that the "Lee" had passed upriver not 25 minutes before. It was then 12:35 on the morning of July 4, 1870.

Leaving Cairo behind, the "Lee" continued on her way. She passed Cape Girardeau, Missouri, at 9:30 in the morning and then, encountering fog, slackened speed. St. Louis lay 150 miles upstream. Captain Cannon sent out his yawl to sound the river in advance of the large steamer and impressed several passengers as lookouts. The "Lee" progressed slowly upstream. She ran aground but was able to free herself without damage.

At 1:30 A.M. the fog lifted and the "Lee," now in a wide stretch of river, resumed her speed. At 11:25 on the morning of July 4, the victorious "Lee" arrived at St. Louis, to be greeted by a cheering throng, brass bands, and general jubilation. She had made the trip from New Orleans in three days, eighteen hours and four minutes, setting a new record.

The "Natchez" left Shepherd's Landing at 6:30 A.M. on July 4, after the fog had receded. She arrived at St. Louis at 5:51 P.M., some five hours and twenty minutes after the "Lee."

The "Lee" was declared the victor, having been the first of the two steamers to arrive at St. Louis. However, Captain Leathers contended that his actual running time had been less than that of Captain Cannon's "Lee." The latter boat lost 36 minutes to repair the malfunction of the hot water pipe it had experienced, and the "Natchez" lost more than seven hours when it tied up at Shepherd's Landing. History, however, has given the victory to Captain John W. Cannon and his steamboat the "Robert E. Lee."

Suggested Readings

Andrist, Ralph K. Steamboats on the Mississippi. New York: American Heritage Press, 1962.

Barthau, Roy L. The Great Steamboat Race. Cincinnati: World, 1952.

Devol, George H. Forty Years a Gambler on the Mississippi. New York: Johnson Reprint, 1892.

Fulkerson, H. S. Random Recollections of Old Time Mississippi. Baton Rouge, La.: Claitor, 1937.

Gould, E. W. Fifty Years on the Mississippi. St. Louis: Nixon-Jones Printing Co., 1889.

"The Great River Race," DeBow's Review, July, 1870.

Jewell, Edward L., ed. Crescent City Illustrated. New Orleans: 1873.

Kane, Harnett. Natchez on the Mississippi. New York: Morrow, 1947.

Kerr, Charles. History of Kentucky. Chicago: American Historical Society, 1922.

Twain, Mark, pseud. Life on the Mississippi. New York: Washington Square Press, 1968.

Way, Frederic, Jr. "Cannon Took a Poke at Leathers," Waterways Journal, July 31, 1943.

________. Pilotin' Comes Natural. New York: Farrar & Rinehart,

1943.
Wellman, Manley Wade. Fastest on the River. New York: Holt, 1937.

THE LAURA D. FAIR CASE (1870-1872)

Laura D. Fair, defendant in one of the most sensational murder trials of the last century, shot her lover, Alexander Parker Crittenden, to death on the San Francisco-Oakland ferry boat "El Capitan" on the afternoon of November 3, 1870.

Born in Holly Springs, Mississippi, at the age of sixteen she married a New Orleans man named Stone. Stone died a year later and Laura then married Tom Grayson. Grayson was a confirmed alcoholic and after six months his wife left him, going with her mother to San Francisco. From there she moved to Shasta, California, where she met William D. Fair, a lawyer, and in February, 1859, the two were married. They moved to Yreka, California, where Fair tried unsuccessfully to practice law. Here, in 1860, their daughter Lillias Lorraine Fair was born. They then moved to San Francisco where, in December, 1861, Fair committed suicide.

Following the death of her third husband Laura Fair operated lodging houses in San Francisco, Sacramento, California, and Virginia City, Nevada. She was involved in money troubles and appeared briefly on the stage.

One of the boarders at her Virginia City establishment was Alexander Parker Crittenden, a successful lawyer and former army officer. Laura Fair promptly fell in love with him although she knew him to be married and the father of seven children.

Some time in 1863 the two became lovers. Crittenden's children and his wife Clara arrived in Virginia City. Laura continued to manage her boarding house and her financial position gradually improved, particularly after she started investing in Comstock Lode mining stocks.

For the next several years Laura and her lover continued their affair, living together at times and seeing each other in hotel rooms. Though professing to love Laura, Crittenden seemed content to remain married to Clara while continuing the unconventional relationship which began in Virginia City and continued in San Francisco and back to Virginia City again. In 1870 Laura married Jesse Snyder, a boarder in her lodging house. This marriage, apparently contracted in order to anger Crittenden, ended in divorce shortly after it was consummated.

Laura Fair obtained a pistol from a San Francisco gunsmith. On Thursday, November 3, 1870, she went to the San Francisco ferry slip and rode across the bay to Oakland on the boat "El Capitan." She knew that Crittenden, also on the boat, planned to meet his wife and family at the Oakland pier, they being scheduled to arrive there by train from the East.

As "El Capitan" started the return trip from Oakland to San Francisco, Laura approached Crittenden, who was sitting on a bench with his wife, fired at him once with her pistol, dropped it on the deck, and walked away. Laura was taken into custody and was charged with murder when Crittenden died 48 hours later. She did not deny shooting her lover or intending to kill him.

The trial, conducted at San Francisco, was sensational in the extreme. Laura Fair pled "not guilty by reason of emotional insanity." The jury found her guilty of murder in the first degree and Judge Samuel H. Dwinelle sentenced her to die by hanging. Her attorneys were able to obtain a stay of execution, which was followed by an appeal to a higher court. A new trial was granted in February, 1872, and, because of several postponements, the woman did not come to trial until September of that year. The jurors at the second trial found her not guilty and Laura Fair went free.

Suggested Readings

Bancroft, Hubert Howe. The History of Nevada, Colorado and Wyoming. San Rafael, Calif.: Bancroft Press, 1888.

DeQuille, Dan, pseud. History of the Big Bonanza. New York: Knopf, 1947.

Fair, Laura D. Wolves in the Fold (pamphlet). San Francisco: 1873.

Fisher, Clement, Jr. Of Walking Beams and Paddle Wheels. San Francisco: Bay Books, 1951.

Harper, Ida H., ed. The Life and Work of Susan B. Anthony. New York: Arno Press, 1969.

Lamott, Kenneth. Who Killed Mr. Crittenden? New York: David McKay, 1963.

Lloyd, B. E. Lights and Shades in San Francisco. San Francisco: Bancroft, 1876.

Lyman, George D. The Saga of the Comstock Lode. New York: Scribner's, 1934.

Morse, John T. Famous Trials. Boston: Little, Brown, 1874.

O'Brien, Robert. "Wolves in the Fold," in Jackson, Joseph Henry, ed. The San Francisco Murders. New York: Duel, Sloan, 1947.

Official Report of the Trial of Laura D. Fair for the Murder of Alex P. Crittenden. San Francisco: San Francisco Cooperative Printing Co., 1871.

Roche, Dr. Philip. The Criminal Mind. New York: Grove Press, 1959.

Shuck, Oscar T. History of the Bench and Bar of California. Los Angeles: Commercial Printing House, 1901.

Stanton, Theodore, and Harriot S. Blatch. Elizabeth Cady Stanton as Revealed in Her Letters. New York: Arno Press, 1921.

Twain, Mark, pseud. Roughing It. Berkeley, Calif.: University of California Press, 1972.

THE TWEED RING (1870-1871)

Political corruption is a thing with which man has been plagued ever since government came into existence. One of the more flagrant examples of this was the Tweed Ring which flourished in New York City during the 1870's. William Marcy Tweed, leader of Tammany Hall, headed a group of city politicians, including Mayor A. Oakey Hall, Comptroller Connolly, City Chamberlain Peter B. Sweeney, and others. In 1870 this group gained control of the city's finances and swindled the treasury of many millions of dollars. Their methods involved such devices as account-raising. Here a businessman who had sold goods or services to the city was required to submit a bill in excess of the amount actually owed him. The conspirators would then arrange to pay him the correct amount and divide the overage among themselves. In one case a bill for $5,000 was submitted as one calling for $55,000. The creditor was paid $5,000 and the members of the Tweed Ring pocketed $50,000. A plasterer is said to have submitted "raised" bills for nearly three million dollars over a nine-month period.

In September, 1870, the New York Times inaugurated an editorial campaign against the Ring. The efforts of Samuel J. Tilden, then chairman of the Democratic State Committee, the cartoons of Thomas Nast which appeared in Harper's Weekly, and the series of exposures in the Times resulted in the arrest of Tweed on October 21, 1871. The jury disagreed at the first trial on December 16, 1871, but on being tried a second time he was convicted and sentenced to imprisonment on Blackwell's Island. On his release in January, 1875, he was arrested by the State of New York in a civil action to recover his stealings, but he escaped and fled to Spain. There he was arrested, returned to New York, and tried once more. Convicted, he was imprisoned again. He died in Ludlow Street jail, New York City, in 1878.

Suggested Readings

Bryce, James. The American Commonwealth. New York: Macmillan, 1914.

Callow, Alexander B., Jr. "The House That Tweed Built," American Heritage Magazine, October, 1965.

Lingley, Charles Ramsdell, and Allen Richard Foley. Since the Civil War. New York: Appleton, 1935.

Lynch, Denis T. "Boss" Tweed: The Story of a Grim Generation. New York: Boni & Liveright, 1927.

Mandelbaum, Seymour J. Boss Tweed's New York. New York: Wiley, 1965.

Myers, Gustavus. The History of Tammany Hall. New York: Burt Franklin, 1967.

Nevins, Allan, and Frank Weitenkampf. A Century of Political Cartoons. New York: Octagon Books, 1975.

Paine, Albert Bigelow. Thomas Nast, His Period and His Pictures.

Gloucester, Mass.: Peter Smith, 1904.
Parkhurst, Charles. Our Fight With Tammany. New York: Scribner's, 1895.
Seitz, Don C. The Dreadful Decade, Detailing Some Phases in the History of the United States from Reconstruction to Resumption, 1869-1879. Westport, Conn.: Greenwood Press, 1968.
_______. Horace Greeley, Founder of the New York Tribune. New York: AMS Press, 1926.
Steinberg, Alfred. The Bosses. New York: Macmillan, 1972.
Stoddard, Theodore. Master of Manhattan. New York: Longmans, Green, 1931.
Werner, M. R. Tammany Hall. Garden City, N.Y.: Doubleday, 1928.

THE BENDER MURDERS (1870-1873)

The Bender family migrated from "somewhere in the east" to the prairies of southeast Kansas in 1869. It is not known where they came from but it is thought that they originated in Germany or Holland, as they spoke English with a Teutonic accent. John "Old Man" Bender, his wife and their children John and Kate built and operated a small combination general store, home and inn. Kate, who was known as "Professor Miss Kate Bender," was a spiritualist who gave lectures in nearby towns and "communicated with the dead." She was, in addition, an accomplished swindler and, like the other members of her family, a murderer.

The Bender inn was so arranged that guests who were served meals there sat on a bench behind which was a canvas curtain. Such guests were killed by being struck on the head from behind the curtain with a heavy hammer. They were then robbed and their bodies buried in an orchard on the premises. A trap door leading to the cellar and a passage cut through to the rear of the house were used for removing the bodies.

Some travelers who stopped at the Bender place were fortunate in not being killed or robbed. A man named Corlew stopped there one day, but refused to stay for supper, thus inadvertently avoiding the bench, canvas curtain, and hammer. He stated later that he had heard "queer scuffling and moaning sounds coming apparently from under the house." Kate Bender had told him that a hog had fallen into the cellar and hurt itself.

Other men passed through the Bender establishment in safety, but often the hammer proved effective and the victims were robbed and buried in the orchard. On March 9, 1873, Dr. William H. York, while on his way home from Fort Scott, Kansas, disappeared from sight. His brother, a colonel at the Fort, became alarmed and instituted a search. The doctor was traced to the Bender inn, where he had taken refuge for the night, and had not been seen since.

The colonel started an investigation which was intensified when the Bender family suddenly vanished, their home showing signs of an extremely hasty departure. No fewer than eleven graves were found in the orchard, and it was shown that those unfortunates who occupied them had died from blows on the head. One, however, a small girl, had apparently been buried alive.

Most of the bodies were identified, one proving to be that of Dr. York. The murdering Benders, who had killed their guests in order to acquire their possessions, were never apprehended, although a wide and intensive search was made for them. They were reported having been seen in Indian Territory. A posse which went in pursuit claimed to have lost the trail but was suspected of having caught up with and executed the four members of the family. This suspicion, however, was never confirmed.

Rumors that the murderous quartet was still alive persisted. Occasionally people thought to be members of the Bender family were arrested but were released when their right identities were established. One such arrest was made in New Orleans. Two women, thought to be Kate Bender and her mother, were brought from Michigan to Kansas for questioning. On November 18, 1889, a preliminary hearing on the charge of murdering Dr. York was held but witnesses could not agree as to whether or not the two Michigan women were members of the Bender family and the charge was dropped.

Suggested Readings

Adleman, Robert H. The Bloody Benders (fiction). New York: Stein & Day, 1970.

Beagle, J. L. "Bender Mound and Mysteries are Recalled," Parsons, Kansas, Daily Republican, March 7, 1926.

"The Benders," Denver News, January 16, 1927.

"The Benders," Kansas Magazine, September, 1886.

"The Benders," Topeka Commonwealth, August 4, 1877.

"The Benders Again," Cherryvale, Kansas, Republican. November 1, 1889.

"Bloody Yarn of 'Hell Benders' Recalled by Col. York's Death," Denver Post, March 11, 1928.

Brigden, Bruce. The Bloody Benders. Cherryvale, Kansas: The Bender Museum, 1966.

Burkholder, Edwin V. "Those Murdering Benders," True Western Adventures, February, 1960.

Conrad, William, and Robert Greenwood. "The Bender Legend," Kansas Magazine, 1950.

Cook, John R. What Became of the Benders? The Port of Buffalo, 1907.

Dick, Leroy F., as told to Jean McEwen. "The Bender Hills Mystery," Parsons, Kansas, Sun, June 9 to August 11, 1934, inclusive.

"The Grave of the Benders," Arkansas Weekly Republican, July 19, 1906.

"The Great Bender Mystery," Frank Leslie's Illustrated Newspaper, January 7, 1873.

Hardy, Allison. Kate Bender: The Kansas Murderess. Girard,

Kansas: Haldeman-Julius Publications, 1944.
Harlow, Alvin F. "Were They the Benders?" in his Murders Not Quite Solved. New York: Messner, 1938.
Harris, John P. "Beautiful Katie," Kansas Magazine, 1936.
Henderson, Mark. "Murder Tavern," The Great West, May, 1968.
James, John T. The Benders of Kansas. Kan-Okla Publishing Co., 1913.
"Light on Benders," Topeka Journal, August 5, 1911.
"Old Bender Caught," Topeka Commonwealth, February 28, 1875.
Ross, Edith Connelly. The Bloody Benders. Wichita, Kansas: Kansas State Historical Society Collections, 1962-68, Vol. XVII.
Rowe, Fayette. "Kate Bender's Fate Still Mystery of Pioneer Kansas," Wichita Eagle Magazine, September 26, 1954.
Shirley, Glenn. "Prairie Vampire: Kate Bender," in his Toughest of Them All. Albuquerque, N.M.: University of New Mexico Press, 1953.
Swallow, Richard. "Arch Murderess of Kansas Still Alive Is Theory," Wichita Eagle, September 4, 1927.
________. "Where Is the Infamous Kate Bender?", Real Detectives, September, 1932.
Symons, Julian. A Pictorial History of Crime. New York: Crown Publishers, 1966.

THE CHICAGO FIRE (1871)

One of the great disasters of United States history, which started on the evening of Sunday, October 8, 1871, was the fire which destroyed much of the city of Chicago before burning itself out the following evening.

The cause of the Chicago fire has never been definitely established. Legend has it that a cow belonging to a Mrs. Patrick O'Leary kicked over a kerosene lantern in a barn at 137 De Koven Street on the city's west side. The O'Leary family, consisting of Patrick, a laborer, his wife Catherina and their five children, lived in a cottage at the De Koven Street address, behind which was a barn where a horse, five cows and a calf were housed. Mrs. O'Leary milked the cows and sold the milk to neighbors.

On the Sunday night in question Dennis Sullivan, a drayman, visited his friends the O'Learys. Around 8:30 Sullivan started home. Halfway down the block he looked back to see "a bright, pulsating glow" in the O'Leary's barn. He rushed back, gave the alarm, and was able to save the calf from the burning building before the heat became unbearable.

In two hours the fire had consumed over 100 acres and was still spreading. The Midwest had suffered a severe drought during the previous summer and the wooden buildings were exceedingly flammable. Streets and vacant lots were littered with combustible material, such as old boxes, discarded lumber and other similar rubbish. The fire department lacked sufficient serviceable equipment to cope with a blaze of any great magnitude. It was undermanned and its members were exhausted from fighting other conflagrations which had broken out in the city. A wind which at times reached gale proportions fanned the flames practically without ceasing. Finally, when the De Koven Street fire was reported, the watchman mistook the location and dispatched an engine company stationed some distance away, rather than one of several which were comparatively nearby.

As the fire spread, looters had a field day. Stores and homes were broken into. "Men and women filled their bellies with whisky and their arms with stolen goods." The authorities were powerless to stop the pillaging. Booty which could not be carried was often tossed into the flames or deliberately destroyed.

The fire eventually burned itself out at Lincoln Park. It was followed by a wave of incendiarism. Seven arsonists, caught in the act, were shot and another was stoned to death by infuriated citizens. Martial law was declared and General Philip H. Sheridan was placed in command.

During the 24 hours that it raged the Chicago fire swept over 2,100 acres, burned more than 17,000 buildings, killed an estimated 500 persons and caused a property loss of over $200,000,000. In addition, many priceless, irreplaceable artifacts were destroyed.

Although Mrs. O'Leary's cow has been accused of starting the Chicago fire, her owner denied that any member of the O'Leary family had gone to the stable after nightfall. She further denied that there had been a lighted lamp on the premises any time during the evening, although the remains of a broken lamp were found in the ruins of the stable.

Another story was that Patrick O'Leary had gone to the barn on the evening of October 8 to obtain some milk for an oyster stew. This he denied. It was also hazarded that the fire may have been started by boys who went to the barn to smoke forbidden pipes and cigars. To this day, however, the origin of the Chicago fire of 1871 is still unknown.

Suggested Readings

Asbury, Herbert. Gem of the Prairie: An Informal History of the Chicago Underworld. New York: Knopf, 1940.

________. "The Great Chicago Fire," in Kartman, Ben, and Leonard Brown, eds. Disaster! New York: Pellegrini & Cudahy, 1948.

Clevely, Hugh. Famous Fires. New York: John Day, 1958.

Colbert, Elias, and Everett Chamberlin. Chicago and the Great Conflagration. New York: Viking Press, 1971.

Cromie, Robert A. The Great Chicago Fire. New York: McGraw-Hill, 1958.

Dreiser, Theodore. Trilogy of Desire; Vol. II, The Titan (fiction). New York: Apollo Editions, 1974.

Farr, Finis. Chicago. New Rochelle, N.Y.: Arlington House, 1973.

Haywood, Charles F. "The Great Chicago Fire," in his General

Alarm. New York: Dodd, Mead, 1967.
Kogan, Herman, and Robert A. Cromie. The Great Fire: Chicago, 1871. New York: Putnam, 1971.
Masters, Edgar Lee. The Tale of Chicago. New York: Putnam, 1933.
Naden, Corinne J. Chicago Fire, 1871. New York: Watts, 1969.
Wagenknecht, Edward. Chicago. Norman, Okla.: University of Oklahoma Press, 1964.

THE GREAT DIAMOND HOAX (1871-1872)

One day in 1871 two prospectors, Philip Arnold and John Slack, deposited in the Bank of California, San Francisco, for safekeeping, several sacks which they said contained rough diamonds. Word of this came to William Chapman Ralston, a wealthy entrepreneur, mine operator, businessman, and cashier and later president of the bank. Some weeks later he interviewed the two prospectors who reluctantly agreed to sell a part of their rights to the desert mine they claimed to have discovered, and which they said was the source of the diamonds.

Before investing, Ralston sent two representatives, one of whom was David C. Colton, a Southern Pacific Railroad official, to inspect the mine. These representatives, conducted by Arnold and Slack, were blindfolded, both going and coming, that they might not disclose the mine's location prematurely. On their return to San Francisco the inspection team reported "acres and acres of precious stones: diamonds, rubies, sapphires, emeralds." Ralston cabled his friend Asbury Harpending, then in London, asking him to be general manager of a company which would be organized to mine the jewels. After some hesitation Harpending agreed.

Arnold and Slack offered to return to their mine, collect "a couple of million dollars worth of diamonds," and turn them over to Ralston as a guarantee of good faith. Ralston agreed and in due course the prospectors delivered to Harpending a sack which, when opened at his home, proved to contain a large assortment of precious stones.

Large sums of money had been loaned to the prospectors. Ralston had determined that a mining company should be organized. Tiffany and Company, New York jewelers, were to appraise the gems produced by Arnold and Slack and legislation covering diamond mining in the United States was to be provided for.

In due course Tiffany appraised a representative sample drawn from the sack of jewels as being genuine and worth about $150,000. A bill was drafted, introduced, and passed through Congress. Known as "Sargent's Mining Bill," it appeared May 18, 1872.

Henry Janin, an outstanding mining engineer of the day, was selected as engineer for the project. He inspected the region where Arnold and Slack claimed to have found the stones and reported that "gems were so plentiful that twenty rough laborers could wash out a million dollars worth of diamonds in a month." On the strength of this Ralston made the prospectors an initial payment of $300,000 and organized the San Francisco and New York Mining and Commercial Company, capitalized at $10,000,000. Twenty-five local businessmen were allowed to participate in the Company, subscribing for $80,000 worth of stock each. Arnold and Slack's interest was bought out for an additional $300,000; altogether they were paid a total of $660,000. They promptly disappeared from the scene.

News of the new diamond mine spread through the world. The London Rothschilds wished to invest in it. Colton resigned his position with the Southern Pacific Railroad to become general manager of the mining company. The board of directors included some of the biggest money men in the west.

Then Ralston received a telegram from Clarence King, a noted geologist and head of the Fortieth Parallel Survey. King had visited the diamond fields and found jewels spread over a wide area. However, he also found rubies and emeralds in forks of trees and in rock crevices "where nature alone could not have placed them." When he uncovered a large diamond which showed evidence of having been cut by hand, he realized that the mine had been "salted." His telegram read, "The alleged diamond fields are fraudulent. Plainly they are salted. The discovery is a gigantic fraud. The Company has been painfully duped."

It was later found that Arnold had bought large quantities of low-grade jewels in London and Amsterdam. Some of the diamonds were identified as South African "niggerheads."

Ralston took sole responsibility for the fiasco and repaid the investors in the Company from his own pocket. John Slack was never heard of again, and it was supposed that he met with foul play. Philip Arnold was traced to Hardin County, Kentucky, and forced to disgorge about $150,000. He kept the balance for himself and when he died in a duel, bequeathed his estate to his relatives.

Suggested Readings

Dana, Julian. The Man Who Built San Francisco: A Study of Ralston's Journey With Banners. New York: Macmillan, 1936.
DeFord, Miriam Allen. "Reconstructed Rebel: Asbury Harpending," in her They Were San Franciscans. Caldwell, Idaho: Caxton Printers, 1941.
Farish, Thomas Edwin. The Gold Hunters of California. Chicago: M. A. Donohue, 1904.
King, Clarence. Helmet of Mambreno. New York: Putnam, 1904.
Lyman, George D. Ralston's Ring. New York: Scribner's, 1937.
MacDougall, Curtis D. Hoaxes. New York: Macmillan, 1940.
Mayer, Robert, comp. San Francisco: A Chronological and Documentary History, 1542-1970. New York: Oceana, 1974.
Smith, H. Allen. The Compleat Practical Joker. Garden City, N.Y.: Doubleday, 1959.

Tilton, Cecil G. William Chapman Ralston, Courageous Builder. Boston: Christopher Publishing House, 1935.
Wilkins, James H., ed. The Great Diamond Hoax, and Other Stirring Episodes in the Life of Asbury Harpending. San Francisco: Barry, 1913.

THE SOUTH IMPROVEMENT COMPANY (1872)

The period from the close of the Civil War in 1865 to the year 1890 saw the proliferation of industrial monopolies and the consolidation of many small concerns into a comparatively few larger ones. The census of 1890 showed that in the previous twenty years the production of manufactured articles had greatly increased, but it also showed that in many industries the number of establishments producing such articles had declined.

After 1865 cutthroat competition between industries was the order of the day. This led to combinations which, businessmen hoped, could achieve monopolistic control of various segments of the industrial world, thus permitting indiscriminate price-setting and insuring greater business profits.

One of the better known examples of combination before 1890 was the Standard Oil Company. In 1862 John D. Rockefeller and Samuel Andrews, the latter the inventor of an inexpensive process for refining crude petroleum, went into the oil business together under the name of Rockefeller and Andrews. In 1870 these two, with Rockefeller's brother William and Henry M. Flagler, who had been associated with Rockefeller since 1865, plus a few others, established the Standard Oil Company of Ohio with a capitalization of $1,000,000.

The business prospered and in 1872 Rockefeller formed the South Improvement Company, an association of the largest oil refiners in Cleveland, Ohio. This Company, headed and controlled by Standard Oil, had as its objective the making of advantageous arrangements with the railroads for transportation facilities. Early in 1872 it signed contracts with the Erie, Pennsylvania, and New York Central Railroads, the three most important in the oil country. By these contracts the railroads were to establish certain freight rates between the crude oil producing regions of Western Pennsylvania and such large refining and shipping centers as Baltimore, Cleveland, New York, Philadelphia, and Pittsburgh. The contracts further specified that the South Improvement Company was to receive substantial rebates, in return for which the railroads were promised the entire freight business of the Company.

Another provision of these contracts was that the railroads were to give the South Improvement Company rebates on all oil shipped by their competitors and furnish it with copies of way-bills on all such shipments each day. While the contracts permitted the railroads to make similar rebates to any concern offering a similar amount of oil freight business, it was extremely unlikely that any other oil-producing company would be able to do this.

Rockefeller then offered competing firms the choice of selling out to him at his evaluation of each firm's worth or being bankrupted. Before the South Improvement Company was dissolved after only three months of corporate life, Rockefeller had acquired 21 of the 26 refineries in Cleveland and controlled one-fifth of the oil-refining business of the country.

Independent oil producers joined forces to fight for survival and after the Oil War of 1872 the railroads were forced to abandon their contracts with the South Improvement Company, whose charter was cancelled by the Pennsylvania legislature. However, the practice of giving rebates continued and the combination, which in 1882 consisted of 39 refiners controlling 90 to 95% of the product, was organized as the Standard Oil Trust. This was declared an illegal monopoly and ordered dissolved by the Ohio Supreme Court the same year. Actual dissolution was not effected until 1899, when Rockefeller established the Standard Oil Company of New Jersey.

Suggested Readings

Cochran, Thomas C., and William Miller. The Age of Enterprise: A Social History of Industrial America. New York: Harper, 1961.

Flynn, John T. God's Gold: The Story of Rockefeller and His Times. Westport, Conn.: Greenwood Press, 1971.

______. Men of Wealth. New York: Simon and Schuster, 1941.

Hidy, Ralph W., and Muriel E. Hidy. Pioneering in Big Business, 1882-1911: History of Standard Oil Company, New Jersey. New York: Harper, 1955.

Jenks, Jeremiah W., and Walter E. Clark. The Trust Problem. New York: Arno Press, 1973.

Josephson, Matthew. The Robber Barons. New York: Harcourt, Brace, 1962.

Kirkland, Edward C. Industry Comes of Age. New York: Holt, Rinehart, 1961.

Lawson, Thomas W. Frenzied Finance. New York: Somerset, 1972.

Lloyd, Henry Demarest. "The Story of a Great Monopoly," Atlantic Monthly, March, 1881.

______. Study in Power: John D. Rockefeller, Industrialist and Philanthropist. New York: Scribner's, 1953.

Rockefeller, John D. Random Reminiscences of Men and Events. New York: Arno Press, 1973.

Seager, Henry R., and Charles A. Gulick, Jr. Trust and Corporation Problems. New York: Arno Press, 1929.

Tarbell, Ida M. History of the Standard Oil Company. New York: McClure, Phillips, 1904.

Williamson, Harold F., and Arnold R. Daum. The American Petroleum Industry. Evanston, Ill.: Northwestern University Press, 1959.

THE CRIME OF '73 (1873)

The so-called "Crime of '73" was the appellation applied to the Coinage Act of February 12, 1873. This legislation, revising the coinage system of the United States, omitted from the coinage the standard silver dollar of 412-1/2 grains and tended to demonetize silver, something bitterly opposed by the silver miners of the West.

Following the Civil War the economic life of the United States developed rapidly, and with this development came an increased demand for additional currency with which to carry out the augmented business transactions. Unless the volume of currency or bank checks increased proportionately with the increase in business, the demand for money would cause its value to rise and prices to fall. Inversely, if the volume of money and checks increased more rapidly than the needs of business required, their value would fall and prices would rise. Neither of these alternatives was desirable, as a substantial change in price level, whether up or down, would harm large groups of people. This situation created difficult financial problems and Congress was subjected to lobbying and pressure by powerful individuals and groups.

For various reasons it did not seem wise or even possible to increase the amount of paper currency. The amount of gold available for new coinage each year was limited and it seemed that silver coins would have to be used to supply the needed increase.

Silver had not been an important medium of exchange except as a subsidiary coin. It was this fact which led to the omission of the silver dollar from the list of authorized United States coins mentioned in the Coinage Act. Certain European countries had reduced their silver coinage, thus throwing large amounts of silver bullion on the international market. Great silver strikes in the western part of the United States had further increased the supply and the price of the metal had fallen. The western silver mine owners, seeking new markets for their product, favored increased coinage of silver. They lobbied for free and unlimited coinage of that metal.

On February 28, 1878, the Bland-Allison Act was passed by Congress. This legislation, which satisfied neither the silver West nor the gold-standard East, represented a compromise between the two factions. It specified that silver dollars should be coined and should be considered legal tender. It further provided that government purchases of silver should not exceed $4,000,000 worth each month and should not be less than $2,000,000 worth per month. This Act was vetoed by President Rutherford B. Hayes but was passed by Congress over his veto.

Suggested Readings

Barrett, Don Carlos. The Greenbacks and Resumption of Specie Payments. Cambridge, Mass.: Harvard University Press,

1931.
Burr, Susan S. Money Grows Up. Publication No. 43. Washington, D.C.: Service Center for Teachers of History, The American Historical Association, 1969.
Commager, Henry Steele, ed. "The Bland-Allison Act," (Doc. No. 299) in his Documents of American History, 8th edition. New York: Appleton, 1968.
_______. "The Crime of '73," (Doc. No. 285) in his Documents of American History, 8th edition. New York: Appleton, 1968.
Fine, Nathan. Labor and Farmer Parties in the United States, 1828-1928. New York: Russell & Russell, 1961.
Friedman, Milton, and Anna J. Schwartz. A Monetary History of the United States. Princeton, N.J.: Princeton University Press, 1963.
Hofstadter, Richard. "Free Silver and the Mind of 'Coin Harvey'," in his The Paranoid Style in American Politics and Other Essays. New York: Random House, 1967.
Laughlin, J. Laurence. The History of Bimetallism in the United States. New York: Appleton, 1897.
Lingley, Charles Ramsdell, and Allen Richard Foley. Since the Civil War. New York: Appleton, 1935.
O'Leary, Paul M. "The Scene of the Crime of 1873 Revisited: A Note," Journal of Political Economy, August, 1960.
Studenski, Paul, and Herman E. Krooss. Financial History of the United States. New York: McGraw-Hill, 1952.
Unger, Irwin. The Greenback Era. Princeton, N.J.: Princeton University Press, 1964.
Williams, Charles R. The Life of Rutherford Birchard Hayes. New York: Da Capo Press, 1970.

THE SALARY GRAB (1873)

During the administration of President Ulysses S. Grant, and particularly during his second term of office, a number of unsavory scandals occurred. General Orville Babcock, his private secretary, was involved in the Whiskey Ring Affair and was also suspected of wrongdoing in connection with the Santo Domingo Negotiations. William Belknap, his Secretary of War, was found to have accepted bribes from a trader at an Indian post. The building of the railroads which followed the Civil War gave various unscrupulous entrepreneurs opportunities for unjust enrichment at the expense of the stockholders and investors. Any number of government officials were involved in questionable, self-serving activities.

These illegal goings-on were by no means confined to Grant's second administration. On March 3, 1873, when his first term as President was drawing to a close, Congress saw fit to pass legislation increasing the salaries of public officials, ranging from the President to the members of the House of Representatives. Adding insult to injury, the salary increases for congressmen were made retroactive, and each would receive $5,000 for the two years just past. The Crédit Mobilier scandal involving public officials had recently come to the attention of the public, as had news of other illegal money-making schemes which benefited the few at the expense of many. News of the "salary grab" and "back pay steal" of the elected officials in the national capital was, to the private citizen, a fresh indication that corruption was rampant in Washington.

Constituents throughout the country contacted their senators and representatives, protesting the increased salaries. Some of the politicians, realizing that they could be voted out of office, returned the increase to the treasury. When the next session opened the law was repealed, except as it applied to the President and the justices of the Supreme Court.

The scandals of the Grant administration turned many voters against the Republican party. In the election of 1876 Rutherford B. Hayes, as a Republican, opposed the Democrat Samuel J. Tilden. In what has been called the "stolen election," Hayes won in a fiercely disputed contest that threatened the country with civil war. Although Hayes received only a minority of the popular votes, a partisan election committee awarded him the office. Skullduggery in government and business continued for many years after the Grant administration, in spite of the efforts of some honest legislators and the writings of the "Muckrakers."

Suggested Readings

Bowers, Claude G. The Tragic Era: The Revolution After Lincoln. Boston: Houghton Mifflin, 1929.

Catton, Bruce. U.S. Grant and the American Military Tradition. Boston: Little, Brown, 1954.

Coolidge, Louis A. Ulysses S. Grant. Edited by John T. Morse, Jr. New York: AMS Press, 1922.

Fuess, Claude M. Carl Schurz, Reformer. New York: Dodd, Mead, 1932.

Hesseltine, William B. Ulysses S. Grant, Politician. New York: Dodd, Mead, 1935.

Josephson, Matthew. The Politicos, 1865-1896. New York: Harcourt, Brace, 1938.

Mayer, George H. The Republican Party, 1854-1966. New York: Oxford University Press, 1967.

Muzzey, David Saville. James G. Blaine: A Political Idol of Other Days. New York: Dodd, Mead, 1934.

Nevins, Allan. Hamilton Fish: The Inner History of the Grant Administration. New York: Dodd, Mead, 1936.

Seitz, Don C. The Dreadful Decade, Detailing Some Phases in the History of the United States from Reconstruction to Resumption, 1869-1879. Westport, Conn.: Greenwood Press, 1968.

Van Deusen, Glyndon G. Horace Greeley, Nineteenth Century Crusader. New York: Hill & Wang, 1964.

White, Leonard D. The Republican Era, 1869-1901. New York: Macmillan, 1958.

THE HORREL-HIGGINS FEUD (1873-1877)

On Friday, March 14, 1873, four state policemen in the town of Lampasas, Texas, were shot to death when they attempted to arrest a cowboy who was, contrary to law, carrying a revolver in a holster. The policemen--Captain Thomas Williams and Privates Wesley Cherry, J. M. Daniels and another private named Melville--had been summoned to town by a group of citizens who hoped they could dissuade high-spirited cowboys from discharging their revolvers at windows and buildings while galloping through town.

No sooner had Captain Williams arrested the gun-carrying cowboy than he and his fellow law enforcers were killed by shots coming from a nearby saloon. The Horrel Brothers--Ben, Bill, Bert, Merrett, Sam, and Tom, all cattlemen--were charged by the inquest jury with the multiple murder.

Following the shooting the Horrel Brothers and their friends fled to the woods. Governor Edmund M. Davis of Texas was asked to send reinforcements, and the following day Adjutant General F. L. Britton and a detachment of ten men arrived at Lampasas. Britton's men were joined by two Minute Man companies which had been hastily organized and by a posse under Sheriff S. T. Denson. After a lengthy search Merrett Horrel and Jerry Scott, the saloon keeper, were found, captured, and jailed. One night shortly afterwards they were freed by a party of men who broke into the building with sledge hammers. The Horrels then took their belongings and cattle and set out for Ruidoso, New Mexico.

The Horrel Brothers were unable to stay out of trouble in their new home. Bill and Ben were killed in fights with New Mexicans and the other brothers decided to return to Texas. Here they stood trial for the shooting of Captain Williams and the other policemen and were acquitted.

John Pinkney Calhoun Higgins, known as "Pink" Higgins, had immigrated from Georgia with his parents and settled in the Lampasas area in 1857, near the Horrel homestead. As a boy he and the Horrel Brothers had been neighbors and friends. However, after they had become young men a quarrel broke out over some steers and Higgins shot and killed Merrett Horrel in the same saloon near which Captain Williams had died. Higgins and his friend Bob Mitchell, who had been present at the shooting, mounted their horses to ride home. Friends of Tom Horrel rode to tell him of what had befallen his brother. A murder charge was filed against Pink Higgins and on June 7, 1877, he and three friends, Bob Mitchell, William R. Wren and B. F. Terry, all heavily armed, rode into town and put up bond. Tom Horrel opened fire on the group from a doorway. The four retaliated with their Winchester rifles. No one was killed although Wren was slightly wounded.

A general free-for-all broke out as friends of both factions joined the fighting. Frank Mitchell, Bob's brother, was killed. The combatants took cover and sniped at one another. Pink Higgins returned with reinforcements and it was not until Captain John Sparks of the Texas Rangers, who happened to be in town, persuaded the leaders to desist that the firing ceased. Sparks put some members of the Horrel faction under guard and induced the Higgins party to return home, following which he released the Horrels.

In response to Captain Sparks' report of the situation, the governor sent a Ranger detachment commanded by Captain N. O. Reynolds to Lampasas. Upon arriving in town, Reynolds, four men of his command, Bob Mitchell, Alonzo Mitchell, and William R. Wren rode to the Horrel home on Sulphur Fork. Here two Horrel brothers were arrested, together with several friends who were present. The Horrel faction was escorted to town where both they and the Higgins men posted bond to keep the peace. From that time on the Horrel-Higgins affair, as a feud, was a thing of the past.

The two remaining Horrel brothers, Sam having left town permanently, were later arrested for allegedly robbing a storekeeper named Vaughn. They were killed in their cells by a mob which broke its way into the jail where they had been confined awaiting trial. Pink Higgins found employment as a bounty hunter for the Spur Ranch in the Southern Panhandle. Eventually he was retired on a pension by his employers. He died of natural causes in 1913.

Suggested Readings

Cutler, James Elbert. Lynch-Law: An Investigation into the History of Lynching in the United States. New York: Negro University Press, 1969.

Douglas, C. L. Famous Texas Feuds. Dallas: The Turner Co., 1936.

Fehrenbach, T. R. Lone Star: A History of Texas and the Texans. New York: Macmillan, 1968.

Gard, Wayne. Frontier Justice. Norman, Okla.: University of Oklahoma Press, 1949.

Raper, Arthur F. The Tragedy of Lynching. Chapel Hill, N.C.: University of North Carolina Press, 1933.

Robson, William A. Civilization and the Growth of Law. New York: Macmillan, 1935.

Seagle, William. The Quest for Law. New York: Knopf, 1941.

Shay, Frank. Judge Lynch: His First Hundred Years. New York: I. Washburn, 1938.

Sonnichsen, C. L. I'll Die Before I'll Run. New York: Harper, 1951.

________. Ten Texas Feuds. Albuquerque: University of New Mexico Press, 1957.

Zane, John Maxey. The Story of Law. Garden City, N.Y.: Garden City Publishing Co., 1927.

THE SLAUGHTER-HOUSE CASES (1873)

The Fourteenth Amendment to the Constitution of the United States became law on July 28, 1868, and the first decision of the Supreme Court involving this Amendment was that given in the Slaughter-House Cases of 1873.

In 1869 the legislature of the state of Louisiana, in order to protect the health of the people of New Orleans, had given a Louisiana corporation a 25-year exclusive right to slaughter cattle within the city's limits, thus forbidding all other persons to build slaughter houses in that area. The corporation, however, was required to permit other persons to use its facilities, for which fixed fees were charged.

Cases were brought before the Louisiana courts involving the possible violation of that part of the Fourteenth Amendment which specifies, "No State shall make or enforce any law which shall abridge the privileges or immunities of citizens of the United States; nor shall any State deprive any person of life, liberty, or property without due process of law...." The United States Supreme Court, by a vote of five to four, upheld the constitutionality of the Louisiana statute.

Justice Samuel F. Miller, in delivering the decision of the Court, first discussed the definition of the word "citizens," as used in the Amendment. The Court found that there is a citizenship of the United States and a citizenship of a state and that these are distinct from each other. The Court declared that the states are forbidden to trespass on those rights possessed by the individual as a federal citizen, be he white or colored. The Amendment was not intended to "constitute this Court a perpetual censor upon all litigation of the States, on the civil rights of their own citizens."

In the Slaughter-House Cases the Court did not believe that "due process of law" had been interfered with by the Louisiana legislation. It was contended that the slaughter-house law violated the clause in the Amendment reading "nor shall any State deny to any person within its jurisdiction the equal protection of the laws." In rebuttal Justice Miller stated, "We doubt very much whether any action of a State not directed by way of discrimination against the Negroes as a class, or on account of their race, will ever be held to come within the purview of this provision."

In summary, the majority opinion of the Supreme Court was that the balance between the states and the national government should be preserved very much as it had been, and that the privileges and immunities applying to state citizenship should rest for their security and protection upon the states where they had theretofore rested.

Suggested Readings

Carson, Hampton L. The History of the Supreme Court. New York: Burt Franklin, 1972.

Collins, Charles W. The Fourteenth Amendment and the States.

New York: Da Capo Press, 1974.
Commager, Henry Steele, ed. "The Constitution of the United States," (Doc. No. 87) in his Documents of American History, 8th edition. New York: Appleton, 1968.
________. "The Slaughter-House Cases," (Doc. No. 284) in his Documents of American History, 8th edition. New York: Appleton, 1968.
Cushman, Robert E. Civil Liberties in the United States. Ithaca, N.Y.: Cornell University Press, 1956.
Haines, Charles G. The American Doctrine of Judicial Supremacy. Los Angeles: University of California Press, 1932.
________. The Role of the Supreme Court in American Government and Politics. Berkeley, Calif.: University of California Press, 1944.
Johnson, Gerald White. The Supreme Court. New York: Morrow, 1962.
Sterne, Horace. "Samuel Freeman Miller, 1816-1890," in Lewis, W. D. Great American Lawyers. Hackensack, N.J.: Fred D. Rothman, 1971.
Strong, Henry. "Justice Samuel Freeman Miller," Annals of Iowa, January, 1894.
Warren, Charles. The Supreme Court in United States History. Boston: Little, Brown, 1923.

THE PANIC OF 1873 (1873)

The Pacific Railway Act of July 1, 1862, authorizing the construction of a transcontinental railroad between Omaha, Nebraska, and San Francisco, California, was the culmination of twenty years of planning. Two years later a second Pacific Railway Act doubled federal land grants and made certain provisions concerning the mortgage on the railroad property that the federal government would hold.

The road was completed on May 10, 1869, symbolized by Governor Leland Stanford of California driving a golden spike into the last railway tie at Promontory Point, Utah. During the next four years some 24,000 miles of new railroad were built, much of it extending into the wilderness well ahead of settlement. Simultaneously, great new stretches of land were being opened and industry was expanding on a huge scale. The net result of these combined factors was a rush of over-speculation which culminated in the Panic of 1873.

The Panic was triggered by the financial failure of Jay Cooke & Co., a Philadelphia banking house. Cooke, founder of the bank that bore his name, had marketed bonds for financing the Civil War and had acted as fiscal agent for the United States Treasury. After the Civil War he had financed the construction of western railroads, particularly the Northern Pacific, by marketing its bonds and thus provided capital for payrolls, purchases of equipment and other expenses. Financial firms, such as Fisk & Hatch, were engaged in similar financing transactions for other railroads and speculative enterprises throughout the country.

A number of events led to the inevitable crash. Devastating fires in Chicago in 1871 and in Boston the following year resulted in the loss of millions of dollars worth of property. In early 1873 the Crédit Mobilier of America was investigated by the government and found to be diverting large sums of railroad money to its promoters and their friends. This created widespread distrust and made investors hesitant about purchasing bonds.

When Jay Cooke & Co. closed its doors on September 18, 1873, the entire financial world was stunned. Bank runs followed, with depositors standing in line, sometimes for hours, waiting to withdraw their funds. Many banks, unable to meet their depositors' demands, failed. Stock prices fell drastically on the exchange and on September 20 it suspended trading, which did not resume for ten days.

Industrial firms, dependent upon the banks, were unable to continue in business. Railroad receiverships, unemployment, industrial paralysis and riots followed. Business did not recover its former prosperity until late in the decade.

Suggested Readings

Bogart, Ernest Ludlow. The Economic History of the United States. New York: Longmans, Green, 1907.

Bremer, C. D. American Bank Failures. New York: Columbia University Press, 1935.

Carman, Harry J. Social and Economic History of the United States. New York: Johnson Reprint, 1934.

Commager, Henry Steele, ed. "The Pacific Railway Act," (Doc. No. 215) in his Documents of American History, 8th edition. New York: Appleton, 1968.

Faulkner, Harold U. American Economic History. New York: Harper & Row, 1960.

Friedman, Milton, and Anna J. Schwartz. A Monetary History of the United States. Princeton, N.J.: Princeton University Press, 1963.

Hoggson, Noble Foster. Epochs in American Banking. New York: John Day, 1929.

Lingley, Charles Ramsdell, and Allen Richard Foley. Since the Civil War. New York: Appleton, 1935.

Nevins, Allan. The Emergence of Modern America, 1865-1878. Saint Clair Shores, Mich.: Scholarly Press, 1971.

Noyes, Alexander D. Forty Years of American Finance. New York: Putnam, 1898.

Studenski, Paul, and Herman E. Krooss. Financial History of the United States. New York: McGraw-Hill, 1952.

White, Henry K. History of the Union Pacific Railway. Clifton, N.J.: Augustus M. Kelley, 1968.

THE CHARLEY ROSS KIDNAPING (1874)

On July 1, 1874, four-year-old Charley Brewster Ross was kidnaped from his home in Germantown, Pennsylvania. He was never seen again.

Young Charley and his six-year-old brother Walter were invited by "two men passing by in a trap" to accompany them to Philadelphia "to buy fireworks." These same two men had driven up, stopped, and talked to the two boys on the 27th of the previous month.

Charley and Walter accepted the invitation and got into the vehicle. En route to Philadelphia Walter was given a quarter and instructed to go into a store and purchase the promised pyrotechnics. When he emerged from the store the trap, the two men and his brother were gone.

Ransom notes, misspelling the boy's name, perhaps intentionally, "charlie buster ros," were received by Charley's father, Christian K. Ross. The first of these indicated that the boy would be returned unharmed if the father did not report the abduction to the police. Other notes followed, demanding $20,000 for the safe return of the child and giving instructions as to how the money should be delivered.

Ross Senior had already reported the kidnaping to the police. The police were unable to locate the boy and the detective agency headed by Allan Pinkerton was called into the case. A reward of $20,000 for the recovery of the missing boy and for the arrest and conviction of his abductors was offered. Reward advertisements portraying an approximate likeness of young Charley and describing him and the two kidnapers were circulated widely.

These advertisements and the activities of the police and the Pinkerton agency produced no results. George W. Walling, chief of the New York police, identified the ransom notes as having been written by a criminal named William Mosher. An effort was made to locate and arrest him but in December, 1874, two burglars were apprehended and fatally shot when they attempted to rob the Brooklyn home of Charles H. Van Brunt. One of the burglars was Joseph Douglas who, before he died, stated that he was one of the two men who had abducted Charley Brewster Ross the previous July. He identified the other kidnaper as Mosher, his fellow-burglar, who had been killed instantly. He did not, however, give any indication of the young boy's fate, and no trace of him has been found to this day.

William Mosher had married the sister of William Westervelt, a former police officer on the New York City force. Mrs. Mosher, when questioned by the police, admitted that she had been privy to the kidnaping but had no knowledge of any of the details, nor did she know where the missing boy was, although she felt that he was still alive.

Westervelt, who seemed to have some connection with the kidnaping, was brought to trial on August 30, 1875. He was found guilty of being an accessory of the kidnapers Mosher and Douglas, and was sentenced to serve seven years' solitary confinement at hard labor. He served his sentence but neither admitted participation in the crime nor disclosed any information concerning Charley Ross' fate.

In 1927 a man calling himself John W. Brown, ill in a Los Angeles hospital, claimed to be the long-lost boy. His claim was disapproved in 1931. In 1939 an Arizona carpenter named Gustav Blair claimed to be Charley Ross. The Superior Court of Phoenix, Arizona, recognized his claim but members of the Ross family refused to acknowledge him as the four-year-old who had been abducted 65 years before.

Suggested Readings

American vs. Italian Brigandage, Philadelphia: Barclay & Co., 1875.
Churchill, Allen. They Never Came Back. Garden City, N.Y.: Doubleday, 1960.
Duke, Thomas S. Celebrated Criminal Cases of America. San Francisco: Barry, 1910.
Horan, James D. The Pinkertons: The Detective Dynasty That Made History. New York: Crown Publishers, 1967.
________, and Howard Swiggett. The Pinkerton Story. New York: Putnam, 1951.
Lavine, Sigmund A. Allan Pinkerton: America's First Private Eye. New York: Dodd, Mead, 1963.
Life, Trial, and Conviction of W. H. Westervelt (pamphlet). Philadelphia: 1879.
McCartney, Clarence. "Charley Ross: The Unforgotten Lost Boy," Ladies' Home Journal, July, 1924.
Messick, Hank, and Burt Goldblatt. Kidnapping: The Illustrated Story. New York: Dial Press, 1974.
Orrmont, Arthur. Master Detective: Alan Pinkerton. New York: Messner, 1965.
Pinkerton, Allan. Criminal Reminiscences and Detective Sketches. Freeport, N.Y.: Books for Libraries, 1970 (Originally published 1878).
Ross, Christian K. Charley Ross, etc. London: Amistead, 1877.
________ The Father's Story of Charley Ross, the Kidnaped Child. Philadelphia: John E. Potter & Co., 1876.
Rowan, Richard Wilmer. The Pinkertons, a Detective Dynasty. Boston: Little, Brown, 1931.
Slattery, William, ed. Abduction: Fiction Before Fact. New York: Grove Press, 1974.
Smith, Edward H. "The Charlie Ross Enigma," in his Mysteries of the Missing. New York: Lincoln MacVeagh-The Dial Press, 1927.
Symons, Julian. A Pictorial History of Crime. New York: Crown Publishers, 1966.
Wallechinsky, David, and Irving Wallace. "Charley Ross," in their The People's Almanac. Garden City, N.Y.: Doubleday, 1975.
Walling, George Washington. Recollections of a New York Chief of Police, With Historical Supplement of the Denver Police. New

York: Caxton Book Concern, Ltd., 1887.
Winslow, Edward W. Crime in a Free Society. Encino, Calif.: Dickenson, 1968.
Zierold, Norman. Little Charley Ross: America's First Kidnapping for Ransom. Boston: Little, Brown, 1967.

THE WHISKEY RING AFFAIR (1874-1876)

The Whiskey Ring Affair was one of the scandals occuring during the administration of President Ulysses S. Grant. For some years prior to 1874 it had been suspected that a ring of revenue officials, collaborating with accomplices in Washington, were in collusion with the distillers to defraud the government of the lawful tax on whiskey. These frauds were discovered in 1875, largely through the efforts of Treasury Secretary Benjamin H. Bristow. Part of the illegal gains was said to have gone into the campaign fund for Grant's reelection, although he was not aware of the source of the money.

Bristow's investigation was hampered by the fact that information concerning his activities was sent out surreptitiously by officials from Washington. However, on May 10, 1875, sixteen distilleries in Chicago, Milwaukee and St. Louis were seized. Some 240 distillers and revenue officials were indicted. Those implicated included the chief clerk in the Treasury Department and General O. E. Babcock, Grant's private secretary and confidential adviser.

Following indictments, trials of the accused began in Jefferson City, Missouri, in October, 1875. Babcock was acquitted in 1876 but did not return to the White House as secretary. Instead he was given the position of Superintendent of Public Buildings and Grounds. Several of the others involved either pled guilty or were convicted despite their pleas of "not guilty," but were later pardoned by the President after a short interval.

Suggested Readings

Anonymous. Some Facts About the Life and Public Services of Benj. H. Bristow (pamphlet). New York: 1876.
Bowers, Claude G. The Tragic Era: The Revolution After Lincoln. Boston: Houghton Mifflin, 1929.
Boynton, H. V. "The Whiskey Ring," North American Review, CXXIII.
Catton, Bruce. U. S. Grant and the American Military Tradition. Boston: Little, Brown, 1954.
Coolidge, Louis A. Ulysses S. Grant. Edited by John T. Morse, Jr. New York: AMS Press, 1922.
Fuess, Carl M. Carl Schurz, Reformer. New York: Dodd, Mead, 1932.
Hesseltine, William B. Ulysses S. Grant, Politician. New York:

Dodd, Mead, 1935.
Josephson, Matthew. The Politicos, 1865-1896. New York: Harcourt, Brace, 1938.
McDonald, General John. Secrets of the Great Whiskey Ring and Eighteen Months in the Penitentiary. St. Louis: W. S. Bryan, 1880.
Nevins, Allan. Hamilton Fish: The Inner History of the Grant Administration. New York: Dodd, Mead, 1936.
Seitz, Don C. The Dreadful Decade, Detailing Some Phases in the History of the United States from Reconstruction to Resumption, 1869-1879. Westport, Conn.: Greenwood Press, 1968.
Willcox, David. "Memorial of Benj. H. Bristow," Annual Report, Association of the Bar of the City of New York, 1897.

THE JUKES AND THE KALLIKAKS (1874 and 1912)

The "Jukes" and the "Kallikaks" were pseudonyms applied to two families studied respectively by the American sociologist Richard Louis Dugdale and the American psychologist Henry Herbert Goddard. The studies were made to determine the hereditary basis, if any, "for such phenomena as mental disorder and feeblemindedness."

Dugdale's study was conducted in 1874 under the auspices of the Prison Association of New York. The report of this study, published in 1875 under the title The Jukes, relates an extraordinary history of the descendants of a backwoodsman identified as "Max." "Max" had two sons, each of whom married one of the "Jukes" sisters. One of these sisters is described in the report as "Margaret, mother of criminals."

Seven hundred and nine of the descendants and blood relations of these unions were traced. Dugdale found that of this total, 208 had, as paupers, received public support, 140 had served time in prison for criminal activities, and a large proportion had been afflicted with mental and/or nervous diseases.

It was calculated that during the 75-year period covered by the study the "Jukes" family had cost New York State more than $1,308,000.

The Dugdale study has been used by a number of researchers in the fields of psychopathology and other social sciences and has precipitated several similar projects. One of the better-known investigations in this area is the one made by Henry Herbert Goddard, mentioned above. His report, published in 1912, is titled The Kallikak Family. Goddard's findings were essentially parallel to those of his predecessor.

Suggested Readings

Adams, Percy W. L. A History of the Jukes Family of Cound, Shropshire, and Their Descendants. Tunstall, England: E. H.

Eardley, 1927.
Chute, B. J. "Jukebox and the Kallikaks" (fiction), in Costain, Thomas B., and John Beecraft, eds. More Stories to Remember. Toronto: Doubleday, 1958.
Dugdale, Richard Louis. Further Studies of Criminals. New York: AMS Press, 1975.
_______. The Jukes, a Study in Crime, Pauperism, Disease and Heredity. New York: AMS Press, 1975. (Originally published 1875.)
Estabrook, Arthur Howard. The Jukes in 1915. Pittsburgh: Carnegie Institute, 1916.
_______, and Charles Benedict Davenport. Nam Family: A Study in Cacogenics. New York: Eugenics Record Office, 1913.
Goddard, Henry H. The Criminal Imbecile. New York: Arno Press, 1915.
_______. Feeble-Mindedness, Its Causes and Consequences. Plainview, N.Y.: Books for Libraries, 1972.
_______. Human Efficiency and Levels of Intelligence. Princeton, N.J.: Princeton University Press, 1920.
_______. The Kallikak Family, a Study in the Heredity of Feeble-Mindedness. New York: Arno Press, 1931.
Putnam, George Haven. Memories of a Publisher, 1865-1915. New York: Putnam, 1915.
Shepard, E. M. The Work of a Social Teacher. New York: 1884.
Winship, Albert Edward. Jukes-Edwards: A Study in Education and Heredity. Harrisburg, Pa.: R. L. Myers, 1900.

THE BEECHER-TILTON SCANDAL AND TRIAL (1875)

The most prominent Congregational minister of the latter part of the 19th century was the Reverend Henry Ward Beecher. He was known to forty million people as a writer, orator, preacher, and ardent abolitionist. The Plymouth Congregational Church of Brooklyn Heights, New York, where he was the "beloved minister," was one of the most fashionable and financially successful in the country.

Beecher, in 1875, was the defendant in a $100,000 adultery suit brought by Theodore Tilton, his closest friend, for allegedly having had illicit sexual intercourse with Tilton's wife, Elizabeth, or "Lib," as she was sometimes called. The trial was one of the most sensational of the decade.

On July 3, 1870, Elizabeth Tilton confessed to her husband that she had been committing adultery with Beecher over a period of approximately two years. Tilton did nothing about the matter until Victoria C. Woodhull and her sister Tennessee Claflin, publishers of Woodhull and Claflin's Weekly, social reformers and advocates of free love, published a vague accusation in the New York World, issue of May 20, 1871. The sisters had apparently learned of the affair from Elizabeth Cady Stanton, a woman's suffrage leader, earlier that month. Beecher was not mentioned specifically in the accusation, and the sisters published it because they were disgusted with the hypocrisy of a leading minister who failed to practice what he so ardently preached.

Theodore Tilton finally swore out a warrant against Beecher on August 20, 1874. In it he charged the minister with having "alienated and destroyed Mrs. Tilton's affections for him" and demanded $100,000 damages.

The trial opened on January 11, 1875, Chief Justice Joseph Neilson presiding. William A. Beach headed the group of lawyers for Tilton, and William M. Evarts and six assistants constituted Beecher's defense staff. Throughout the trial Beecher, through his attorneys, attempted to save his own reputation at the expense of all others concerned, including Elizabeth Tilton.

When the scandal first became general knowledge Beecher had defended himself before a committee representing Plymouth Church. Tilton, who had testified before this committee, covered essentially the same ground in court. He had, for some fifteen years, been an intimate friend of Beecher, who had secured a position for him as an editor of Henry Bowen's newspapers, the *Union* and the *Independent*.

Many letters had been exchanged between the Tiltons, Beecher and others, 57 such being produced as exhibits. On November 2, 1872, a story had appeared in *Woodhull and Claflin's Weekly* charging Beecher with adultery with Elizabeth Tilton. Francis D. Moulton, a friend of the Tiltons and of Beecher, had been asked by Mrs. Tilton to act as an intermediary and did so, as he later testified.

The trial dragged on, being characterized by long outbursts of oratory from the battery of attorneys present. Beecher as a witness did his case more harm than good. He begged the question no less than 894 times, giving such replies as "I don't know" and "I cannot remember." More than two million words were spoken or introduced as written evidence.

Nearly six months after the trial began Justice Neilson charged the jury and a week later, on July 2, 1875, he dismissed it as the members were unable to come to an agreement.

Beecher brought a charge of criminal libel against Tilton but, on reflection, decided to drop it. He returned to his pastorate at Plymouth Congregational Church where he discharged all those who had not sided with him in the trial. His congregation remained, in general, loyal to him but he spent much time away from the pulpit engaged in lecture tours. He died of a cerebral hemorrhage on March 8, 1887.

Elizabeth Tilton lived quietly in Brooklyn with her mother following the trial. A letter she wrote to her legal counsel in 1878 confessing that the charge of adultery brought by her husband was true, though published widely in newspapers throughout the country, failed to revive the scandal. She died, a blind recluse, in 1897.

Suggested Readings

Abbott, Austin. Official Report of the Trial of Henry Ward Beecher. New York: McDivitt, Campbell, 1875.
Abbott, Lyman. Henry Ward Beecher. New York: Somerset, 1903.
Aymar, Brandt, and Edward Sagarin. "Reverend Henry Ward Beecher," in their A Pictorial History of the World's Great Trials. New York: Crown Publishers, 1967.
Beecher, William C., and the Rev. Samuel Scoville. A Biography of Henry Ward Beecher. New York: Charles L. Webster, 1888.
Hibben, Paxton. Henry Ward Beecher: An American Portrait. New York: Doran, 1927.
The Independent. Scrapbooks of clippings relating to the Beecher-Tilton Trial. New York: New York Public Library, 1975.
Knox, Thomas W. The Life and Works of Henry Ward Beecher. New York: Harper, 1887.
Rourke, Constance M. Trumpets of Jubilee. New York: Harper, 1927.
Shaplen, Robert. Free Love and Heavenly Sinners: The Story of the Great Henry Ward Beecher Scandal. New York: Knopf, 1954.
Stowe, Lyman Beecher. Saints, Sinners, and Beechers. Indianapolis: Bobbs-Merrill, 1934.
Theodore Tilton vs. Henry Ward Beecher, Action for Criminal Conspiracy ... Verbatim Report by the Official Stenographer. New York: McDivitt, Campbell, 1875.
Williamson, Francis P. Preacher and His Accusers: A Complete History of the Great Controversy. Philadelphia: Flint & Company, 1874.

THE MASON COUNTY WAR (1875)

One of the most bloody feuds in Texas history was known as the "Mason County War." It erupted in 1875.

The cause of the feud was cattle theft, something which plagued the American West for over half a century. It started in Mason County when Lige Baccus, his cousin, and three companions were arrested and lodged in jail by Sheriff John Clark. They were charged with stealing cattle belonging to a group of German ranchers.

The Germans decided to make an example of the five. A mob, not waiting to determine the innocence or guilt of the men, descended upon the jail, broke down the door, took the prisoners out and started to hang them. In this the mob was only partially successful; Sheriff Clark and some volunteers, together with Captain Dan W. Roberts of the Texas Rangers, were able to rescue three of the alleged cattle rustlers.

The next day one of the rescued men shot one of the members of the mob and then left town. The lynching and retaliatory shooting brought about the formation of two factions: the German cattle families of Mason County against the cattlemen of the outlying districts.

A cattle driver named Tim Williamson employed by Charles Lemberg was arrested by deputy sheriff John Worley on a charge of cattle theft. As Lemberg, Williamson and Worley were riding to town so that Lemberg might post bond for his employee, the three men were ambushed by a sizable party of horsemen. Williamson was shot dead, as was his horse. The other two men made their escape.

One of Tim Williamson's friends was Scott Cooley, a cattleman. Cooley resolved to avenge Williamson's death, as did a number of other friends, including Mose and John Beard, George Gladden, and John Ringgold. This group planned to attack those Germans they felt were involved in the killing, but before they could take definite action a shooting affray occurred between Gladden and Mose Beard and Sheriff Clark's posse. Both men attempted to escape. Beard was killed but Gladden's life was spared because of the entreaties of his friend Charlie Keller.

The feud spread and grew in intensity, with killings being met with counter-killings. Men were shot from ambush and people answering a call at their doors were killed without warning. John Ringgold was indicted for murder but the charge was dismissed. John Beard shot Dan Hoerster, a prominent German rancher and merchant. Deputy Sheriff Worley, while helping a neighbor, Dr. Harkett, dig a well, was shot by Scott Cooley who then departed, realizing that the Texas Rangers would soon be after him. However, he had at one time been a member of that organization and fifteen Rangers, under the command of Captain Roberts, resigned rather than pursue one of their own.

Other assassinations followed. Gladden was charged with killing a man named Peter Barder. He was tried, convicted, sentenced to 99 years in prison, and shortly thereafter pardoned. Sheriff Clark left the region, never to return.

The Mason County War finally came to an end when the members of the Cooley faction realized that the Texas Rangers meant business and could and would put down any further insurrections. With Cooley gone from the area the Rangers no longer felt any personal friendship or loyalty to him.

Cooley returned to Blanco County where he had formerly lived. There, shielded from the law by his friends, he died of a cerebral hemorrhage a few years later.

Suggested Readings

Bierschwale, Margaret. "Mason County, Texas, 1845-1870," Southwestern Historical Quarterly, April, 1944.

Cutler, James Elbert. Lynch-Law: An Investigation into the History of Lynching in the United States. New York: Negro University Press, 1969.

Douglas, C. L. Famous Texas Feuds. Dallas: The Turner Co., 1936.
Eilers, Kathryn Burford. A History of Mason County, Texas. Unpublished M.A. thesis, University of Texas, 1939.
Fehrenbach, T. R. Lone Star: A History of Texas and the Texans. New York: Macmillan, 1968.
Gard, Wayne. Frontier Justice. Norman, Okla.: University of Oklahoma Press, 1949.
Gillett, James B. Six Years With the Texas Rangers. Austin, Tex.: Von Boeckmann-Jones, 1921.
Harris, William Foster. The Look of the Old West. New York: Viking Press, 1955.
Hunter, J. Marvin. "Brief History of Early Days in Mason County," Frontier Times, November, 1928, March, 1929.
Raper, Arthur F. The Tragedy of Lynching. Chapel Hill, N.C.: University of North Carolina Press, 1933.
Roberts, Captain Dan W. Rangers and Sovereignty. San Antonio: Wood Printing & Engraving Co., 1914.
Robson, William A. Civilization and the Growth of Law. New York: Macmillan, 1935.
Seagle, William. The Quest for Law. New York: Knopf, 1941.
Shay, Frank. Judge Lynch: His First Hundred Years. New York: I. Washburn, 1938.
Sonnichsen, C. L. Billy King's Tombstone. Caldwell, Idaho: Caxton Printers, 1942.
________. "The Hoodoo War," in his Ten Texas Feuds. Albuquerque: University of New Mexico Press, 1957.
________. I'll Die Before I'll Run. New York: Harper, 1951.
Ward, Don. Cowboys and Cattle Country. New York: American Heritage Press, 1961.
Webb, Walter Prescott. The Texas Rangers. Boston: Houghton Mifflin, 1935.
Zane, John Maxey. The Story of Law. Garden City, N.Y.: Garden City Publishing Co., 1927.

THE MOLLY MAGUIRES (1875-1877)

The Molly Maguires was a society formed by the anthracite coal miners of eastern Pennsylvania about 1854. It was modeled after a similar organization in Ireland from which it took its name. It had as its object the planning and execution of a campaign of physical violence against the mine owners and the state and municipal police whom the owners controlled.

The society, also sometimes called the "Sleepers," grew in both numbers and strength during and after the Civil War, and with this growth came an increase in violent demonstrations against the mine owners.

In 1875 a strike of miners, engineered by the Molly Maguires, precipitated the hiring of the detective Allan Pinkerton by Franklin Benjamin Gowen, president of the Philadelphia and Reading

Coal & Iron Company. Pinkerton was to infiltrate the society and, by espionage, destroy it. Pinkerton sent James McParlan, an Irish-American operative in his employ, to join the society and accumulate evidence against its members.

McParlan, under the alias "James McKenna," was accepted by the Molly McGuires as a criminal fleeing from the law. In 1876 and 1877, after more than two and a half years, he was able to present evidence in court which resulted in the conviction of many members of the society for various crimes, including murder. The back of the society was broken and shortly thereafter it ceased to exist.

McParlan continued in the service of Allan Pinkerton as a detective and labor spy until his death in 1919.

Suggested Readings

Bimba, Anthony. The Molly McGuires. New York: International Publishers, 1932.

Broehl, Wayne G., Jr. The Molly McGuires. Cambridge, Mass.: Harvard University Press, 1964.

Coleman, J. Walter. The Molly McGuire Riots. Richmond: Garrett & Massie, 1936.

Dewees, Franklin P. The Molly McGuires. Philadelphia: Lippincott, 1877.

Doyle, Sir Arthur Conan. The Valley of Fear (fiction). New York: Doran, 1914.

Heaps, Willard Allison. Riots, U.S.A., 1765-1970. New York: Seabury Press, 1970.

Horan, James D. Desperate Men. New York: Putnam, 1949.

________. The Pinkertons: The Detective Dynasty That Made History. New York: Crown Publishers, 1967.

________, and Howard Swiggett. The Pinkerton Story. New York: Purnam, 1951.

Hynd, Alan. "The Murdering Sleepers," in Klein, Alexander. The Double Dealers. Philadelphia: Lippincott, 1958.

Lavine, Sigmund A. Allan Pinkerton: America's First Private Eye. New York: Dodd, Mead, 1963.

Lucey, Ernest W. The Mollie McGuires of Pennsylvania, or Ireland in America. London: George Bell & Sons, 1882.

Orrmont, Arthur. Master Detective: Allan Pinkerton. New York: Messner, 1965.

Pinkerton, Allan. Criminal Reminiscences and Detective Sketches. Freeport, N.Y.: Books for Libraries, 1970. (Originally published 1878.)

________. The Mollie McGuires and the Detectives. New York: G. W. Dillingham, 1877.

________. Professional Thieves and the Detective. New York: G. W. Carleton & Co., 1881.

________. Strikers, Communists, Tramps and Detectives. New York: G. W. Carleton & Co., 1878.

Pinkerton, Robert A. "Detective Surveillance of Anarchists," North American Review, November, 1901.

Rowan, Richard Wilmer. The Pinkertons: A Detective Dynasty. Boston: Little, Brown, 1931.
Schlegel, Marvin W. Ruler of the Reading: The Life of Franklin B. Gowen. Harrisburg: Archives Publishing Co., 1947.
Wallechinsky, David, and Irving Wallace. "The Molly Maguires," in their The People's Almanac. Garden City, N.Y.: Doubleday, 1975.

THE STOLEN ELECTION (1876)

The Presidential election of 1876 was fraught with political corruption, a carryover from the administration of Ulysses S. Grant. The Republican candidate, Rutherford B. Hayes, had polled 4,035,924 popular votes and the Democratic candidate, Samuel J. Tilden, had polled 4,287,670, thus winning a popular majority of over a quarter of a million votes. Tilden, however, had only 184 uncontested electoral votes; Hayes had 166, which left Tilden short of a majority in the electoral college. In addition, the four electoral votes of Florida, the eight of Louisiana and seven of South Carolina, for a total of 19, were all disputed.

Charges were made by the Republicans that the Democrats had achieved popular majorities in the three southern states by intimidation of Negro voters and that, had such voters not been threatened, they would have voted Republican. The first of these charges was true, the second mere conjecture.

Federal troops, occupying the South and supporting the administration in Washington, enabled the Republicans to set up electoral bodies which cast votes for Hayes. In Oregon the Democrats retaliated by removing a Hayes elector on technical grounds, replacing him with a Tilden man. The resulting dispute ended in a deadlock which it was vitally necessary should be resolved, as Grant's term was drawing to a close.

On January 29, 1877, legislation was enacted by Congress which provided for a commission, known as the Electoral Commission of 1877. This Commission consisted of five senators, five representatives and four justices of the Supreme Court, the justices to choose an independent justice so as to bring their total number to five. Justice David Davis, selected as the independent, had been chosen by the Illinois legislature as U. S. Senator and refused to serve on the commission. A substitute for Davis, nominally non-partisan but actually Republican, was designated and the net result was that the commission, when finally completed, had a membership which outnumbered the Democrats eight to seven.

The Electoral Commission, on March 2, 1877, formally announced that Hayes had been elected. Tilden accepted the decision, realizing that to do otherwise could precipitate a second Civil War, but always maintained that he had wrongfully been deprived of the election.

Suggested Readings

Barnard, Harry. Rutherford B. Hayes and His America. Indianapolis: Bobbs-Merrill, 1954.
Bigelow, John, ed. Letters and Literary Memorials of Samuel J. Tilden. Port Washington, N.Y.: Kennikat Press, 1971.
Buchanan, Lamont. Ballot for Americans. New York: Dutton, 1956.
Clubb, Jerome M., and Howard W. Allen, eds. Electoral Change and Stability in American Political History. New York: Free Press, 1971.
Congressional Quarterly Service. Presidential Candidates from 1788 to 1964, Including Third Parties, 1832-1964, and Popular Electoral Vote: Historical Review. Washington, D.C.: Government Printing Office, 1964.
Dunning, William A. Reconstruction: Political and Economic. New York: Harper, 1907.
Haworth, Paul L. The Hayes-Tilden Election. Indianapolis: Bobbs-Merrill, 1927.
Lingley, Charles Ramsdell, and Allen Richard Foley. Since the Civil War. New York: Appleton, 1935.
Nevins, Allan. Abram S. Hewitt, With Some Account of Peter Cooper. New York: Harper, 1935.
Porter, Kirk H., and Donald Bruce Johnson, comps. National Party Platforms, 1840-1960. Urbana, Ill.: University of Illinois Press, 1956.
Rhodes, James F. History of the United States from the Compromise of 1850. New York: Macmillan, 1893-1906.
Richardson, Leon Burr. William E. Chandler, Republican. New York: Dodd, Mead, 1940.
Robinson, Lloyd. The Stolen Election: Hayes versus Tilden, 1876. Garden City, N.Y.: Doubleday, 1968.
Roseboom, Eugene H. A History of Presidential Elections, from George Washington to Richard M. Nixon. New York: Macmillan, 1970.
Schlesinger, Arthur M., Jr., ed. History of American Presidential Elections, 1789-1968. New York: Chelsea House, 1971.
Stampp, Kenneth M. The Era of Reconstruction. New York: Knopf, 1965.
Williams, Charles R. The Life of Rutherford Birchard Hayes. New York: Da Capo Press, 1970.
Woodward, C. Vann. Reunion and Reaction: The Compromise of 1877 and the End of Reconstruction. Boston: Little, Brown, 1951.

CUSTER'S LAST STAND (1876)

Following the completion of the first transcontinental railroad in 1869, some 24,000 miles of new lines were laid throughout the West within the next four years. Many of these lines were built in Indian territory despite the violent protests of the aborigines. It was often necessary, in building the railroads across the plains, to provide armed men to protect the construction workers from Indian attacks.

With the railroads came farmers and settlers, which spelled the doom of the native Indian. A gold strike in the Black Hills on the Sioux reservation precipitated agitation to remove the Indians to make room for the white miners and settlers.

Inroads on tribal territories led to the Sioux War of 1876, in which the Indians attempted to regain the Black Hills from which they had been removed.

George Armstrong Custer, an American general officer who had served in the Civil War, was attached to the Seventh Cavalry Regiment and assigned to Indian fighting. In 1873 he was ordered to Dakota Territory to protect railroad surveyors and gold miners. After three years of skirmishing it was determined to crush the Indians by a three-way envelopment. One of the groups participating in the maneuver was commanded by General Alfred H. Terry, Custer's commanding officer. Custer, ordered to scout in advance of the main force, located a detachment of Sioux which, underestimating their number, he attacked, contrary to the orders of his superior. Led by Sitting Bull, Gall, and Crazy Horse, the Indians completely destroyed the entire center column of the cavalry, including Custer and 264 of his men. The battle, called the Battle of Little Big Horn, is more popularly known as "Custer's Last Stand." It was fought on June 25, 1876.

Custer Battlefield National Monument, a national monument in Big Horn County, Montana, marks the site of the encounter, one of the last important Indian resistances in the United States.

Suggested Readings

Ambrose, Stephen E. Crazy Horse and Custer. Garden City, N.Y.: Doubleday, 1975.

Britt, Albert. Great Indian Chiefs. New York: Whittlesey House, 1938.

Cooke, David C. Fighting Indians of the West. New York: Dodd, Mead, 1954.

Dines, Glen. Crazy Horse. New York: Putnam, 1966.

Downey, Fairfax. Indian-Fighting Army. New York: Scribner's, 1941.

Garst, Shannon. Crazy Horse, Great Warrior of the Sioux. Boston: Houghton Mifflin, 1950.

________. Custer, Fighter of the Plains. New York: Messner, 1944.

Gessner, Robert. Massacre: A Survey of Today's American Indian. New York: Da Capo Press, 1972.

Graham, W. A. The Custer Myth: A Source Book of Custeriana. Harrisburg, Pa.: The Stackpole Co., 1953.

Grant, Bruce. American Indians, Yesterday and Today. New York: Dutton, 1960.

Henry, Will, pseud. Custer's Last Stand. Philadelphia: Chilton

Books, 1966.
Heuman, William. Custer, Man and Legend. New York: Dodd, Mead, 1968.
Johnson, D. M. Warrior for a Lost Nation: A Biography of Sitting Bull. Philadelphia: Westminster Press, 1969.
Lindquist, G. E. E., et al. The Indian in American Life. New York: Friendship Press, 1944.
Meadowcroft, Enid La Monte. The Story of Crazy Horse. New York: Grosset & Dunlap, 1954.
Miles, Nelson A. Serving the Republic. Plainview, N. Y.: Books for Libraries, 1974.
Miller, David Humphreys. Custer's Fall. New York: Duel, Sloan, 1957.
Monaghan, Jay. Custer: The Life of General George Armstrong Custer. Boston: Little, Brown, 1959.
Rachlis, Eugene. Indians of the Plains. New York: American Heritage, 1960.
Reynolds, Quentin. Custer's Last Stand. New York: Random House, 1951.
Roland, Albert. Great Indian Chiefs. New York: Crowell, 1966.
Sandoz, Mari. Crazy Horse, The Strange Man of the Oglalas. New York: Knopf, 1942.
Seymour, Flora W. The Story of the Red Man. Plainview, N. Y.: Books for Libraries, 1929.
Stewart, Edgar I. Custer's Luck. Norman: University of Oklahoma Press, 1955.
Utley, Robert M. Custer and the Great Controversy. Los Angeles: Westernlore Press, 1962.
Vaughn, Jesse W. The Reynolds Campaign on Powder River. Norman: University of Oklahoma Press, 1961.
Vestal, Stanley. Sitting Bull, Champion of the Sioux. Norman: University of Oklahoma Press, 1957.
_______. Warpath and Council Fire: The Plains Indians' Struggle for Survival in War and in Diplomacy, 1851-1891. New York: Random House, 1948.
Wellman, Paul I. The Indian Wars of the West. Garden City, N. Y.: Doubleday, 1954.
Wissler, Clark. The American Indian. Gloucester, Mass.: Peter Smith, 1922.
_______. Indians of the United States. New York: Doubleday, 1940.

THE SAN ELIZARIO SALT WAR (1877)

A natural salt deposit in a chain of waterless lakes in the Guadalupe Mountains, some ninety miles east of El Paso, Texas, was the cause of the so-called San Elizario Salt War, which raged in the year 1877. For decades residents of the area had mined the salt, which was sufficiently pure to be used without refining.

As an ownerless product of nature it was considered free for the taking.

This was the situation until the middle 1870's when Charles Howard, a lawyer, arrived in El Paso. He entered politics and became district judge. In his official capacity he declared the salt deposits as of the public domain and therefore open to claim. This he did after staking out such a claim in his own name. Assisted by his associate, Justice of the Peace Gregoria N. García, he then proceeded to levy a charge on each cartload of salt taken from what had legally become his property.

The natives, who considered the salt a free gift of nature, protested. Madonia Gandara and José Marie Juarez, two Mexicans, announced their intention of taking the salt without paying for it. Howard and García had them arrested "on suspicion." An armed mob of Mexicans led by Visto Salcido and Léon Granillo released the prisoners from jail and then sought out Howard and García. These two were, for their protection, arrested by the local authorities and placed under guard in the local jail. The mob stormed the jail, demanding that the two officials be turned over to it. Louis Cardis, a local citizen, and the Reverend Pierre Bourgad, priest of the parish, dissuaded the mob from its purpose and the mob dispersed. Cardis, though a friend of Howard, felt that he had acted illegally in the matter of the salt claim and said so publicly.

Howard remained in jail until he agreed to relinquish all claims to the salt deposit, post bond to guarantee such relinquishment, and leave the area permanently. García was to resign his post as justice of the peace. Howard agreed, posted bond and departed for New Mexico. On October 10 he returned to San Elizario and shot Louis Cardis to death. He then left for New Mexico, pursued by a band of Mexican horsemen. Safe in his new home he was kept advised of developments in the salt situation by his agent McBride, whom he had instructed to remain on the job.

McBride advised Howard that salt was being removed from the mines and Howard resolved to return to San Elizario once more. He sought and received protection from the Texas Rangers, a small detachment of which rode with him on his return trip. News of Howard's homecoming had been received in San Elizario and a band of Mexicans, numbering almost 400 men, gathered to await him. These Mexicans made no secret of their intentions: to kill Howard on sight.

Ranger Lieutenant J. B. Tays and twelve men joined Howard and his party at Ysleta, Texas, after Tays had learned of the formation of the armed Mexican mob. The following day trouble broke out between the Mexicans and the Rangers. Charles E. Ellis, one of Howard's bondsmen, was lassoed and dragged to his death behind a Mexican's horse. The Rangers barricaded an adobe house at San Elizario and occupied it. The Mexicans laid siege to it but neither side could gain an advantage. The mob, reinforced with additional men, demanded that Howard be surrendered to it. Lieutenant Tays refused.

A shooting battle followed. Men on both sides were killed. The Mexicans tunneled into a nearby store, looted it and used it as an attacking point.

The fighting continued for three days and nights, after which Howard surrendered and delivered himself to the Mexicans. Through a trick the Mexicans persuaded the Rangers to surrender. A Mexican firing squad, commanded by Desiderio Apodaca, then executed Howard, McBride, and John G. Atkinson, another of Howard's bondsmen.

The San Elizario Salt War was over. The Mexican authorities promised to "see that justice was done," but few arrests were made and only a small portion of the property taken from the looted store was returned. As in earlier days, salt was once again taken from the Guadalupe Mountains without payment being made to Charles Howard or anyone else.

Suggested Readings

Cutler, James Elbert. Lynch-Law: An Investigation into the History of Lynching in the United States. New York: Negro Universities Press, 1969.

Douglas, C. L. Famous Texas Feuds. Dallas: The Turner Co., 1936.

Fehrenbach, T. R. Lone Star: A History of Texas and the Texans. New York: Macmillan, 1968.

Gard, Wayne. Frontier Justice. Norman, Okla.: University of Oklahoma Press, 1949.

Gillett, James B. Six Years With the Texas Rangers. Austin, Texas: Von Boeckmann-Jones, 1921.

Harris, William Foster. The Look of the Old West. New York: Viking Press, 1955.

Hughes, Annie E. The Beginning of Spanish Settlement in the El Paso District. Berkeley, Calif.: University of California Press, 1914.

MacCallum, Esther Darbyshire. The History of St. Clement's Church, El Paso, Texas, 1870-1925. El Paso: McMath, 1925.

Mills, W. W. El Paso: A Glance at Its Men and Contests for the Last Few Years. Austin, Texas: Republican Office, 1871.

_______. Forty Years at El Paso, 1858-1898. Privately printed, 1901.

Raper, Arthur F. The Tragedy of Lynching. Chapel Hill, N.C.: University of North Carolina Press, 1933.

Rippey, James Fred. The United States and Mexico. New York: Crofts, 1931.

Roberts, Captain Dan W. Rangers and Sovereignty. San Antonio: Wood Printing & Engraving Co., 1914.

Robson, William A. Civilization and the Growth of Law. New York: Macmillan, 1935.

Seagle, William. The Quest for Law. New York: Knopf, 1941.

Shay, Frank. Judge Lynch: His First Hundred Years. New York: I. Washburn, 1938.

Sonnichsen, C. L. "The El Paso Salt War," in his Ten Texas Feuds. Albuquerque: University of New Mexico Press, 1957.

_______. I'll Die Before I'll Run. New York: Harper, 1951.
Twitchell, Ralph Emerson. The Leading Facts of New Mexican History. Cedar Rapids, Ia.: Torch Press, 1911-1917.
Ward, Charles Francis. The Salt War of San Elizario (1877). Unpublished M.A. thesis, University of Texas, 1932.
Ward, Don. Cowboys and Cattle Country. New York: American Heritage Press, 1961.
Webb, Walter Prescott. The Texas Rangers. Boston: Houghton Mifflin, 1935.
White, Owen P. Out of the Desert: The Historical Romance of El Paso. El Paso: McMath, 1924.
Zane, John Maxey. The Story of Law. Garden City, N.Y.: Garden City Publishing Co., 1927.

THE MEMPHIS PLAGUE (1878)

In 1878 the city of Memphis, Tennessee, was not only virtually bankrupt but also filthy. Wells and cisterns were polluted and slops and garbage were disposed of by dumping them in gutters or streams of running water. Because of the poor financial condition of the city's exchequer, maintenance of the few sewage disposal facilities which did exist had been woefully neglected. Pools of stagnant water made excellent breeding grounds for the malaria-carrying mosquitoes aedes calopus. Under such conditions it was inevitable that a disastrous epidemic of malaria, or yellow fever as it was called, should strike Memphis and the surrounding area.

The first reported victim of the disease was a Mrs. Bionda who, with her husband, operated a small restaurant on the river front. She passed away on August 13, 1878. On August 14 twenty-two new cases of malaria were reported and citizens began to flee the city. This was similar to the hegira of August, 1873, when an epidemic of the mosquito-borne disease had caused some 25,000 Memphians to leave. Before the 1878 exodus had run its course, over half the population had departed, 17,000 natives contracted the disease and over 5,000 persons died of it.

Memphis resembled a ghost city. Burial carts made their rounds and the cry, "Bring out your dead!," was heard, just as it had been in medieval England during the days of the Black Plague. Annie Cook, proprietress of a well known brothel, discharged her girls and used her premises as a hospital for yellow fever patients. She contracted the disease and died on September 11.

Natives who were financially unable to travel further established a tent camp near the city. This was occupied by several thousand persons and was patrolled by both white and Negro militia. Looting and carousing also abounded. Dead and dying people were robbed. Many of the living hesitated to approach the dead, and decomposing bodies were allowed to remain on the streets untouched, often for several days.

Charitable institutions and persons contributed aid to the stricken area. Donations of money came in from all parts of the country. The Howard Association spent more than half a million dollars and sent 2,900 nurses to administer to the needs of the sick.

The epidemic lasted until mid-October when frost covered the South and a breeze, blowing the mosquitoes away, followed. It was determined that 75% of the whites remaining in Memphis had died; the mortality rate of the Negroes was about 10%.

The yellow fever scourge was not confined to Memphis. Greenville, Mississippi, New Orleans, Louisiana, and various cities and towns in the Mississippi Valley also suffered from it. Over 4,000 persons died in New Orleans alone.

It was not until Major Walter Reed, Aristides Agramonte, James Carroll, and Jesse Lazear proved, at the turn of the century, that malaria is carried by mosquitoes which have stung a person already suffering from the disease, that effective control measures could be initiated and the disease virtually stamped out.

Suggested Readings

Capers, Gerald M. *The Biography of a River Town: Memphis, Its Heroic Age.* Chapel Hill, N.C.: University of North Carolina Press, 1939.

Carter, Hodding. "Bring Out Your Dead," in Kartman, Ben, and Leonard Brown, eds. *Disaster!* New York: Pellegrini & Cudahy, 1948.

_______. *Lower Mississippi.* New York: Rinehart, 1942.

De Kruif, Paul. *Microbe Hunters.* Edited by Harry G. Grover. New York: Harcourt, Brace, 1932.

Eberle, Irmengarde. *Modern Medical Discoveries.* New York: Crowell, 1958.

Finlay, Carlos E. *Carlos Finlay and Yellow Fever.* New York: Oxford University Press, 1940.

Gibson, John M. *Physician to the World.* Durham, N.C.: Duke University Press, 1950.

Hill, Ralph N. *Doctors Who Conquered Yellow Fever.* New York: Random House, 1957.

McCrocklin, James H. *Garde D'Haiti, 1915-1934.* Menasha, Wis.: George Banta Co., 1955.

Sullivan, Mark. *Our Times, The United States, 1900-1925.* New York: Scribner's, 1928-1935.

Wood, Laura N. *Walter Reed, Doctor in Uniform.* New York: Messner, 1943.

THE HATFIELD-McCOY FEUD (1878-1890)

Webster defines the word "feud" as "a bitter, continuous hostility, especially between two families, clans, etc." One of the better known feuds which flourished in the West Virginia-Kentucky border hill country in the last century was that between the Hatfield and the McCoy families.

The original cause of the animosity between the two families is not known, although some people are of the opinion that it grew out of or was, in a way, an extension of the Civil War. Anderson "Devil Anse" Hatfield had served in the Confederate Army as a captain and Randall McCoy, his neighbor, had joined the Union forces where he remained an enlisted man for the duration.

The two men were released from military service when the war ended in 1865 and returned to their homes. Hatfield acquired land and by 1878, when the feud really started, owned some 70,000 acres. He farmed, hunted wild game, and raised swine. He also fathered thirteen children. Randall McCoy owned and operated a farm a few miles away from that of Devil Anse.

In 1878 a dispute concerning the ownership of a hog claimed by both McCoy and Floyd Hatfield, one of Devil Anse's kinsmen, occurred. McCoy brought suit to recover the animal and the court decided in his favor.

Some historians feel that the incident of the lawsuit over the hog started the Hatfield-McCoy feud, which was to continue for more than a decade and cost a score of lives. Others suggest that the trouble originated when, in 1863, Devil Anse as a Confederate army officer shot and killed Harmon McCoy, a Union soldier, in one of the skirmishes. It is reasonably sure, however, that there was already bad feeling between the two families when Jonse Hatfield, the eighteen-year-old girl-chasing eldest son of Devil Anse, fell in love with Rosanna, the daughter of Randall McCoy. He met her in 1880 at a local election, seduced her that evening and took her home to his family.

Rosanna McCoy remained at Jonse Hatfield's home, though they were never married, Devil Anse refusing to permit his son to marry a McCoy, although he had no objection to their living together in his house. In due course Jonse tired of Rosanna and turned his romantic attention to Mary Stafford, who was being courted by Rosanna's brother Tolbert McCoy. Tolbert warned Jonse to stay away from Mary. Rosanna, though pregnant, went back to her parents for a brief stay, then returned to the Hatfield home and finally went to live with an aunt.

Other trouble was brewing. Bud McCoy, a relative of Randall, attempted to kill Bob Hatfield, Devil Anse's cousin. Bob tried to get a warrant for Bud's arrest but failed as there were McCoys in political office.

Jonse Hatfield began courting Nancy McCoy, Rosanna's cousin, but also occasionally visited Rosanna at the home of her aunt. One evening three of Randall McCoy's sons surprised the two, seized Jonse, tied his hands behind his back and led him away. Warned by Rosanna, Devil Anse and several men of his family rode off into the night and rescued Jonse before his captors could do him harm.

Rosanna had her illegitimate child. That Fall the baby died and his mother passed away soon after. Jonse then married Nancy McCoy.

Other incidents followed. Bill Staton, a member of the Hatfield clan, was shot and killed by Sam McCoy as the result of a quarrel. Ellison Hatfield, commissioned a deputy sheriff and assigned to locate and bring in Sam McCoy and his brother Paris who had been involved in the quarrel with Staton, captured the latter. Both brothers were, in due course, brought to trial and Sam McCoy was acquitted on grounds of self-defense.

In 1882 at the August elections a fight broke out between members of the two warring families. Ellison Hatfield was stabbed to death. In turn, Tolbert, Phemar and Little Randall McCoy were captured by members of the Hatfield family and, on instructions from Devil Anse, shot to death.

The McCoys sought legal redress for the murder of the three men but while summonses for 23 members of the Hatfield clan were issued, no other action was taken by the authorities. A reward of $500 was offered for the apprehension of Devil Anse, "dead or alive." He, convinced that his son Jonse's sister-in-law was a spy, arranged to have her flogged one night when her husband, Bill Daniels, was away from home. She died of the beating.

Trouble developed between Jonse Hatfield and his wife Nancy, she accusing him of having been implicated in the flogging of her sister. Jeff McCoy was shot and killed by Cap Hatfield, then a deputy sheriff, for the murder of Fred Woolford, a mail carrier.

So the trouble went on. Killing followed killing, with each family avenging the death of one or more of its members by adding to the numbers already killed. On New Year's Eve, 1887, a group of raiders from the Hatfield family, led by Devil Anse, attacked the cabin of Randall McCoy. Allifair McCoy, aged fifteen, was shot to death by Cotton Top Mounts and the cabin was set afire.

A gun battle between the two factions resulted in the deaths of over twenty feuding men. Wall Hatfield was arrested. Jonse Hatfield, tired of the constant turmoil, left for Washington State, there to settle down. His wife did not accompany him but instead set up housekeeping with Frank Phillips.

Cotton Top Mounts, killer of Allifair and Little Randall McCoy, was tried for and convicted of murder. In 1890 he was hanged. Cap Hatfield was sent to jail in Huntington, West Virginia, for a murder he committed in 1896. With the aid of tools provided by friends he managed to cut his way out of his cell and escaped.

The Hatfield-McCoy feud had about run its course. Both families entered the coal mining business. Devil Anse "got religion" and he and the remaining members of his family were baptized. Instead of being enemies the two families stood together, united against the unions which sought to organize the family-operated mines. In 1921, the year of Devil Anse's death, an industrial war which had no connection with the feud resulted in the shooting death of Sid Hatfield, a thirty-year-old coal miner and one of Devil Anse's descendants.

Suggested Readings

Burns, J. A. The Crucible: A Tale of the Kentucky Feuds (fiction). Oneida, Ky.: The Oneida Institute, 1928.
Coates, Harold Wilson. Stories of Kentucky Feuds. Cincinnati: Holmes Coal Sales Co., 1942.
Dibble, R. F. "Devil Anse," American Mercury, May, 1925.
________. "Devil Anse," in French, Joseph Lewis, ed. A Gallery of Old Rogues. New York: A. H. King, 1931.
Donnelly, Shirley. The Hatfield-McCoy Feud Reader. Parsons, W. Va.: McClain Printing Co., 1971.
Hatfield, G. Elliott. The Hatfields. Stanville, Ky.: Big Sandy Valley Historical Society, 1974.
Jones, Virgil C. The Hatfields and the McCoys. Chapel Hill, N. C.: University of North Carolina Press, 1948.
Kroll, Harry Harrison. Their Ancient Grudge (fiction). Indianapolis: Bobbs-Merrill, 1946.
Spivak, J. L. Devil's Brigade: The Story of the Hatfield-McCoy Feud. New York: Brewer, 1930.
Thomas, Jean. "Romeo and Juliet," in her Big Sandy. New York: Holt, 1940.
________. "Romeo and Juliet in the Mountains," Etude, June, 1948.

THE WHITE RIVER MASSACRE (1879)

The uprising of a small band of Ute Indians at White River, Colorado, on September 29, 1879, was the direct result of certain white "land-grabbers" appropriating the Indians' rich lands for themselves. Historians consider it a shocking example of cupidity and of ignoring the rights of the original Americans. The "White River Massacre" which followed was the natural result of such greed.

Nathan Meeker, a former newspaper editor and novelist, had been appointed Indian agent at the White River Ute Agency in Northern Colorado in 1877. An idealist, he hoped to found a Utopian community and help the Indians improve their lot, abandon their native ways, and live as did the white man.

Meeker arrived at his new post in May, 1878, and his wife Arvilla and their youngest daughter Josie joined him shortly thereafter. He found the Utes to be led by Chief Ouray, who felt that his people should live in peace with the white man, modify their wasteful hunting economy, and "sell off bits of land as required by events." In 1868 Ouray had negotiated a treaty by which the Utes had been given forever a large tract of land on the Colorado western slope. In 1873 a portion of this land had been returned to the federal government but the Utes were still the richest Indian tribe in the nation.

Unfortunately for Ouray the state's white population had increased to the point where more land was demanded by them and such land-grabbers as senator Henry Moore Teller and his associates hoped to seize the Ute's lands, consigning the Indians to army-guarded desert camps.

Teller's group proceeded to accuse the Indians falsely of various outrages, thus destroying their reputation. Two Indians, Chief Jack and Chief Douglas, resented Teller's tactics and felt that Nathan Meeker, a Teller appointee, planned to steal the Indians' land and destroy their way of life.

Following his arrival at White River, Meeker explained to Jack, Douglas, and the Indian medicine man Johnson his plans for them. He intended, he said, to enable them to live as the white man did, engaging in modern farming, sleeping in beds, using privies, wearing underwear, and attending the Agency school. Chief Jack was not convinced that these plans were feasible and felt that his people would be exploited by the whites.

Matters came to a head when, in the spring of 1879, Frederick W. Pitkin, who had been elected governor of Colorado on a "Utes-must-go" platform, set about to fulfill his campaign promises. He saw to it that anti-Ute propaganda was published in Denver newspapers. Senator Teller forced Chief Ouray to sell a parcel of prime farm land for a comparatively small sum, which the buyer refused to pay. Forest fires, common in dry spring weather, were blamed on deliberate arson by the Utes.

The Indians ceased to cooperate with Meeker at the Agency. Eventually he mailed a list of complaints to the Indian Office. A physical encounter between Meeker and Johnson occurred, following which Meeker, though not badly hurt, telegraphed to Washington, asking that troops be sent in as he feared for his life and for those of the other whites at the Agency.

A detachment of cavalry, under the command of Major Thomas T. Thornburgh was dispatched from Rawlins, Wyoming. At Milk Creek, Colorado, a group of Indians was encountered and a battle ensued, in which Thornburgh, eleven of his men and many of the Utes were killed. Six days later the survivors were rescued by relief troops from Rawlins.

On Monday, September 29, 1879, occurred the White River Massacre. News of the battle between the soldiers from Rawlins and the Utes had been brought to the Agency by an Indian messenger on horseback. The messenger spoke to Chief Douglas. Immediately some of Douglas' men fired on the whites, killing two of them. Arvilla Meeker, her daughter Josie, Flora Ellen Price, the wife of Meeker's plowman, the latter's two children and Frank Dresser, an employee at the Agency, hid in an adobe milk house. The Indians set some of the log buildings on fire.

When evening fell the whites left the milk house. Dresser attempted to run for help but was shot and fatally wounded. Arvilla Meeker was wounded by a bullet which grazed her thigh.

The body of Nathan Meeker was found by his wife a hundred yards from the house. He had been shot in the side of the head. In all, eight men, in addition to Meeker, were killed. The two children and three women were held as hostages by the Indians for 23 days before they were rescued. The women testified that, during their captivity, they were raped and generally mistreated.

The aftermath of the Massacre was eminently satisfactory to the land-grabbers. Although only twenty Utes had actually participated in the action, all 700 of them were penalized in that money owed them by the government was paid to relatives of the white victims. The 1868 treaty rights of the Utes were voided and they were banished to a barren area in Utah.

Suggested Readings

Bancroft, Hubert Howe. The History of Nevada, Colorado and Wyoming. San Rafael, Calif.: Bancroft Press, 1888.
Coman, Katherine. Economic Beginnings of the Far West. Clifton, N.J.: Augustus M. Kelley, 1967.
Gessner, Robert. Massacre: A Survey of Today's American Indian. New York: Da Capo Press, 1972.
Goodale, Elaine. "On the Indian Question," Hartford Courant, October 16, 1885.
Grant, Bruce. American Indians, Yesterday and Today. New York: Dutton, 1960.
Paxson, Frederic L. The History of the American Frontier, 1763-1893. Dunwoody, Ga.: Norman S. Berg, 1967.
_______. The Last American Frontier. New York: Cooper Square, 1970.
Seymour, Flora W. Indian Agents of the Old Frontier. New York: Octagon Books, 1973.
_______. The Story of the Red Man. Plainview, N.Y.: Books for Libraries, 1929.
Sprague, Marshall. "The Bloody End of Meeker's Utopia," American Heritage Magazine, October, 1957.
Swanton, John Reed. Indian Tribes of North America. Washington, D.C.: Smithsonian Institute Press, 1952.
Tebbel, John, and Keith Jennison. The American Indian Wars. New York: Bonanza Books, 1960.
Vestal, Stanley. Warpath and Council Fire: The Plains Indians' Struggle for Survival in War and in Diplomacy. New York: Random House, 1948.
Wissler, Clark. The American Indian. Gloucester, Mass.: Peter Smith, 1922.
_______. Indians of the United States. New York: Doubleday, 1940.

THE SINGLE TAX MOVEMENT (1879-1897)

The Single Tax Movement enjoyed a brief popularity in the latter part of the 19th century. It was promulgated by its creator Henry George, an American economist, social reformer, writer, and politician. He publicized the Movement in his book, Progress and Poverty which appeared in 1879, and also in other writings and in public lectures.

George desired to "correct," as he saw it, the concentration of land ownership which followed railroad development, particularly in the state of California. He advocated a single tax on land because he believed that the privilege of land ownership was greatly abused. As a measure of social reform primarily and revenue only incidentally, the Single Tax was designed to decentralize ownership, eliminate private profit from the sale of land, and redistribute wealth.

The incidence of the Single Tax would be on the land owner, would remain where the single-taxer wanted it and would not be diffused throughout the community or shifted to consumers. Again, it was assumed by the single-taxer that land is a "gift of nature," and that landlords receive rent from it only because of the scarcity of the better grades of land. Such income, said George, is not the result of real effort, is unearned and undeserved. Consequently it should revert to society "through the channel of the Single Tax."

In essence, Henry George held that land values represent monopoly power, and that the entire tax burden should be laid on land, thus freeing industry from taxation and equalizing opportunities by destroying monopoly power.

Following the publication of Progress and Poverty George lectured and published magazine articles concerning his theory. In 1886 he was monimated for Mayor of New York City by various labor and liberal groups, opposing the Tammany Hall candidate, Abram Stevens Hewitt, and the Republican candidate, Theodore Roosevelt. Hewitt was victorious, although George made a strong showing. He ran again for Mayor in 1897 and died the same year.

The single-tax movement is today virtually a dead issue. Economists consider the concepts upon which it is based unsound. Adherence to the program is, at the present time, confined to a small minority of students of tax problems.

Suggested Readings

Aaron, Daniel. Men of Good Hope. New York: Oxford University Press, 1951.

Chamberlain, John. Farewell to Reform. New York: Quadrangle/New York Times Co., 1965.

Commager, Henry Steele, ed. "The Single-Tax," (Doc. No. 313) in his Documents of American History, 8th edition. New York: Appleton, 1968.

De Mille, Anna G. Henry George, Citizen of the World. Edited by Don C. Shoemaker. Westport, Conn.: Greenwood Press, 1972.

Geiger, George R. The Philosophy of Henry George. Westport, Conn.: Hyperion Press, 1975.

George, Henry. The Complete Works of Henry George. New York: AMS Press, 1906-11.

________. Our Land and Land Policy (pamphlet). 1871.

________. Progress and Poverty. New York: Robert Schalkenbach Foundation, 1954. (Originally published in 1879.)

George, Henry, Jr. The Life of Henry George. New York: Robert Schalkenbach Foundation, 1900.
Post, Louis F. The Prophet of San Francisco. New York: Gordon Press, 1970.
Rose, Edward J. Henry George. Boston: Twayne, 1968.
Young, Arthur N. The Single Tax Movement in the United States. Clifton, N.J.: Augustus M. Kelley, 1916.

THE SELDEN PATENT (1879-1912)

On May 8, 1879, George Baldwin Selden, an American patent lawyer and inventor, filed an application at the United States Patent Office for what he claimed to be "the first automobile." On November 5, 1895, he was advised by the Bureau of Patents that his application had been approved, the necessary search and other formalities complied with, and a patent, No. 549,160, was granted.

Selden's patent covered every essential feature of the modern gasoline-propelled automobile. It included the idea of using an internal combustion engine as a source of propulsion and combining the engine with a clutch in the power train. It also covered the idea of a reducing gear by which the propelling wheels revolved at a lesser speed than the engine.

Several automobile manufacturing companies took out licenses under the Selden patent. Selden never manufactured automobiles as did Alexander Winton, Elwood Haynes, Charles E. Duryea, Henry Ford, and other pioneers in the field. In 1899 he sold out his rights under the 1895 patent on a royalty basis. This sale ultimately realized him $200,000.

Protracted litigation followed Selden's sale of his patent rights. The Ford Motor Company, established in 1903, and several other automobile manufacturing companies refused to pay royalties. With each car sold Ford included a guarantee to the purchaser that any penalties assessed by a court as a result of a lawsuit involving the Selden patent would not be passed on to the purchaser.

In 1911 Ford gained the court decision that "its four-cycle engine was of a different fundamental type from the Selden engine," and the matter was permanently resolved when the patent rights expired in 1912.

Suggested Readings

Bolton, Sarah K. Lives of Poor Boys Who Became Famous. New York: Crowell, 1939.
Burlingame, Roger. Henry Ford. New York: Knopf, 1954.
________. Inventors Behind the Inventor. New York: Harcourt, Brace, 1947.
Caldwell, Cyril Cassidy. Henry Ford. New York: Messner, 1947.

Clancy, Louise B. The Believer: The Life of Mrs. Henry Ford. New York: Coward-McCann, 1960.
Epstein, Ralph C. The Automobile Industry: Its Economic and Commercial Development. New York: Arno Press, 1972.
Fanning, Leonard M. Men, Money and Automobiles: The Story of an Industry. Cleveland: World, 1969.
Flynn, John T. Men of Wealth. New York: Simon & Schuster, 1941.
Ford, Henry, and Samuel Crowther. My Life and Work. Garden City, N.Y.: Doubleday, 1922.
Gilbert, Miriam. Henry Ford, Maker of the Model T. Boston: Houghton Mifflin, 1962.
Glasscock, C. B. The Gasoline Age. Indianapolis: Bobbs-Merrill, 1937.
Hagedorn, Hermann. Americans: A Book of Lives. New York: John Day, 1946.
Janeway, Elizabeth. The Early Days of Automobiles. New York: Random House, 1956.
Marquis, Samuel S. Henry Ford: An Interpretation. Boston: Little, Brown, 1923.
Morgan, W. J. "Selden Patent War Near End," New York Globe and Commercial Advertiser, December 24, 1910.
Nevins, Allan, with the collaboration of Frank Ernest Hill. Ford: The Times, The Man, The Company. New York: Scribner's, 1954.
Neyhart, Louise Albright. Henry Ford, Engineer. Boston: Houghton Mifflin, 1950.
Simonds, William A. Henry Ford: His Life, His Work, His Genius. Indianapolis: Bobbs-Merrill, 1943.
Sullivan, Mark. Our Times, The United States, 1900-1925. New York: Scribner's, 1928-1935.
Sward, Keith. The Legend of Henry Ford. New York: Holt, Rinehart, 1948.

THE MOREY AND MURCHISON LETTERS (1880 and 1888)

The device of the forged or spurious letter, used to discredit a political opponent in an election campaign, is not unknown in American politics. In the campaign of 1880 such a letter, purporting to have been written by the Republican Presidential candidate General James A. Garfield, was circulated in the Pacific states. This letter, expressing opposition to the restriction of Chinese immigration and addressed to a "Mr. Morey," was a forgery. Complete exposure in the short time remaining before election day was impossible and while the letter injured Garfield's chances in the West, he and his Vice-Presidential candidate Chester A. Arthur were victorious over the Democrats General Winfield Scott Hancock and William English, running for the office of President and Vice President respectively.

In the election of 1888 the Republicans used a spurious letter to discredit their Democratic opponent in a manner reminiscent of the Democrats' "Morey Letter." This was a missive, supposedly from a Charles F. Murchison, a naturalized American of English birth, but actually a fictitious character, sent to Lord Lionel Sackville Sackville-West, the British minister in Washington. This letter requested the minister's opinion "as to whether President Grover Cleveland's hostile policy in a recent controversy with Canada had been adopted for campaign purposes and whether after election the President would be more friendly towards England." The British Minister indiscreetly replied to the letter, stating that it was his opinion that Cleveland "would show a conciliatory spirit toward Great Britain," and recommended that the fictitious Murchison vote the Democratic ticket as advantageous to British interests. The correspondence between Lord Sackville-West and "Murchison" was withheld until shortly before the election and was then published in newspapers and on handbills which were given a wide distribution. The Republicans declared that the Democrat Cleveland was the "British Candidate." Cleveland at first overlooked the incident but was subjected to such pressure that he eventually dismissed the minister. The British government, in turn, refused to fill the vacancy until there was a change of American administration.

In the following Presidential election Benjamin Harrison and Levi P. Morton were elected President and Vice President, defeating Cleveland and Allen G. Thurman. Cleveland, however, was reelected President in 1892, along with Adlai E. Stevenson as Vice President. In this election the Republicans Benjamin Harrison and Whitelaw Reid were defeated.

Suggested Readings

Cleveland, Grover. Presidential Problems. Plainview, N.Y.: Books for Libraries, 1975.

Hoar, George F. Autobiography of Seventy Years. New York: Somerset, 1903.

Muzzey, David Saville. James G. Blaine: A Political Idol of Other Days. New York: Dodd, Mead, 1934.

Nevins, Allan. Grover Cleveland: A Study in Courage. New York: Dodd, Mead, 1932.

Porter, Kirk H., and Donald Bruce Johnson, comps. National Party Platforms, 1840-1960. Urbana, Ill.: University of Illinois Press, 1956.

Rhodes, James F. History of the United States from the Compromise of 1850. New York: Macmillan, 1893-1906.

Roseboom, Eugene H. A History of Presidential Elections, from George Washington to Richard M. Nixon. New York: Macmillan, 1970.

Russell, Charles Edward. Blaine of Maine: His Life and Times. New York: Cosmopolitan Book Corp., 1931.

Schlesinger, Arthur M., Jr., ed. History of American Presidential Elections, 1789-1968. New York: Chelsea House, 1971.

Smith, Theodore C. The Life and Letters of James Abram Gar-

field. Hamden, Conn.: Shoe String Press, 1968.
Stanwood, Edward. James Gillespie Blaine. Edited by John T. Morse, Jr. New York: AMS Press, 1905.

THE ASSASSINATION OF PRESIDENT GARFIELD (1881)

The assassination of President James A. Garfield at the Baltimore & Potomac Railway station in Washington, D.C., on July 2, 1881 by Charles J. Guiteau, a disappointed office-seeker, led to reforms in the hiring policies of the federal government. It also led to the passage of the country's first civil service law, the Pendleton Act of January 16, 1883.

On the morning of July 2 Garfield, who had planned a vacation with his family, traveled from the White House to the railroad depot with several members of his cabinet who had come to see him off. One of the members of the party was Secretary of State James G. Blaine.

Arriving approximately ten minutes before train time, Garfield and Blaine chatted in the station waiting room. Suddenly two pistol shots rang out and Garfield collapsed, bleeding profusely and, according to witnesses, crying out, "My God, what is this?"

Blaine rushed to Garfield's aid. Police and doctors were quickly called and police officer Patrick Kearney, on guard at the station, arrested the assailant, 39-year-old Charles J. Guiteau, who made very little effort to escape.

Dr. Smith Townsend, the District of Columbia health officer, had been summoned. He found the President lying on the lobby floor, in shock. He administered first aid and ordered Garfield carried to an office upstairs and laid on a mattress. Blaine sent for Garfield's friend Dr. D. W. Bliss, who examined the President and determined that one bullet had only grazed his arm but the other had entered his back near the spine. Garfield was in pain but fully conscious.

Under the circumstances the projected vacation was, of course, cancelled. Lucretia, Garfield's wife, then in New Jersey recovering from an attack of malaria, was asked by telegraph to return to Washington at once. At 10:45 a police ambulance took the President home to the White House where he was put to bed.

Guiteau, meanwhile, had been incarcerated in the Washington jail. Mobs formed and it was necessary to send federal troops to prevent a disturbance. Mrs. Garfield arrived at the White House that evening and was appalled when physicians told her her husband had only a few hours to live. Garfield, however, lingered on for several weeks. He had rallied somewhat and expressed a wish to go to Elberon, New Jersey, where he felt the sea air would be beneficial to him.

In August he and his family were moved to the new location but he passed away on September 19. An autopsy showed that the immediate cause of his death was not Guiteau's bullet but a huge blood clot which had formed within the bullet track and eventually broke into the abdominal cavity.

Guiteau's trial was opened on November 14, 1881, Judge Walter S. Cox presiding. An investigation of his background disclosed that he was a religious fanatic, inexpert lawyer, debt-dodger, syphilitic, lecturer, writer on religious subjects, and unsuccessful political office seeker. He had arrived in Washington on March 5, 1881, where he managed to see Garfield and asked to be appointed ambassador to Austria or else consul at Paris. Garfield courteously refused his request, and after he had pestered former President Ulysses S. Grant, Secretary of State James G. Blaine and Vice President Chester A. Arthur, he was barred from the White House.

In May Guiteau wrote a threatening letter to Garfield. This letter was not answered. On June 8 he purchased the British .44 caliber "Bull Dog" pistol with which he shot Garfield at the railroad station. When he was arrested a letter, written in anticipation of the assassination, was found on his person. This letter read, in part, "I presume the President was a Christian, and that he will be happier in Paradise than here.... His death was a political necessity.... I am going to jail."

The trial was a spectacle. Guiteau was defended by his brother-in-law George Scoville, who attempted to plead insanity for his client, but to no avail. While in prison Guiteau started to write his memoirs.

On January 25, 1882, the jury, after deliberating for less than an hour, found the accused guilty. He was sentenced to hang and the sentence was carried out on June 30, 1882. As the trap was sprung the condemned man was reciting a poem which he had written in his cell that morning.

Suggested Readings

Bryce, James. The American Commonwealth. New York: Macmillan, 1914.

Commager, Henry Steele, ed. "Pendleton Act," (Doc. No. 308) in his Documents of American History, 8th edition. New York: Appleton, 1968.

Fish, Carl R. The Civil Service and the Patronage. New York: Russell & Russell, 1963.

Hoyt, Edwin P. James A. Garfield. Chicago: Reilly & Lee, 1964.

Muzzey, David Saville. James G. Blaine: A Political Idol of Other Days. New York: Dodd, Mead, 1934.

Potter, John Mason. Plots Against the Presidents. New York: Astor, 1968.

Robertson, Archie. "Murder Most Foul," American Heritage Magazine, August, 1964.

Severn, Bill. Teacher, Soldier, President: The Life of James A. Garfield. New York: I. Washburn, 1964.

Smith, Theodore C. The Life and Letters of James Abram Garfield. Hamden, Conn.: Shoe String Press, 1968.

Symons, Julian. A Pictorial History of Crime. New York:

Crown Publishers, 1966.
Thomas, Harrison C. The Return of the Democratic Party to Power in 1884. New York: AMS Press, 1919.

THE SHOOTOUT AT THE O.K. CORRAL (1881)

The city of Tombstone, in southeast Arizona, mushroomed shortly after a rich silver strike was made there in 1878. Prior to this the "stealing of horses and cattle was fairly commonplace," and gangs of rustlers, such as the one headed by Newman "Old Man" Clanton, were much in evidence. As this thievery was profitable, various officials such as John Behan, sheriff of Cochise County, were, in all probability, engaged in it.

Old Man Clanton's gang included his three sons: Joseph "Ike," Phineas "Finn" and William "Billy" Clanton, and the outlaw William Brocius Graham, better known as Curly Bill. Others of the Clanton gang were John Ringgold, known as Johnny Ringo, and Tom and Frank McLowry, also spelled McLoury.

Sheriff Behan was appointed in 1881. A rival candidate for the position was Wyatt Berry Stapp Earp, a professional gambler, card sharp, bigamist, church deacon, buffalo hunter, confidence man, extrovert, and peace officer. Wyatt Earp, his brothers Newton, Morgan, Warren, Virgil, and James, as well as his friend John Henry "Doc" Holliday, arrived in Tombstone in 1879. A series of robberies followed, in which Earp and his followers were suspected of being implicated. Virgil Earp had been appointed city marshal and so was in a position to deputize his brothers.

Bad feeling grew between the Behan, Earp, and Clanton factions, and the Clanton gang found itself being blamed by the others for various illegal happenings. While reliable witnesses later told widely divergent stories concerning the origin of the feud, it is generally accepted that the Clantons knew that the Earps were behind much of the crime in Tombstone and for that reason the Earps felt that the Clantons should be eliminated.

The famous "Shootout at the O.K. Corral" occurred on the afternoon of October 26, 1881. Billy and Ike Clanton and the McLowry Brothers had come to town for supplies. Wyatt, Morgan and Virgil Earp, together with Doc Holliday, were in town. When they saw the Clantons and the McLowrys, who had gone to the corral to get their horses, they opened fire on them. Accounts differ as to whether or not the Earps first warned the others that they were under arrest.

In the shooting, the McLowrys and Billy Clanton were killed. Ike Clanton was wounded in the shoulder. Virgil and Morgan Earp were seriously injured and Doc Holliday was grazed by a bullet.

As a result of the shooting, Virgil Earp was relieved from his office of city marshal. The three dead men were buried following a highly publicized funeral. Feeling ran high against the Earps and on March 18, 1882, Morgan Earp was assassinated.

After further trouble, including the shooting of Frank Stilwell by Wyatt Earp, the latter and his gang left Tombstone.

Suggested Readings

Bartholomew, Ed Ellsworth. Wyatt Earp, 1879 to 1882, The Man and the Myth. Toyahvale, Texas: Frontier Book Co., 1964.

Boyer, Glenn D. The Suppressed Murder of Wyatt Earp. San Antonio: Naylor, 1967.

Breakenridge, William. Helldorado. Boston: Little, Brown, 1928.

Burns, Walter Noble. Tombstone: An Iliad of the Southwest. Garden City, N. Y.: Doubleday, 1927.

Chilton, Charles. The Book of the West. Indianapolis: Bobbs-Merrill, 1962.

Clum, John P. It All Happened in Tombstone. Edited by John Gilchriese. Flagstaff, Ariz.: Northland Press, 1965.

Coolidge, Dane. Fighting Men of the West. Plainview, N. Y.: Books for Libraries, 1968.

Duffen, William A. "Notes on the Earp-Clanton Feud," Arizoniana, Fall, 1960.

Faulk, Odie B. Tombstone: Myth and Reality. New York: Oxford University Press, 1972.

Hall-Quest, Olga W. Wyatt Earp, Marshal of the Old West. New York: Ariel Books, 1956.

Hamill, Lloyd, and Rose Hamill. Tombstone Picture Gallery. Glendale, Calif.: Western Americana Press, 1960.

Harris, William Foster. The Look of the Old West. New York: Viking Press, 1955.

Holbrook, Stewart H. Wyatt Earp, U. S. Marshal. New York: Random House, 1956.

Lake, Stuart N. The Life and Times of Wyatt Earp. Boston: Houghton Mifflin, 1956.

Lyon, Peter. "The Wild, Wild West," American Heritage Magazine, August, 1960.

Martin, Douglas D. The Earps of Tombstone. Tombstone, Ariz.: Tombstone Epitaph, 1959.

Miller, Nyle H., and Joseph W. Snell. Great Gunfighters of the Kansas Cowtowns 1867-1886. Lincoln, Neb.: University of Nebraska Press, 1963.

Myers, John M. Doc Holliday. Lincoln, Neb.: University of Nebraska Press, 1973.

Olsson, John Olof. Welcome to Tombstone. London: Elek Books, 1956.

Raine, William MacLeod. Famous Sheriffs and Western Outlaws. New York: Popular Library, 1929.

Rosa, J. G. Gunfighter: Man or Myth? Norman, Okla.: University of Oklahoma Press, 1969.

Tilghman, Zoe A. Spotlight: Bat Masterson and Wyatt Earp as U. S. Marshals. San Antonio: Naylor, 1960.

Walters, Lorenzo D. Tombstone's Yesterdays. Tucson: Acme Printing Co., 1928.

Ward, Don. Cowboys and Cattle Country. New York: American

Heritage Press, 1961.
Waters, Frank. The Earp Brothers of Tombstone. New York: Clarkson N. Potter, 1960.

THE DEATHS OF BILLY THE KID AND JESSE JAMES (1881 and 1882)

Two notorious American outlaws of the last century were William Bonney, better known as "Billy the Kid," and Jesse Woodson James, also called "The Robin Hood of Missouri" and "Mr. Thomas Howard." Both men died violent deaths while comparatively young.

Billy the Kid was born in New York City in 1859. Little is known of his early life. He came west, possibly to escape the consequences of some crime he had committed. In Silver City, New Mexico, as a juvenile delinquent, he was arrested for petty thievery. At Fort Grant, Arizona Territory, he killed his first victim, a blacksmith named F. P. "Windy" Cahill, following an argument.

A fugitive from justice, he fled to Mesilla, New Mexico, where he worked as a cattle rustler and horse thief. In 1877 he became involved in the so-called Lincoln County war, a struggle between the forces of cattle barons John Tunstall and Alexander McSween and their competitors, the Murphy-Dolan-Riley ranching and trading combine. The Kid joined the Tunstall-McSween group, legend having it that he was devoted to Tunstall. When the latter was murdered by a Murphy-Dolan-Riley posse, precipitating the war between the two factions, the Kid supposedly resolved to avenge his friend's death.

In the struggle that followed Tunstall's death, killing followed killing. Billy the Kid is said to have done away with 21 men, one for each year of his life. Federal troops were brought in from Fort Stanton. Alexander McSween was shot to death in July, 1878, which virtually ended the war. The Kid then turned to stealing horses and cattle. He was under indictment for the murder of Sheriff William Brady. Following secret meetings with Governor Lew Wallace of New Mexico he surrendered to the authorities and was placed in the Lincoln jail, having agreed to turn state's evidence against other killers. He broke arrest, was caught, tried and convicted of Brady's murder, and sentenced to hang. He obtained a pistol, shot two guards, and escaped from jail again, hiding out in the Fort Sumner area.

Patrick F. "Pat" Garrett, sheriff of Lincoln County, with deputies John W. Poe and Kip McKinney, came to Fort Sumner looking for the Kid. On the night of July 14, 1881, they went to the home of Pete Maxwell to seek information. Garrett entered the house to talk to Maxwell, who had gone to bed. While Garrett was there the Kid came in. He saw a figure sitting near the head of Maxwell's bed. The light was dim. He asked, "Who is there?"

Sheriff Garrett, recognizing the voice, shot twice, the first bullet killing the Kid, the second missing.

Jesse Woodson James, in the course of his career as a "Missouri ruffian, murderer, bank robber, train robber and American demigod ... and leader of a gang of thugs, in fifteen years held up eleven banks, three stages, one county fair and one payroll messenger, in the process looting some $200,000 and killing at least sixteen men."

During the Civil War Jesse, together with his elder brother Frank, served with William C. Quantrill, leader of a troop of Confederate irregulars. When the war was over the James brothers, Cole and Jim Younger, and possibly Jim Reed united in a gang devoted to robbing and killing. By the year 1874 the James gang was notorious throughout the Middle West. The gubernatorial campaign of that year was concerned with suppressing outlawry so that "capital and immigration could once again enter our state" [of Missouri]. The gang, however, continued its activities unchecked.

Eventually the Younger Brothers were arrested in 1876, following an unsuccessful bank robbery attempt at Northfield, Minnesota. They were replaced in the James gang by the two Ford Brothers, Charles and Bob.

The robbing of railroad trains by the James gang led to the offering of a $5,000 reward for the capture of either Jesse or Frank. This reward was offered at the suggestion of Allan Pinkerton, the detective whose agency had been hired to protect railway trains and mail coaches from the many outlaw gangs engaged in plundering them. Bob Ford resolved to collect the reward. Jesse James was then hiding out in St. Joseph, Missouri, under the name of Thomas Howard. On April 3, 1882, shortly before the gang was scheduled to set off on a bank-robbing expedition, Jesse noticed that a picture of his horse Skyrocket hanging on the wall was slightly askew. He stepped up on a chair to straighten it and Bob Ford shot him in the back of the head, killing him instantly.

Ford met a violent death in 1891 when he was shot and killed by Edward O. "Red" Kelly, a western outlaw.

Frank James surrendered to the authorities in 1883. At a farcical trial he was acquitted and spent the rest of his life quietly and respectably on his father-in-law's Missouri farm. He died in 1915.

A rumor sprang up that Jesse James had not been killed in 1882 but had escaped, his "murder being staged." As late as 1941 an old man, claiming to be the Missouri outlaw, appeared on the scene. His story was that he had adopted the alias "Frank Dalton" and had lived under it for more than sixty years.

A grave, said to be the final resting place of Jesse James, is to be found near the site of the Kearney Baptist Church, Missouri. The inscription on the tombstone reads:

In Loving Remembrance of My Beloved Son
JESSE JAMES
Died April 3, 1882
Aged 34 Years, 6 Months, 28 Days
Murdered by a Traitor and Coward Whose
Name is Not Worthy to Appear Here

Suggested Readings

Adams, Ramón F. A Fitting Death for Billy the Kid. Norman, Okla.: University of Oklahoma Press, 1960.

Adler, Alfred. "Billy the Kid: A Case Study in Epic Origins," Western Folklore, April, 1951.

Avery, Delos. "The Life and Death of Billy the Kid," Chicago Sunday Tribune, August 27, 1944.

Billy the Kid. Character ballet (1 act). Music: Aaron Copland; choreography, Eugene Loring; Book, Lincoln Kirstein; scenery and costumes, Jared French. In Krakova, Rosalyn. New Borzoi Book of Ballet. New York: Knopf, 1956.

Billy the Kid: Las Vegas Newspaper Accounts of His Career, 1880-1881. Waco, Texas: W. M. Morrison, 1958.

Breihan, Carl W. The Complete and Authentic Life of Jesse James. New York: Fell, 1953.

_______. The Day Jesse James Was Killed. New York: Fell, 1961.

Brent, William. The Complete and Factual Life of Billy the Kid. New York: Fell, 1964.

Burns, Walter Noble. The Saga of Billy the Kid. Garden City, N.Y.: Doubleday, 1926.

Chapman, Arthur. "Billy the Kid--A Man All 'Bad'," Outing Magazine, April, 1905.

Corle, Edwin. Billy the Kid. New York: Sloane, 1953.

Croy, Homer. Jesse James Was My Neighbor. New York: Duell, Sloan, 1949.

Cunningham, Eugene. "The Kid Still Rides," New Mexico Magazine, March, 1935.

Dacus, Joseph A. Illustrated Lives and Adventures of Frank and Jesse James, and the Younger Brothers, the Noted Western Outlaws. St. Louis: N. D. Thompson, 1882.

Donald, Jay. Outlaws of the Border. Chicago: Coburn, 1882.

Doughty, Francis W. Old King Brady and Billy the Kid (fiction). New York: Frank Tousey, 1890.

Dykes, J. C. Billy the Kid: The Bibliography of a Legend. Albuquerque: University of New Mexico Press, 1952.

Fable, Edmund, Jr. Billy the Kid, the New Mexican Outlaw; or, The Bold Bandit of the West! (fiction). Denver: Denver Publishing Co., 1881.

Ferguson, Harvey. "Billy the Kid," American Mercury, May, 1925.

Fishwick, Marshall. "Billy the Kid: Faust in America," Saturday Review, October 11, 1952.

Garrett, Pat F. The Authentic Life of Billy the Kid. Norman, Okla.: University of Oklahoma Press, 1954. (Originally written in 1882.)

Hamlin, William Lee. The True Story of Billy the Kid. Caldwell, Idaho: Caxton Printers, 1959.

Henry, Will, pseud. Death of a Legend. New York: Random House, 1954.

Horan, James D. "The Rise and Fall of Jesse James," in his

Desperate Men: Revelations from the Sealed Pinkerton Files. Garden City, N.Y.: Doubleday, 1962.
Hough, Emerson. "Billy the Kid: The True Story of a Western 'Bad Man'," Everybody's Magazine, September, 1901.
Hunt, Frazier. The Tragic Days of Billy the Kid. New York: Hastings House, 1956.
Huntington, George. Robber and Hero: The Story of the Raid on the First National Bank of Northfield, Minn. ... in 1876. Minneapolis: Ross & Haines, 1895.
James, Jesse E. Jesse James, My Father: The First Only True Story of His Adventures Ever Written. New York: Fell, 1957.
Jenardo, Don, pseud. The True Life of Billy the Kid. New York: Frank Tousey, 1881.
Keleher, William A. Violence in Lincoln County. Albuquerque: University of New Mexico Press, 1957.
Love, Robertus. The Rise and Fall of Jesse James. New York: Putnam, 1926.
Lyon, Peter. "The Wild, Wild West," American Heritage Magazine, August, 1960.
Otero, Miguel. The Real Billy the Kid. New York: R. R. Wilson, 1936.
Poe, John W. The Death of Billy the Kid. Boston: Houghton Mifflin, 1933.
Raine, William MacLeod. "Billy-the-Kid," Pacific Monthly, July, 1908.
Settle, William A. Jesse James Was His Name. Columbia, Mo.: University of Missouri Press, 1966.
Siringo, Charles A. Riata and Spurs. Boston: Houghton Mifflin, 1927.
Sonnichsen, C. L., and William V. Morrison. Alias Billy the Kid. "I Want to Die a Free Man...." Albuquerque: University of New Mexico Press, 1955.
Steckmesser, Kent Ladd. "The Outlaw: Billy the Kid," in his The Western Hero in Story and Legend. Norman, Okla.: University of Oklahoma Press, 1965.
Triplett, Frank. The Life, Times, and Treacherous Death of Jesse James. Edited by Joseph Snell. Chicago: Swallow Press, 1970.

THE HILL-SHARON LAWSUIT (1881-1887)

Sarah Althea Hill, better known as Althea Hill, was an American adventuress. In 1880, when she was in her late twenties, she met an extremely wealthy widower, Senator William Sharon, who was then sixty. Sharon offered to give her financial advice and she visited him in his San Francisco office several times. Here he allegedly attempted to make love to her and, when repulsed, proposed marriage.

Sharon's proposal was accepted. He told her that for personal reasons he wished the marriage kept secret. Under

California law, he said, they could marry simply by agreeing in writing to do so. He dictated a declaration of marriage which she wrote out and which they both signed.

Althea moved into the Grand Hotel in San Francisco, which was connected by a bridge to the Palace Hotel, Sharon's residence. For the next year the marriage, though entered into in a most unusual manner, progressed satisfactorily. Then Sharon accused Althea of revealing his business secrets to others and demanded that she sign a paper stating that she was not Mrs. Sharon. This she refused to do, even for a consideration of $500 which was to be paid her each month. Eventually she accepted a lump payment of $7,500 plus some personal notes, for which she signed a release of claims. Sharon then had the manager of the Grand Hotel notify her to vacate her room. When she persisted in remaining, Sharon had the outer door removed from its hinges and the carpet taken up.

The girl then sought the assistance of Mammy Pleasant, a Negro entrepreneur, businesswoman, voodoo practitioner and opportunist. Althea told Mammy Pleasant of her "marriage" to Sharon and, after a short reconciliation with the senator, consulted George Tyler, Mammy's lawyer. Tyler called in, as assistant counsel, several prominent lawyers, including Judge David S. Terry, a "tall, strong, fighting man from Texas." This was the same Terry who, in 1859, challenged Senator David Broderick, a San Francisco politician, to a duel in which Broderick was killed.

In her lawsuit against Sharon, Althea claimed marriage and sought damages. Sharon, in turn, claimed blackmail and that the marriage document submitted as evidence by the plaintiff was a forgery. Althea then brought a second suit, seeking a divorce and a property settlement. For six years these two suits went from court to court.

Judge Jeremiah Sullivan of the Superior Court of San Francisco, before whom the second case was tried without a jury, granted Althea Hill a divorce and alimony of $2,500 per month. Sharon promptly appealed and petitioned in the federal court to have the declaration of marriage cancelled as a forgery.

An official examiner was appointed to take testimony from witnesses and report to the court. This occupied the next six months, during which time Althea "sneered at the lawyers, the witnesses, and the examiner, and took to carrying a pistol in the courtroom."

Justice Stephen Johnson Field, a United States Supreme Court justice, was on circuit in California and was assigned to assist Judge Lorenzo Sawyer. Althea's case first came before him in 1885. When she displayed her pistol in court Field gave orders that she would, in the future, refrain from carrying small arms in the examiner's presence. This case resulted in the court's judgment that the declaration of marriage was a forgery, a finding contrary to that of Judge Sullivan. Sharon had died in 1885, before this judgment had been given, and on January 7, 1886, Althea Hill and David Terry were married, Terry believing that Sullivan's order had died with Sharon and not bothering to appeal it.

Sharon's children, through their attorneys, persuaded the California Supreme Court to reduce Althea's alimony from $2,500 to $500 a month. They then filed a petition in federal court to revive a federal order on the girl to surrender her marriage contract, an order which she had, until that time, ignored. Justice Field heard the petition and he, together with Judges Lorenzo Sawyer and George Sabin, revived the order against Althea. While Field was reading the decision the girl provoked a scene in the courtroom, and she and her husband Terry were promptly expelled. Terry was carrying a bowie knife and Althea had a pistol in her handbag. Both were found guilty of contempt of court and sentenced to prison for short terms.

Subsequent appeals to the United States Supreme Court and to the California Supreme Court failed. Eventually the state court agreed with the federal court that Judge Sullivan's finding in Althea's favor had not been supported by the evidence. Terry and his wife made threats against the life of Field and the latter was advised to acquire a bodyguard. The United States marshal appointed David Neagle to protect Field.

On August 14, 1889, Field and Neagle were returning by train from Los Angeles where Field had been trying a case. Around midnight Terry and Althea boarded the train at Fresno, where they had been living. Neagle, who saw them come aboard, informed Field, who was unconcerned. Neagle remained on guard for the rest of the night. In the morning the train stopped at Lathrop, near Stockton, where the passengers were to breakfast at the station restaurant. Field and Neagle entered the eating house and were shortly followed by Terry and Althea. Seeing Field, Althea turned and dashed back to the train. Terry slapped Field viciously in the face. Neagle sprang to his feet and, as Terry reached--or searched--for his bowie knife, fired twice, killing Terry instantly. Althea rushed into the room with an open handbag. This, when examined, was found to contain a loaded pistol. At Stockton she swore out a warrant for the arrest of Field and Neagle for the murder of her husband.

The governor of California ordered Field to be freed from arrest. Neagle's case was ultimately ruled on by the United States Supreme Court which held that Neagle, in protecting Field, "had been acting in pursuance of a law of the United States," and thus was not guilty of murder.

In 1892 Mammy Pleasant had Sarah Althea Hill committed to an asylum for the insane and she died there 45 years later.

Suggested Readings

Articles in Daily Examiner (San Francisco) beginning September 4, 1884 and August 15, 1889.

Bancroft, Hubert Howe. The History of California. San Rafael, Calif.: Bancroft Press, 1888.

________. The History of Nevada, Colorado and Wyoming. San Rafael, Calif.: Bancroft Press, 1888.

Emrich, Duncan, ed. Comstock Bonanza. New York: Vanguard

Press, 1950.
Field, Stephen J. Personal Reminiscences of Early Days in California. Privately printed, 1893.
Hagan, H. H. "A California Saga," Commercial Law League Journal, November, 1929.
Lewis, Oscar. The Silver Kings. New York: Knopf, 1947.
Lyman, George D. Ralston's Ring. New York: Scribner's, 1937.
_______. The Saga of the Comstock Lode. New York: Scribner's, 1934.
Maccracken, Brooks W. "Althea and the Judges," American Heritage Magazine, June, 1967.
O'Meara, James. Stephen J. Field Arrested for Conspiracy and Murder of the Hon. David S. Terry (pamphlet), 1889.
Potts, C. S. "David S. Terry," Southwest Review, April, 1934.
Swisher, Carl B. Stephen J. Field, Craftsman of the Law. Chicago: University of Chicago Press, 1969.

THE "BLUEBEARD" HOCH MURDERS (1881-1905)

Johann Hoch, a German-American amatory adventurer, married and then either deserted or poisoned between 43 and 50 women prior to his trial, conviction and death by hanging. In all cases he married in order to obtain his various wives' money and property for himself.

It is not certain that this Bluebeard's name was Johann Hoch. He used at least twenty aliases in the course of his infamous career, which began in 1881 and lasted until 1905.

Hoch, as we shall call him, had left his native Germany for America in 1888, having married and buried one wife, married and deserted a second, and then, on the boat, courted and married a third woman who died two months later.

Arriving in Chicago, Hoch proceeded to court, marry and either desert or poison a series of wives, mostly recently widowed German-speaking elderly women of some property. His 24-year career included marriages in such widely separated cities as New York and San Francisco.

In 1895 Hoch appeared in Wheeling, West Virginia where, under the name of Jacob Huff, he opened a saloon and courted and won the hand of Mrs. Caroline Hoch, a widow with nine hundred dollars in savings, a house, and a $2,500 insurance policy on her life. Mrs. Hoch, before her remarriage, discussed the matter with her pastor, the Reverend Herman C. A. Haass of the St. Matthew's German Lutheran Church. Reverend Haass took a dim view of the contemplated nuptials, as he had been consulted by several other property-owning widows of his parish and had learned that Huff had proposed marriage to all of them. Nevertheless he reluctantly performed the ceremony on April 18, 1895. On June 14 the bride became ill. Haass was summoned and saw the husband administer a white powder to his wife. On June 15 she died.

Reverend Haass suspected Huff of poisoning his recent bride but could not prove his suspicions. Early in July, 1895, a pile of clothing, a watch containing a picture of Huff, and a suicide note were found on the bank of the Ohio River, near Wheeling. It appeared that Huff--or Hoch--had committed suicide. However, Reverend Haass was advised that Jacob Huff, or Johann Hoch, as he was now calling himself, had been seen in Zanesville, Ohio, after the evidence of apparent suicide was discovered.

Feeling that his original suspicions were now confirmed, Haass attempted to locate Huff-Hoch. An avid newspaper reader, he scanned the stories in the several English and German language papers to which he subscribed, and occasionally came across a notice of the death or abandonment of a recently re-married widow whose husband had disappeared, taking her cash assets with him.

In 1898 Reverend Haass read a story in a Chicago newspaper concerning one Martin Dotz who had been arrested for bigamy and swindling. He wrote to Captain Luke Colleran of the Chicago police, stating that he suspected "Dotz" of being the man he felt had murdered Mrs. Caroline Hoch. With his letter he enclosed a copy of the picture which Johann Hoch had left in the watch on the bank of the Ohio River.

Chicago Police Inspector George Shippy confronted "Martin Dotz" with the picture and "Dotz" acknowledged that it was indeed a picture of him. Further investigations by Shippy revealed that "Dotz" had been married to several women, all whom had died suddenly shortly after their marriages.

On November 14, 1898, Mrs. Caroline Hoch's body was exhumed. It was found that her vital organs had been removed, thus making it impossible to determine whether or not she had died by poison. On June 30, 1900, Hoch was released from prison in Chicago where he had served a term for bigamy, and was then taken to Wheeling. The authorities there, however, could not prove murder by poison and Hoch was freed for lack of evidence. Upon being freed he went to Argos, Indiana, where he married again. His new wife and her daughter traveled with him to Chicago, where the two women disappeared.

Between 1898 and 1905 Hoch married, robbed and murdered or deserted several other women. The body of Marie Welker Hoch, one of his late wives, was exhumed on January 22, 1905, and found to be heavily impregnated with arsenic. That murder had been committed was obvious and the case became an overnight sensation. Hoch's picture appeared in newspapers throughout the country and friends and relatives of his many wives, as well as ex-wives who had escaped death, told their stories to reporters and to the police.

Hoch had left Chicago. He appeared in New York, calling himself Henry Bartels. Renting a room in a boarding house operated by Mrs. Catherine Kimmerle, a German widow, he shortly proposed marriage to her. Mrs. Kimmerle saw Hoch's picture in the *New York American*, recognized him as "Henry Bartels," and notified the police.

That night Hoch was arrested at his boarding house. The police officers found a hollow fountain pen in his possession. This pen contained fifty-eight grains of arsenic. He was extradited to

Chicago where he was tried for the murder of Mrs. Marie Welker Hoch. He was found guilty and sentenced to be hanged. The sentence was carried out at the Chicago county jail on February 23, 1906.

Suggested Readings

Abrahamsen, David, M.D. The Murdering Mind. New York: Harper & Row, 1973.

Bromberg, Walter. Mold of Murder: A Psychiatric Study of Homicide. New York: Grune & Stratton, 1961.

Catton, Joseph. Behind the Scenes of Murder. New York: Norton, 1940.

Gribble, Leonard. They Had a Way With Women. New York: Roy Publishers, 1967.

Guttmacher, M. S. The Mind of the Murderer. Freeport, N.Y.: Books for Libraries, 1960.

Jesse, F. Tennyson. Murder and Its Motives. London: Harrap, 1952.

Reinhardt, James Melvin. The Psychology of Strange Killers. Springfield, Ill.: Thomas, 1962.

Schutzer, A. I. "The Lady Killer," American Heritage Magazine, October, 1964.

Wilson, Colin, and Patricia Pitman. Encyclopedia of Murder. New York: Putnam, 1962.

Wolfgang, Marvin E., comp. Studies in Homicide. New York: Harper & Row, 1967.

THE HILLMON TRIALS (1882-1903)

The Hillmon trials involved life insurance claims made by Sallie Quinn Hillmon, alleged widow of John W. Hillmon, a cattle herder who may or may not have been shot to death on Crooked Creek near Medicine Lodge, Kansas, on March 17, 1879.

Hillmon had married Sallie in that year. Levi Baldwin, Sallie's cousin and Hillmon's friend, a cattleman of Lawrence, Kansas, indicated that he would finance a cattle ranch for the newlyweds if Hillmon could locate a suitable one somewhere in the southwest. Prior to setting out on a search for a ranch Hillmon, at Baldwin's request, insured his life with two insurance companies for a total of $10,000. Shortly thereafter he took out, from another company, a third policy, this one for $5,000, and also had himself vaccinated against smallpox. He and a man named John H. Brown left in search of a ranch and on March 17 Hillmon was accidentally shot and killed by Brown, or at least it was so claimed.

Sallie Hillmon and Levi Baldwin claimed that the body was Hillmon's. The insurance companies, however, declined to pay on the policies, claiming that the body of someone else had been substituted in an attempt to bilk them. Following an inquest Brown had had the body buried in the graveyard at Medicine Lodge. At the insistence of representatives of the insurance companies it was exhumed and the representatives denied that it was that of Hillmon. In all, three inquests were held and the body was examined, measured and photographed. Brown, Sallie Hillmon and several others identified the body as Hillmon's. The insurance companies, however, persistently maintained that Hillmon was still alive and the body was that of another. At the third inquest the coroner's jury found that the body was that of an unknown man feloniously shot by Brown.

State Senator W. J. Buchan entered the case as attorney for Brown. Brown decided to turn state's evidence and Buchan drafted a "confession" which Brown signed. This document stated, in part, that Brown and Hillmon had enlisted a stranger whom they called "Joe" to travel with them and work on the ranch once it had been located and purchased. Brown stated further that Hillmon had shot "Joe" and then, after changing clothes with him, "had started north with Joe's valise." Brown later repudiated his "confession."

Further investigation by the insurance companies revealed that a man named Frederick Adolph Walters, a resident of Fort Madison, Iowa, who had gone to Crooked Creek to seek his fortune, had disappeared. Accustomed to write to his sister and his sweetheart regularly, his letters had suddenly ceased. It was claimed by the insurance companies that the body said to be that of Hillmon was actually that of Walters.

In due course the matter came to court and before it was finally resolved six separate trials were held and on two occasions appeals were made to the United States Supreme Court. The first two trials, held in 1882 and 1885, resulted in hung juries. The third trial (1888) resulted in a finding for Sallie Quinn Hillmon. This, however, was reversed when appealed to the Supreme Court in 1892. The fourth and fifth trials, held between 1892 and 1897, also resulted in hung juries. Trial No. 6 was a victory for Mrs. Hillmon and, in 1903, was appealed once more to the Supreme Court. Again the findings of the lower court were reversed.

In 1899 one of the insurance companies, Mutual Life Insurance Company of New York, gave up the battle and paid off on its policy. The New York Life Insurance Company had already given in and settled with the widow, and at last the Connecticut Mutual Life Insurance Company submitted and paid. In all, after 21 years of litigation John W. Hillmon's "widow" received a total of $35,700 from the three companies, including accumulated interest.

The truth of the Hillmon cases is as debatable today as it was almost a century ago. Frederick Adolph Walters was never seen again and many theories and opinions have been put forward in connection with "The Case of the Anonymous Corpse."

Suggested Readings

Cahn, William. _Matter of Life and Death: The Connecticut Mutual Story_. New York: Random House, 1970.

"Five Trials of an Insurance Case," New York Times, April 5, 1896, p. 2.
"The Hillmon Insurance Case," New York Times, January 9, 1895, p. 11.
"Hillmon Tells His Story," New York Times, January 2, 1894, p. 1.
Maccracken, Brooks W. "The Case of the Anonymous Corpse," American Heritage Magazine, June, 1968.
McNeal, T. A. When Kansas Was Young. New York: Macmillan, 1922.
Roenigk, Adolph, ed. Pioneer History of Kansas. Lincoln, Kans.: A. Roenigk, 1933.
"Says He Can Produce Hillmon," New York Times, September 15, 1895, p. 5.
Stone, Mildred F. Since 1845: A History of the Mutual Benefit Life Insurance Company. New Brunswick, N.J.: Rutgers University Press, 1957.
Wigmore, John Henry. The Principles of Judicial Proof. Boston: Little, Brown, 1931.

THE PERALTA LAND GRANT SWINDLE (1883-1895)

James Addison Reavis, one of the most audacious swindlers in America's history, chose as his victim the federal government of the United States. He laid claim to a tract of land approximately as large as the combined states of New Hampshire and New Jersey.

In 1871 he met Dr. George M. Willing, a petty swindler, whose specialty was the making of spurious Spanish land-grant claims. Willing operated in the American Southwest, where such claims were often recognized in American courts and settled on a quit-claim basis. Genuine claims were frequent enough to make it a waste of time to dispute well-documented false ones, even though the documents might later turn out to be forged.

Willing had mapped out a large land grant in the Arizona Territory. Like many other grants made by the King of Spain, the boundaries of this one were exceedingly vague. Although Willing planned only a quit-claim settlement, Reavis became interested in acquiring the land for himself by establishing a valid claim which would stand up in court.

Willing died in Prescott, the territorial capital, where he had gone to file his claim on the property. Reavis, through a subterfuge, obtained from Willing's effects a deed to the grant, made over to a third party whose name had been left blank. He then proceeded to acquaint himself with the ramifications of international law, the Spanish language as it was written and spoken in the 18th century, the manner in which Spanish and Mexican records were filed, recorded and preserved, and as much about land grants and all that concerned them as any man then living. All this was in preparation for his gigantic land claim which, he realized, would be bitterly contested by government lawyers and officials. His efforts in this direction came to be known as the "Peralta Land Grant Swindle."

Reavis produced a long line of noble Spanish ancestors whose family name was Peralta. Documents were found in Mexico and Spain corroborating the claim that King Ferdinand VI of Spain had, in return for certain valuable services, granted Don Miguel de Peralta de la Cordoba a huge section of land in what is present-day Arizona and New Mexico. Reavis placed his own name on the deed he had obtained from Willing's effects and, to explain why Willing had transferred the property to him, obtained a bill of sale from Willing's widow.

In 1883 he filed his claim in the office of the Surveyor-General of Arizona Territory, using the name he had assumed, James Addison Peralta-Reavis. The previous year he had adopted an orphan half-breed girl named Sophia Treadway. He showed, by producing appropriate documents, that she was actually the granddaughter of Don Miguel de Peralta de la Cordoba.

Money began to pour in. Reavis collected from the railroads which had built across "his land," from farmers who feared they might lose their prosperous farms and from mining companies in the area. He sold franchises to these various operators and with the proceeds embarked on a life of luxury and high living.

His adopted daughter, now known as Doña Sophia Loreto Micaela Maso y Peralta de la Cordoba, he claimed as having become his wife. Evidence of this was a contract of marriage dated 1882 but actually prepared in 1887. Thus Reavis strengthened his claim, adding marriage to the woman who was rightfully entitled to inherit the land to his purchase from Dr. Willing's widow, as evidenced by the bill of sale she had signed.

Until 1890 Reavis prospered. In that year Royal A. Johnson, Surveyor-General of the Arizona Territory, found certain discrepancies in Reavis' claim and pronounced the claim a fraudulent one. Reavis replied with a $10,000,000 lawsuit, claiming that the federal government had wrongfully given away to others property belonging to his wife.

The United States Government sent special agents to Mexico and Spain to examine the documents of which Reavis had produced certified copies. Without exception the documents were found in the archives where they were said to be. However, on close examination by experts they turned out to be forgeries, planted by Reavis with the assistance of bribed custodians and government officials.

It was found that Don Miguel de Peralta de la Cordoba had never existed, nor had the assortment of ancestors and descendants, copies of whose portraits Reavis had produced. Doña Sofia Loreto Micaela Maso y Peralta de la Cordoba was found to be the posthumous daughter of John A. Treadway, a squaw man.

Reavis' case was lost. In 1896 he was tried for conspiracy, convicted, and sentenced to six years in prison, of which he served almost two years. Eventually his wife divorced him for non-support.

Suggested Readings

Bancroft, Hubert Howe. The History of Arizona and New Mexico. San Francisco: Bancroft, 1889.
Bechdolt, F. R. When the West Was Young. New York: Century, 1922.
Brief of Argument in the Peralta Grant Case. San Francisco: Bancroft, 1884.
Cookridge, E. H., pseud. The Baron of Arizona. New York: John Day, 1967.
Crane, H. "He Fooled a Nation," Sunday Referee, October 21, 1937.
Dodge, I. F. Our Arizona. New York: Scribner's, 1929.
Farish, G. H. The History of Arizona. San Francisco: Filmer, 1915.
Fugate, F. L. The Spanish Heritage of the Southwest. El Paso: Western Press, 1952.
Hoffman, O. Reports on Land Cases. San Francisco: Numa Hubert, 1892(?).
Hopkins, R. C. Muniments of Title of the Barony of Arizona and Translation Into English. San Francisco: Bancroft, 1893.
Johnson, Royal A. Report of the Surveyor-General Upon the Alleged Peralta Grant. Phoenix: Arizona Gazette Book and Job Office, 1890.
Kelland, Clarence Budington. "The Red Baron of Arizona," Saturday Evening Post, October 11, 1947.
Leighton, P. "The Baron of Arizona," Everybody's Magazine, May, 1951.
Lockwood, Frank C. Arizona Characters. Los Angeles: Times-Mirror, 1928.
McClintock, J. H. Arizona. Chicago: Clarke, 1916.
Myers, John M. "The Prince of Swindlers," American Heritage Magazine, August, 1956.
Powell, Donald M. The Peralta Grant: James Addison Reavis and the Barony of Arizona. Norman, Okla.: University of Oklahoma Press, 1960.
Reynolds, M. G. Spanish and Mexican Land Laws. St. Louis: Buxton, 1895.
Strover, W. "The Story of the Red Baron," Arizona Magazine, September, 1919.
Tipton, W. M. "The Prince of Imposters," Land of Sunshine, February, 1891; March, 1891.

THE HAYMARKET SQUARE RIOT (1886)

The tragic event known as the "Haymarket Square Riot" occurred in the Haymarket, a square in Randolph Street, Chicago, on May 4, 1886. The affair was the outgrowth of a violent strike and riot at the McCormick Reaper Works in Chicago, in which several participants had been killed by the police the day before.

A group of international anarchists called the May 4 meeting to protest police violence. While the meeting was in session police attempted to disperse the assembled crowd and a riot resulted, during which a bomb was thrown. Seven policemen were killed and 27 were injured.

Eight anarchists, all of whom had attended the meeting, were arrested and charged with being accessories to the bombing. All had, on previous occasions, publicly advocated violence against law enforcement officers.

The eight were duly tried and convicted. Of the eight, seven were sentenced to death and one to life imprisonment. Ultimately four were hanged, one committed suicide, one had his sentence reduced to fifteen years' imprisonment, and the death sentences of the other two were commuted to life terms in prison.

Governor John Peter Altgeld of Illinois pardoned the three who were in prison in 1893, primarily on the grounds that no evidence connecting any of them with the throwing of the bomb had been presented at their trials.

Suggested Readings

Adamic, Louis. Dynamite: The Story of Class Violence in America. Gloucester, Mass.: Peter Smith, 1959.

Altgeld, John Peter. "The Chicago Martyrs: Reasons for Pardoning Fielden, Neebe and Schwab," in his The Complete Works of John Peter Altgeld. Montclair, N.J.: Patterson Smith, 1976.

Browne, Waldo R. Altgeld of Illinois: A Record of His Life and Work. New York: B. W. Heubsch, 1924.

Cook, Roy. Leaders of Labor. Philadelphia: Lippincott, 1966.

Darrow, Clarence. The Story of My Life. New York: Scribner's, 1932.

David, Henry. The History of the Haymarket Affair. New York: Russell & Russell, 1958.

Gary, Joseph E. "The Chicago Anarchists of 1886," Century Magazine, April, 1893.

Gompers, Samuel. Seventy Years of Life and Labour. Clifton, N.J.: Augustus M. Kelley, 1966.

Gurko, Miriam. Clarence Darrow. New York: Crowell, 1965.

Heaps, Willard Allison. Riots, U.S.A., 1765-1970. New York: Seabury Press, 1970.

Iman, Raymond S., and Thomas W. Koch. Labor in American Society. Chicago: Scott, Foresman, 1965.

Johnsen, Julia E., comp. The Closed Shop. New York: Wilson, 1942.

Lawson, John D., ed. American State Trials, 1569-1920. Wilmington, Del.: Scholarly Resources, 1972.

Lens, Sidney. Unions and What They Do. New York: Putnam, 1968.

Litwack, Leon F., ed. The American Labor Movement. Englewood Cliffs, N.J.: Prentice-Hall, 1962.

Meltzer, Milton. Bread--and Roses: The Struggle of American

Labor. New York: Knopf, 1967.
Mordell, Albert. Clarence Darrow, Eugene V. Debs and Haldeman-Julius. Girard, Kansas: Haldeman-Julius Publications, 1950.
Morris, Richard B. "Ordeal by Jury," in his Fair Trial. New York: Knopf, 1952.
Stone, Irving. Clarence Darrow for the Defense. Garden City, N.Y.: Doubleday, 1941.
Weinberg, Arthur, ed. Attorney for the Damned. New York: Simon & Schuster, 1957.
Werstein, Irving. Pie in the Sky: An American Struggle, the Wobblies and Their Times. New York: Delacorte, 1969.

THE BLIZZARD OF '88 (1888)

One of the most devastating snowstorms in the history of New York rendered that city virtually helpless from the morning of Monday, March 12, 1888 until midday of Sunday, March 18.

The storm, known for years as the "Blizzard of '88" and the "Goliath of New York Blizzards," was completely unexpected. On the morning of the 12th the weather prediction was "clearing and colder, preceded by light snow." By the time this forecast was published in the morning papers, two feet of snow had fallen and the temperature had dropped to fifteen degrees. A gale was blowing at forty miles per hour and the steadily falling barometer was to reach a low of 29.62.

The blizzard, which had started as a comparatively gentle "Dakota Storm," increased in violence and raged up the Atlantic seaboard. In short order all areas within a thirty-mile radius of New York City were paralyzed and the heavy snowfall covered cities as far away as Baltimore, Philadelphia, Pittsburgh and Washington, D.C. A large portion of the state of New Jersey lay prostrate.

Great drifts of snow covered New York City. Icy winds severed telephone poles and communication lines were snapped. People were blown through the air, to be buried in the deep snow, some to die there. Buggies and wagons were overturned on the streets.

All transportation was halted. Trains, unable to buck the huge piles of snow on the tracks, ceased running. Passengers stranded on elevated trains were brought down to the streets below on ladders.

Ships in New York Harbor were flying distress signals. No fewer than 27 vessels were washed ashore and William A. W. Stewart's yacht "Cythera," on her way to Bermuda, disappeared and was never seen again.

By Tuesday, March 13, some of the telephone and telegraph lines had been repaired and communication by wire had been partially restored. The following day saw the storm abated but huge snow drifts still prevented trains from entering the city and supplies of food, coal, milk, and other necessities were running dangerously low. What items were available were being sold by profiteers at double and triple their normal prices.

By Thursday the sun had appeared once more and things were getting back to normal as New York dug itself out. By Sunday, March 18, the danger was past. The storm, following the coastline, had lost much of its intensity. It passed through New England and eventually dissipated itself in Canada.

The Blizzard of '88 caused property damage reaching into the millions. Lives lost were estimated at over 400. It was not until December 26, 1947, that a comparable storm again buried America's largest city in drifts of snow.

Suggested Readings

Battan, Louis Joseph. The Nature of Violent Storms. Garden City, N.Y.: Doubleday, 1961.
Lane, Frank Walter. The Elements Rage. Philadelphia: Chilton, 1965.
Sloane, Eric. The Book of Storms. New York: Duell, Sloan, 1956.
Smith, L. P. Weather Studies. Oxford: Pergamon Press, 1966.
Trewartha, Glenn T. The Earth's Problem Climates. Madison, Wis.: University of Wisconsin Press, 1961.
_______. An Introduction to Climate. New York: McGraw-Hill, 1968.
Turkel, Stephen. "The Blizzard of 1888," in Kartman, Ben, and Leonard Brown, eds. Disaster! New York: Pellegrini & Cudahy, 1948.
Van Straten, Florence W. Weather or Not. New York: Dodd, Mead, 1966.
Werstein, Irving. The Blizzard of '88. New York: Crowell, 1960.
Winchester, James H. Hurricanes, Storms, Tornadoes. New York: Putnam, 1968.

THE FOUNDING OF HULL HOUSE (1889)

Hull House, a social settlement, was established by Jane Addams and Ellen Gates Starr in 1889. Located in the city of Chicago, Illinois, Hull House was designed primarily as a welfare agency for needy families, but was also intended to provide recreational facilities for slum children, in order to reduce juvenile delinquency. Other objects were to assist foreign-born residents of Chicago to learn the English language and acquire American citizenship.

When founded, Hull House, the first social settlement of its kind in the United States, was confined to a single building. Eventually it expanded until it occupied no fewer than thirteen buildings and became the leading and one of the largest such institutions in

America. Today its facilities include a social service center, a theater in which dramatic performances may be given, a day nursery, a well-equipped gymnasium, recreation and meeting rooms, workshops of various kinds, a music school, and classrooms in which adult education classes are held.

Since 1912 the Joseph T. Bowen Country Club has been operated in conjunction with Hull House. This is a summer camp for children and is located at Waukeegan, Illinois.

Jane Addams' settlement is financed by donations from private citizens and grants by other social-welfare agencies. It has, over the years, provided countless opportunities for underprivileged Chicagoans, both juvenile and adult.

Suggested Readings

Abbott, Edith. "The Hull House of Jane Addams," Social Science Review, September, 1952.

Addams, Jane. The Excellent Becomes the Permanent. Plainview, N. Y.: Books for Libraries, 1932.

______. Twenty Years at Hull House. New York: Macmillan, 1910.

Bolton, Sarah K. Lives of Girls Who Became Famous. New York: Crowell, 1942.

Conway, Jill. "Jane Addams: An American Heroine," Daedalus, Spring, 1964.

Davis, Allen F. "Jane Addams vs. the Ward Boss," Journal of the Illinois State Historical Society, Autumn, 1960.

Farrell, John C. Beloved Lady: A History of Jane Addams' Ideas on Reform and Peace. Baltimore: John Hopkins University Press, 1967.

Hagedorn, Hermann. Americans: A Book of Lives. New York: John Day, 1946.

James, Edward T., ed. "Jane Addams," in his Notable American Women. Cambridge, Mass.: The Belknap Press of Harvard University Press, 1971.

Lasch, Christopher, ed. The Social Thought of Jane Addams. New York: Bobbs-Merrill, 1965.

Linn, James W. Jane Addams: A Biography. Westport, Conn.: Greenwood Press, 1968.

Scott, Anne F. "Saint Jane and the Ward Boss," American Heritage Magazine, December, 1960.

THE OKLAHOMA LAND RUSH (1889)

In 1803 the area now including the present state of Oklahoma, with the exception of the extreme western Panhandle strip north of Texas, was acquired by the United States from France as part of the Louisiana Purchase. In 1834 it became the Indian Territory, which it remained until after the Civil War. After the War the Indian nations occupying the Territory were forced to cede the western half to the United States as a home for other Indian tribes. Desirable land was becoming scarce and much of the land in the Territory was unoccupied. White men were forbidden by law to settle there, but this law was frequently disregarded.

In 1885 President Rutherford B. Hayes received authorization from Congress to renegotiate with the Creek and Seminole Indians for the purpose of opening these unoccupied lands for settlement. These negotiations were completed in 1889, and at noon on April 22 of that year the land was opened to the public for settlement.

The blast of a cavalry bugle was the signal that any settler who so desired could legally enter the previously forbidden Territory and stake out his claim. The mob of settlers rushed over the line on foot, on horseback, and in buggies and wagons. Trains were jammed with people, tools, furniture, and even portable houses, carried in from Texas, Nebraska, and Kansas.

By the evening of the first day a stretch of prairie had become the city of Guthrie, with a population of 10,000 persons, and Oklahoma had a population of 50,000.

Additional lands were opened to settlement from 1891 through 1906. In 1907 Oklahoma was admitted to the Union as the 46th state.

Suggested Readings

Alley, John. City Beginnings in Oklahoma Territory. Norman, Okla.: University of Oklahoma Press, 1939.

Buchanan, James, and Edward E. Dale. A History of Oklahoma. Evanston, Ill.: Northwestern University Press, 1935.

Dale, Edward E. Oklahoma, The Story of a State. Evanston, Ill.: Northwestern University Press, 1949.

________, and Jesse L. Rader. Readings in Oklahoma History. Evanston, Ill.: Northwestern University Press, 1930.

________, and Morris L. Wardell. History of Oklahoma. New York: Harper, 1948.

Ferber, Edna. Cimarron (fiction). New York: Bantam Books, 1963.

Fisher, Aileen Lucia. Cherokee Strip: The Race for Land. New York: Aladdin, 1956.

Foreman, Grant. A History of Oklahoma. Norman, Okla.: University of Oklahoma Press, 1942.

Gideon, D. C. Indian Territory, Descriptive, Biographical, and Genealogical ... With a General History of the Territory. New York: Harper, 1901.

Gittinger, Roy. The Formation of the State of Oklahoma, 1803-1906. Norman, Okla.: University of Oklahoma Press, 1939.

Hodge, Frederick W., ed. Handbook of American Indians North of Mexico. East Saint Clair Shores, Mich.: Scholarly Press, 1910.

McReynolds, Edwin C. Oklahoma: The Story of Its Past and Present. Norman, Okla.: University of Oklahoma Press, 1961.

Rainey, George. The Cherokee Strip. Guthrie, Okla.: Merrill, 1933.
Royce, Charles C. "Indian Land Cessions in the United States," Bureau of American Ethnology, Eighteenth Annual Report. Washington, D.C.: 1900.
Stewart, Dora A. The Government and Development of Oklahoma Territory. Oklahoma City: Oklahoma City University Press, 1933.
Thoburn, Joseph B., and Muriel H. Wright. Oklahoma: A History of the State and Its People. New York: Knopf, 1929.
Tinkle, Lon. The Story of Oklahoma. New York: Random House, 1962.

THE JOHNSTOWN FLOOD (1889)

The site of Johnstown, Pennsylvania, was settled in 1791 by Joseph Jahns, a Swiss, and about 1800 the town, named for him, was laid out. In 1831 it was incorporated as a borough and in 1889 as a city.

Johnstown lies in a deep valley and on May 31, 1889 it was inundated when the dam across the south fork of the Conemaugh River, twelve miles east of the city, broke following unusually heavy rains. The waters of Conemaugh Lake, impounded by the dam, poured into the city at approximately twenty miles per hour.

The lake was three miles long and varied from sixty to seventy feet in depth, and the waters released from it submerged Johnstown and completely destroyed seven other towns nearby. Over 2,200 persons were drowned and property damage was estimated at twelve million dollars. Seven hundred and seventy-nine unidentified victims of the flood are buried in Grandview Cemetery at Johnstown.

The city was soon rebuilt. In 1936 another flood cost the lives of about 25 persons and inflicted much property damage. Subsequently U.S. Army Engineers constructed a flood control system costing over eight million dollars which, it is hoped, will prevent the recurrence of similar catastrophes.

Suggested Readings

Baker, George W., and Dwight W. Chapman, eds. Man and Society in Disaster. New York: Basic Books, 1962.
Barrows, Harold Kilbirth. Floods, Their Hydrology and Control. New York: McGraw-Hill, 1948.
Barton, Allen H. Communities in Disaster. Garden City, N.Y.: Longmans-Green, 1961.
Briggs, Peter. Rampage: The Story of Disastrous Floods, Broken Dams and Human Fallibility. New York: Donald McKay, 1973.
Chamberlin, Jo. "Johnstown Remembers," in Kartman, Ben, and Leonard Brown, eds. Disaster! New York: Pellegrini &

Cudahy, 1948.
Dacy, Douglas C., and Howard Kunreuther. The Economics of Natural Disasters. New York: Free Press, 1969.
Dempsey, David K. Flood. New York: Ballantine Books, 1956.
Dolson, Hildegarde. Disaster at Johnstown: The Great Flood. New York: Random House, 1965.
Garrison, W. B. Disasters That Made History. Nashville: Abingdon, 1973.
Hewitt, Ronald. From Earthquake, Fire and Flood. New York: Scribner's, 1957.
Sutton, Ann, and Myron Sutton. Nature on the Rampage. Philadelphia: Lippincott, 1962.
Wallechinsky, David, and Irving Wallace. "The Johnstown Flood," in their The People's Almanac. Garden City, N.Y.: Doubleday, 1975.

AROUND THE WORLD WITH NELLIE BLY (1889-1890)

One of the outstanding female journalists of her day was Elizabeth Cochrane Seaman, better known by her pseudonym, "Nellie Bly." She served on the staff of such newspapers as the Pittsburgh Dispatch, the New York World, and the New York Journal. In 1888 she spent ten days in an insane asylum on Blackwell's Island in order to gather material for a series of articles on the treatment of inmates. This series appeared later under the title "Ten Days in a Madhouse."

Nellie Bly was clearly not "a woman who wrote on scented paper with violet ink." She specialized in sensationalism and journalistic stunts which would increase the circulation of the newspaper of which she happened to be a staff member at the time.

Jules Verne's best selling novel, Around the World in Eighty Days, published in 1873, inspired the "female reporter" and her editors to determine whether the eighty-day record established by the fictional Phileas Fogg and his manservant Passepartout could be bettered. On November 4, 1889, she set out from Hoboken, New Jersey, on the first stage of her round-the-world tour. She traveled by commercial transportation "without an umbrella, which was considered not quite nice."

She crossed the Atlantic and the Mediterranean, proceeding to Aden, Colombo, Hong Kong, Tokyo, and San Francisco. She traveled by steamer, train, ricksha and sampan.

On January 25, 1890, gun salutes at New York welcomed the returning "daring young female reporter." She had made her trip around the world in the unbelievable time of 72 days, six hours and eleven minutes. She described her record-breaking journey in the publication Nellie Bly's Book: Around the World in Seventy-two Days.

The trip was exotic and exciting, and while the woman was abroad the World kept the reading public's attention focused on her progress. She "seemed to embody the romance of journalism, the lure of travel, and the pluck of the American girl."

Songs and dances were dedicated to the "young reporter." Games, clothes, and toys were named after her, and parades were held in her honor.

"Nellie Bly" married Robert L. Seaman, a New York businessman, in 1895. After his death in 1910 she abandoned journalism to manage her late husband's iron manufacturing company. Her business activities proved unsuccessful and she joined the staff of the New York Journal in 1919. She died in 1922.

Suggested Readings

Baker, Nina. Nellie Bly, Reporter. New York: Scholastic Book Service, 1972.

Cochrane, Elizabeth. Nellie Bly's Book: Around the World in Seventy-Two Days. New York: 1890.

Graves, Charles P. Nellie Bly, Reporter for the "World." Scarsdale, N.Y.: Garrard Publishing Co., 1970.

Hahn, Emily. Around the World With Nellie Bly. Boston: Houghton Mifflin, 1959.

Jakes, John. "Elizabeth Cochrane," in his Great Women Reporters. New York: Putnam, 1969.

James, Edward T., ed. "Elizabeth Cochrane Seaman," in his Notable American Women. Cambridge, Mass.: The Belknap Press of Harvard University Press, 1971.

Juergens, George. Joseph Pulitzer and the New York "World." Princeton, N.J.: Princeton University Press, 1966.

Kunitz, Stanley J., and Howard Haycraft. "Elizabeth Cochrane Seaman," in their American Authors, 1600-1900. New York: H. W. Wilson, 1938.

Logie, Iona Robertson. Careers for Women in Journalism. Scranton, Pa.: International Textbook, 1938.

Noble, Iris. Nellie Bly, First Woman Reporter. New York: Messner, 1956.

Rittenhouse, Mignon. The Amazing Nellie Bly. Plainview, N.Y.: Books for Libraries, 1956.

Ross, Isabel. Ladies of the Press. New York: Arno Press, 1974.

Schwanberg, W. A. Pulitzer. New York: Scribner's, 1967.

Seitz, Don C. Joseph Pulitzer, His Life and Letters. New York: AMS Press, 1970.

Shuler, Marjorie. A Passenger to Adventure. New York: Appleton, 1939.

Verne, Jules. Around the World in Eighty Days (fiction). Translated by George M. Towle. New York: Dodd, Mead, 1956. (Originally published 1873.)

THE MAFIA IMBROGLIO (1890)

The Mafia, sometimes also known as the "Black Hand," an Italian secret society, originated in Sicily in the 19th century. Without any centralized organization, it consisted of a large number of small bands of criminals, each of which was autonomous in its own district. These bands engaged in cattle stealing, extortion, kidnaping for ransom and, by employing terrorist methods against peasant voters, obtained political offices which enabled them to influence police officials and obtain weapons legally.

Late in the 19th century the Italian government took vigorous steps to suppress the Mafia and many of its members came to America. They settled in large metropolitan centers and continued their criminal activities in the new country.

In October, 1890, David C. Hennessy, chief of the New Orleans, Louisiana police, was murdered. The circumstances surrounding the crime indicated that it had been committed by members of the Mafia. Eleven Italians were arrested. Of the eleven, five were held for trial and the others were tried at once. Three were acquitted and three, following trials which resulted in hung juries, were to be tried a second time. A mob, believing that there had been a serious miscarriage of justice, seized and killed the eleven Italians.

In retaliation the Italian government demanded the trial and punishment of the persons involved in the killings and protection for all the Italians in New Orleans. It also demanded an indemnity for the victims of the mob.

These demands were not met, and the Italian government withdrew its minister. A grand jury investigation in New Orleans resulted in an excusing of the participants in the killing, none of whom was ever brought to trial.

The situation was complicated by the fact that the American government, while dealing with foreign countries on a diplomatic basis, was virtually powerless to act in a situation requiring judicial action by a state. Eventually diplomatic activity resulted in an expression of regret on the part of the United States and the payment of $24,000.

Suggested Readings

Evans, Oliver W. New Orleans. New York: Harper, 1959.

Gage, Nicholas, comp. Mafia, U.S.A. New York: Playboy Press, 1972.

Hess, Genner. Mafia and Mafiosi: The Structure of Power. Lexington, Mass.: Lexington Books, 1973.

Joey, pseud., with David Fisher. Killer: Autobiography of a Hit Man for the Mafia. New York: Playboy Press, 1973.

Lewis, Norman. Honored Society: A Searching Look at the Mafia. New York: Putnam, 1964.

Perisco, Joseph E. "Vendetta in New Orleans," American Heritage Magazine, June, 1973.

Puzo, Mario. The Godfather (fiction). New York: Putnam, 1969.
Salerno, Ralph. Crime Confederation: Cosa Nostra and Allied Operations in Organized Crime. Garden City, N.Y.: Doubleday, 1969.
Sondern, Frederick, Jr. Brotherhood of Evil: The Mafia. New York: Farrar & Rinehart, 1959.
Talese, Gay. Honor Thy Father. New York: World, 1971.
Teresa, Vincent Charles, with Thomas C. Renner. My Life in the Mafia. Garden City, N.Y.: Doubleday, 1973.
Train, Arthur. Courts, Criminals and the Camorra. New York: Arno Press, 1912.

THE SHERMAN ANTI-TRUST ACT AND THE RULE OF REASON (1890 and 1911)

Following the Civil War the growth of trusts and corporations and the attendant malpractices of big business led to an increasing demand for federal government regulation. Such demands were incorporated in the platforms of both the Democratic and Republican parties in 1888.

On December 4, 1889, Senator John Sherman introduced a bill providing for the regulation of trusts. This bill was the first federal act ever passed which attempted to regulate the corporate giants. It was written, in its final form, primarily by senators George Franklin Edmunds and George Frisbie Hoar, and was couched in general and often ambiguous language. It was intended, primarily, to protect trade and commerce from unlawful restraints and monopolies. This legislation is known as the Sherman Antitrust Act.

A number of questions concerning the Act arose. These had to be answered by court decisions. One such question, resolved by the "Rule of Reason," was this: Did the Act prohibit all contracts in restraint of trade between the states, or only unreasonable restraints?

Two early anti-trust suits brought before the Supreme Court involved the American Tobacco Company and the Standard Oil Company. In May, 1911, the Court found both of these companies guilty of "combining to restrain and monopolize trade" and ordered "a dissolution of the conspiring elements into separate, competing units." At this time the Court also answered some of the questions which had arisen in connection with the Sherman Anti-trust Act. The phraseology of Section I appears to forbid restraints of all kinds, which was in line with the previous decisions of the Court. The so-called "Rule of Reason" resulted from the American Tobacco and Standard Oil cases, in which it was declared that "only those restraints were forbidden that were unreasonable."

As a result, the attention of some of the persons who were opposed to trusts was focused on the obiter dictum rather than on the decisions themselves.

Chief Justice Edward Douglass White read the Court's decision. Associate Justice John Marshall Harlan's dissenting opinion agreed with the decision but condemned the _obiter dictum_. He felt that the exact words of the law forbade _every_ contract, not just those in which the forbidden restraints _were_ "unreasonable."

The dissolution of the two companies concerned did not seem to benefit the public particularly, and the fact that the prices of Standard Oil stocks immediately rose on the exchanges where they were listed indicated that the Supreme Court decision was inconsequential.

Suggested Readings

Burton, Theodore E., and John T. Morse, Jr. _John Sherman_. Boston: Houghton Mifflin, 1906.

Clark, John D. _Federal Trust Policy_. Boston: Houghton Mifflin, 1931.

Commager, Henry Steele, ed. "The Rule of Reason," (Doc. No. 375) in his _Documents of American History_, 8th edition. New York: Appleton, 1968.

_______. "The Sherman Anti-Trust Act," (Doc. No. 320) in his _Documents of American History_, 8th edition. New York: Appleton, 1968.

_______. "United States v. Socony-Vacuum Oil Co.," (Doc. No. 530) in his _Documents of American History_, 8th edition. New York: Appleton, 1968.

Ely, Richard T. "The Founding and Early History of the American Economic Association," _American Economic Review_, XXVI, Supp. 141, 1936.

Letwin, William L. _Law and Economic Policy in America_. New York: Random House, 1956.

Seager, Henry R., and Charles A. Gulick, Jr. _Trust and Corporation Problems_. New York: Arno Press, 1929.

Sherman, John. _Recollections of Forty Years in the House, Senate and Cabinet: An Autobiography_. New York: Greenwood Press, 1968.

Taft, William Howard. _The Anti-Trust Act and the Supreme Court_. Millwood, N.Y.: Kraus Reprint, 1914.

Thorelli, Hans B. _The Federal Antitrust Policy_. Baltimore: Johns Hopkins University Press, 1955.

Warren, Charles. _The Supreme Court in United States History_. Boston: Little, Brown, 1923.

Whitney, Simon N. _Antitrust Policies: American Experience in Twenty Industries_. New York: Twentieth Century Fund, 1958.

THE WOUNDED KNEE INCIDENTS (1890 and 1973)

In February, 1973, history repeated itself to some degree when Indians at Wounded Knee, South Dakota, resisted the white man as did their Sioux ancestors in 1890. The 1973 siege was not the wholesale massacre of the earlier day but two Indians, Frank Clearwater and Lawrence Lamont, were killed in gun battles.

South Dakota was originally acquired by the United States from France as a part of the 1803 Louisiana Purchase. Until 1870 settlement of the region was slow, but when gold was discovered in the Black Hills in 1874 gold seekers began to flock there.

The federal government attempted to keep settlers out of the area until an agreement could be negotiated with the Sioux Indians. The Indians refused to cede their land, further gold strikes were made, and eventually armed warfare broke out between the Indians and the whites. General George Armstrong Custer and his command were wiped out in an encounter at Little Big Horn in 1876. In 1890/91 another revolt at Wounded Knee resulted in the death of Sitting Bull and the massacre of 300 Sioux men, women, and children. The revolt was crushed by American soldiers commanded by General Nelson A. Miles.

In February, 1973, armed supporters of the American Indian Movement seized control of the town of Wounded Knee. This takeover emanated from a bitter internal tribal struggle between the American Indian Movement (A.I.M.) supporters and the followers of Richard Wilson who had been appointed Tribal Council President by the Bureau of Indian Affairs. The A.I.M. insurgents, numbering approximately 200, hoped, by the takeover, to dramatize their grievances. They demanded that Wilson be removed from his position and that a Senate investigation be made into alleged mishandling of Indian affairs by both the B.I.A. and the federal government. This takeover resulted in a veritable parade of government officials to the town. South Dakota senators James Abourezk and George McGovern put in appearances but bureaucratic bickering and intermittent gunplay hampered negotiations.

A few short-lived peace agreements were achieved at meetings held in such informal places as tepees and a school house, but these were quickly voided by quarrels over disarmament and other aspects of the controversy. Eight shootouts occurred and a United States marshal was wounded.

On May 5, 1973, a lasting peace was achieved and the 120 occupiers surrendered. This surrender was with the understanding that government investigation into Indian affairs would be made and that Washington officials would be present at a series of treaty talks.

During the 1973 siege Wounded Knee suffered severe property damage and heavy property losses. About fifteen insurgents were arrested following the surrender while other A.I.M. leaders had either escaped or been captured earlier.

Suggested Readings

"Behind a Modern-Day Indian Uprising: Sioux Militancy," U.S. News and World Report, March 12, 1973.

Brown, Dee. Bury My Heart at Wounded Knee. New York: Holt, Rinehart, 1970.

Collier, P. "Wounded Knee: The New Indian War," Ramparts,

June, 1973.
Coman, Katherine. *Economic Beginnings of the Far West.* Clifton, N.J.: Augustus M. Kelley, 1967.
Dilley, R. "Standoff at Wounded Knee," *Christian Century*, May 9, 1973.
Eastman, Elaine Goodale. "The Ghost Dance War and Wounded Knee Massacre of 1890-1891," *Nebraska History*, 26, 1945.
"The Future of the Indian Question," *North American Review*, January, 1891.
Gessner, Robert. *Massacre: A Survey of Today's American Indian.* New York: Da Capo Press, 1972.
Greene, Jerome A. "The Sioux Land Commission of 1889: Prelude to Wounded Knee," *South Dakota History*, Winter, 1970.
McGregor, James H. *The Wounded Knee Massacre.* Baltimore: Wirth Bros., 1940.
Miles, Nelson A. *Serving the Republic.* Plainview, N.Y.: Books for Libraries, 1974.
"Not With a Bang," *Newsweek*, May 21, 1973.
Paxson, Frederic L. *The History of the American Frontier, 1763-1893.* Dunwoody, Ga.: Norman S. Berg, 1967.
________. *The Last American Frontier.* New York: Cooper Square, 1970.
Pfaller, Father Louis. "The Indian Scare of 1890," *North Dakota History*, Spring, 1972.
"Return to Wounded Knee," *Newsweek*, March 12, 1973.
Schell, Herbert S. *The History of South Dakota.* Lincoln, Neb.: University of Nebraska Press, 1961.
Schultz, T. "Bamboozle Me Not at Wounded Knee," *Harper's*, June, 1973.
Seymour, Flora W. *The Story of the Red Man.* Plainview, N.Y.: Books for Libraries, 1929.
Smith, Rex Alan. *Moon of Popping Trees.* New York: Reader's Digest Press, 1975.
"Suspenseful Show of Red Power: Sioux Protest," *Time*, March 19, 1973.
Tebbel, John, and Keith Jennison. *The American Indian Wars.* New York: Bonanza Books, 1960.
Vestal, Stanley. *Sitting Bull, Champion of the Sioux.* Norman, Okla.: University of Oklahoma Press, 1957.
________. *Warpath and Council Fire: The Plains Indians' Struggle for Survival in War and in Diplomacy, 1851-1891.* New York: Random House, 1948.
Wissler, Clark. *The American Indian.* Gloucester, Mass.: Peter Smith, 1922.
________. *Indians of the United States.* New York: Doubleday, 1940.
"Wounded Knee: Just a Prelude?," *U.S. News and World Report*, March 26, 1973.

THE CARLYLE HARRIS MORPHINE MURDER (1891)

In 1889 Carlyle Harris, a 21-year-old medical student, while vacationing at Ocean Grove, New Jersey, met 18-year-old Helen Potts and her family. The girl fell in love with him but he, while attracted to her as he had been to countless other girls, regarded her only as a possible conquest. Helen's mother forbade her daughter to become engaged to "a medical student and only in his second year at college." In February, 1890, the two were secretly married in New York, under false names.

Shortly thereafter Harris lost interest in his bride, who had become pregnant. He bungled an attempt to perform an abortion and the girl's mother, alarmed at her daughter's ailing health, sent her to her uncle, a physician, in Scranton, Pennsylvania. The uncle diagnosed the cause of her condition and performed a proper abortion with the assistance of another Scranton doctor. Helen told her uncle of her marriage to Harris but the latter denied its legality, claiming to be married to someone else, whom he never identified.

In New York Helen's mother resolved to make her daughter's marriage legal. She consulted an attorney and Harris, under pressure, signed an affidavit affirming his marriage. The mother then arranged to send her daughter to Comstock School for Young Ladies in New York City. Harris' affections for Helen had completely cooled but the girl's mother insisted that the two be married "in a Christian way" on February 8, 1891, the first anniversary of their secret marriage.

On January 20 Harris visited Ewen McIntyre's drug store. There he had the pharmacist fill a prescription for 27 grains of quinine and one grain of morphine, to be made into six capsules. Each capsule was to contain 1/6 of a grain of morphine and 4-1/2 grains of quinine. The written prescription which Harris gave the druggist was signed "C. W. H., Student."

Two days later Helen complained of a headache and Harris, who had been visiting her, gave her the capsules and told her to take one each night. He then left town. When he returned on January 31 he persuaded the girl to take the fourth and last remaining capsule in the box. That night she died and the box which had contained the capsules was found in her room. It was traced to Harris, who was summoned to the scene and eventually charged with murder by morphine poisoning. He had removed two of the capsules from the box and later turned one of them over to the coroner. At his trial the prosecution contended that he had taken out two of the capsules in order to protect himself if arrested, and had unloaded one of the capsules he had given Helen and substituted three grains of morphine, a lethal dose, for the quinine in it.

Reporters on the staff of the New York World became suspicious and conducted an investigation. Helen's body was exhumed and an autopsy revealed the presence of morphine but not quinine in the stomach.

Recorder Frederic Smyth presided over Carlyle Harris' trial which commenced in January, 1892, and lasted three weeks. The prosecution was handled by Francis L. Newman, and William Travers Jerome acted as defense counsel. Medical experts testified for both sides. Harris did not take the stand in his own defense.

The prosecution's evidence was all circumstantial. Recorder Smyth, in charging the jury, stated correctly that circumstantial evidence is legal evidence. An hour and twenty minutes after the jury retired it returned to the courtroom and rendered its verdict: guilty of murder in the first degree.

Harris was condemned to die in the electric chair. Through his attorneys he appealed the sentence. Dissatisfied with the performance of Jerome and his associates, he engaged William F. Howe and his law partner Abraham Henry Hummel to represent him. The Court of Appeals unanimously upheld the lower court's verdict. In an opinion read by Judge John G. Gray, the jurists stated that "there was not a single fact in the record to help out the presumption of the defendant's innocence." Governor Roswell P. Flower of New York State refused to intervene and on May 8, 1893, Carlyle Harris was strapped into the chair and electrocuted. His last words were, "I have no further motive for concealment, and I desire to state that I am absolutely innocent of the crime for which I am to be executed."

Suggested Readings

Abrahamsen, David, M.D. The Murdering Mind. New York: Harper & Row, 1973.

Collins, Ted, ed. "Carlyle Harris," in his New York Murders. New York: Duell, Sloan, 1944.

Gribble, Leonard. "The Secret Bride," in his They Had a Way With Women. New York: Roy Publishers, 1967.

Ledyard, Hope, pseud. The Judicial Murder of Carlyle Harris. New York: Privately printed for Mrs. Frances McCready Harris, 1893.

Morris, Richard B. "The Case of the Morphine Murder," in his Fair Trial. New York: Knopf, 1952.

O'Brien, Frank M. Murder Mysteries of New York. New York: F. W. Payson, 1932.

Pearson, Edmund. "The Sixth Capsule, or Proof by Circumstantial Evidence," in his Masterpieces of Murder. Boston: Little, Brown, 1963.

Reinhardt, James Melvin. The Psychology of Strange Killers. Springfield, Ill.: Thomas, 1962.

Rovere, Richard H. Howe & Hummel: Their True and Scandalous History. New York: Farrar, Straus, 1947.

Smith, Edward H. Famous American Poison Mysteries. London: Hurst & Blackett, 1927.

Train, Arthur. "The Fall of Hummel," Cosmopolitan, May, June, 1908.

Wellman, Francis L. The Art of Cross-Examination. New York:

Macmillan, 1962.
______. Gentlemen of the Jury: Reminiscences of Thirty Years at the Bar. New York: Macmillan, 1924.

THE JOHNSON COUNTY INVASION (1892)

The cattle boom of the American West began about the year 1880. Prior to that time the cattle industry was largely dominated by ranchers with small herds. These operators in due course found themselves in competition with large cattle companies controlling vast acreage and dealing in cattle by the hundreds of thousands.

The boom vanished during the extremely severe winter of 1886/87. Heavy snows and bitterly cold weather rendered it impossible for many animals to survive and herd losses averaged fifty per cent. The beef barons administering the large cattle companies with headquarters in Cheyenne, Wyoming, anticipating heavy financial losses, blamed their depleted cattle herds not only on the freezing weather but also on thieves. This situation degenerated into what one observer called "the bitter conflict that has ranged incessantly between large and small owners."

The large cattle companies belonged to an organization known as the Wyoming Stock Growers Association. Before the Johnson County Invasion of April, 1892, at least six persons, including Jim Averell and his paramour, Cattle Annie Watson, were hanged or shot as suspected cattle thieves. These killings were instigated by the Association. Many cowboys were blackballed by this same group and were unable to find employment with any of the larger cattle ranches.

By 1890 a group of blackballed cowboys had settled in Johnson County, Wyoming, and were in the business of rustling cattle from the large outfits, on the theory that "if a dog is given a bad name he will steal." In 1892 a group calling itself the Northern Wyoming Farmers and Stockgrowers Association announced that it planned to hold an independent roundup. This was against state law and contrary to the wishes of the Wyoming Stock Growers Association. One of the foremen of the roundup was Nathan D. Champion, a Texas cowhand and reputedly an expert with the six-shooter.

It was then that the members of the Wyoming Stock Growers Association determined to invade Johnson County with the avowed purpose of exterminating certain persons it considered objectionable. Seventy names were placed on a "death list," including that of Johnson County sheriff "Red" Angus. This plan was known to Acting Governor Amos W. Barber and to United States Senators Joseph M. Carey and Francis E. Warren. None of these men offered any objection. The Wyoming Stock Growers Association imported 25 hired gunfighters from Texas, making up, with its own members, a group of fifty. The expedition was headed by Major Wolcott, and Sam T. Clover, a Chicago reporter, accompanied it, hoping to obtain a story for his employer, the Herald.

On April 5, 1892, the invaders entrained at Cheyenne, making an overnight run to Casper, Wyoming. Here they mounted horses and headed for Buffalo, the county seat of Johnson County. Before reaching this objective they learned that Champion and some other cowboys were at a cabin a few miles away.

Proceeding to the cabin, the invaders waited. Two trappers had spent the night there and as these men emerged they were taken prisoner. Then Nick Ray, one of the "wanted" cowboys, appeared at the cabin door and was immediately shot down. In spite of flying bullets Champion, the only other occupant of the cabin, dragged Ray inside. Ray died that morning and the invaders set fire to the cabin and shot and killed Champion when he attempted to escape from the burning building. His body was left on the snow with a card pinned to his shirt, reading "Cattle thieves, beware!"

Next day the invaders, warned that the citizens of Buffalo were up in arms to avenge the deaths of Ray and Champion, barricaded themselves at the TA Ranch. Here they were besieged by an impromptu army of over three hundred cowboys led by Sheriff "Red" Angus.

Word of the situation had been sent to the governor who, in turn, got through to President Benjamin Harrison. The President authorized the use of federal troops from Fort McKinney to "suppress the insurrection." The troops, commanded by Colonel J. J. Van Horn, arrived at the TA Ranch and accepted the surrender of Major Wolcott and his invading army, which was taken to the fort.

None of the invaders was ever brought to justice although they did "pay an admitted $100,000 as the price of the invasion, counting legal expenses but not mentioning the illegal." The two trappers, witnesses to the shooting of Ray and Champion, were hustled out of the state and, because of legal technicalities, could not be subpoenaed. In due course the Johnson County Invasion faded into history.

Suggested Readings

Atherton, Lewis Eldon. The Cattle Kings. Bloomington, Ind.: Indiana University Press, 1961.

Brayer, Garnet M. and Herbert O. Brayer. American Cattle Trails, 1540-1900. Bayside, N.Y.: Pioneer Trails Assn., 1952.

Brown, Dee. Trail Driving Days. New York: Scribner's, 1952.

Chilton, Charles. The Book of the West. Indianapolis: Bobbs Merrill, 1962.

Dale, Edward Everett. Cow Country. Norman, Okla.: University of Oklahoma Press, 1965.

Frink, Maurice. When Grass Was King. Boulder, Colo.: University of Colorado Press, 1956.

Gann, Walter. Tread of the Longhorns. San Antonio: Naylor, 1949.

Harris, William Foster. The Look of the Old West. New York: Viking Press, 1955.

McCoy, Joseph Geiting. Historic Sketches of the Cattle Trade of

the West and Southwest. Kansas City: Ramsey, Millet & Hudson, 1874.
Mercer, Asa Shinn. The Banditti of the Plains, or the Cattlemen's Invasion of Wyoming--The Crowning Infamy of the Ages. Cheyenne, Wyo.: Mercer's Print Shop, 1894.
_______. Big Horn Country: Wyoming, the Gem of the Rockies. Seattle: Shorey, 1967.
Rollinson, John K. Wyoming Cattle Trails: History of the Migration of Oregon-Raised Herds to Mid-Western Markets. Caldwell, Idaho: Caxton Printers, 1948.
Rosa, J. G. Gunfighter: Man or Myth? Norman, Okla.: University of Oklahoma Press, 1969.
Smith, Helena Huntington. "The Johnson County War," American Heritage Magazine, April, 1961.
Von Richthofen, Walter Baron. Cattle Raising on the Plains of North America. Norman, Okla.: University of Oklahoma Press, 1964.
Ward, Don. Cowboys and Cattle Country. New York: American Heritage Press, 1961.

THE HOMESTEAD STRIKE (1892)

The Homestead Strike of 1892 was one of the most violent in American industrial history. Homestead, Pennsylvania, a borough of Allegheny County, is situated on the Monongahela River, six miles southeast of the center of Pittsburgh. A major industry in the area was and is the production of steel.

In 1881 the Homestead Steel Works were built. These were acquired by the firm of Carnegie, Phipps and Company in 1883. Andrew Carnegie was the principal owner of the company and Henry Clay Frick was Chairman and Managing Head.

In 1892 the company reduced wages paid to the workers and refused to recognize or negotiate with the Iron and Steel Workers' Union. As a result the workers went out on strike. The company employed 300 armed guards, members of Allan Pinkerton's detective agency, to protect their property.

On July 6, 1892, a riot occurred in which a number of men were killed and wounded. The steel company had prepared two barges to transport the Pinkerton guards to the Homestead wharf. The barges were to be towed by two steam tugs, the "Little Bill" and the "Tide."

At 4:00 P.M. the strikers sighted the flotilla coming towards them and a huge crowd gathered on the river banks. The "Tide" developed engine trouble and the "Little Bill" was pulling both barges loaded with Pinkerton's armed strike-breakers.

To this day the question of which side fired the first shot has never been decided. Gunfire broke out, however, and a good many people were hit and several killed. The "Little Bill" cast off from the barges and headed for Braddock, leaving the Pinkerton men behind. The shooting, temporarily suspended, resumed later.

Sixty Pinkertons attempted to land and were driven back to their barges. Attempts to set the barges on fire by pouring oil on the water and igniting it were unsuccessful. At 5:00 in the afternoon the Pinkertons, trapped on the barges, capitulated. They were permitted to disembark, their weapons and ammunition were seized and the barges set ablaze. The Pinkertons, now prisoners, were held in an impromptu jail and were treated with great brutality by their captors.

At midnight on July 7 the Pinkerton men were removed to Pittsburgh where those who needed medical attention were hospitalized. On July 12, the governor called out the state militia and put the borough under martial law. On July 23 Alexander Berkman, a Russian-born anarchist, shot and stabbed Frick in his Pittsburgh office. Berkman's sympathies were with the strikers and Frick represented to him everything that was evil in industrial management.

In spite of Berkman's savage attack Frick recovered. The strike ended in a victory for the company and the Homestead Steel Works were reopened in November.

Suggested Readings

Cook, Roy. Leaders of Labor. Philadelphia: Lippincott, 1966.

Harvey, George. Henry Clay Frick. New York: Scribner's, 1928.

Heaps, Willard Allison. Riots, U.S.A., 1765-1970. New York: Seabury Press, 1970.

Hendrick, Burton J. The Life of Andrew Carnegie. Garden City, N.Y.: Doubleday, 1932.

Horan, James D. The Pinkertons: The Detective Dynasty That Made History. New York: Crown Publishers, 1967.

________, and Howard Swigett. The Pinkerton Story. New York: Putnam, 1931.

Iman, Raymond S., and Thomas W. Koch. Labor in American Society. Chicago: Scott, Foresman, 1965.

Johnsen, Julia E., comp. The Closed Shop. New York: Wilson, 1942.

Lavine, Sigmund A. Allan Pinkerton: America's First Private Eye. New York: Dodd, Mead, 1963.

Lens, Sidney. Unions and What They Do. New York: Putnam, 1968.

Lingley, Charles Ramsdell, and Allen Richard Foley. Since the Civil War. New York: Appleton, 1935.

Litwack, Leon F., ed. The American Labor Movement. Englewood Cliffs, N.J.: Prentice-Hall, 1962.

Meltzer, Milton. Bread--and Roses: The Struggle of American Labor. New York: Knopf, 1967.

Orrmont, Arthur. Master Detective: Allan Pinkerton. New York: Messner, 1965.

Pinkerton, Allan. Criminal Reminiscences and Detective Sketches. Freeport, N.Y.: Books for Libraries, 1970. (Originally published 1878.)

________. Professional Thieves and the Detective. New York:

G. W. Carleton & Co., 1881.
_______. Strikers, Communists, Tramps and Detectives. New York: G. W. Carleton & Co., 1878.
Pinkerton, Robert A. "Detective Surveillance of Anarchists," North American Review, November, 1901.
Rowan, Richard Wilmer. The Pinkertons, a Detective Dynasty. Boston: Little, Brown, 1931.
Wallechinsky, David, and Irving Wallace. "Assassinations--Henry Clay Frick," in their The People's Almanac. Garden City, N.Y.: Doubleday, 1975.
Wilson, Margaret Barclay. A Carnegie Anthology. Privately printed, 1915.
Winkler, John K. Incredible Carnegie. Plainview, N.Y.: Books for Libraries, 1931.
Wolff, Leon. Lockout: The Story of the Homestead Strike of 1892: A Study of Violence, Unionism and the Carnegie Steel Empire. New York: Harper, 1965.

THE LIZZIE BORDEN CASE (1892-1893)

"Lizzie Borden, with an ax
Gave her mother forty whacks.
When she saw what she had done
She gave her father forty-one."

This bit of doggerel, sung to the tune of the then-popular song, "Ta-ra-ra Boomdeay," does not jibe with the "not guilty" verdict of the jury which sat at her 1893 trial for the alleged ax-murder of her stepmother and her father in the year 1892.

The Borden family consisted of Andrew J. Borden, a bank president and successful businessman; his second wife Abby; and Lizzie and Emma Borden, his unmarried daughters by his first marriage. The Bordens lived at 92 Second Street, Fall River, Massachusetts. The other occupant of the house was Bridget Sullivan, a maid-of-all-work who had been employed by Andrew Borden for two years and nine months.

On the morning of August 4, 1892, John V. Morse, Abby Borden's brother, who had been an overnight guest at the Borden home, rose early and was in the sitting room by 6:00 A.M. Bridget Sullivan appeared shortly thereafter and started to prepare breakfast. Abby Borden joined her brother, as did Andrew Borden. After breakfast Andrew departed for his bank, after inviting Morse to return for the noon meal. Bridget, having washed the breakfast dishes, prepared to clean the windows as she had been instructed to do. Lizzie Borden had not as yet come down for breakfast, and her older sister Emma was away visiting friends in Fairhaven, Massachusetts.

Andrew Borden arrived at the bank at 9:30, at about which time his wife Abby was murdered with an ax in the guest room of her home. The autopsy showed that she had been struck no fewer than nineteen times.

Returning home at 10:40 A.M., Andrew Borden was admitted by Bridget Sullivan who had been washing windows on the south side of the house. She locked the front door after Borden came inside. Lizzie, who had arisen by this time, had a brief conversation with her father who then lay down on his sitting room sofa. His daughter supposedly went to the barn loft where she "searched for some sinkers." Bridget Sullivan, who went to her attic room to rest, later testified that she had heard the city hall clock strike the hour of eleven.

Lizzie Borden, returning to the house at approximately 11:08, found the body of her father, the victim of an ax murderer, lying on the sofa, his skull split open. She screamed and Bridget, from the head of the stairs, asked, "What's the matter?" Lizzie told her that her father had been killed and sent Bridget to bring Dr. Seabury W. Bowen. This was at 11:12 A.M.

Dr. Bowen was not at his office and a neighbor, Mrs. Adelaide B. Churchill, was sent by Bridget to bring Miss Alice E. Russell, Lizzie's best friend. Miss Russell came to the Borden home at once. The police were notified and it was after their arrival that the mutilated body of Abby Borden was found in the guest room by Bridget Sullivan. Like her husband she was the victim of an ax murderer. The coagulation of the blood caused medical examiner Dr. William A. Dolan to conclude that she had been murdered prior to her husband.

John V. Morse returned to the Borden home. The police, in searching the house, located four hatchets in the cellar. However, no evidence which could connect them with the two murders was found.

A preliminary hearing was held before Judge Joseph C. Blaisdell on August 22, and Lizzie Borden, who had been arraigned following the inquest, was returned to jail at Taunton, Massachusetts, where she was being held. On December 2 the grand jury of Bristol County returned indictments against Lizzie Borden and on May 8, 1893, she was taken to New Bedford, Massachusetts, and arraigned before Judge J. W. Hammond of the Supreme Court to plead on the indictments. Her plea was "not guilty," and the trial date was set for June 5.

Lizzie Borden was represented by attorneys Andrew Jennings and former governor George D. Robinson. The trial was presided over by three judges, Chief Justice Albert Mason, Associate Justice Caleb Blodgett, and Justice Dewey.

The trial lasted thirteen days. What evidence was introduced by the prosecution was purely circumstantial. Contradictory stories were told by various witnesses concerning a Bedford cord dress which Lizzie had burned in the kitchen stove. The prosecution claimed that it had been spattered with blood when the murders were committed and its burning involved the destruction of evidence. The defense claimed that the dress had been disposed of as it had been smeared with paint when the Borden home had been painted some time before. The various police officers involved told conflicting stories. Hannah Reagan, matron of the police station where Lizzie had been held, gave evidence under oath which was contradicted by her sister Emma, also under oath. Lizzie did not take the stand, saying only, "I am innocent. I leave it to my counsel to speak for me." A drug clerk testified that Lizzie had tried to purchase some hydrocyanic acid a few days previous and that he had refused to sell it to her. His testimony was uncorroborated.

Following closing arguments and the charging of the jury by Justice Dewey the jury retired and, after a short deliberation, returned a verdict of "not guilty."

Subsequently the Borden sisters sold the house on Second Street and moved to a new residence on Frank Street in Fall River. Here, in Maplecroft, their new home, they lived until 1905, when Emma, having quarreled with her sister, moved out. Lizzie remained there until 1927, when she died following a serious operation.

The Borden murders aroused much speculation. They have furnished material for serious studies, for novels, plays, an opera and even a ballet. While acquitted by a court of law, Lizzie Borden is thought by some to have been guilty of murder, although one student of the crime suggested that the killer might have been the servant, Bridget Sullivan. Another puts forth the theory that the murders may have been committed by an unknown homicidal maniac who gained entrance to the house during the night.

Suggested Readings

Aymar, Brandt, and Edward Sagarin. "Lizzie Borden," in their *A Pictorial History of the World's Great Trials*. New York: Crown Publishers, 1967.

DeMille, Agnes, choreographer; music by Morton Gould. *Fall River Legend*. American ballet (prologue, 8 scenes), based on the true story of Lizzie Borden. First produced at Metropolitan Opera House, New York City, April 22, 1948.

Elmslie, Kenward, librettist; music by Jack Beeson. *Lizzie Borden*. American opera (3 acts), based on the true story of Lizzie Borden. First produced at City Center, New York City, April 1, 1965.

Gross, Gerald. "A Postscript to 'The Final Word': The Pearson-Radin Controversy Over the Guilt of Lizzie Borden," in Pearson, Edmund. *Masterpieces of Murder*. Boston: Little, Brown, 1963.

Hart, Harold H. "The Borden Murder Case," in his *From Bed to Verse*. New York: Hart Publishing Co., 1966.

Ketchum, Richard M. "Fall River Legend," *American Heritage Magazine*, April, 1964.

Kunstler, William H. "Murder in Hatred: The Commonwealth of Massachusetts versus Lizzie Borden," in his *First Degree*. New York: Oceana Publications, 1960.

Lincoln, Victoria E. *Private Disgrace: Lizzie Borden by Daylight*. New York: Putnam, 1967.

Lustgarden, Edgar. "Lizzie Borden," in his *Verdict in Dispute*.

New York: Scribner's, 1950.
________. "The Lizzie Borden 'Axe Murder' Case," in Rubenstein, Richard E., ed. Great Courtroom Battles. Chicago: Playboy Press, 1973.
Pearson, Edmund. "The Final Word; The End of the Borden Case," in his Masterpieces of Murder. Boston: Little, Brown, 1963.
________. Five Murders: With a Final Note on the Borden Case. New York: Crime Club, 1928.
________. Studies in Murder. Garden City, N.Y.: Doubleday, 1924.
________. The Trial of Lizzie Borden. Garden City, N.Y.: Doubleday, 1937.
Pearson, Mrs. Edmund. "A Postscript to 'The Pearson-Radin Controversy'," in Pearson, Edmund. Masterpieces of Murder. Boston: Little, Brown, 1963.
Porter, Edwin H. The Fall River Tragedy--A History of the Borden Murders. Fall River, Mass.: G. H. R. Buffinton, Publisher, 1893.
Radin, Edwin D. Lizzie Borden--The Untold Story. New York: Simon & Schuster, 1961.
Reach, James. "The Myth of Lizzie Borden," in Boucher, Anthony, pseud., ed. The Quality of Murder. New York: Dutton, 1962.
Symons, Julian. A Pictorial History of Crime. New York: Crown Publishers, 1966.
Wallechinsky, David, and Irving Wallace. "The Lizzie Borden Case," in their The People's Almanac. Garden City, N.Y.: Doubleday, 1975.

POLLOCK VS. FARMERS' LOAN AND TRUST COMPANY (1894)

The constitutionality of the income tax was established with the addition of the Sixteenth Amendment to the Constitution of the United States on February 25, 1913. Prior to this the legality of this tax was open to question and was the subject of an 1894 Supreme Court decision in the case of Pollock vs. Farmers' Loan and Trust Company.

This was not the first income tax decision to be made by the Court. It was generally assumed that this question had been settled by previous decisions, particularly in the 1880 case of Springer vs. United States, in which the law had been upheld.

The Wilson Tariff Bill of 1894 had included provision for such a tax. In the Pollock case, arguments against the tax were pressed vigorously for economic and social reasons as well as on constitutional grounds. Important financial interests, vitally concerned with the issue, lobbied against it and hired the best available legal talent to defeat it.

Article I, Section 2 of the Constitution deals with "direct" taxes and states that "direct Taxes shall be apportioned among the several States ... according to their respective Numbers." Consequently, if an income tax is a direct tax, it must be apportioned among the states according to population. Those who framed the Constitution did not define the word "direct," nor could they have, for no such tax existed in either the United States or England when the Constitution was drawn up. This placed the Supreme Court in the ambiguous position of defining a word which those who framed the Constitution could not define. The general thinking hitherto had been to regard the income tax as indirect and therefore constitutional, even though it was not apportioned according to population.

The Pollock case was heard on two separate occasions. On the first hearing Justice Howell Edmunds Jackson was ill and so did not participate. The remaining justices voted four to four. At the second hearing Jackson dissented from the majority opinion while Justice David Josiah Brewer reversed himself, voting with the majority. By the narrow margin of five to four the income tax provision of the Wilson-Gorman Act was declared null and void.

It was this decision, which aroused bitter dissatisfaction, that led to the passage of the Sixteenth Amendment. This Amendment reads, "The Congress shall have power to lay and collect taxes on income, from whatever source derived, without apportionment among the several States and without regard to any census or enumeration."

Suggested Readings

Boudin, Louis B. Government by Judiciary. New York: William Goodwin, 1932.

Carson, Hampton L. The History of the Supreme Court. New York: Burt Franklin, 1972.

Commager, Henry Steele, ed. "The Constitution of the United States," (Doc. No. 87) in his Documents of American History, 8th edition. New York: Appleton, 1968.

________. "Pollock v. Farmers' Loan and Trust Company," (Doc. No. 333) in his Documents of American History, 8th edition. New York: Appleton, 1968.

Goode, Richard. Individual Income Tax. Washington, D.C.: Brookings Institution, 1964.

Johnson, Gerald White. The Supreme Court. New York: Morrow, 1962.

McCloskey, Robert G. The American Supreme Court. Chicago: University of Chicago Press, 1960.

Nevins, Allan. Grover Cleveland: A Study in Courage. New York: Dodd, Mead, 1932.

Seligman, Edwin R. The Income Tax. Clifton, N.J.: Augustus M. Kelley, 1914.

Surface, William E. Inside Internal Revenue. New York: Coward-McCann, 1967.

Warren, Charles. The Supreme Court in United States History. Boston: Little, Brown, 1923.

COXEY'S ARMY (1894)

One of the more severe business and economic depressions in American history was that of 1893. Bank failures, reduced production of such commodities as coal and iron, bankruptcies of business firms, falling prices on the stock and commodities exchanges, a shortage of coin and paper currency and other characteristics of a financial panic led inevitably to mass unemployment and hardship among the wage earners.

"General" Jacob Sechler Coxey, an American politician of Massillon, Ohio, proposed to end the depression of 1893 by a public works program to be undertaken by federal and local authorities. He specified that the Secretary of the Treasury should issue a non-interest bearing 25-year bond, to be retired at the rate of 4% per year. He specified further that $500,000,000 should be provided by the federal government for the construction of roads throughout the nation. This construction, he proposed, would be under the direction of the Secretary of War; for an eight-hour work-day each workman should be paid not less than $1.50, while "teams and labor" should be paid not less than $3.50 for a day's work.

Starting on March 25, 1894, Coxey led an "army" of unemployed workmen from Massillon to Washington, D.C., arriving there on April 28. At the capital he and his followers, on May 1, attempted to stage a demonstration. This led to his arrest. He was released on June 10, after being nominated in Ohio for Congress. Later he was a candidate for various public offices, from Mayor of Massillon to President of the United States. He was elected to the office of mayor and served in that capacity from 1931 through 1933.

Suggested Readings

Commager, Henry Steele, ed. "Coxey's Program," (Doc. No. 332) in his Documents of American History, 8th edition. New York: Appleton, 1968.

Coxey, Jacob S. The Big Idea (broadside). Massillon, Ohio: April 16, 1946. Published by the author.

________. The Coxey Plan (pamphlet). Massillon, Ohio: April, 1914. Published by the author.

Gray, Wood. Hidden Civil War. New York: Viking Press, 1942.

Howard, Oliver Curtis. "The Menace of Coxeyism," North American Review, June, 1894.

McMurray, Donald L. Coxey's Army: A Study of the Industrial Army Movement of 1894. New York: AMS Press, 1970.

Milton, George Fort. Abraham Lincoln and the Fifth Column. New York: Vanguard Press, 1942.

Nye, Russel B. A Baker's Dozen: Thirteen Unusual Americans. East Lansing, Mich.: Michigan State University Press, 1956.

Stead, William T. "The Coxey Crusade," Review of Reviews, April, 1894.

________. "Coxeyism," Review of Reviews, July, 1894.

Tragle, Henry I. Coxey's Army. New York: Grossman, 1974.
Vallandigham, James. The Life of Clement L. Vallandigham. Baltimore: 1872.
Vincent, Henry. The Story of the Commonweal. New York: Arno Press, 1970.

THE PULLMAN STRIKE (1894)

In 1880 George Mortimer Pullman, American industrialist and inventor of the sleeping car used on railroads, founded the industrial town of Pullman, Illinois, near Chicago. This was a typical company town. The houses, largely sub-standard, were rented to employees at high rates. The employees were required to trade at the company-owned store where the prices charged were exorbitant. Those employees who protested or who refused to comply with Pullman rules were summarily discharged.

In 1893 and 1894 the American economy suffered a general depression. The Pullman Palace Car Company reduced the wages of its workmen approximately 25% but did not reduce the prices charged at the company store or the rents charged for company-owned houses. The salaries paid to the higher company officials remained untouched.

A committee of workmen requested a return to the former wage scale, but this was refused. When three members of the committee were laid off, the employees struck. In June, 1894, the American Railway Union, under the leadership of Eugene V. Debs, sided with the striking workmen, and the General Managers' Association, comprising the officials of 24 railroads entering Chicago, sided with the Pullman Company.

Violence followed. Tracks were torn up, railroad cars were tipped over and railroad property was destroyed. Ultimately the federal government entered the situation after complaints reached Washington that the United States mails and interstate commerce were being interfered with.

The Constitution requires that the United States protect states from domestic violence on the application of the legislature or of the executive if the legislature is not in session. The President is also empowered to use federal force to execute federal laws. Governor John Peter Altgeld of Illinois protested President Grover Cleveland's decision to send federal troops to Chicago and asked that they be withdrawn. His request was ignored, the President stating that the mail service was being obstructed and that this was sufficient justification for the use of federal troops.

On July 2 the United States District Court issued a blanket injunction forbidding all persons to interfere with the operation of trains. Debs and other union officers continued to direct the strike, for which they were arrested and imprisoned for contempt of court.

The combination of federal troops and the loss of their union leaders made the strikers realize that their cause was a lost one, and they surrendered.

Later investigations by a commission appointed by President Cleveland showed that the Pullman Company had exploited its employees, that there had been no destruction of property in the town of Pullman, and that the Company had refused to arbitrate the dispute in spite of requests that they do so by public officials. The commission also found that the Managers' Association had armed and paid 3,600 deputy marshals who, under the direction of the Managers, acted as both United States officers and railroad employees.

Suggested Readings

Browne, Waldo R. Altgeld of Illinois: A Record of His Life and Work. New York: B. W. Huebsch, 1924.

Buder, Stanley. Pullman: An Experiment in Industrial Order and Community Planning. New York: Oxford University Press, 1967.

Cleveland, Grover. Presidential Problems. Plainview, N.Y.: Books for Libraries, 1975.

Commager, Henry Steele, ed. "The Altgeld-Cleveland Controversy," (Doc. No. 334) in his Documents of American History, 8th edition. New York: Appleton, 1968.

________. "In Re Debs," (Doc. No. 336) in his Documents of American History, 8th edition. New York: Appleton, 1968.

________. "U.S. v. Debs et al.," (Doc. No. 335) in his Documents of American History, 8th edition. New York: Appleton, 1968.

Cook, Roy. Leaders of Labor. Philadelphia: Lippincott, 1966.

Darrow, Clarence. The Story of My Life. New York: Scribner's, 1932.

Dos Passos, John. "Lover of Mankind," in his The 42nd Parallel. New York: Random House, 1930.

Ginger, Ray. Altgeld's America: The Lincoln Ideal vs. Changing Realities. New York: Funk & Wagnalls, 1958.

________. The Bending Cross: A Biography of Eugene Victor Debs. New Brunswick, N.J.: Rutgers University Press, 1949.

Heaps, Willard Allison. Riots, U.S.A., 1765-1970. New York: Seabury Press, 1970.

Iman, Raymond S., and Thomas W. Koch. Labor in American Society. Chicago: Scott, Foresman, 1965.

Johnsen, Julia E., comp. The Closed Shop. New York: Wilson, 1942.

Lens, Sidney. Unions and What They Do. New York: Putnam, 1968.

Lindsey, Almont. The Pullman Strike. Chicago: University of Chicago Press, 1942.

Lingley, Charles Ramsdell, and Allen Richard Foley. Since the Civil War. New York: Appleton, 1935.

Litwack, Leon F., ed. The American Labor Movement. Englewood Cliffs, N.J.: Prentice-Hall, 1962.

Meltzer, Milton. Bread--and Roses: The Struggle of American Labor. New York: Knopf, 1967.

Nevins, Allan. Grover Cleveland: A Study in Courage. New York: Dodd, Mead, 1932.
Rayback, Joseph G. A History of American Labor. New York: The Free Press, 1959.
Stone, Irving. Clarence Darrow for the Defense. New York: Doubleday, 1941.
Taft, Philip. Organized Labor in American History. New York: Harper, 1964.
Warne, Colston E., ed. The Pullman Boycott of 1894. Boston: Heath, 1955.

THE HOLMES-MUDGETT MURDERS (1894)

On September 4, 1894, a body, thought to be that of B. F. Perry, an American business swindler, was discovered in Perry's Philadelphia office by a carpenter named Smith. Smith notified the authorities who determined at the subsequent inquest that Perry "had died of congestion of the lungs caused by the inhalation of flame or chloroform."

The Philadelphia branch of the Fidelity Mutual Life Association then received a letter from Jephtha D. Howe, a St. Louis attorney, stating that Perry's real name was Benjamin F. Pitezel, and that he had been insured for $10,000 by Fidelity's St. Louis office the previous November. Howe offered to bring members of Pitezel's family to Philadelphia to identify the remains.

The insurance company contacted one Harry Howard Holmes of Wilmette, Illinois, who, their investigations showed, had known Pitezel in Chicago. Holmes was asked to help with the identification of Perry/Pitezel's body and answered that he would be happy to do so provided his expenses were paid. On September 23, in Philadelphia, he identified the body as that of Pitezel. The same day Howe and Alice Pitezel, the fifteen-year-old daughter of the dead man, arrived and confirmed the identification. Howe explained that Pitezel had taken the name of Perry "owing to financial difficulties." The insurance company paid Howe as the legal representative of Pitezel's widow the sum of $9,175 and gave Holmes ten dollars as expenses.

The insurance company presumed that with the payment of the claim the matter was closed. However, Marion Hedgspeth, a convicted train robber who was serving time in a St. Louis prison, revealed that a fellow-prisoner named H. M. Howard had told him of a plan to swindle an insurance company to the extent of $10,000. Hedgspeth stated further that he had been offered $500 to recommend an attorney who would participate in the enterprise, that he had recommended Jephtha D. Howe, that Howe had told him that "the deal went off smoothly," and that he had not received any part of the promised $500. This was in October, 1894, after Pitezel had been found dead and after Howard had been released from prison.

The insurance company called in Allan Pinkerton, the famous detective, to investigate Holmes. Holmes was arrested in Boston on November 17. Following his arrest he admitted that he had attempted to swindle the insurance company. He stated that Pitezel was involved, that the dead body found in Perry/Pitezel's office had been obtained from a physician whom he did not identify, and that the conspirators planned that, following the payment of the insurance money, Pitezel would disappear.

Pitezel's wife traveled to Boston where, when questioned by the police, she denied any complicity in the fraud. She was anxious for news concerning her husband and three children. Alice had come to Philadelphia with Howe to identify her father's body. The other two children, Howard and Nellie, had been taken from St. Louis by Holmes, ostensibly to join Alice who, Holmes had told Pitezel's wife, was in Covington, Kentucky.

The case became complicated. After months in jail Holmes was tried for and convicted of fraud. Sentence was suspended and he remained in jail while an intense search was made for the Pitezel children who, Holmes stated, had gone to England with one Minnie Williams, a friend of his.

Detective Frank P. Geyer traced the movements of Holmes prior to his arrest. His odyssey, following the path of Holmes, took him to Indianapolis, Cincinnati, Detroit, and Toronto, in which cities Holmes, accompanied by three children, had stayed. He also learned that Holmes had murdered them, two by gas and one probably beaten to death.

It was determined that Holmes was actually Herman Webster Mudgett and that he had, in the course of various illegal transactions, used a number of other aliases, including "H. M. Howard" while incarcerated in the jail at St. Louis where he had confided in Marion Hedgspeth. He had three wives, each of whom he had married under a different name. It was found that in 1893 he had constructed, in Chicago, at the intersection of 63rd and Wallace Streets, a "murder castle." This extraordinary building contained secret staircases, airtight chambers with false ceilings, a room lined with asbestos "to stifle screams," a shaft through which bodies could be dropped to the cellar, a crematory and vats of quicklime. Detectives exploring the cellar found human skulls and teeth. A trunk and a piece of watch chain found at the "murder castle" were identified as having belonged to Minnie Williams who had lived there briefly with him. Other evidence linked Holmes/Mudgett with Pitezel in a fraudulent real estate transaction involving property belonging to Miss Williams.

The multiple murderer was tried, convicted, and sentenced to death by hanging. The sentence was carried out on May 7, 1896.

Suggested Readings

Abrahamsen, David, M.D. The Murdering Mind. New York: Harper & Row, 1973.

Asbury, Herbert. Gem of the Prairie: An Informal History of the Chicago Underworld. New York: Knopf, 1940.

Brannon, William T. "The Anatomical Practice of Dr. H. H. Holmes," in Boucher, Anthony, pseud. The Quality of Murder. New York: Dutton, 1962.
Duke, Thomas S. Celebrated Criminal Cases of America. San Francisco: Barry, 1910.
Gribble, Leonard. "The Castle of Secrets," in his They Had a Way With Women. New York: Roy Publishers, 1967.
Guttmacher, M. S. The Mind of the Murderer. Freeport, N.Y.: Books for Libraries, 1960.
Holmes, Harry Howard, pseud. Holmes' Own Story (pamphlet). Philadelphia, 1895.
Horan, James D. The Pinkertons: The Detective Dynasty That Made History. New York: Crown Publishers, 1967.
________, and Howard Swigett. The Pinkerton Story. New York: Putnam, 1931.
Irving, H. B. "The Mysterious Mr. Holmes," in his A Book of Remarkable Criminals. New York: Doran, 1918.
Lavine, Sigmund A. Allan Pinkerton: America's First Private Eye. New York: Dodd, Mead, 1963.
Logan, Guy B. H. "America's Mass Murderers," in his Rope, Knife and Chair. New York: Duffield, 1930.
Orrmont, Arthur. Master Detective: Allan Pinkerton. New York: Messner, 1965.
Pinkerton, Allan. Criminal Reminiscences and Detective Sketches. Freeport, N.Y.: Books for Libraries, 1970. (Originally published 1878.)
________. Professional Thieves and the Detectives. New York: G. W. Carleton & Co., 1881.
Reinhardt, James Melvin. The Psychology of Strange Killers. Springfield, Ill.: Thomas, 1962.
Rowan, Richard Wilmer. The Pinkertons, a Detective Dynasty. Boston: Little, Brown, 1931.
Symons, Julian. A Pictorial History of Crime. New York: Crown Publishers, 1966.

COIN'S FINANCIAL SCHOOL (1894)

During the last decades of the 19th century the silver question was one which occupied the attention of virtually everyone from "the man in the street" to the legislators in Washington. Various Acts affecting the coinage of silver had been passed, including the Coinage Act of February 12, 1873, the Bland-Allison Act of February 28, 1878, and the Sherman Silver Purchase Act of July 14, 1890. The matter of silver vs. gold constituted an important issue in the platforms of both the Democratic and Republican Presidential conventions of 1896.

Citizens of the South and West favored a policy of free and unlimited coinage of silver; those of the East preferred a gold standard monetary system.

One of the more influential propaganda publications advocating the free coinage of silver was William H. Harvey's book, Coin's Financial School, which appeared in 1894. This was an emotional appeal to its readers, and economists and students of monetary policy found a number of flaws in its logic. It submitted arguments for the increased use of silver and "brought forward objections which were triumphantly demolished."

The cartoon-type illustrations made points favoring the author's arguments. One such cartoon depicted Senator John Sherman and President Grover Cleveland digging out part of a foundation of a house which had been built on a stable basis of gold and silver. The portion they were removing was labeled "silver." Again, an illustration showed a one-legged man, representing the state of the country without the free coinage of the two metals. Still another picture showed New York and New England capitalists milking a cow which was being fed by Western farmers.

Suggested Readings

Bell, Ovid. "'Silver Dick' Bland," Missouri Historical Society Bulletin, Vol. 9, 1953.

Commager, Henry Steele, ed. "The Bland-Allison Act," (Doc. No. 299) in his Documents of American History, 8th edition. New York: Appleton, 1968.

________. "Bryan's Cross of Gold Speech," (Doc. No. 342) in his Documents of American History, 8th edition. New York: Appleton, 1968.

________. "The Crime of '73," (Doc. No. 285) in his Documents of American History, 8th edition. New York: Appleton, 1968.

________. "Repeal of the Sherman Silver Purchase Act," (Doc. No. 328) in his Documents of American History, 8th edition. New York: Appleton, 1968.

________. "The Sherman Silver Purchase Act," (Doc. No. 321) in his Documents of American History, 8th edition. New York: Appleton, 1968.

Destler, Chester M. American Radicalism, 1865-1901. New York: Quadrangle/The New York Times Co., 1966.

Harvey, William H. Coin's Financial School. Edited by Richard Hofstadter. Cambridge, Mass.: Harvard University Press, 1975. (First published, 1894.)

Hofstadter, Richard. "Free Silver and the Mind of 'Coin Harvey'," in his The Paranoid Style in American Politics and Other Essays. New York: Random House, 1967.

Lewis, E. G. Contributions of John Sherman to Public and Private Finance. (Unpublished Ph.D. dissertation, University of Illinois, 1932.)

Nevins, Allan. Grover Cleveland: A Study in Courage. New York: Dodd, Mead, 1932.

Porter, Kirk, and Donald Bruce Johnson, comps. National Party Platforms, 1840-1960. Urbana, Ill.: University of Illinois Press, 1956.

Randall, James G. "John Sherman and Reconstruction," Mississippi

Valley Historical Review, December, 1932.
Robinson, Edgar E. The Evolution of American Political Parties. New York: Johnson Reprint, 1971.
Sherman, John. Recollections of Forty Years in the House, Senate, and Cabinet: An Autobiography. New York: Greenwood Press, 1968.
Sullivan, Mark. Our Times, The United States, 1900-1925. New York: Scribner's, 1928-1935.

THE DURRANT MURDERS (1895)

The Emmanuel Baptist Church in San Francisco was the scene of two gruesome murders in April, 1895. The murderer was William Henry Theodore Durrant, a schizophrenic medical student and assistant Sunday School superintendent. His victims were Blanche Lamont, a 21-year-old high school girl, and Minnie Flora Williams, a housemaid. Both girls were friends of Durrant.

On the afternoon of April 3, Durrant met Blanche Lamont as she left school. Together they rode on a cable car to the Emmanuel Baptist Church where, at approximately 4:30, they were seen going inside. Half an hour later George King, the church organist, arrived to find Durrant composed but pale and looking extremely tired. Questioned, Durrant told King that he had repaired a leaky gas jet and had inhaled some of the fumes.

When Blanche Lamont failed to return home her family instituted a search. Her nude body was found in the belfry of the church. She had been strangled.

On April 12 Minnie Flora Williams left her place of employment to meet Durrant. They were seen entering the church together by a man named Zengler. This man saw Durrant leave the church alone shortly afterwards. Minnie Williams' knife-slashed body was found behind a door in the church library the following day by a woman who had come to put up some Easter decorations.

Durrant was eventually arrested and charged with both murders. He protested his innocence until the last. Nevertheless, he was tried, found guilty of murder and hanged on January 7, 1898.

Suggested Readings

Boucher, Anthony, pseud. "The Demon in the Belfry; The Case of Theodore Durrant--The Legends," in Jackson, Joseph Henry, ed. San Francisco Murders. New York: Duell, Sloan, 1947.
Duke, Thomas S. Celebrated Criminal Cases of America. San Francisco: Barry, 1910.
"Durrant Greatly in Evidence," New York Times, July 25, 1895, p. 8.
"Durrant's Trial Will Begin Today," New York Times, July 22, 1895, p. 1.

Fanning, Pete. Great Crimes of the West. San Francisco: Barry, 1929.
Gribble, Leonard. "The Teen-age Nudes," in his They Had a Way With Women. New York: Roy Publishers, 1967.
Leach, Harold, comp. The Crime of the Century, or the Mystery of Emanuel Baptist Church, San Francisco. San Francisco: Yosemite Publishing Co., 1895.
Logan, Guy B. H. Great Murder Mysteries. London: Paul, 1931.
McDade, Thomas M., comp. The Annals of Murder. Norman, Okla.: University of Oklahoma Press, 1961.
"No Change of Venue for Durrant," New York Times, July 26, 1895, p. 5.
Peixotto, Edgar D. Report of the Trial of Henry Theodore Durrant. Detroit: Collector Publishing Co., 1899.
Symons, Julian. A Pictorial History of Crime. New York: Crown Publishers, 1966.
Taussig, Jacob. Letter F., or, Startling Revelations in the Durant Case. New York: N. Savier & Co., 1895.
Teilhet, Hildegarde. "The Demon in the Belfry; The Case of Theodore Durrant--The Facts," in Jackson, Joseph Henry, ed. San Francisco Murders. New York: Duell, Sloan, 1947.

THE "CROSS OF GOLD" SPEECH (1896)

The "Cross of Gold" speech was delivered by William Jennings Bryan, delegate from Nebraska to the Democratic National Convention at Chicago, on July 8, 1896. It is considered to be the outstanding speech of his career and "one of the most notable in the history of American oratory." It "dramatized Bryan before the Convention and did much to bring about his already probable nomination" for the office of President of the United States. It was the concluding speech in the debate on the adoption of the Democratic platform.

In 1896 the silver question was an important economic and political issue. William McKinley, the Republican candidate, ran on a platform which included a gold monetary standard and which opposed the free coinage of silver. This allied him with the eastern businessmen and alienated the western advocates of the coinage of silver.

When the Democrats held their convention at Chicago the situation was dominated by the silver advocates. Bryan captivated the assembled delegates with his famous "Cross of Gold" speech. In it he contended that a gold standard would tend to enrich the already wealthy businessman and that farmers, miners, and workingmen were businessmen as much as were "the few financial magnates who, in a back room, corner the money of the world." Bryan insisted that the money contest was between the "idle holders of idle capital and the struggling masses who produce the capital." He concluded by saying, "You shall not press down upon the brow of labor this crown of thorns, you shall not crucify mankind upon a cross of gold."

Following this stirring oration Bryan was nominated for President on the fifth ballot. The Democratic platform as finally adopted committed that party to free and unlimited coinage of silver and gold at the existing legal ratio. By this, sixteen parts of silver by weight were declared equal to one part of gold in the minting of coins.

The Republican National Convention, held at St. Louis, Missouri, nominated William McKinley on a platform advocating the gold standard and high tariffs. The victorious McKinley became the 25th President of the United States.

Suggested Readings

Bryan, William Jennings. The First Battle. Port Washington, N.Y.: Kennikat Press, 1970.

________, and Mary B. Bryan. The Memoirs of William Jennings Bryan. Philadelphia: Winston, 1925.

Coletta, Paolo E. William Jennings Bryan. Lincoln, Neb.: University of Nebraska Press, 1969.

Commager, Henry Steele, ed. "Bryan's Cross of Gold Speech," (Doc. No. 342) in his Documents of American History, 8th edition. New York: Appleton, 1968.

________. "The Democratic Platform of 1896," (Doc. No. 343) in his Documents of American History, 8th edition. New York: Appleton, 1968.

________. "The Republican Platform of 1896," (Doc. No. 341) in his Documents of American History, 8th edition. New York: Appleton, 1968.

Congressional Quarterly Service. Presidential Candidates from 1788 to 1964, Including Third Parties, 1832-1964, and Popular Electoral Vote: Historical Review. Washington, D.C.: Government Printing Office, 1964.

Dos Passos, John. "The Boy Orator of the Platte," in his The 42nd Parallel. New York: Random House, 1930.

Glad, Paul W. McKinley, Bryan and the People. Philadelphia: Lippincott, 1964.

Harvey, William H. Coin's Financial School. Edited by Richard Hofstadter. Cambridge, Mass.: Harvard University Press, 1975. (First published, 1894.)

Hibben, Paxton. The Peerless Leader: William Jennings Bryan. New York: Russell & Russell, 1929.

Hofstadter, Richard. "Free Silver and the Mind of 'Coin Harvey'," in his The Paranoid Style in American Politics and Other Essays. New York: Random House, 1967.

Johnson, Alexander, and James A. Woodburn. American Orations. Plainview, N.Y.: Books for Libraries, 1896.

Laughlin, J. Laurence. The History of Bimetallism in the United States. New York: Appleton, 1897.

Leech, Margaret. In the Days of McKinley. New York: Harper, 1959.

Lingley, Charles Ramsdell, and Allen Richard Foley. Since the Civil War. New York: Appleton, 1935.
Porter, Kirk H., and Donald Bruce Johnson, comps. National Party Platforms, 1840-1960. Urbana, Ill.: University of Illinois Press, 1956.
Porter, Robert P. The Life of William McKinley, Soldier, Lawyer, Statesman. Cleveland: World, 1896.
Roseboom, Eugene H. A History of Presidential Elections, from George Washington to Richard M. Nixon. New York: Macmillan, 1970.
Schlesinger, Arthur M., Jr., ed. History of American Presidential Elections, 1789-1968. New York: Chelsea House, 1971.
Sullivan, Mark. Our Times, The United States, 1900-1925. New York: Scribner's, 1928-1935.
Werner, M. R. Bryan. New York: Harcourt, Brace, 1929.

THE CONWAY KIDNAPING (1897)

John Conway, the young son of Michael J. Conway, a railroad train dispatcher, was kidnaped from his home in Albany, New York, on Monday, August 16, 1897. His mother learned that John had been kidnaped when a small boy delivered to her a verbose, misspelled, ungrammatical letter signed "The Captain of the Gang." The writer of the letter demanded three thousand dollars ransom and gave detailed instructions concerning the manner in which the money was to be paid.

The boy had been playing in front of his home when last seen. His father, upon being shown the ransom note, immediately contacted the Albany police. Two detectives were dispatched to search the area and question the neighbors. Their search yielded nothing.

In accordance with the instructions in the note, a man was sent to the meeting place indicated, but he left there a dummy package of paper rather than the cash demanded. Unfortunately, the police accompanied him to the scene and revealed their presence by carrying lighted lanterns. No one came to pick up the package.

The police devoted the following day to further searching. This brought no results. A newspaper reporter, John F. Farrell, hoping for a scoop, interviewed the boy's father. He learned, among other things, that Joseph M. Hardy, Conway's brother-in-law, had, from time to time, borrowed small sums of money from him. On one occasion he had demanded one thousand dollars, a demand which Michael Conway refused. Failing to receive the money, Hardy had made vague threats.

Farrell investigated Hardy. He learned that his financial affairs were chaotic and that he was a close friend of H. G. Blake, "a man of no very definite social grade, means of livelihood, or character." Further investigation disclosed the fact that on the morning of August 16, a man resembling Blake had rented and signed for a horse and buggy. The handwriting on the receipt resembled that of Blake but the name was different. It was assumed that Blake had rented the horse and buggy under a fictitious name.

Hardy, following questioning by the reporter, was allowed to return home, believing that he was under no suspicion. Blake was taken to the newspaper office by the reporter and was offered $2,500 for information which would lead to the recovery of young John Conway. He was shown a wallet of money and agreed to take the reporter to the place where the boy was hidden. It seemed obvious that he was one of the kidnapers.

That night Blake, two reporters and two police officers, the last disguised as drivers, set out in a horse-drawn buggy. They proceeded to the woods north of Albany where they discovered John Conway by a campfire. With him was a masked man armed with a pistol.

The policemen exchanged shots with the masked man. The boy was rescued and placed in the buggy, and the reporters and police officers drove rapidly away. They were not followed.

Young John Conway was returned to his parents unharmed. He said that while playing in front of his house on Monday morning he had been accosted by a stranger who enticed him into a waiting buggy. He was driven into the country, spent one night in a deserted cabin, and another in a church before his captors set up the camp in the woods from which he was rescued.

Joseph M. Hardy was arrested. Blake, located in Schenectady, New York, where he had fled, was also arrested. Neither would identify the masked man who had fired on the rescuing party in the woods.

In New York City William N. Loew, a lawyer, notified the newspapers that he believed he could contribute some valuable information concerning the kidnaping. He had, a month previously, acted as attorney for Albert Warner in a case involving the attempted blackmail of one Bernard Myers. Warner was also an attorney. He told Loew that he was interested in a plan to organize kidnaping on a commercial scale. At the time Loew dismissed Warner's tale as nonsense, but later felt that he should tell the authorities what Warner had told him.

In Albany Blake, under questioning, admitted that he was an old schoolmate of Warner's and that Warner had been involved in the kidnaping. Warner promptly disappeared from his New York office and a nationwide hunt was made for him. He was eventually located in Riley, Kansas, working as a farmhand under the name of George Johnson. He was arrested and returned to Albany where he was ultimately tried and sentenced to fifteen years' imprisonment. Hardy and Blake were also tried and given identical sentences of fourteen and one half years.

Suggested Readings

"Albany's Abduction Case," New York Times, August 21, 1897, p. 1.

"Albany's Kidnapped Boy," New York Times, August 18, 1897, p. 1.

"Boy Abducted in Albany," New York Times, August 17, 1897, p. 1.
Cohen, Bruce J. Crime in America. Itasca, Ill.: Peacock, 1970.
Hibbert, Christopher. Roots of Evil: A Social History of Crime and Punishment. New York: Funk & Wagnalls, 1968.
"Kidnapper Warner's Fate," New York Times, December 24, 1897, p. 5.
"Kidnapping Victims: Tragic Aftermaths," Saturday Evening Post, April, 1976.
Messick, Hank, and Burt Goldblatt. Kidnapping: The Illustrated History. New York: Dial Press, 1974.
Slattery, William, ed. Abduction: Fiction Before Fact. New York: Grove Press, 1974.
Smith, Edward H. "The Stolen Conway Boy," in his Mysteries of the Missing. New York: Lincoln Macveagh-The Dial Press, 1927.
"Stolen Child Recovered," New York Times, August 20, 1897, p. 1.
Sutherland, Edwin H. On Analyzing Crime. Edited by Karl Schussler. Chicago: University of Chicago Press, 1972.
Winslow, Robert W. Crime in a Free Society. Encino, Calif.: Dickenson, 1968.

THE DE LÔME LETTER (1897-1898)

When William McKinley, 25th President of the United States, took office in 1897, Cuba was in revolt against Spain. Grover Cleveland, McKinley's predecessor, was loath to interfere in the internal affairs of other nations but the Cuban situation was affecting American investments in that island. The barbaric cruelty of the Cuban insurrectionists under Máximo Gómez y Baez and the Spanish military under General Valeriano Weyler horrified the American people. The private citizens of Cuba, caught between the warring factions, were experiencing an appalling loss of life, health, and property.

McKinley was more responsive to public opinion than Cleveland had been, and he was aware that the American people were coming more and more to favor intervention in Cuba. His Republican party had, in its platform, committed itself to Cuban independence through American action.

Early in 1898 two events occurred which angered the United States and which helped lead to the Spanish-American War the same year. One was the destruction of the battleship "Maine" in Havana Harbor and the other was the de Lôme Letter.

The "Maine" sinking is discussed elsewhere in this book. The de Lôme Letter was written by Señor Enrique Dupuy de Lôme, Spanish minister to the United States, to Don José Canalejas, a personal friend in Havana. The letter, in the Spanish language, was a private communication not intended for publication.

Written in December, 1897, on the letterhead of the Legation of Spain, it was posted from Washington, D.C., where it had been composed. It was stolen from the Havana post office by an insurgent spy and held by the Cuban junta in the United States until it could be used most effectively.

The de Lôme letter was given to the press of February 9, 1898, and a New York newspaper published it in translation. It spoke of McKinley as "weak and a bidder for the admiration of the crowd, besides being a would-be politician (politcastro) who tries to leave a door open behind himself while keeping on good terms with the jingoes of his party."

The de Lôme letter further indicated that the writer felt "it would be very advantageous to take up, if only for effect, the question of commercial relations and to have a man of some prominence sent hither [to Washington, D.C.] in order that I may make use of him here to carry on a propaganda among the Senators and others in opposition to the junta and to try to win over the refugees...."

When this letter was published de Lôme realized that his usefulness as Spanish Minister to the United States was at an end and immediately cabled his resignation. This was promptly accepted by the Spanish government.

Suggested Readings

Alger, Russell A. The Spanish-American War. Plainview, N.Y.: Books for Libraries, 1901.

Beard, Charles A., and Mary R. Beard. A Basic History of the United States. New York: Doubleday, 1944.

Chadwick, French E. The Relations of the United States and Spain: The Spanish-American War. New York: Russell & Russell, 1968.

Commager, Henry Steele, ed. "The De Lôme Letter," (Doc. No. 345) in his Documents of American History, 8th edition. New York: Appleton, 1968.

Craven, Avery O., Walter Johnson and F. Roger Dunn. A Documentary History of the American People. Boston: Ginn, 1951.

Latané, John H. America as a World Power, 1897-1907. Saint Clair Shores, Mich.: Scholarly Press, 1971.

Leech, Margaret. In the Days of McKinley. New York: Harper, 1959.

Millis, Walter. The Martial Spirit. New York: Viking Press, 1965.

Olcott, Charles S. The Life of William McKinley. New York: AMS Press, 1916.

Porter, Robert P. The Life of William McKinley, Soldier, Lawyer, Statesman. Cleveland: World, 1896.

Wilkerson, Marcus M. Public Opinion and the Spanish-American War. New York: Russell & Russell, 1967.

THE GRANDFATHER CLAUSE (1898)

Following the Civil War the American Negro was a citizen of civil and political importance. Legally and constitutionally his position seemed to be impregnable. However, in the mind of the southern white, the black man was an inferior being and one who should be "kept in his place." Such organizations as the Ku Klux Klan, the Terry Terribles and the Byram Bulldozers were organized to keep the Negroes subjugated and subdued. These organizations resorted to intimidation and violence to enforce their ideas of white supremacy.

As a result of this situation, Congress passed the Enforcement Laws of 1870/71, generally called the "Force Acts." These laws laid heavy penalties on individuals "who should prevent citizens from exercising their constitutional political powers," primarily the right to vote. Eventually these Enforcement Laws were found to be unconstitutional and, in 1883, in the Supreme Court case of United States vs. Harris, they were voided on a technicality.

In the South the movement to eliminate the Negro from active participation in politics continued. In 1890 the state of Mississippi adopted a new constitution which specified certain prerequisites to the voting privilege. One of these was the payment of all taxes which were legally demanded of the citizen during the two preceding years. Another provided that, after January 1, 1892, "every voter must be able to read any section of the state constitution or give an interpretation of it when read to him." This made it certain that the ignorant Negro, when examined on this point by a white examiner, would have little chance of passing the educational test. In 1898 the state of Louisiana inaugurated the "Grandfather Clause." This device, which was subsequently used by several other southern states, allowed citizens to vote provided they had that right before January 1, 1867. Descendants of citizens qualifying under this clause were also entitled to vote, regardless of their educational or property qualifications. It was obvious that, as no Negroes had voted in that state prior to the date specified, they were effectively disfranchised. As a result of the "Grandfather Clause" the Negro vote dwindled away to negligible proportions.

The "Grandfather Clause" remained in effect until the year 1915. Then the Supreme Court declared it unconstitutional on the grounds that its only intention was to evade the provision of the Fifteenth Amendment which states, "The right of citizens of the United States to vote shall not be denied or abridged by the United States or by any State on account of race, color, or previous condition of servitude."

Suggested Readings

Chalmers, David M. Hooded Americanism: The First Century of the Ku Klux Klan, 1865-1965. Garden City, N.Y.: Doubleday,

1965.
Channing, Edward, B. Hart and Frederick Jackson Turner. Guide to the Study and Reading of American History. Norwood, Pa.: Norwood Editions, 1912.
Commager, Henry Steele, ed. "The Constitution of the United States," (Doc. No. 87) in his Documents of American History, 8th edition. New York: Appleton, 1968.
Dye, Thomas R. The Politics of Equality. Indianapolis: Bobbs Merrill, 1971.
Fleming, Walter L. Documentary History of Reconstruction. Gloucester, Mass.: Peter Smith, 1906-1907.
Henderson, L. J., comp. Black Political Life in the United States. Edited by L. J. Henderson, Jr. San Francisco: Chandler, 1972.
Keech, W. R. The Impact of Negro Voting. Chicago: Rand McNally, 1968.
Lewison, Paul. Race, Class and Party. New York: Russell & Russell, 1963.
Lingley, Charles Ramsdell, and Allen Richard Foley. Since the Civil War. New York: Appleton, 1935.
Mabry, William Alexander. The Negro in North Carolina Politics Since Reconstruction. Durham, N.C.: University of North Carolina Press, 1940.
Moon, Henry Lee. Balance of Power: The Negro Vote. Garden City, N.Y.: Doubleday, 1948.
Randel, William P. The Ku Klux Klan: A Century of Infamy. Philadelphia: Lippincott, 1965.

THE DESTRUCTION OF THE "MAINE" (1898)

In the late 19th century the island of Cuba, in the Caribbean Sea, was Spain's most valuable possession in the Western Hemisphere. Her attitude toward her colony had long been one of exploitation, which was the cause of numerous riots and rebellions. American statesmen from Thomas Jefferson on had considered the Cuban question. In 1859 the Senate Committee on Foreign Relations had indicated that the United States should regard the acquisition of Cuba as desirable.

Cuba and Spain had engaged in a barbarous war, the "Ten Years' War," from 1868 to 1878. The inhumane treatment of Cubans by the Spaniards had horrified the American people. American trade rights were interfered with and American citizens had been killed by the Spaniards.

War seemed imminent after the steamer "Virginius," sailing under American registry, was captured by the Spaniards and her crew shot. Diplomatic relations, however, staved off war between Spain and the United States until 1898, when the Spanish-American War was fought.

Insurrection broke out again in 1895, and cruelties were perpetrated by both sides. Máximo Gómez y Baez, the Cuban patriot and general, conducted guerrilla warfare, hoping to exhaust the Spaniards and bring about American intervention. General Valeriano Weyler, governor-general and General-in-Chief of the Spanish army, retaliated with vicious atrocities against the Cuban working classes. In addition to the internal disturbances in Cuba, American investments in and trade with Cuba were being interfered with and American property in Cuba was being destroyed.

At 9:40 P.M. on February 15, 1898, the American battleship "Maine," lying in Havana Harbor, was destroyed and sunk by an explosion. Two hundred and sixty men and officers were lost. Captain Charles Dwight Sigsbee, commander of the vessel, asked that popular judgment of the disaster, which shocked the world, be suspended until a report could be made by a court of inquiry.

Examinations were conducted by both the American and Spanish authorities. These, when completed, resulted in different conclusions. The American court's findings were that the "Maine" had been destroyed by a submarine mine, the detonation of which, in turn, had caused the partial explosion of two or more of her powder magazines. No evidence was found that any individual was directly responsible for the explosion. The Spanish court of inquiry came to the conclusion that the sinking was due solely to an explosion in the "Maine's" magazines.

American feelings, which agreed with the findings of the American court, ran high, and "Remember the Maine!" became a pro-war catch phrase of the day.

On April 11 President William McKinley, in spite of having received official notice that the Spanish government had "ordered a suspension of hostilities in Cuba in order to prepare and facilitate peace," sent a message to Congress indicating that the matter of the Cuban situation "was now with that body." Congress formally declared war on Spain on April 25, 1898; the war was won by the United States and terminated in December of that year.

In 1911 the wreck of the sunken "Maine" was raised. An examination of the hulk substantiated the findings of the American court of inquiry in 1898. It was thought that the explosion may have been perpetrated by a Cuban sympathizer who desired American intervention in the Cuban situation.

It has also been suggested that the "Maine" explosion may have been caused by a faulty mechanism within the vessel rather than by the deliberate act of a saboteur.

Suggested Readings

Alger, Russell A. The Spanish-American War. Plainview, N.Y.: Books for Libraries, 1901.

Bonis, Andrew, and Cris Johnson. Remember the Maine. New York: Crossman, 1973.

Chadwick, French E. The Relations of the United States and Spain: The Spanish-American War. New York: Russell & Russell, 1968.

Chidsey, Donald Barr. The Spanish-American War. New York: Crown, 1971.

"Destruction of the War Ship Maine Was the Work of an Enemy," New York Journal, Feb. 17, 1898, p. 1, col. 1.
Freidel, Frank. The Splendid Little War. Boston: Little, Brown, 1958.
Latané, John H. America as a World Power, 1897-1907. Saint Clair Shores, Mich.: Scholarly Press, 1971.
Leckie, Robert. Wars of America. New York: Harper, 1968.
Leech, Margaret. In the Days of McKinley. New York: Harper, 1959.
Mahan, Alfred T. Lessons of the War with Spain. Plainview, N.Y.: Books for Libraries, 1899.
Millis, Walter. The Martial Spirit. New York: Viking Press, 1965.
Olcott, Charles S. The Life of William McKinley. New York: AMS Press, 1916.
Pratt, Julius W. Expansionists of 1898. Baltimore: Johns Hopkins University Press, 1936.
Pringle, Henry F. Theodore Roosevelt: A Biography. New York: Harcourt, Brace, 1956.
Sargent, Herbert H. The Campaign of Santiago de Cuba. Plainview, N.Y.: Books for Libraries, 1907.
Schley, Winfield Scott. Forty-Five Years Under the Flag. New York: Somerset, 1904.

THE ROUGH RIDERS (1898)

Theodore Roosevelt, the American statesman, author, editor, sportsman, politician and amateur soldier who was to become the 26th President of the United States, was Assistant Secretary of the Navy when the Spanish-American War was declared on April 25, 1898.

This war was, at first, fought at sea, both in the Philippine Islands and off the island of Cuba. On May 19 the Spanish Admiral Pascual Cervera y Topete had, with his naval force, occupied Santiago Bay on the eastern end of Cuba. This bay is between four and five miles long and is reached through a narrow entrance channel. Commodore Winfield Scott Schley, second in command to Admiral William T. Sampson, headed the American squadron which set up a blockade outside the bay.

Santiago Harbor was comparatively easy for the Spaniards to defend, owing to elevated positions existing at the mouth of the channel. With the Spanish fleet in the bay and the American blockade surrounding the channel which Cervera would have to take in order to escape, Sampson settled down to await developments.

It became obvious that, if Santiago were to be captured, a joint movement of both the army and navy would be required. In addition to 50,000 regular army soldiers, over 200,000 volunteers had been called for. Theodore Roosevelt and his friend Colonel Leonard Wood, the latter an American physician, army officer and colonial administrator, organized the First U.S. Volunteer

Cavalry, popularly known as the "Rough Riders." This group was composed largely of athletes from eastern colleges, cowboys, Indians, ranchers, and would-be adventurers.

Commanded by Colonel Wood and Lieutenant Colonel Roosevelt, the regulars, the Rough Riders and a few other volunteers were sent to Tampa, Florida, and Chickamauga Park, Georgia, for training. General Nelson A. Miles commanded that portion of the army which ultimately landed in Havana, and General William R. Shafter commanded the Santiago expedition.

On June 14, 1898, 16,887 officers and men set sail from Tampa and eight days later began to land at Daiquiri, sixteen miles east of Santiago.

General Henry W. Lawton, commanding an infantry division, moved west and seized Siboney. General Joseph Wheeler, a cavalry commander, defeated a Spanish force at Las Guasimas on June 24.

It was at this point that the American armies met serious difficulties. San Juan Hill, on the left of the American line, overlooked the country towards the east. El Caney, a fortified village held by a small Spanish force, was on the right, and the country between was a roadless jungle with only a few narrow trails. General Lawton, engaging Spanish troops at El Caney, suffered a loss of more than 400 killed and wounded before driving the Spaniards out of their well-entrenched position.

San Juan Hill was equally well defended. The terrain was rough, the heat intense, and much tropical growth made maneuvering difficult. On July 1, 1898, Roosevelt led his Rough Riders up the Hill, at one point becoming separated from them and returning to them only with great difficulty. Regular troops fought their way up the opposite side of the hill, which was taken by the Americans. The Spaniards fell back to Santiago.

On July 3 Admiral Cervera attempted to leave Santiago Bay. Admiral Sampson was away at a conference with General Shafter, and Commodore Schley directed an action which resulted in the loss of the Spanish fleet. On July 16 Santiago surrendered to the Americans.

With the war won, Roosevelt and his Rough Riders wished to return home. He and his men sent a "round robin" to President William McKinley requesting that they be relieved from active duty. Shortly thereafter, following return to the United States, the Rough Riders were disbanded.

Suggested Readings

Alger, Russell A. The Spanish-American War. Plainview, N.Y.: Books for Libraries, 1901.

Chadwick, French E. The Relations of the United States and Spain: The Spanish-American War. New York: Russell & Russell, 1968.

Chidsey, Donald Barr. The Spanish-American War. New York: Crown, 1971.

Dos Passos, John. "The Happy Warrior," in his Nineteen-Nineteen.

New York: Random House, 1931.
Freidel, Frank. The Splendid Little War. Boston: Little, Brown, 1958.
Hagedorn, Hermann. Leonard Wood: A Biography. Millwood, N.Y.: Kraus Reprint, 1931.
Jones, Virgil C. Roosevelt's Rough Riders. Garden City, N.Y.: Doubleday, 1971.
Latané, John H. America as a World Power, 1897-1907. Saint Clair Shores, Mich.: Scholarly Press, 1971.
Leckie, Robert. Wars of America. New York: Harper, 1968.
Leech, Margaret. In the Days of McKinley. New York: Harper, 1959.
Lingley, Charles Ramsdell, and Allen Richard Foley. Since the Civil War. New York: Appleton, 1935.
Mahan, Alfred T. Lessons of the War With Spain. Plainview, N.Y.: Books for Libraries, 1899.
Matthews, Herbert Lionel. Cuba. New York: Macmillan, 1964.
Miles, Nelson A. Serving the Republic. Plainview, N.Y.: Books for Libraries, 1974.
Millis, Walter. The Martial Spirit. New York: Viking Press, 1965.
Olcott, Charles S. The Life of William McKinley. New York: AMS Press, 1916.
Pringle, Henry F. Theodore Roosevelt: A Biography. New York: Harcourt, Brace, 1956.
Roosevelt, Theodore. Autobiography. Edited by Wayne Andrews. New York: Octagon Books, 1973.
________. The Rough Riders. Williamstown, Mass.: Corner House, 1971.
Sargent, Herbert H. The Campaign of Santiago de Cuba. Plainview, N.Y.: Books for Libraries, 1907.
Schley, Winfield Scott. Forty-Five Years Under the Flag. New York: Somerset, 1904.

THE ANNEXATION OF HAWAII (1898-1900)

When the Hawaiian Islands were discovered by the English navigator and explorer Captain James Cook in 1778, political sovereignty was divided between four native monarchs. By 1810 Kamehameha I, King of the Island of Hawaii, had conquered the other rulers and became sole monarch of the entire group of islands. His dynasty lasted until 1872, in which year Kamehameha V, the last of his direct descendants, passed away and the kingdom was torn by political strife.

Missionaries from New England had, since the year 1820, done much to educate the natives and convert them to Christianity. In 1840 constitutional rule was adopted and in 1852 the royal government was liberalized. The constitutional movement, which was supported primarily by foreigners who favored the annexation of the kingdom by the United States, lasted through the reigns of Kings

Lunalilo and Kalakaua, and came to a head in 1893 when Queen Liliuokalani was deposed in January of that year.

President Benjamin Harrison favored annexation of the Islands by the United States, and on February 13, 1893, recommended to the Senate that such annexation be immediately accomplished. Grover Cleveland, who did not favor annexation, succeeded Harrison as President and appointed James H. Blount special commissioner to Hawaii. Blount's investigation convinced him that American interests had been responsible for the revolution that overthrew Queen Liliuokalani and established a United States protectorate. He ordered the American flag lowered and the protectorate ended, an action supported by Cleveland. The latter, on December 18, 1893, in a message to the Senate and House of Representatives, withdrew the treaty of annexation which had been proposed during Harrison's administration.

Sanford B. Dole, an American lawyer and head of the revolutionary provisional government, proclaimed on July 4, 1894, the Hawaiian Republic, which was recognized by the United States. Dole became President, which position he held until 1898. The movement for annexation continued, and on July 7, 1898, the necessary legislation was approved by the American Congress. On June 14, 1900, the Hawaiian Islands were formally constituted as the Territory of Hawaii, with Dole becoming the first governor.

In 1940 a protracted movement for admission of the Territory as a State of the American Union was endorsed by the Hawaiian electorate. In March, 1959, Congress approved legislation granting statehood to Hawaii, and the formal proclamation of statehood was made by President Dwight D. Eisenhower on August 21 of that year.

Suggested Readings

Callahan, James M. American Relations in the Pacific and Far East, 1874-1900. New York: Cooper Square, 1901.

Chambers, H. E. Constitutional History of Hawaii. New York: Johnson Reprint, 1973.

Cleveland, Grover. Presidential Problems. Plainview, N.Y.: Books for Libraries, 1975.

Commager, Henry Steele, ed. "The Annexation of Hawaii," (Doc. Nos. 330 and 348) in his Documents of American History, 8th edition. New York: Appleton, 1968.

________. "Cleveland's Withdrawal of Treaty for Annexation of Hawaii," (Doc. No. 331) in his Documents of American History, 8th edition. New York: Appleton, 1968.

Daws, Gavan. Shoal of Time: A History of the Hawaiian Islands. New York: Macmillan, 1968.

Dewey, David R. National Problems, 1885-1897. Westport, Conn.: Greenwood Press, 1968.

Foster, John W. American Diplomacy in the Orient. New York: Da Capo Press, 1970.

Kuykendall, Ralph S., and A. Grove Day. Hawaii: A History from Polynesian Kingdom to American State. Englewood

Cliffs, N.J.: Prentice-Hall, 1961.
Lee, William Storrs. The Islands. New York: Holt, 1966.
Leech, Margaret. In the Days of McKinley. New York: Harper, 1959.
Lewis, Oscar. Hawaii, Gem of the Pacific. New York: Random House, 1954.
Lingley, Charles Ramsdell, and Allen Richard Foley. Since the Civil War. New York: Appleton, 1935.
Moore, Charles Forrest. Parade of the Presidents. New York: W. E. Rudge, 1928.
Moore, John B. American Diplomacy: Its Spirit and Achievements. Norwood, Pa.: Norwood Editions, 1905.
Nevins, Allan. Grover Cleveland: A Study in Courage. New York: Dodd, Mead, 1932.
Olcott, Charles S. The Life of William McKinley. New York: AMS Press, 1916.
Porter, Robert P. The Life of William McKinley, Soldier, Lawyer, Statesman. Cleveland: World, 1896.
Stone, Adrienne. Hawaii's Queen Liliuokalani. New York: Messner, 1947.
Sullivan, Mark. Our Times: The United States, 1900-1925. New York: Scribner's, 1928-1935.

THE CLARKE KIDNAPING (1899)

In May, 1899, Mrs. Arthur Clarke of New York City hired a young woman calling herself Carrie Jones as nurse for her twenty-months-old daughter Marion.

On Sunday, May 21, the nurse took the little girl for a visit to Central Park in a wicker carriage. At noon the baby's father set off for a stroll in the park. There he found his daughter's wicker carriage. It was empty and little Marion and her nurse were not to be seen.

The father contacted a policeman, who advised him not to worry and to return home. "The nurse," he said, "has probably got lost in the winding drives of the park and will, no doubt, find her way home shortly."

Arthur Clarke went home. At two o'clock he returned to the park and consulted a captain of police, with the same results. He went home a second time and at four o'clock a boy delivered a note to him. The note, addressed to Mrs. Clarke and signed "Three," indicated that both baby and nurse were "in safe hands," and that further communication could be expected. The note warned Mrs. Clarke not to say anything to the newspapers or the police.

In spite of the warning in the note, Arthur Clarke notified the police that his daughter had been abducted. A gigantic search for the missing baby and her nurse was immediately started. Hundreds of people furnished the police with information which, when run down, proved valueless. However, a Mrs. Cosgriff, proprietress of a New York boarding house, stated that she had rented a room to two women and a small girl on the evening of May 21 and that the three had spent the night there, leaving the following day. One of the women had mentioned a town with a name ending in "burg." Mrs. Cosgriff's description of the baby led Mrs. Clarke to declare that it was her missing daughter.

A search of towns with the "burg" termination--Pittsburgh, Plattsburg, Williamsburg, Petersburg and the like--was instigated, with detectives being sent to each. A girl named Jones was arrested in Pennsylvania but turned out to have no connection with the kidnaping.

No further communication was received by the Clarkes. On Thursday, June 1, Mrs. Ada B. Corey, postmistress at St. John, New York, reported that a little girl and a woman giving her name as "Beauregarde" had picked up some letters at the post office. William Charleston, deputy sheriff of Rockland County, who received Mrs. Corey's information, was able to locate the little girl and the Beauregarde woman before they could leave town and he followed them to a farmhouse belonging to Frank Oakley near Sloatsburg, New York. Here he learned that the woman had been living at the Oakley farm with her husband and that ten days previously she had been joined by another woman and a baby girl.

Charleston seized the woman, the child, and the woman's husband. Arthur Clarke, summoned to the scene, identified the baby as his missing daughter. The woman, however, was not Carrie Jones. When questioned she gave her name as "Mrs. George Beauregarde," but later stated that her correct name was "Mrs. James Wilson."

Mrs. Wilson told a story of a couple, unknown to her, who had brought the infant to her and paid her to board it for the summer. James Wilson, arrested at the farmhouse, disclaimed all knowledge of the affair other than what his wife had told him.

Freddy Lang, the boy who had brought the note signed "Three" to the Clarke house, was located by the police. He identified Mrs. Wilson as the woman who had handed him the note and a five-cent piece and asked him to deliver it to Mrs. Clarke. Mrs. Cosgriff identified Mrs. Wilson and Marion Clarke as two of the three who had spent the night in her boarding house. The Wilsons remained mute.

On June 2, 1899, a reporter located "Carrie Jones" at the home of an aunt. He learned that her true name was Bella Anderson and that she had formerly been employed as a waitress at the Mills Hotel in New York. Here she had met Mr. and Mrs. George Beauregarde Barrow, the true names of the arrested pair. They had suggested a kidnap plot to her, her part being merely to find a position as nurse to the child of some wealthy family, seize the child and turn it over to them. For this Bella Anderson was to receive one half of whatever ransom was collected.

Bella Anderson had advertised for a place as a child's nurse, using the name "Carrie Jones." Mrs. Arthur Clarke had employed her and she, in turn, had taken little Marion Clarke to Central Park and delivered her to Mrs. Barrow.

The three kidnapers were tried and convicted. George Beauregarde Barrow was sentenced to serve fourteen years and ten months in prison. Bella Anderson was sentenced to serve four years and Mrs. Barrow received a sentence of twelve years and ten months. It was suspected that the Barrows may have had accomplices in the kidnap plot but this was never proven.

Suggested Readings

"Clark Baby Not Yet Found," New York Times, May 24, 1899, p. 1.
Cohen, Bruce J. Crime in America. Itasca, Ill.: Peacock, 1970.
"Evidence Against Baby Kidnappers," New York Times, June 5, 1899, p. 1.
Hibbert, Christopher. Roots of Evil: A Social History of Crime and Punishment. New York: Funk & Wagnalls, 1968.
"Kidnapping Victims: Tragic Aftermaths," Saturday Evening Post, April 1976.
"Marion Clark Kidnapped," New York Times, May 23, 1899, p. 1.
"Marion Clarke Found at Last," New York Times, June 2, 1899, p. 1.
Messick, Hank, and Burt Goldblatt. Kidnapping: The Illustrated History. New York: Dial Press, 1974.
'No Clue to Clarke Baby," New York Times, May 30, 1899, p. 2.
Slattery, William, ed. Abduction: Fiction Before Fact. New York: Grove Press, 1974.
Smith, Edward H. "The Kidnappers of Central Park," in his Mysteries of the Missing. Lincoln MacVeagh-The Dial Press, 1927.
Sutherland, Edwin H. On Analyzing Crime. Edited by Karl Schussler. Chicago: University of Chicago Press, 1972.
Winslow, Robert W. Crime in a Free Society. Encino, Calif.: Dickenson, 1968.

THE RICE MURDER (1900)

William Marsh Rice was an American merchant and philanthropist. Born in Springfield, Massachusetts, in 1816, he had migrated to Texas where, in 1838, he opened a store in the city of Houston. In due course he developed a large exporting, importing, and retail business and accumulated great wealth.

Following the death of his wife he moved to New York. Here he engaged Charles F. Jones as his secretary and leased an apartment on Madison Avenue. Before long he found himself involved in bitter litigation over the validity of his late wife's will and, without ever meeting the man, incurred a hearty dislike for Albert T. Patrick, an attorney representing his legal adversaries.

Rice had endowed an educational institute in Houston, which later came to be known as Rice Institute. Jones, his secretary, felt that he would be a more deserving recipient of the Rice fortune than the institute and was well aware that Rice had made a will leaving virtually his entire estate to the school.

One evening Patrick called on Rice in connection with the lawsuit and Rice refused to see him. He did, however, meet Jones, and the two conceived a scheme whereby Rice's fortune could be channeled into their pockets.

Jones typed a new will leaving the bulk of the Rice estate to Patrick and eliminating the bequest to the school. Patrick forged Rice's name to this spurious document. It was understood that, following Rice's death and the probating of his estate, Patrick would pay Jones the sum of $10,000 a year.

The conspirators realized that if Rice were to leave his fortune to Patrick, a complete stranger, suspicions would be aroused, embarrassing questions asked and an investigation made. They therefore set about to show, by means of forged personal letters supposedly written by Rice to Patrick, that the two were old and close friends. Checks were forged to show that the signatures on them and on the will were identical. A typewritten order was produced, instructing that securities and cash in several safe deposit boxes be turned over to Patrick.

Rice was then in his eighties. However, his health was excellent for a man of his age, and he decided to invest several million dollars in a Texas mill. It was then that Jones and Patrick acted. One night in the year 1900 Jones murdered Rice by placing a chloroform-soaked towel over his face. A death certificate, indicating that Rice had died of natural causes, was issued and the body was immediately cremated.

Patrick, greedy to obtain immediate possession of a large sum of cash, instructed Jones to issue several checks, totaling $250,000, to which Rice's name was forged. On one check Jones carelessly typed "Abert" instead of "Albert T. Patrick." The bank where this check was presented telephoned and asked that Rice confirm personally the validity of the check. As Rice was dead this obviously was not possible. Jones panicked. Both men were arrested and charged, first with forgery, then with murder.

Jones turned state's evidence. In March, 1902, Patrick was sentenced to be executed but the sentence was commuted to life imprisonment. He was released from confinement in 1912 and died in 1940.

Suggested Readings

Abrahamsen, David, M.D. The Murdering Mind. New York: Harper & Row, 1973.

Busch, Francis X. They Escaped the Hangman. Indianapolis: Bobbs-Merrill, 1953.

Collins, Ted, ed. "Rice-Patrick Case," in his New York Murders. New York: Duell, Sloan, 1944.

"Forgery Charge Hearing against Patrick and Jones before Magistrate Brann; Jones Confesses to District Attorney," New York Times, October 16, 1900, p. 2.

"Investigation into the Affairs of the Late Millionaire; Patrick and Jones Suspected," New York Times, September 28, 1900, p. 2.

Morris, Sylvia Stallings, ed. William Marsh Rice and His Institute.

Houston: Rice Institute, 1972.
O'Brien, Frank M. Murder Mysteries of New York. New York: F. W. Payson, 1932.
Pearson, Edmund. "The Firm of Patrick & Jones," in his Masterpieces of Murder. Boston: Little, Brown, 1963.
_______. Five Murders: With a Final Note on the Borden Case. New York: Crime Club, 1928.
"Professor Witthaus's Report of Analysis for Poison," New York Times, October 28, 1900, p. 1.
Reinhardt, James Melvin. The Psychology of Strange Killers. Springfield, Ill.: Thomas, 1962.
Sanders, Bruce. "The Case of the Missing Letters," in his They Caught These Killers. New York: Roy Publishers, 1968.
"Second Will to Be Offered for Probate," New York Times, October 31, 1900, p. 14.
"Suspicious Death," New York Times, September 26, 1900, p. 1.
Symons, Julian. A Pictorial History of Crime. New York: Crown Publishers, 1966.
Train, Arthur. "The Patrick Case Complete," American Magazine, May, 1907.
_______. True Stories of Crime from the District Attorney's Office. New York: McKinly, Stone & MacKenzie, 1908.

THE CUDAHY KIDNAPING (1900)

One of the wealthiest families of Omaha, Nebraska in the year 1900 was that of the millionaire meat packer Edward A. Cudahy. On the evening of December 18 of that year he sent his fifteen-year-old son Edward A. Cudahy, Jr., popularly known as "Eddie" Cudahy, to deliver a number of magazines to the home of Dr. Fred Rustin, a friend. Young Eddie Cudahy performed the errand, but when he did not return home as expected, his father notified the Omaha police.

A city-wide search was instituted, but without success. The following morning a note, signed "Jack" and indicating that Eddie Cudahy was being held for a $25,000 ransom, was reported having been received by the boy's father. This turned out to be a "police fake," and the newspaper accounts of it were in error; such a note was never delivered.

A genuine ransom note, dated December 19, 1900, was received by Cudahy, Sr. This communication, which was unsigned, demanded $25,000 in gold as the price of Eddie Cudahy's freedom. It contained detailed instructions concerning the delivery of the money.

Although the police advised Cudahy, Sr. not to comply with the instructions contained in the note, the father arranged to withdraw the required sum from the First National Bank. That evening he drove in his buggy to the designated rendezvous and left the money beside a lantern tied to a tree by the roadside, as directed in the anonymous ransom letter. He then drove home and, shortly after midnight, Eddie Cudahy walked into his home. He was unharmed.

Young Eddie stated that after he had delivered the magazines to Dr. Rustin, he had been accosted by two men who, calling him "Eddie McGee" and claiming to be police, had carried him off in a buggy. The two had bandaged his eyes and mouth, tied his arms, and taken him to a two-story house. He realized that he had been kidnaped.

He remained blindfolded in an upstairs room during his captivity. One of the men went away and the other remained to guard him. Finally the first man returned and young Eddie, still blindfolded, was taken to the buggy and driven to a point near his home, where he was released.

The police instigated a search for a two-story house or cottage and eventually located the one in which the boy had been held. This was at 3604 Grover Street, Omaha. The police also located the men who had sold a horse and buggy to a "suspicious looking stranger."

Eventually enough clues were assembled to implicate one of Edward Cudahy, Sr.'s former employees. He was Pat Crowe, a one-time butcher, later a criminal, convict, and jail-breaker. He had been seen in Omaha on the day of the kidnaping and had since disappeared.

A world-wide hunt for Crowe followed but he was not located. His accomplice, James Callahan, however, was apprehended, brought to trial, and acquitted. In 1906, five years after the Cudahy kidnaping, Crowe offered to surrender himself to the authorities provided all rewards for his capture, totaling $55,000, were withdrawn. These terms were met and in February, 1906, Crowe was tried and, surprisingly enough, acquitted.

Following his trial Crowe abandoned his life of crime. He became "a kind of peregrine reformer, lecturer, pamphleteer, and semi-mendicant."

Suggested Readings

Cohen, Bruce J. Crime in America. Itasca, Ill.: Peacock, 1970.

Crowe, Patrick T. His Story, Confessions and Reformation. New York: 1906.

________, as told to Thomas Regan. Spreading Evil. New York: Branwell, 1927.

"Cudahy Kidnapper Is Fully Identified," New York Times, December 31, 1900, p. 1.

Hibbert, Christopher. Roots of Evil: A Social History of Crime and Punishment. New York: Funk & Wagnalls, 1968.

"Joy in the Cudahy Home," New York Times, December 21, 1900, p. 1.

"Kidnappers' House Located," New York Times, December 22, 1900, p. 1.

"Kidnapping Victims: Tragic Aftermaths," Saturday Evening Post, April, 1976.

Messick, Hank, and Burt Goldblatt. Kidnapping: The Illustrated History. New York: Dial Press, 1974.
"Omaha Boy Held for $25,000 Ransom," New York Times, December 20, 1900, p. 1.
"Omaha Police Confident," New York Times, December 23, 1900, p. 6.
Slattery, William, ed. Abduction: Fiction Before Fact. New York: Grove Press, 1974.
Smith, Edward H. "Eddie Cudahy and Pat Crowe," in his Mysteries of the Missing. New York: Lincoln MacVeagh-The Dial Press, 1927.
Sutherland, Edwin H. On Analyzing Crime. Edited by Karl Schussler. Chicago: University of Chicago Press, 1972.
Winslow, Robert W. Crime in a Free Society. Encino, Calif.: Dickenson, 1968.

THE MUCKRAKERS (1900-1906)

Following the Civil War, great industrial and economic expansion took place in the United States. Many great American fortunes were founded between 1865 and 1900 and such entrepreneurs as John D. Rockefeller, Andrew Carnegie, Henry Clay Frick, Jay Gould, Edward H. Harriman, George Pullman and others became enormously wealthy. In achieving their wealth they did not hesitate to use such devices as monopolies and trusts, and with these devices came a series of notorious malpractices. The attitude of many business magnates, as expressed by one of them, was "the public be damned!"

As early as 1889 it was realized that federal regulation of growing trusts and corporations was needed. Such legislation as the Sherman Anti-trust Act of 1890 and the Clayton Act of 1914 attempted to regulate big business and "protect trade and commerce against unlawful restraints and monopolies," but many abuses continued.

The first decade of the 20th century saw the rise of the "muck-rake magazines," although Henry Demarest Lloyd had attacked the Standard Oil Company in the Atlantic Monthly twenty years earlier. These publications exposed business chicanery and political double-dealing in a manner not shared by the newspapers of the period. The term "muckraker," referring to the authors of such exposés, was coined by President Theodore Roosevelt in 1906. It was inspired by the man with the muck-rake in John Bunyan's Pilgrim's Progress.

The journalists who wrote and published material of this sort included Henry Demarest Lloyd, known as "The First of the Muckrakers," Ida M. Tarbell, Lincoln Steffens, Upton Sinclair, Ray Stannard Baker, Samuel Hopkins Adams, Samuel Sidney McClure, and Mark Sullivan, among others. Periodicals in which these articles appeared were numerous, including Collier's, McClure's Magazine, Munsey's Magazine, American Magazine, and Everybody's

Magazine. Many of these articles, which often ran serially, were later published in book form.

Attacks on the activities of railroads in politics, the fixing of prices by such organizations as the sugar trust, and the adulteration of food and drugs were quite profitable for the magazines that published them. One such muck-rake publication increased its circulation three-fold between 1901 and 1906. Another increased its circulation four-fold and still another no less than seven-fold.

Suggested Readings

Aaron, Daniel. Men of Good Hope. New York: Oxford University Press, 1951.

Baker, Ray Stannard. The New Industrial Unrest: Reasons and Remedies. New York: Arno Press, 1920.

Bryce, James. The American Commonwealth. New York: Macmillan, 1914.

Chalmers, David M. The Muckrake Years. Cincinnati: Van Nostrand, 1974.

_______. The Social and Political Ideas of the Muckrakers. Plainview, N.Y.: Books for Libraries, 1964.

Commager, Henry Steele, ed. "The Clayton Anti-Trust Act," (Doc. No. 403) in his Documents of American History, 8th edition. New York: Appleton, 1968.

_______. "The Concentration of Wealth," (Doc. No. 407) in his Documents of American History, 8th edition. New York: Appleton, 1968.

_______. "The Rule of Reason," (Doc. No. 375) in his Documents of American History, 8th edition. New York: Appleton, 1968.

_______. "The Sherman Anti-Trust Act," (Doc. No. 320) in his Documents of American History, 8th edition. New York: Appleton, 1968.

_______. "United States v. Socony-Vacuum Oil Co.," (Doc. No. 530) in his Documents of American History, 8th edition. New York: Appleton, 1968.

Dell, Floyd. Upton Sinclair: A Study in Social Protest. New York: Doran, 1927.

Filler, Louis. Crusaders for American Liberalism. New York: Harcourt, Brace, 1939.

Lloyd, Henry Demarest. "The Story of a Great Monopoly," Atlantic Monthly, March, 1881.

Lyon, Peter. Success Story: The Life and Times of S. S. McClure. New York: Scribner's, 1963.

Reiger, C. C. The Era of the Muckrakers. Chapel Hill, N.C.: University of North Carolina Press, 1932.

Sinclair, Upton. The Autobiography of Upton Sinclair. New York: Harcourt, Brace, 1962.

_______. The Brass Check. New York: Arno Press, 1970.

_______. The Jungle. Cambridge, Mass.: Robert Bentley, 1971.

_______. The Money Changers. Boston: Gregg Press, 1969.

Steffens, Lincoln. The Autobiography of Lincoln Steffens. New York: Harcourt, Brace, 1931.

______. The Shame of the Cities. New York: Sagamore Press, 1957.
______. Upbuilders. Seattle: University of Washington Press, 1969.
Sullivan, Mark. The Education of an American. New York: Johnson Reprint, 1970.
______. Our Times, The United States, 1900-1925. New York: Scribner's, 1928-1935.
Swados, Harvey, ed. Years of Conscience: The Muckrakers, an Anthology of Reform Journalism. Cleveland: World, 1962.
Tarbell, Ida M. History of the Standard Oil Company. New York: McClure, Phillips, 1904.
Weinberg, Arthur. The Muckrakers. New York: Simon & Schuster, 1961.
Wilson, Harold S. McClure's Magazine and the Muckrakers. Princeton, N.J.: Princeton University Press, 1970.

THE CAPTURE OF EMILIO AGUINALDO (1901)

In 1896, shortly before the Spanish-American War, the Filipinos, led by General Emilio Aguinaldo, a Chinese-Tagalog, had risen against the Spanish government. They wished to secure more liberal treatment at the hands of the Spaniards and wished also to eliminate the influence of the Catholic priests from politics.

In 1891 several secret societies had been organized to act against the Spanish authorities. The most prominent of these was the Liga Filipina (Philippine League), founded by Dr. José Rizal, a Manila oculist. Another was the Katipunan (Union), which was established to secure complete independence from Spain by open revolt. Dr. Rizal was betrayed to the Spaniards on August 19, 1896, and was executed in December of that year.

Emilio Aguinaldo, leader of the rebel forces, was at first successful, but his rebellion was broken in 1897. Following the signing of the Pact of Biac-na-bito, guaranteeing Spanish reforms within three years, he and his rebel association withdrew from the islands, as specified by the Pact. However, with the Spanish-American War of 1898, Aguinaldo returned to Malolos, Luzon, and proclaimed himself president of an independent Philippine republic. Spain was defeated in the war, but Aguinaldo continued his struggle, this time against the American forces, insisting on the right of the natives to constitute the independent republic they had planned.

On February 4, 1899, hostilities began at Manila when an American sentry was provoked into firing on a Filipino patrol. The insurgents then resorted to guerrilla warfare. Aguinaldo, headquartered in Palanan, a remote village in the jungles of Northern Luzon, directed a "hit and run" warfare. Then, on February 8, 1901, General Frederick Funston, an American brigade commander, learned that a small group of rebels had voluntarily surrendered to the Americans. One of these was Cecilio Segismundo, a messenger who was carrying an order to General Ubano Lacuna, a Filipino insurrectionist leader. This message disclosed Aguinaldo's whereabouts and ordered Lacuna to reassign 200 rebel troops to the Palanan headquarters.

Funston suggested a scheme to his military superior, General Arthur MacArthur. By substituting loyal Filipino troops for the rebels ordered by Aguinaldo he planned to proceed to Palanan and capture the rebel chief. Funston, together with four other American officers, would accompany the expedition, posing as captured American soldiers.

Eighty-five loyal Filipino soldiers were selected to make the trip. They and the five American officers boarded the frigate "Vicksburg" on March 6, 1901, and sailed northward from Manila Bay. On March 14, in the midst of a dark, rainy, tropical night, they landed at Casiguran Bay, approximately 110 miles south of Palanan.

Posing as rebels, the invaders obtained some food for the trip to Palanan from the officials of a local village. On March 17 the party started overland. There was only enough food for half rations and the trek through the jungle was one of hardship.

On March 23 the detachment arrived at the village of Dinundungan where, on orders from Aguinaldo, the American "prisoners" were to be detained. Funston arranged to have a message with Aguinaldo's forged signature sent to Dinundungan ordering that the Americans be brought to Palanan, and the five officers were thus able to rejoin their troops.

As Funston and his men approached Palanan they opened fire on the Aguinaldo forces. Aguinaldo, hearing the gunfire, stepped to the window of his headquarters building and looked out. He was promptly seized by Hilario Tal Placiso, one of Funston's soldiers.

Aguinaldo's rebel forces surrendered to the invaders. Returned to Manila, Aguinaldo issued a proclamation ordering his followers to lay down their arms and abandon the insurrection. He swore allegiance to the United States. Sporadic warfare, however, continued under the leadership of other insurgent Filipino generals until 1902 when the last of these surrendered. Emilio Aguinaldo then retired to private life but in 1935 was a candidate for the office of President of the Philippine Commonwealth. He and Bishop Gregorio Aglipay were defeated in the election by Manuel Quezon.

Suggested Readings

Blount, James H. The American Occupation of the Philippines, 1898-1912. New York: Putnam, 1913.

Editors of the Army Times. "The Capture of Aguinaldo," in their The Tangled Web. Harrisburg, Pa.: Army Times Publishing Co., 1963.

Forbes, W. Cameron. The Philippine Islands. Wilmington, Del.: Scholarly Resources, 1928.

Funston, Frederick. Memories of Two Wars. New York: Scribner's, 1911.

Grunder, Garel A., and William E. Livezey. The Philippines and

the United States. Norman, Okla.: University of Oklahoma Press, 1957.
Kuhn, Delia, and Ferdinand Kuhn. The Philippines Yesterday and Today. New York: Holt, Rinehart, 1966.
Leech, Margaret. In the Days of McKinley. New York: Harper, 1959.
Mellis, Walter. The Martial Spirit. New York: Viking Press, 1965.
Morgan, H. Wayne. America's Road to Empire. New York: Wiley, 1965.
Pratt, Julius W. America's Colonial Experiment. New York: Prentice-Hall, 1950.
Reyes, José S. Legislative History of America's Economic Policy Toward the Philippines. New York: AMS Press, 1923.
Roosevelt, Nicholas. The Philippines: A Treasure and a Problem. New York: AMS Press, 1970.
Storey, Moorfield, and Marcial P. Lichauco. The Conquest of the Philippines by the United States, 1898-1925. Plainview, N.Y.: Books for Libraries, 1970.
Wolff, Leon. Little Brown Brother. Garden City, N.Y.: Doubleday, 1961.

THE McCORMICK MYSTERY (1901)

Was ten-year-old Willie McCormick kidnaped? Was he murdered? Or did he meet his unfortunate end as the result of a ghastly accident? It is known that he disappeared from his home in the Highbridge section of New York City on March 27, 1901, and that his body was found in the muddy waters of Cromwell Creek six weeks later, but these questions have never been answered with certainty.

Willie McCormick left his home on the evening in question to attend church. The last person known to see him alive was Mrs. Tierney, the mother of a playmate who lived in the neighborhood. This woman spoke to him briefly, telling him that her boy would not be able to attend church.

When Willie's two elder sisters, who had attended church services, returned home they reported that their brother had not been there. William McCormick, the boy's father, alarmed at his son's unexplained absence, searched the neighborhood, but without success. Next morning he communicated with the police, who expressed the opinion that young Willie had merely run away from home, an opinion in which his distraught parents did not concur.

The story of the boy's disappearance became headline material in the newspapers, and the possibility that he might have been kidnaped was suggested. Father J. A. Mullin, the family's priest, a man of wealth, when appealed to, offered a reward of $10,000 for the apprehension of the kidnapers and the return of the boy. Additional rewards were offered by Michael McCormick, Willie's uncle, by Oscar Willgerodt, a well-to-do neighbor, and by a local restaurateur and family friend. Altogether the rewards totaled $19,000, but no trace of the boy was forthcoming. Various suspects were apprehended, questioned, and acquitted. Two gypsy girls were interrogated but were found to have no connection with the missing boy.

Then a ransom note, signed "Kid," was received by Willie McCormick's father. This note, scrawled on cheap paper, seemed to have been written by some semi-literate person. It demanded the sum of $200 for the boy's return and gave instructions concerning the payment of the money. The police considered this communication the work of "some mental defective" and instructed McCormick to attend the rendezvous as directed but to leave a bundle of worthless paper rather than the requested currency.

McCormick, following police instructions, deposited a package containing only paper at the designated spot near the waterfront. However, no one came to claim it although detectives, hidden in the area, kept it under surveillance for several days and nights.

Ten days later McCormick received a second missive from "Kid." In this he was censured for communicating with the police and demand was made for not $200 but $2,000 ransom. Instructions were given concerning the place where this money should be left. Again, in accordance with instructions from the police, Willie's father left a dummy package and again no one appeared to pick it up. The third and last letter from "Kid" upbraided the father for again consulting the police and stated that Willie McCormick had been taken to England.

Pat Sheedy, a gambler who had acted as intermediary between criminals and the police on previous occasions, was brought into the case by Michael McCormick. Sheedy was authorized to pay the kidnapers $5,000 upon the return of the missing boy, with a guarantee that no prosecution would be pressed and no questions would be asked. Sheedy, though well connected with the underworld, was unable to learn anything of Willie McCormick's fate.

On May 10 John Garfield, a bridge tender, found the body of a boy floating in Cromwell Creek. Half an hour later the body was identified as that of Willie McCormick. The autopsy revealed that the body had been in the water for an extended period but it was not possible to determine whether it had been there for as long as six weeks, the full length of time since the disappearance. The body was badly decomposed and it could not be ascertained whether the boy had been choked to death or had died by drowning. No bones were broken.

The writer of the letters signed "Kid" was never apprehended. To this day the circumstances surrounding the death of young Willie McCormick remain shrouded in mystery.

Suggested Readings

"Big Reward for McCormick," New York Times, April 17, 1901, p. 2.

Cohen, Bruce J. Crime in America. Itasca, Ill.: Peacock, 1970.

"Funeral of Willie McCormick," New York Times, May 14, 1901, p. 3.
Hibbert, Christopher. Roots of Evil: A Social History of Crime and Punishment. New York: Funk & Wagnalls, 1968.
"Kidnapping Now Feared by McCormick Family," New York Times, April 4, 1901, p. 2.
"Kidnapping Victims: Tragic Aftermaths," Saturday Evening Post, April, 1976.
"May Be Willie McCormick," New York Times, April 5, 1901, p. 6.
Messick, Hank, and Burt Goldblatt. Kidnapping: The Illustrated History. New York: Dial Press, 1974.
"Missing Boy Seen on a Bay Ridge Car," New York Times, April 3, 1901, p. 1.
" 'Pat' Sheedy to Find Willie McCormick," New York Times, April 19, 1901, p. 7.
Slattery, William, ed. Abduction: Fiction Before Fact. New York: Grove Press, 1974.
Smith, Edward H. "The Mystery at Highbridge," in his Mysteries of the Missing. New York: Lincoln MacVeagh-The Dial Press, 1927.
Sutherland, Edwin H. On Analyzing Crime. Edited by Karl Schussler. Chicago: University of Chicago Press, 1972.
Winslow, Robert W. Crime in a Free Society. Encino, Calif.: Dickenson, 1968.

THE ASSASSINATION OF PRESIDENT McKINLEY (1901)

On September 4, 1901, William McKinley, 25th President of the United States, arrived at the Pan-American Exposition at Buffalo, New York. Here he was to give a speech on behalf of reciprocal trade agreements. September 5 had been designated President's Day and McKinley was then to address the public at the Esplanade shortly before noon.

The President made his speech as scheduled. Then he and his wife, Ida Saxton McKinley, were taken on a conducted tour of the Exposition. He also attended a luncheon at the New York State Building and a card reception at the Government Building. Mrs. McKinley had luncheon with the Board of Women Managers of the Exposition.

In the late afternoon McKinley and his wife were escorted to the home of John G. Milburn, President of the Exposition, where they were house guests. That evening they watched a lavish fireworks display.

On the morning of September 6 the President, his party, and a number of diplomats and distinguished guests rode to Niagara Falls on a special train. This visit completed, they returned to the Exposition, where a public reception had been arranged. This was to be held in the Temple of Music at four o'clock. Mrs. McKinley, being tired, had decided not to attend the reception and was driven from the Exposition station to the Milburn house. Having half an hour to spare, McKinley stopped for refreshments at the Mission Building. Then he, Milburn, and George Bruce Cortelyou, his secretary, drove to the Temple of Music in a victoria, dismounted, and entered. The President was guarded by three secret service men, a special detachment of Exposition police, several Buffalo detectives, and eleven soldiers. Also on guard were a number of plain clothes policemen stationed throughout the crowd.

McKinley stood at the corner of a dais with Milburn standing at his left in order to introduce members of the public to the President. A huge crowd had gathered and those who were to shake hands with McKinley were to proceed in single file down a narrow aisle to the dais.

Leon Czolgosz, a Polish-American anarchist and follower of the anarchist Emma Goldman, had arrived at the Temple of Music earlier and stood near the head of the line of persons waiting to meet the President. The day was swelteringly hot and many of those present were fanning themselves with palm leaf fans and wiping their brows with handkerchiefs. In his right pocket Czolgosz carried a short-barreled Iver Johnson .38 caliber revolver under a large cotton handkerchief. The line of greeters proceeded toward the President, who shook hands with each person in turn.

At seven minutes past four Czolgosz stood before McKinley. He pulled his hand from his pocket and fired two shots through the handkerchief at the President. He was knocked down, pinioned, and dragged to the center of the hall. McKinley was assisted to a chair.

Police reinforcements shouldered their way through the crowd. The Temple of Music was cleared and Czolgosz was taken to an inner office. A motor ambulance removed the President to the Exposition emergency hospital. Here it was found that one bullet had merely grazed a rib but the other had penetrated the stomach causing a serious abdominal wound.

McKinley was moved to the Milburn home and medical specialists were called in to attend him. At first it was reported that his condition was "satisfactory," and later that he was "improving." However, late in the afternoon of Saturday, September 14, 1901, President McKinley passed away.

Leon Czolgosz was tried for murder, found guilty, and sentenced to death. He was executed at Auburn Prison, New York, on the following October 29.

Suggested Readings

Beer, Thomas. Hanna. New York: Octagon Books, 1973.
Donovan, Robert J. "The Man Who Didn't Shake Hands," in his The Assassins. New York: Harper, 1955.
Glad, Paul W. McKinley, Bryan and the People. Philadelphia: Lippincott, 1964.
Leech, Margaret. In the Days of McKinley. New York: Harper, 1959.

Morgan, H. Wayne. William McKinley and His America. Syracuse: Syracuse University Press, 1963.
Olcott, Charles S. The Life of William McKinley. New York: AMS Press, 1916.
Porter, Robert P. The Life of William McKinley, Soldier, Lawyer, Statesman. Cleveland: World, 1896.
Potter, John Mason. Plots Against the Presidents. New York: Astor, 1968.
Sievers, H. J., ed. William McKinley, 1843-1901: Chronology, Documents, Bibliographical Aids. Dobbs Ferry, N.Y.: Oceana, 1970.
Sullivan, Mark. Our Times, The United States, 1900-1925. New York: Scribner's, 1928-1935.
Symons, Julian. A Pictorial History of Crime. New York: Crown Publishers, 1966.
Wallechinsky, David, and Irving Wallace. "Assassinations--William McKinley," in their The People's Almanac. Garden City, N.Y.: Doubleday, 1975.

THE PLATT AMENDMENT (1903)

Following the Spanish-American War it was considered of vital importance to establish a policy dealing with permanent relations between the United States and Cuba. Spain had withdrawn from the island of Cuba in 1898 and in that year the Teller Resolution of April 19 stated that the United States disclaimed any intention of exercising "sovereignty, jurisdiction, or control over said Island except for the pacification thereof, and asserts its determination that when it is accomplished to leave the government and control of that Island to its people."

General Leonard Wood, military governor of the island, arranged for a convention to draw up a form of government. The constitution provided by this convention in November, 1900, contained no provision for future relations with the United States. The American Senate was unwilling to acquiesce in this situation, and therefore, under the leadership of Senator Orville H. Platt, passed the so-called "Platt Amendment." This, which was actually a series of amendments, was added to the Army Appropriation Bill of March 2, 1901. The fifth amendment was drafted by General Wood and the others were the work of Elihu Root, then Secretary of War.

The Cuban convention added these amendments to the Cuban Constitution as an appendix and they were incorporated into a treaty with the United States in May, 1903.

The Platt Amendment provided that Cuba would enter into no foreign agreements contrary to the interests of the United States and granted the United States the right to intervene in Cuban affairs if this became necessary in terms of keeping order. It also placed restrictions on the Cuban public debt and indicated that Cuba should sell or lease necessary coaling stations to the United States.

Cuba, satisfied that the United States had no intention of meddling in her internal affairs, incoporated the Platt Amendment into her final constitution. On May 20, 1902, U.S. control of the island of Cuba was formally relinquished. The Platt Amendment was abrogated by treaty on May 31, 1934.

Suggested Readings

Beals, Carleton. The Crime of Cuba. New York: Arno Press, 1970.

Chapman, Charles E. History of the Cuban Republic. New York: Octagon Books, 1970.

Commager, Henry Steele, ed. "The Abrogation of the Platt Amendment," (Doc. No. 491) in his Documents of American History, 8th edition. New York: Appleton, 1968.

________. "The Platt Amendment," (Doc. No. 360) in his Documents of American History, 8th edition. New York: Appleton, 1968.

Coolidge, Louis A. An Old-Fashioned Senator, Orville H. Platt of Connecticut. Port Washington, N.Y.: Kennikat Press, 1974.

"Cuba and the Platt Amendment," Foreign Policy Association: Information Service Reports. April 3, 1929.

Fitzgibbon, Russell H. Cuba and the United States, 1900-35. New York: Russell & Russell, 1964.

Guggenheim, Harry F. The United States and Cuba. New York: Arno Press, 1970.

Hagedorn, Hermann. Leonard Wood: A Biography. Millwood, N.Y.: Kraus Reprint, 1931.

Hill, Leonard C. Roosevelt and the Caribbean. New York: Russell & Russell, 1965.

Jenks, Leland H. Our Cuban Colony: A Study in Sugar. New York: Arno Press, 1970.

Leech, Margaret. In the Days of McKinley. New York: Harper, 1959.

Porter, Robert P. The Life of William McKinley, Soldier, Lawyer, Statesman. Cleveland: World, 1896.

Root, Elihu. Military and Colonial Policy of the United States. New York: AMS Press, 1970.

THE SAN FRANCISCO GRAFTERS (1901-1911)

Political graft in American cities is by no means new. From earliest times certain opportunists have acquired control of various metropolitan areas and managed them for their own aggrandizement.

One of the more notorious graft rings was the one headed by Abraham Ruef in San Francisco at the start of the 20th century. Ruef was a brilliant and ambitious attorney who preferred to operate behind the scenes rather than in the open. Accordingly, he selected Eugene Schmitz, leader of the orchestra at the Columbia Theater, to be his front man and puppet.

In 1901 Ruef, by pulling various political strings, was able to have Schmitz nominated and elected Mayor of San Francisco, a feat which he duplicated in 1905. Ruef became "legal counsel" for business interests and others who wished political favors, collected his fees and, in turn, passed a portion of them on to Schmitz and the members of the board of supervisors who were little more than henchmen. Vice flourished. Houses of prostitution and gamblers were allowed to operate without interference provided they paid the necessary tribute to Ruef and his associates. Large corporations, such as the Home Telephone Company, paid "legal fees" for franchises and legislation favorable to them, as did the United Railroads.

In due course Fremont Older, managing editor of the San Francisco Bulletin, determined that civic reform was called for. In 1905 Schmitz was reelected Mayor in spite of the editorial opposition of the Bulletin. In 1907 the third of three attempts on Older's life failed. A warrant for his arrest was sworn out in Los Angeles and he was apprehended by two deputies and two detectives and placed on a train bound for that city. It was planned that he be shot "while trying to escape." He was recognized by a San Francisco attorney who gave the alarm and saved him.

Older realized that the graft ring was so strongly entrenched that outside help to overcome it was required. He enlisted the aid of James D. Phelan, former Mayor of San Francisco, and Rudolph Spreckels, a prominent and wealthy local citizen. Attorney Francis J. Heney was engaged to help with the prosecution and William J. Burns, a well-known detective, was also hired to participate in the struggle.

William H. Langdon, the district attorney appointed by Ruef, and who had turned out to be honest, appointed Heney as his assistant. This precipitated a situation in which Ruef ordered Schmitz to discharge Langdon, following which Schmitz was to appoint Ruef to the office. Langdon, however, remained in office after Older had revealed the order in the Bulletin and an angry crowd had gathered at the court house where the judge was to rule on the dismissal.

The prosecution got under way but made little progress until two minor members of the graft ring defected. These two contrived with the prosecutors to entrap certain members of the board of supervisors who were involved in the proffering of bribes in connection with an ordinance regulating skating rinks. Ultimately confessions were obtained from most of the seventeen members of the board of supervisors.

District Attorney Langdon dismissed the old grand jury and a new one was impaneled. Ruef and Schmitz were promptly indicted for mulcting brothels. Ruef failed to appear in court, for which he was arrested. After much urging, under promise of immunity he made a full confession. Schmitz was tried for extortion, found guilty, and sentenced to prison. He was, however, freed by the district court of appeals on a technicality.

Francis J. Heney attempted to prosecute the highly placed business executives who had paid bribes. Unfortunately for him, public opinion turned against him. It was one thing to prosecute "men from the lower social strata," but something altogether different to seek to destroy the "best people," such as wealthy members of early pioneer families and top-level business executives.

Tirey L. Ford who, as the agent of Patrick Calhoun, head of the United Railroads, had bribed Ruef with $200,000, was tried three times. In spite of ample evidence of guilt, the jury disagreed in one trial; in the other two he was acquitted. Each case involved the bribing of a different supervisor.

Abraham Ruef was eventually tried for accepting bribes. While he had been granted immunity in return for his confession, he had persisted in partly repudiating it and insisting that all payments made to him had been legal fees only. During the course of this trial a prospective juror, declared ineligible to serve because of his prison record, shot and wounded Heney. The latter's place was taken by attorney Hiram Johnson, later governor of and United States senator from California.

The trial continued. Ruef was found guilty and was sentenced to serve fourteen years in prison. He was the only member of the San Francisco graft ring to actually go to jail.

Langdon refused to consider further service as district attorney and a former football hero, Charles M. Fickert, whose liaison with the grafters was well known, was elected. Fickert refused to continue any of the cases still pending against the California businessmen.

William P. Lawlor, the judge who had presided in several of the cases, ordered the prosecution to continue but was overruled by the court of appeals. The remaining indictments were quashed.

Ruef served four years and seven months of his sentence and was then paroled. He entered the real estate business and ultimately died a bankrupt in 1936. Schmitz remained in San Francisco, stayed in politics and, ironically, served several terms on the board of supervisors.

Suggested Readings

Bean, Walton E. "Boss Ruef, the Union Labor Party and the Graft Prosecution in San Francisco, 1901-1911," Pacific Historical Review, November, 1948.

________. Boss Ruef's San Francisco. Berkeley, Calif.: University of California Press, 1952.

Bliven, Bruce. "The Boodling Boss and the Musical Mayor," American Heritage Magazine, December, 1959.

DeFord, Miriam Allen. "The Last of the Tolstoyans," in her They Were San Franciscans. Caldwell, Idaho: Caxton Printers, 1941.

Gatlin, Dana. "Great Cases of Detective Burns: How Abe Ruef Confessed," McClure's Magazine, February, 1911.

Hoag, Edwin. American Cities: Their Historical and Social

Development. Philadelphia: Lippincott, 1969.
Hynd, Alan. In Pursuit: The Cases of William J. Burns. Camden, N.J.: Nelson, 1968.
Inglis, William. "For the Kingdom of California: True Story of San Francisco's Civil War between the Grafters and the Elaborate Forces of the Prosecution," Harper's Weekly, May 23, May 30, June 6, June 13, 1908.
Irwin, Will. "They Who Strike in the Dark," American Magazine, April, 1909.
McKee, Irving. "The Background and Early Career of Hiram Warren Johnson, 1866-1910," Pacific Historical Review, February, 1950.
Mowry, George E. The California Progressives. New York: Quadrangle/The New York Times Co., 1963.
Older, Fremont. My Own Story. New York: Macmillan, 1926.
Olin, Spencer G., Jr. California Prodigal Sons: Hiram Johnson and the Progressives. Berkeley, Calif.: University of California Press, 1968.
Palmer, Frederick. "Abe Ruef of the Law Offices," Collier's Weekly, January 12, 1907.
Raudebaugh, Charles. "San Francisco: The Beldam Dozes," in Allen, Robert S., ed. My Fair City. New York: Vanguard Press, 1947.
Records of the San Francisco Graft Prosecution. Bancroft Library, University of California, Berkeley, California.
Ruef, Abraham. "The Road I Traveled," San Francisco Bulletin (serially), 1912.
Steffens, Lincoln. The Autobiography of Lincoln Steffens. New York: Harcourt, Brace, 1931.
_______. The Shame of the Cities. New York: Sagamore Press, 1957.
_______. "William J. Burns, Intriguer," American Magazine, April, 1908.
Steinberg, Alfred. The Bosses. New York: Macmillan, 1972.
Thomas, Lately, pseud. A Debonaire Scoundrel. New York: Holt, Rinehart, 1962.
Wells, Evelyn. Fremont Older. New York: Appleton, 1936.

THE ANTHRACITE COAL STRIKE (1902)

Prior to 1899 the anthracite coal miners of Pennsylvania considered themselves the exploited victims of the coal mining companies which employed them. They were required to trade at the company-operated stores where prices were exorbitantly high. Miners who made their purchases elsewhere were discharged. Coal operators required the miners to load 3,000 pounds of coal and accept pay for 2,000 pounds. Wages were low, and unskilled immigrant labor from Hungary, Italy, and Poland which competed with the American miners for jobs kept them low. Someone observed that "many of the companies earned money not only by mining coal but by mining miners."

As the workers in the coal fields had no organization of sufficient strength to negotiate with the owners or force any changes in their working conditions they were, as individuals, helpless to better their lot. In 1899 the United Mine Workers of America sent organizers into the Pennsylvania coal regions and in 1900 a strike was called. An agreement was reached with the mine operators and work was resumed. However, the settlement was not satisfactory to either side. The mine operators indicated that they stood ready to deal with their employees on an individual basis but would not negotiate with a union or other labor organization.

On May 12, 1902, the miners struck again. About 147,000 men left their jobs, remaining out until October 23. The total loss to both miners and operators was $100,000,000.

Coal became in short supply and public feeling, with winter approaching, bordered on panic. The sentiment of the people swung more and more to the side of the miners. George Frederick Baer, president of the Philadelphia and Reading Company, unwisely referred to himself and his fellow mine operators in a letter as "those Christian men to whom God in His infinite wisdom has given control of the property interests of the country." Although Baer later denied writing such a letter, a photographic copy of it was published and given wide distribution. This incident obviously did the mine owners a good deal of harm and tremendously weakened their case in the public eye.

The Anthracite Coal Strike became a matter of national importance. President Theodore Roosevelt, concerned with the suffering which would ensue from a heatless winter, called a conference of miners and operators to meet with him in Washington, D.C. on October 3, 1902. The conference was held and the operators indicated that they resented Roosevelt's interference. They requested that federal troops be sent to the coal fields "to keep the miners in order," and suggested that suits be brought against the miners by the Attorney General under the Sherman Anti-trust Law. John Mitchell, president of the United Mine Workers of America, suggested arbitration and pledged the miners to accept it, but the owners refused.

President Roosevelt then resolved to use pressure on the owners. He arranged for former President Grover Cleveland to chair an investigation commission and he sent Elihu Root, Secretary of War, to discuss the matter with the financier J. P. Morgan. Morgan agreed that Roosevelt should appoint a commission. He succeeded in getting the mine operators to agree, but with the proviso that Cleveland not serve. Roosevelt then appointed Judge George Gray as chairman and the miners promptly returned to work.

The commission made its investigation and, concluding that neither side was completely right, made an award that satisfied most of the complaints of both parties. The miners received a ten per cent wage increase and it was determined that further disputes should be referred to a six-man Board of Conciliation, three members of which would represent the miners and the other three would represent the mine owners. This essentially gave the miners the collective bargaining which they had demanded.

Suggested Readings

Commager, Henry Steele, ed. "President Roosevelt and the Trusts," (Doc. No. 356) in his Documents of American History, 8th edition. New York: Appleton, 1968.
Cook, Roy. Leaders of Labor. Philadelphia: Lippincott, 1966.
Cornell, Robert J. The Anthracite Coal Strike of 1902. Washington, D. C.: Catholic University of America Press, 1957.
Crandall, Allen. The Man from Kinsman. Sterling, Colo.: Published by the author, 1965.
Darrow, Clarence. The Story of My Life. New York: Scribner's, 1932.
Foner, Philip S. History of Labor Movements in the United States. New York: International Publishers, 1964.
Ghent, William J. Our Benevolent Feudalism. New York: Macmillan, 1902.
Gluck, Elsie. John Mitchell, Miner. New York: John Day, 1929.
Gompers, Samuel. Seventy Years of Life and Labour. Clifton, N. J.: Augustus M. Kelley, 1966.
Gurko, Miriam. Clarence Darrow. New York: Crowell, 1965.
Harrison, Charles Yale. Clarence Darrow. New York: Jonathan Cape and Harrison Smith, 1931.
Heaps, Willard Allison. Riots, U. S. A., 1765-1970. New York: Seabury Press, 1970.
Iman, Raymond S., and Thomas W. Koch. Labor in American Society. Chicago: Scott, Foresman, 1965.
Johnsen, Julia E., comp. The Closed Shop. New York: Wilson, 1942.
Lens, Sidney. Unions and What They Do. New York: Putnam, 1968.
Litwack, Leon F., ed. The American Labor Movement. Englewood Cliffs, N. J.: Prentice-Hall, 1962.
Meltzer, Milton. Bread--and Roses: The Struggle of American Labor. New York: Knopf, 1967.
Roosevelt, Theodore. Senate Reports, 58th Congress, special session. Document No. 6 (serial No. 4556), The Reports of the President's Commission.
Stern, Gerald E., ed. Gompers. Englewood Cliffs, N. J.: Prentice-Hall, 1971.
Stone, Irving. Clarence Darrow for the Defense. New York: Doubleday, 1941.
Taft, Philip. The A. F. of L. in the Time of Gompers. New York: Harper, 1957.
Weinberg, Arthur, ed. Attorney for the Damned. New York: Simon & Schuster, 1957.
Weinstein, James. The Corporate Ideal and the Liberal State. Boston: Beacon Press, 1968.
Wiebe, Robert H. "The Anthracite Strike of 1902: A Record of Confusion," Mississippi Valley Historical Review, September, 1961.
________. Businessmen and Reform. Cambridge, Mass.: Harvard University Press, 1962.

THE KITTY HAWK FLIGHTS (1903)

For countless centuries man has watched birds in flight and sought to emulate them. In the late 19th century such aviation pioneers as Octave Chanute, Samuel Pierpont Langley, Otto Lilienthal, Charles Matthews Manly, and others experimented with gliders and machine-driven heavier-than-air flying machines. The research and experiments of these pioneers of flight resulted in the discovery of a number of principles of aerodynamics. However, it was not until December 17, 1903, that Orville Wright and his brother Wilbur succeeded in making a flight in a biplane of their own design and construction which raised itself into the air by its own power, sailed forward without a reduction in speed, and landed at a point as high as the one from which it started.

The first historic flight of the Wright Brothers, in which their plane traveled a distance of 120 feet in twelve seconds, was made on the sand dunes of Kitty Hawk, North Carolina. Orville Wright was at the controls and Wilbur ran alongside the plane, steadying it.

Three other flights were made that day, with the brothers alternating at the controls. The last flight was the longest. With Wilbur Wright as pilot the machine traveled 852 feet in 59 seconds.

Later in the day the machine was caught by a gust of wind, turned over and smashed.

Further experimentation by the Wright Brothers and others, such as Glenn Hammond Curtiss, resulted in many improvements in the airplane. Eventually what had once been a dream and later a source of ridicule became an accepted part of modern life. The original Wright machine was restored and today is on permanent exhibition at the National Air Museum at Washington, D. C.

Suggested Readings

Dos Passos, John. "The Campers at Kitty Hawk," in his _The Big Money_. New York: Random House, 1936.

Duke, Neville Frederick, ed. _The Saga of Flight_. New York: John Day, 1961.

Hyde, Margaret Olroyd. _Flight Today and Tomorrow_. New York: Whittlesey House, 1962.

Kelley, Fred C. _Miracle at Kitty Hawk: The Letters of Wilbur and Orville Wright_. New York: Farrar, Straus, 1951.

________. _The Wright Brothers: A Biography Authorized by Orville Wright_. New York: Farrar, Straus, 1951.

Lewellyn, John Bryan, and Irwin Shapiro. _The Story of Flight_. New York: Golden Press, 1959.

McFarland, Marvin W. _The Papers of Wilbur and Orville Wright_. New York: McGraw-Hill, 1953.

Maizlish, I. L. _Wonderful Wings: A Story of Aviation_. Evanston, Ill.: Row, 1941.

Miller, Francis Trevelyan. _The World in the Air: The Story of Flying in Pictures_. New York: Putnam, 1930.

Morris, Lloyd R., and Kendall Smith. Ceiling Unlimited. New York: Macmillan, 1953.
Reynolds, Quentin. The Wright Brothers, Pioneers of American Aviation. New York: Random House, 1950.
Smith, Henry L. Airways. New York: Knopf, 1942.
________. Airways Abroad. Madison, Wis.: University of Wisconsin Press, 1950.
Stevenson, Augusta. Wilbur and Orville Wright: Boys With Wings. Indianapolis: Bobbs-Merrill, 1959.
Stever, H. G., et al. Flight. New York: Time Magazine, 1965.
Sutton, Felix. We Were There at the First Airplane Flight. New York: Grosset & Dunlap, 1960.
Walsh, John. One Day at Kitty Hawk. New York: Crowell, 1975.
Whitehouse, Arch. The Early Birds. Garden City, N.Y.: Doubleday, 1965.

THE IROQUOIS THEATER FIRE (1903)

The most disastrous conflagration in theatrical history occurred in Chicago, Illinois, on the afternoon of Wednesday, December 30, 1903, when fire of an unknown origin broke out in the Iroquois Theater. Five hundred and eighty-eight men, women and children died in the ensuing panic, and many others were injured.

Eddie Foy, an American vaudevillian and entertainer, was appearing at the theater in the extravaganza "Mr. Bluebeard." The theater was new and supposedly fireproof. The production was of such a nature as to appeal to children and there were many of them in the audience. All the 1,600 seats in the theater had been sold and in addition there were several hundred standees. Foy had brought his six-year-old son Brian to see the performance and, unable to find a seat for him in the audience, had placed him on a stool in the wings.

As Foy was putting on the costume he wore in the second act he heard an unusual noise outside. He ran from his dressing room, located his son, and also found two stage hands attempting to put out a fire in the flimsy scenery backstage. One was using a fire extinguisher, the other was beating the flames back with a stick.

Brian Foy was sent outside in the custody of a stage hand. His father ran to the stage where the chorus was singing "In the Pale Moonlight." The orchestra had ceased to play. Stepping to the footlights, the actor told the chorus to leave and instructed the orchestra to play.

Out front the theater audience was streaming toward the exits. Foy entreated it to remain calm, begging the people not to lose their heads. Tragedy might have been averted had not the asbestos curtain become entangled with a heavy electric light reflector and stopped twelve feet above the stage floor. The scenery, now ablaze, crashed onto the stage and over the orchestra pit, creating a panic.

In the subsequent investigation it was found that the exit doors were locked and bodies, pushed from behind, had piled up inside the doors. Most of the casualties were children. Since the fire had started backstage, the theater suffered little damage. Only one performer, an aerialist who had been waiting high in the wings to do her act, was killed. Foy and his son survived, the former continuing on the stage until his death in 1928. Brian Foy became a motion picture producer.

Suggested Readings

Banner, Hubert S. "The Iroquois Theater Calamity," in his Calamities of the World. London: Hurst & Blackett, Ltd., 1932.

Chicago's Awful Theater Horror, Told by the Survivors. Chicago: Monarch Books, 1904.

Everett, Marshall, pseud. The Great Chicago Theater Disaster ... The Complete Story Told by the Survivors. Chicago: Publishers Union of America, 1904.

Foy, Eddie. Clowning Through Life. New York: Dutton, 1928.

________, and Alvin F. Harlow. "The Iroquois Theater Fire," in Kartman, Ben, and Leonard Brown, eds. Disaster! New York: Pellegrini & Cudahy, 1948.

Haywood, Charles F. General Alarm. New York: Dodd, Mead, 1967.

"Lest We Forget"; Chicago's Awful Theater Horror, by the Survivors and Rescuers, with an Introduction by Bishop Fallows. Chicago: Memorial Publications, 1904.

Morris, John V. "A New Age: Motors and Two Platoons: Triangle Shirtwaist Fire," in his Fires and Firefighters. Boston: Little, Brown, 1953.

Northrop, H. D. The World's Greatest Calamities: The Baltimore Fire and the Chicago Theater Horror. Chicago: National, 1904.

Samuels, Charles, and Louise Samuels. Once Upon a Stage. New York: Dodd, Mead, 1974.

Wallechinsky, David, and Irving Wallace. "The Iroquois Theater Fire," in their The People's Almanac. Garden City, N.Y.: Doubleday, 1975.

THE NORTHERN SECURITIES CASE (1904)

On December 3, 1901, in his first message to Congress, President Theodore Roosevelt addressed himself to the growing problem of business trusts, proposing not their prohibition but regulation. His policy was illustrated by the federal government's 1904 suit against the Northern Securities Company which the J. P. Morgan and James J. Hill interests had formed to control major railroads in the West. It was in this suit that Roosevelt made his first bid for popularity as a "trust buster."

The Northern Securities Company was a holding company, incorporated in the state of New Jersey, and was established to acquire and hold a majority of the stock of the Northern Pacific and Great Northern Railway Companies which, in turn, controlled the Chicago, Burlington, and Quincy Railroad. Early in 1901 Edward H. Harriman, president of the Union Pacific Railway, had made an unsuccessful effort to get a controlling interest in the Northern Pacific. It was then that Hill and Morgan joined forces and set up the Northern Securities Company.

Northern Securities was capitalized at $400,000,000, a sum sufficiently great to make it highly improbable that anyone could or would purchase a controlling interest in its securities in the open market. In addition, it was authorized by its charter to own shares of other companies engaged in such activities as coal mining, which would be of value to the railroads, which used coal-burning locomotives.

The organizers of the Northern Securities Company felt that they were in no way violating anti-trust legislation such as the Sherman Anti-trust Law and felt further that, in view of the decision of the Supreme Court in the Knight case, they were on safe legal ground.

President Roosevelt, however, asked Attorney General Philander C. Knox to investigate the legality of the new holding company. As a result, proceedings were instituted against the Company and in 1904, by the slim margin of five to four, the government's contention was upheld that Northern Securities was a combination in restraint of trade and therefore illegal. The Company was ordered dissolved.

Roosevelt was extremely pleased with the outcome, as the Court's decision indicated that future prosecutions made under the Sherman Act might run the gauntlet of the courts.

In all, the Roosevelt administration filed suits against 43 other corporations and achieved victories over the tobacco and oil trusts.

Suggested Readings

Beard, A. E. S. "Empire Builder," in her _Our Foreign-born Citizens_. New York: Crowell, 1958.

Burt, Olive. _The Story of American Railroads_. New York: John Day, 1969.

Clark, George T. _Leland Stanford_. Palo Alto, Calif.: Stanford University Press, 1931.

Commager, Henry Steele, ed. "Northern Securities Company v. United States," (Doc. No. 363) in his _Documents of American History_, 8th edition. New York: Appleton, 1968.

________. "President Roosevelt and the Trusts," (Doc. No. 356) in his _Documents of American History_, 8th edition. New York: Appleton, 1968.

________. "The Sherman Anti-Trust Act," (Doc. No. 320) in his _Documents of American History_, 8th edition. New York: Appleton, 1968.

Corey, Lewis. The House of Morgan. New York: AMS Press, 1930.

Daggett, Stuart. Chapters on the History of the Southern Pacific. New York: Ronald Press, 1922.

Flynn, John T. Men of Wealth. New York: Simon & Schuster, 1941.

Fogel, Robert William. The Union Pacific Railroad: A Case in Premature Enterprise. Baltimore: Johns Hopkins Press, 1960.

Griswold, Wesley S. A Work of Giants. New York: McGraw-Hill, 1962.

Heilbroner, Robert L., et al. In the Name of Profit. Garden City, N.Y.: Doubleday, 1972.

Howard, Robert W. The Great Iron Trail. New York: Putnam, 1962.

Josephson, Matthew. The Robber Barons. New York: Harcourt, Brace, 1962.

Kennan, George. The Life of E. H. Harriman. Plainview, N.Y.: Books for Libraries, 1922.

Lewis, Oscar. The Big Four. New York: Knopf, 1938.

McCague, James. Moguls and Iron Men: The Story of the First Transcontinental Railroad. New York: Harper, 1964.

Meyer, Balthasar H. "History of the Northern Securities Case," Madison, Wis., Bulletin of the University of Wisconsin, Vol. 1, 1906.

________. Railway Legislation in the United States. New York: Arno Press, 1973.

Noyes, Alexander D. Forty Years of American Finance. New York: Putnam, 1898.

Pringle, Henry F. Theodore Roosevelt: A Biography. New York: Harcourt, Brace, 1956.

Pyle, Joseph G. The Life of James J. Hill. Gloucester, Mass.: Peter Smith, 1917.

Riegel, Robert E. The Story of Western Railroads. Gloucester, Mass.: Peter Smith, 1926.

Roosevelt, Theodore. Autobiography. Edited by Wayne Andrews. New York: Octagon Books, 1973.

Salt, Harriet. Mighty Engineering Feats. Philadelphia: Penn, 1937.

Steffens, Lincoln. The Autobiography of Lincoln Steffens. New York: Harcourt, Brace, 1931.

Trottman, Nelson S. History of the Union Pacific: A Financial and Economic Survey. New York: Ronald Press, 1923.

Tutorow, Norman E. Leland Stanford, Man of Many Careers. Menlo Park, Calif.: Pacific Coast, 1971.

Wood, Florence Dorothy. Long Eye and the Iron Horse: A Biography of Grenville Dodge and the Union Pacific Railroad. New York: Criterion, 1966.

THE "GENERAL SLOCUM" DISASTER (1904)

Inadequate safety precautions and personnel attempting to perform tasks for which they were not qualified were directly responsible for the fire which destroyed the Knickerbocker Steamship Company's boat "General Slocum" on June 15, 1904.

The scene of the disaster was the East River, New York Harbor. St. Mark's Kindergarten was having its annual outing. Reverend G. F. C. Haas, pastor of St. Mark's, his family, as well as 400 children of the kindergarten and their parents plus some other passengers were aboard. Altogether, 2,358 persons were taking the cruise. The "General Slocum," commanded by Captain W. H. Schaick, was a fast paddle-wheel-driven excursion boat only thirteen years old.

When the trip started the steamer headed north. As it was passing 83rd Street one of the passengers, a fourteen-year-old boy named Frank Perditsky, saw smoke coming from the hold. He ran to the pilot house and informed Captain Schaick, who is alleged to have replied, "Shut up and mind your own business!"

Other craft in the harbor saw the smoke rising from the "General Slocum" and sounded warning whistles. The steamer disregarded these whistles and continued on her way. Off 110th Street some passengers noticed smoke pouring from a companionway. Captain Schaick, again told of the smoke, sent mate Edward Flanagan to investigate. Flanagan ran to the scene of the blaze and reported to the captain through a speaking tube. The fire had burned through a partition into the engine room.

Captain Schaick sounded the alarm gong. The passengers stampeded and dozens were trampled. Although there were several docks nearby where the "Slocum" could have been landed, the captain chose to head for Riker's Island, some distance away. The crew attempted to extinguish the blaze but the fire hoses proved faulty and split under the pressure of the water within them. The crew abandoned the fire hoses and fled for safety.

In virtually no time the flames spread, trapping passengers aft, forward, and below. Frank Conklin, chief engineer, abandoned the engine room, and was closely followed by the engine room crew. The steamer continued to forge ahead.

The life preservers, fastened to the walls with wire, proved to be rotten and fell apart when passengers attempted to put them on. The ten lifeboats, completely inadequate for the number of passengers aboard the burning steamer, also proved to be wired in their places.

The panic increased. Suddenly all three decks collapsed, sending hundreds of passengers to death, either in the flames or else by sliding into the river. The "Slocum" proceeded ahead, eventually striking a rock in the channel between Riker's and North Brother Islands. It began to sink and a fire boat was able to play streams of water on it and extinguish the flames.

In all, 1,021 persons lost their lives in the disaster. A coroner's jury conducted an intensive investigation into the causes of the "Slocum" tragedy. It was learned that laxity began with the owners of the ship, the Knickerbocker Steamship Company, and the United States Steamboat Inspection Service in the New York district. Safety inspections had been superficial. The life preservers had not been thoroughly examined since the steamer had been first commissioned. Henry Lundberg, who claimed to have made an inspection prior to the ship's last trip, was found to have had virtually no experience in inspecting steamships. Edward Flanagan, the mate, did not hold the required license for his position and the crew members were an assortment of dock workers who knew nothing of running a ship or of emergency procedures at sea.

The coroner's jury found that eleven men should be held under bond on charges of first degree manslaughter. These were Captain Schaick, Mate Flanagan, Inspector Lundberg and Commodore John A. Pease of the Knickerbocker Steamship Company, as well as the Company's president, secretary, and five stockholders.

In due course Schaik was sentenced to a ten-year prison term for criminal negligence. The others succeeded in escaping responsibility for the deaths of the "Slocum's" passengers.

The "General Slocum" tragedy precipitated reforms in marine safety which still apply today.

Suggested Readings

Armstrong, Warren. "The General Slocum," in his _Fire Down Below_. New York: John Day, 1968.
Boesen, Victor. "Holiday Holocaust," _Coronet_, March, 1945.
________. "Holiday Holocaust," in Kartman, Ben, and Leonard Brown, eds. _Disaster!_ New York: Pellegrini & Cudahy, 1948.
Clevely, Hugh. _Famous Fires_. New York: John Day, 1958.
"The 'General Slocum' an Unlucky Craft," _New York Times_, June 16, 1904, p. 6.
" 'General Slocum' Disaster," _Harper's Weekly_, June 25, 1904.
" 'General Slocum' Relief Fund," _Charities_, September 3, 1904.
Hoehling, Adolph A. _Great Ship Disasters_. New York: Cowles, 1971.
Mielke, Otto. _Disaster at Sea_. New York: Fleet, 1958.
Morris, John V. "The Devil Goes on a Sunday School Picnic," in his _Fires and Firefighters_. Boston: Little, Brown, 1953.

THE NAN PATTERSON MURDER CASE (1904-1905)

In New York City on the morning of June 4, 1904, Caesar Young, an Anglo-American gambler, bookmaker and racehorse owner, was shot and killed in a hansom cab occupied by himself and Nan Patterson, his showgirl mistress. The girl became the defendant in two sensational murder trials, both of which resulted in hung juries.

Young was married and apparently was sincerely in love with his wife Margaret. She knew of her husband's affair with Nan Patterson and, understandably, disapproved of it. She desired him to break off the liaison and return to her. Young was tiring of his mistress and had previously attempted to induce her to go to London without him. This she had declined to do.

Margaret Young succeeded in persuading her husband to take a joint vacation trip to Europe. They were to sail on the "Germanic" on June 4. That morning Young telephoned Nan at her hotel and asked her to meet him at 59th Street. She kept the appointment. Young had been drinking but they entered a nearby saloon and he consumed a glass each of whiskey and brandy. This was followed by more drinks at another saloon. They then hailed and entered a hansom cab. Young instructed the driver to proceed to a hat store where he purchased a new hat. Then he reentered the cab which proceeded to Bleecker Street and West Broadway, where he had additional drinks.

After the two came out of the saloon they got back into the cab. Shortly thereafter Frederick Michaels, the cab driver, heard the sound of a shot. Nan immediately opened the trapdoor in the roof of the cab and said, "For God's sake, drive to a drugstore!" Michaels did so and the druggist directed him to the Hudson Street Hospital. Young was dead on arrival, having been shot by a pistol which, still warm, was found in his right hand coat pocket.

Nan Patterson was held, at first as a material witness; later she was charged with murder in the first degree.

The pistol with which Young had been shot was traced to a New York pawn shop operated by Hyman Stern. Stern was unable to identify the man who had purchased it from him or the woman who had accompanied the purchaser.

Nan Patterson's sister Julia was married to J. Morgan Smith, "a big florid man, addicted to liquor and gambling." Nan was extremely fond of her sister and frequently gave her and her husband money which she had obtained from Young, the Smiths being perpetually in dire financial straits. When the murder pistol was traced to Stern's pawnshop the Smiths promptly left town and were apprehended some months later.

Nan was defended in both trials by Abraham Levy, a prominent New York criminal lawyer. His defense tactics consisted of emphasizing the "reasonable doubt" that his client had murdered her lover. He pointed out that other possibilities existed: Young could have committed suicide or the pistol could have been fired accidentally. In spite of excellent courtroom performances by the prosecutors, neither jury was able to agree. Following the second trial District Attorney William Travers Jerome, realizing that a third trial could also be unavailing, requested that Nan be discharged on her own recognizance. Judge John W. Goff granted the motion and Nan Patterson was a free woman. Following a brief engagement in vaudeville she dropped from sight.

What actually happened inside the hansom cab that June morning will never be known. Newman Levy, in his book <u>The Nan Patterson Case</u>, furnishes a plausible theory concerning Caesar Young's death. Levy rules out the possibility of murder on the grounds that

Nan Patterson regarded Young as her "meal ticket" and, as Al Smith once observed, "One does not shoot Santa Claus." Levy also feels that Young was not a suicide, that he was tiring of his mistress and was psychologically incapable of taking his own life.

Levy's hypothesis is that J. Morgan Smith and his wife Julia, who had enjoyed Young's largesse, were a loath to see him cast Nan off. Once this occurred their source of money would be gone. Smith conceived the idea of obtaining a revolver and inducing Nan to threaten to kill herself unless Young agreed not to abandon her.

Smith, according to Levy's theory, obtained a pistol from the pawnbroker Hyman Stern and gave it to Nan, who had it in her handbag on the fatal morning. In the hansom cab she produced the pistol and Young attempted to take it away from her, whereupon the weapon was discharged accidentally. The girl, says Levy, then placed it in Young's coat pocket where it was later found.

Suggested Readings

Bromberg, Walter. Mold of Murder: A Psychiatric Study of Homicide. New York: Grune & Stratton, 1961.

Guttmacher, M. S. The Mind of the Murderer. Freeport, N.Y.: Books for Libraries, 1960.

Jesse, F. Tennyson. Murder and Its Motives. London: Harrap, 1952.

Kunstler, William H. "Murder in Jealousy: The State of New York versus Nan Patterson," in his First Degree. New York: Oceana Publications, 1960.

Lester, David, and Gene Lester. Crime of Passion: Murder and the Murderer. Chicago: Nelson Hall, 1975.

Levy, Newman. The Nan Patterson Case. New York: Simon & Schuster, 1959.

McDade, Thomas M., comp. The Annals of Murder. Norman, Okla.: University of Oklahoma Press, 1961.

Reinhardt, James Melvin. The Psychology of Strange Killers. Springfield, Ill.: Thomas, 1962.

Sparrow, Gerald. Women Who Murder. New York: Abelard, 1970.

Wolfgang, Marvin E., comp. Studies in Homicide. New York: Harper & Row, 1967.

THE "COMELY WIDOW" MURDERS (1904-1908)

In 1900 Mrs. Merrell Sorenson, a Norwegian-American small-time actress, was widowed when her husband died at their newly-acquired farm near La Porte, Indiana. The two had moved to the farm from Chicago following the loss of their home by fire, for which the insurance company paid them $2,000. Sorenson's death may have been caused by arsenic administered by his wife, although this was never proved. The widow collected $3,000 on his life insurance policy.

Shortly after Sorenson's death the widow, whose first name was Belle, met and married Peter Gunness on condition that he take out $4,000 worth of life insurance. This he did. In 1904 a meat chopper "accidentally fell from a shelf and killed him." Belle Gunness was the only witness to the alleged accident.

It was then that the twice-widowed woman began to insert advertisements in various newspapers. A typical one went as follows:

> Personal. Comely young widow, but lonesome, owning a fine farm near La Porte, Indiana, wishes to make the acquaintance of a respectable gentleman of substantial means. Object, matrimony. No letters considered unless writer is willing to become personally acquainted at the earliest opportunity.

These advertisements drew hundreds of replies. Those considered "suitable" were answered; the others were ignored. A "suitable" reply was one from a man who had "substantial means," was willing to bring his "substantial means" with him, and was so situated that his disappearance would go virtually unnoticed. At the outset of this correspondence the "comely widow" purchased a barrel of quicklime, telling the dealer that she intended to use it for cleaning out cesspools.

In due course Ole Lindboe of Chicago, a bachelor, visited the La Porte farm. He brought $200 with him and met the widow, who was anything but comely. He also met her three adopted children. After some weeks he mysteriously disappeared. His fur coat, gold ring and watch were later sold by Belle Gunness.

The next visitor was Eric Gerhalt who came with $2,500 in cash and vanished four months later. Then came John Moos, who also disappeared after a few weeks. Ray Lamphere was next. The widow took a liking to him and he stayed on. He became suspicious concerning the fate of her previous visitors, and when John Alden appeared in response to one of her advertisements and subsequent correspondence, Lamphere was sent on an overnight errand. He returned ahead of schedule to find the widow dragging Alden's dead body out of the parlor. She stated that Alden had attacked her and that she had killed him with an ax in self-defense. The two cut Alden's body into small pieces and buried them in the garden.

Several other men came to the La Porte farm, perhaps twenty in all. The last of these was Andrew C. Helgelin of Aberdeen, South Dakota. He brought, in addition to cash, some securities which Belle Gunness persuaded him to sell. He was killed and buried.

Axel Helgelin, Andrew's brother, concerned that his letters to Andrew went unanswered, made an investigation. Before he could come to La Porte the farmhouse caught fire and on the night of April 27, 1908, burned to the ground. The bodies of Belle Gunness and her three adopted children were found and, on inspection, it was learned that the four had been battered to death before the place caught fire.

Lamphere was arrested. He told what he knew of the widow's activities. Excavations were made on the farm and the dismembered remains of twelve bodies, sprinkled with quicklime, were found. Axel Helgelin was able to identify one corpse as that of his brother.

Lamphere was charged with murder by arson, tried and sentenced to twenty years' confinement. He died in prison.

It was generally supposed that Lamphere had killed Belle Gunness in order to prevent her from killing him. The money she had obtained from her victims was never accounted for. None of it was ever traced to Lamphere and it is possible that it was destroyed when the farmhouse burned.

Suggested Readings

Abrahamsen, David, M.D. The Murdering Mind. New York: Harper & Row, 1973.

Bromberg, Walter. Mold of Murder: A Psychiatric Study of Homicide. New York: Grune & Stratton, 1961.

Catton, Joseph. Behind the Scenes of Murder. New York: Norton, 1940.

Guttmacher, M. S. The Mind of the Murderer. Freeport, N.Y.: Books for Libraries, 1960.

"Indiana's Murder Farm," Harper's Weekly, May 30, 1908.

Jesse, F. Tennyson. Murder and Its Motives. London: Harrap, 1952.

Logan, Guy B. H. "America's Comely Widow," in his Rope, Knife and Chair. New York: Duffield, 1930.

Reinhardt, James Melvin. The Psychology of Strange Killers. Springfield, Ill.: Thomas, 1962.

Sparrow, Gerald. Women Who Murder. New York: Abelard, 1970.

Symons, Julian. A Pictorial History of Crime. New York: Crown Publishers, 1966.

THE LIFE INSURANCE INVESTIGATION (1905-1906)

During and after the administration of President Ulysses S. Grant national political and business ethics sank to a very low level. Such legislation as the Sherman Anti-trust Act of 1890, the result of some of the more notorious malpractices of big business, stemmed from a growing demand for the regulation of trusts by the federal government. This Act spawned a huge number of lawsuits and further legislation, such as the Clayton Anti-trust Act of 1914.

As a result of the popular demand for a cessation of business abuses as publicized by such "muckrakers" as Lincoln Steffens, Upton Sinclair, and Ida M. Tarbell, a number of investigations were launched. Those discovered in the meat packing industry led to the enactment of pure food and drug acts. Others led to similar legislation and regulation.

The insurance industry found its business methods being looked into. Early in 1905 charges against the Equitable Life Assurance Company were published in various newspapers. These charges were taken up by the New York legislature and an investigation committee, known as the Armstrong Commission, was appointed to examine the affairs of life insurance companies doing business in New York State. Charles Evans Hughes, then a New York attorney, was retained by the Commission as legal counsel.

The results of the investigation substantiated the charges published in the newspapers and shocked the American public. Excessive salaries were paid to company officers. Many unreasonable expenses were incurred and the rights of policy holders were disregarded. As if these abuses were not enough, it was also discovered that several of the larger insurance companies had apportioned certain geographical areas to themselves and, within these areas, had spent huge sums to bribe politicians and office-holders to pass legislation favorable to them. Large sums of money, on which spurious records or no records at all were kept, were spent for this purpose.

The Armstrong Commission's final report on the investigation condemned these various abuses and included a program of reform legislation, which was put into effect.

Laws regulating political campaign expenditures had long been in existence but, prior to the insurance investigation, had been flagrantly disregarded. As a result of the report of the Armstrong Commission, New York, in 1906, passed laws forbidding corporations to make contributions for political purposes. In 1907 Congress passed similar legislation dealing with federal campaigns, and subsequently most states passed laws restricting the use of campaign funds.

Suggested Readings

Armstrong-Hughes Committee. Testimony Before a Sub-Committee of the Committee on Privileges and Elections. United States Senate, 62nd Congress, 2nd Session, pursuant to Senate Resolution 79 (Clapp Report).

Commager, Henry Steele, ed. "The Clayton Anti-Trust Act," (Doc. No. 403) in his Documents of American History, 8th edition. New York: Appleton, 1968.

________. "The Sherman Anti-Trust Act," (Doc. No. 320) in his Documents of American History, 8th edition. New York: Appleton, 1968.

Fellowes, Gordon. Insurance Racketeers. London: Allen, 1935.

Gollin, James. Pay Now, Die Later: What's Wrong with Life Insurance, A Report of Our Biggest and Most Wasteful Industry. New York: Random House, 1966.

Gudmundsen, John. The Great Provider: The Dramatic Story of Life Insurance in America. South Norwalk, Conn.: Industrial Publications Co., 1959.

Henderson, Ralph. The Grim Truth About Life Insurance. New York: Putnam, 1957.

Hughes, Charles Evans. The Autobiographical Notes of Charles Evans Hughes. Edited by David J. Danelsic and others. Cambridge, Mass.: Harvard University Press, 1973.
May, Earl Chapin. A Study of Human Security: The Prudential. Garden City, N.Y.: Doubleday, 1950.
Morrow, Ralph E. Insurance Surveys. Indianapolis: Rough Notes Co., 1951.
Perkins, Dexter. Charles Evans Hughes and American Democratic Statesmanship. Boston: Little, Brown, 1956.
Pusey, Merlo John. Charles Evans Hughes. New York: Macmillan, 1951.
Report of the Legislative Insurance Investigating Committee (10 vols.), 1905-1906.

THE SAN FRANCISCO EARTHQUAKE AND FIRE (1906)

The city of San Francisco, California is situated on the San Andreas fault, which stretches from Southern California northwest for six hundred miles. The term "fault," as used geologically, signifies "a line of fracture in rock formations accompanied by displacement of the rock mass at either side of the line." Movement accompanying faulting of the earth's crust is usually slow but can occur suddenly, causing severe earthquakes. Such was the case when a sudden movement along 270 miles of the San Andreas fault caused the San Francisco earthquake which was followed by a devastating fire.

At twelve minutes and six seconds past five o'clock on the morning of April 18, 1906, while most of the populace was asleep, San Francisco and certain nearby peninsula towns were hit by the most disastrous earthquake in the history of North America. Intermittent shocks continued for one minute and five seconds. The earthquake itself caused comparatively little damage, but broke water mains, shook down chimneys, and severed electrical connections. Shortly after the last shock was felt, no fewer than sixteen separate fires were raging in various areas south of Market Street. No water was available to extinguish them because of the broken mains, and the fire department found itself helpless against the flames.

By noon the spreading conflagration had devastated more than a square mile of territory, and in the afternoon the fire crossed Market Street at Third and Kearney. It spread north and west, gutting the business and financial districts, Chinatown and the Barbary Coast.

Refugees fled the town, going by ferry to the East Bay cities of Berkeley, Oakland, and Alameda, or by any means of transportation they could muster to the peninsula cities on the south. Others camped in Golden Gate Park and on the beaches west of the city. Buildings were dynamited in an effort to confine the fire and prevent it from spreading still further.

After three days the raging conflagration, which had reached Van Ness Avenue, burned itself out. Destroyed were 28,188 buildings in an area of more than four square miles, about a quarter of which were north of Market Street. Three hundred and fifteen people were known to have died as a result of the catastrophe. Of the 352 reported missing, only a handful were ever found, and countless numbers were made homeless. The value of the property destroyed is estimated at about $400,000,000.

Suggested Readings

Asbury, Herbert. The Barbary Coast. New York: Knopf, 1933.
Beebe, Lucius, and Charles Clegg. San Francisco's Golden Era: A Picture Story of San Francisco Before the Fire. Berkeley, Calif.: Howell-North Books, 1960.
Bronson, William. The Earth Shook, The Sky Burned. Garden City, N.Y.: Doubleday, 1959.
Bryce, James. The American Commonwealth. New York: Macmillan, 1914.
Clevely, Hugh. Famous Fires. New York: John Day, 1958.
Dolan, Edward F. Disaster 1906: The San Francisco Earthquake and Fire. New York: Messner, 1967.
Haywood, Charles F. "Disaster at San Francisco," in his General Alarm. New York: Dodd, Mead, 1967.
Hodgson, John H. Earthquakes and Earth Structure. Englewood Cliffs, N.J.: Prentice-Hall, 1964.
Jordan, David Starr, ed. The California Earthquake of 1906. San Francisco: A. M. Robertson, 1907.
Lane, Frank Walter. Elements Rage. Philadelphia: Chilton, 1965.
Lewis, Oscar. California Heritage. New York: Crowell, 1949.
________. San Francisco: Mission to Metropolis. Berkeley, Calif.: Howell-North Books, 1966.
Riesenberg, Felix, Jr. Golden Gate. New York: Knopf, 1940.
________. "San Francisco, City of Courage," in Kartman, Ben, and Leonard Brown, eds. Disaster! New York: Pellegrini & Cudahy, 1948.
Thomas, Gordon, and Max M. Witts. San Francisco Earthquake. New York: Stein & Day, 1971.

THE MURDER OF STANFORD WHITE (1906)

Stanford White was the most famous American architect of his day. Among the buildings designed by him were Madison Square Presbyterian Church, Madison Square Garden, the Century Club, Metropolitan Club, and Players Club, all in New York City. He also designed several buildings for the University of Virginia and Battle Monument at the United States Military Academy at West Point.

In addition to his architectural activities White was a sensualist and man-about-town who, though married, had maintained several mistresses, one of whom was Evelyn Nesbit, later Evelyn Nesbit Thaw. This girl, born in Tarentum, Pennsylvania, had, at fourteen, been an artist's model and then a performer in the musical hit "Florodora." Irvin S. Cobb, journalist, humorist, and short story writer, described her as "the most exquisitely lovely human being I ever looked at." She met and, in 1905, married Harry K. Thaw, the irresponsible playboy heir to a Pittsburgh railroad and coke fortune.

On the evening of June 25, 1906, Thaw and his wife, with two male guests, were attending the opening of the musical production "Mamzelle Champagne" at the roof theater of Madison Square Garden. Thaw suddenly rose, walked quickly to a table at which Stanford White was seated and shot him three times with a pistol, killing him.

Six months later Thaw was on trial for murder, a trial which became one of the most sensational of the decade. Evelyn Nesbit Thaw's testimony was lurid in the extreme. She indicated that before her marriage she had been White's mistress. He, according to her sworn testimony, had arranged to meet her, taken her to his New York apartment one night, plied her with drugged champagne, and seduced her.

Subsequent to becoming his mistress, she said, White was in the habit of placing her naked in a red velvet swing in his apartment and "pushing her so high her feet touched a Japanese parasol which hung from the ceiling." In 1903 she made two premarital trips to Europe with Thaw but returned alone from the first one to resume her relationship with White. Thaw, she said, was in the habit of beating her with a rawhide whip and engaging in other sadistic practices.

Prior to her marriage, she testified, she had told Thaw of her relationship with White. Thaw became infuriated at any mention of White's name, referring to him as "the Bastard" and "the Beast."

William Travers Jerome, district attorney, was unable to secure a conviction in the first of Thaw's two trials. Delphin M. Delmas, Thaw's defense counsel, pictured him as a wronged husband "avenging his wife's honor by killing her seducer." The jury disagreed and nine months later a second trial was held. The jury here found Thaw "not guilty by reason of insanity." The judge committed him to Matteawan State Hospital for the Criminal Insane, declaring him to be manic-depressive and dangerous to the public safety.

Thaw spent much of the rest of his life in asylums. He was released in 1924 and lived in semi-seclusion, except for occasional unconventional escapades which brought him into contact with the police. He died in 1947.

Evelyn Nesbit Thaw divorced her husband soon after the trials were over. Her later years were filled with squabbles with the Thaw family over money, a few appearances at second-rate night clubs and cabarets, suicide attempts and affairs with various men. She led a purposeless life until her death in a convalescent home at the age of 82.

Suggested Readings

Atwell, Benjamin H. The Great Harry Thaw Case. Chicago: Laird & Lee, 1907.
Baldwin, Charles. Stanford White. New York: Dodd, Mead, 1931.
Collins, Frederick L. Glamorous Sinners. New York: R. Long and R. R. Smith, 1932.
Frank, H. "The Thaw-White Tragedy," Arena, September, 1906.
Goldsmith, G. "I Remember McKim, Mead and White," American Institute of Architects Journal, April, 1950.
Gribble, Leonard. Justice? Stories of Famous Modern Trials. New York: Abelard, 1971.
Jessel, George. Essay in Manhattan (poem). New York: Holt, 1961.
Ketchum, Richard M. "Faces from the Past," American Heritage Magazine, June, 1969.
_______. "The Girl in the Red Velvet Swing," American Heritage Magazine, June, 1969.
Langford, Gerald. The Murder of Stanford White. Indianapolis: Bobbs-Merrill, 1962.
Lavine, Sigmund A. Famous American Architects. New York: Dodd, Mead, 1967.
MacKenzie, F. A. The Trial of Harry K. Thaw. London: G. Bles, 1928.
Nesbit, Evelyn. Prodigal Days: The Untold Story. New York: Messner, 1934.
Saarinen, A. "The Splendid World of Stanford White," Life, September 16, 1966.
Samuels, Charles. "The Dementia Americana of Harry Thaw," in Boucher, Anthony, pseud. ed. The Quality of Murder. New York: Dutton, 1962.
Sanders, Bruce. "First Night Murder," in his Murder in Big Cities. New York: Roy Publishers, 1962.
Symons, Julian. A Pictorial History of Crime. New York: Crown Publishers, 1966.
"Thaw Murders Stanford White," New York Times, June 26, 1906, p. 1.
Train, Arthur. My Day in Court. New York: Scribner's, 1939.

THE HAYWOOD TRIAL (1907)

The murder trial of William D. ("Big Bill") Haywood, Charles Moyer, and George Pettibone was precipitated by the assassination of Frank R. Steunenberg, former Governor of Idaho.

The defendants were officials and members of the Western Federation of Miners which had, for fourteen years, fought the western mine owners in an effort to correct the intolerable conditions under which they were forced to labor. Steunenberg, as governor had, in 1899, called in federal troops to help break a strike, and for this he was killed by an infernal machine rigged to the gate of his home in Caldwell, Idaho.

A professional dynamiter and murderer known as Harry Orchard had confessed to the crime, implicating Haywood, Moyer, Pettibone, and Jack Simpkins, another Federation member, as instigators. Simpkins disappeared but the other three were kidnaped and illegally brought from Denver, Colorado to Boise, Idaho. James McParlan, a Pinkerton detective and labor spy, was instrumental in obtaining Orchard's confession. Haywood, Moyer, and Pettibone were charged with murder. As Orchard's unsupported statement was not legally sufficient to convict, a corroborating witness was needed. Orchard had implicated a homesteader named Steve Adams as an accomplice, and Adams was arrested in Oregon and brought to Boise. Here he was persuaded by McParlan and Orchard to sign a suitable confession.

Clarence Darrow, a prominent criminal lawyer, entered the case as defense attorney. On Darrow's instructions, Adams repudiated his confession and was promptly charged with another murder, that of an obscure claim jumper. Adams' trial was held at Wallace, Idaho, beginning in February, 1907. It lasted three weeks and resulted in a hung jury. Under these circumstances Adams could not legally testify in the case of Haywood, Moyer, and Pettibone, which was to follow.

The Haywood Trial, as it came to be called, was held in Boise. The prosecution attempted to show Orchard as a reformed character who repented his early transgressions and wished to make amends. Darrow produced over a hundred witnesses to disprove Orchard's statements. He showed further that Orchard, as a miner who had suffered as a result of Steunenberg's actions while Governor of Idaho, had reason to hate him, and as a thief, murderer, poisoner, kidnaper, and perjurer was the type of man who would not hesitate to do away with an enemy by violent means.

The trial, which opened in May, 1907, lasted over two months. Haywood was found not guilty. Steve Adams was tried a second time and Darrow, although suffering from an infected ear, defended him once more. This trial, like the first Adams trial, ended with a jury which could not agree.

George Pettibone was tried, with Orchard the primary witness against him. Like Haywood, he was acquitted. The state of Idaho dropped the case against Moyer, and Steve Adams was extradited to Colorado, tried on still another charge, found not guilty, and released.

Harry Orchard was tried for murder and sentenced to death. The sentence was commuted to life imprisonment and Orchard remained in the Boise penitentiary for the remainder of his life.

Suggested Readings

Busch, Francis X. Prisoners at the Bar. Indianapolis: Bobbs-Merrill, 1952.
Cook, Roy. Leaders of Labor. Philadelphia: Lippincott, 1966.
Crandall, Allen. The Man from Kinsman. Sterling, Colo.: Published by the author, 1965.
Darrow, Clarence. The Story of My Life. New York: Scribner's, 1932.
Dos Passos, John. "Big Bill," in his The 42nd Parallel. New York: Random House, 1930.
Gurko, Miriam. Clarence Darrow. New York: Crowell, 1965.
Harrison, Charles Yale. Clarence Darrow. New York: Jonathan Cape and Harrison Smith, 1931.
Haywood, William D. The Autobiography of Big Bill Haywood. New York: International, 1966.
Iman, Raymond S., and Thomas W. Koch. Labor in American Society. Chicago: Scott, Foresman, 1965.
Johnsen, Julia E., comp. The Closed Shop. New York: Wilson, 1942.
Lens, Sidney. Unions and What They Do. New York: Putnam, 1968.
Litwack, Leon F., ed. The American Labor Movement. Englewood Cliffs, N.J.: Prentice-Hall, 1962.
Meltzer, Milton. Bread--and Roses: The Struggle of American Labor. New York: Knopf, 1967.
Ravitz, Abe C., and James N. Primm. The Haywood Case. San Francisco: Chandler Publishing Co., 1963.
Rayback, Joseph G. A History of American Labor. New York: The Free Press, 1959.
Stone, Irving. Clarence Darrow for the Defense. Garden City, N.Y.: Doubleday, 1941.
Weinberg, Arthur, ed. Attorney for the Damned. New York: Simon & Schuster, 1957.
Werstein, Irving. Pie in the Sky: An American Struggle, the Wobblies and Their Times. New York: Delacorte, 1969.

THE GENTLEMEN'S AGREEMENT (1908)

In 1854 Commodore Matthew C. Perry of the American navy visited Japan and obtained permission for American ships to take on coal and provisions at two Japanese ports. Subsequently Townsend Harris, as first Consul General to Japan, negotiated commercial treaties and obtained other privileges. For many years thereafter relations between the two countries were extremely friendly.

During the first years of the 20th century, however, friction began to develop. The Russo-Japanese War, which ended in 1905, had been brought to a peaceful settlement at a conference held at Portsmouth, New Hampshire. This conference had been arranged by President Theodore Roosevelt at the secret request of the Japanese government and the Emperor of Germany. The Japanese people expected a heavy indemnity and when this did not materialize they blamed Roosevelt and the American people.

When San Francisco was severely damaged by the earthquake and fire of April, 1906, Japan contributed to a relief fund. However, Americans in the West, particularly in California, had objected to the presence of Japanese in that state and the Board of Education had segregated Japanese children from American children in the public schools. This act resulted in much indignation on the part of the Japanese.

The most effective remedy for the objections of the Californians to the presence of Japanese in their state was the terminating of Japanese immigration. The Japanese, however, refused to agree to a treaty similar to the one which restricted the entry of Chinese. It was then that the "Gentlemen's Agreement" of 1908 was entered into. Baron Kogoro Takahira, Japanese ambassador at Washington, D.C., and Elihu Root, American Secretary of State, arranged that the Japanese government would undertake to prevent the emigration of Japanese laborers to the United States.

In 1882 certain restrictive limitations on immigration had been imposed. Following World War I it was realized that unless quotas were established and enforced, the United States would be subjected to a deluge of refugees from war-torn Europe, something to which organized labor strenuously objected. In 1921 Congress passed the Emergency Quota Act, and in 1924 the Immigration Quota Law, more restrictive than the 1921 Act, was also passed. One provision of the 1924 Law concerned the exclusion of the Japanese who had hitherto been debarred by the "Gentlemen's Agreement" of 1908.

Suggested Readings

Bailey, Thomas Andrew. Theodore Roosevelt and the Japanese-American Crisis. Gloucester, Mass.: Peter Smith, 1934.

Chu, Daniel, and Samuel Chu. Passage to the Golden Gate. Garden City, N.Y.: Doubleday, 1967.

Commager, Henry Steele, ed. "Japanese Immigration--The Gentlemen's Agreement," (Doc. No. 367) in his Documents of American History, 8th edition. New York: Appleton, 1968.

Dennett, Tyler. Americans in Eastern Asia. Wilmington, Del.: Scholarly Resources, 1922.

________. Roosevelt and the Russo-Japanese War. Gloucester, Mass.: Peter Smith, 1958.

Griffis, William E. Townsend Harris, First American Envoy to Japan. Plainview, N.Y.: Books for Libraries, 1895.

Gulick, Sidney L. The American Japanese Problem. New York: Jerome S. Ozer, 1971.

Hornbeck, Stanley K. Contemporary Politics in the Far East. New York: Arno Press, 1970.

Iyenaga, Toyokichi, and Kenosake Sato. Japan and the California Problem. Wilmington, Del.: Scholarly Resources, 1972.

McKenzie, R. D. Oriental Exclusion. San Francisco: R&E

Research Associates, 1970.
Morse, Hosea B., and Harley F. MacNair. Far Eastern International Relations. New York: Russell & Russell, 1967.
Roosevelt, Theodore. Autobiography. Edited by Wayne Andrews. New York: Octagon Books, 1973.
Root, Elihu. "The Real Question Under the Japanese Treaty and the San Francisco Board Resolution," Proceedings of the American Society of International Law, 1907.

THE WHITLA KIDNAPING (1909)

One of the more sensational kidnaping cases of the 20th century was that of eight-year-old Willie Whitla, who was abducted from the East Ward School in Sharon, Pennsylvania.

On the morning of March 18, 1909, a strange man driving a horse and buggy appeared at the school and told Wesley Sloss, the janitor, that Willie's father, James P. Whitla, a wealthy attorney, had sent for him. Sloss got the boy from his classroom and turned him over to the stranger.

Willie Whitla failed to come home for lunch and his parents thought he might have accepted an invitation to eat his midday meal at the home of a friend. This idea was dispelled when that afternoon an anonymous letter stating that the boy had been kidnaped and demanding $10,000 ransom was delivered to the Whitla home. The envelope containing the letter was addressed in Willie Whitla's hand; the letter itself had been written by someone else.

The ransom note directed that the message, "A. A. Will do as you requested. J. P. W.," be inserted in the personal columns of the Cleveland Press, Youngstown Vindicator, Indianapolis News, and Pittsburgh Dispatch.

In spite of the fact that the anonymous letter warned against James Whitla's notifying the police of the kidnaping, the alarm was given and a search was conducted in the eastern part of the country. While the fact of the kidnaping could not be concealed, the boy's father and his wealthy uncle Frank M. Buhl had no confidence in the ability of the police to locate the missing boy. Consequently they did not divulge the contents of the ransom note or their plans to recover young Willie Whitla. The requested message was placed in the four newspapers indicated and a second note, containing additional instructions, was promptly received. Following the directions in the second note, the boy's father, accompanied by a private detective, went to Cleveland and from there to Ashtabula, Ohio. He left $10,000 at a designated spot in Flatiron Park and returned to his hotel to wait for his son. The boy failed to appear and early in the morning James Whitla was informed that the local police had been watching the park, thus scaring off the abductors. No one had appeared to claim the money.

James Whitla returned to Sharon. There he received a third note instructing him to go to Cleveland where, at Dunbar's Drug Store, he would find a letter addressed to "William Williams," which letter would contain further instructions. Evading the police, the father went to Cleveland where he picked up the promised note at the designated drug store. This directed him to go to a candy store operated by a Mrs. Hendricks, where the ransom was to be delivered. This was to be carefully packaged and left for a "Mr. Hayes" who would call for it.

Following these instructions, James Whitla was given a note by Mrs. Hendricks. This told him to go to the Hollenden Hotel where he was to wait for his son. At the hotel he waited and shortly after eight o'clock Willie Whitla, accompanied by two older boys who had found him on a streetcar, entered the lobby and was reunited with his father. He was tired but unharmed.

His story was that he had been taken to Cleveland in a buggy by a man who told him he was complying with his father's request to take him to a hospital in order to avoid a smallpox epidemic. In Cleveland he was taken care of by a "Mr. and Mrs. Jones." After his sojourn at the "hospital" "Mr. Jones" put him on a streetcar and told him to go to the Hollenden Hotel where his father was waiting for him.

The police then instituted an all-out search for the kidnapers. The druggist at Dunbar's Drug Store and Mrs. Hendricks, proprietress of the candy store, were interrogated and declared they were not aware that they had in any way been connected with the Whitla kidnaping, nor were the police able to establish any such connection. Mrs. Hendricks was able to furnish a description of the man who left the note for James Whitla and returned later to pick up the package of money.

The Cleveland police located an apartment building on Prospect Avenue which turned out to be the one in which Willie Whitla had been held. It was learned that a furnished apartment in this building had been rented to a man and a woman answering to the descriptions furnished by the boy.

A substantial reward for the apprehension of the kidnapers had been offered and many people, hoping to collect it, were on the lookout. One evening shortly after Willie Whitla's return, a man and a woman entered Patrick O'Reilly's saloon on Ontario Street in Cleveland and ordered drinks, the man paying with a new five dollar bill. He then ordered "drinks for the house," paying with another new similar bill. In a short time he had spent six such five dollar bills. This aroused O'Reilly's suspicions. He notified the police and the man and woman were arrested, taken to a precinct station, and questioned. When searched, almost all the ransom money was found on the woman's person.

The prisoners were James H. Boyle and his wife, Helen McDermott Boyle. They were charged with kidnaping, tried, convicted, and sentenced to long prison terms. Boyle attempted to implicate Harry Forker, Willie Whitla's uncle, in the plot. He told a long, rambling story of blackmail, in which he was the blackmailer and Forker was the victim. When Forker, according to Boyle's story, could not raise the sum demanded by Boyle, Forker suggested a plan to kidnap his nephew. Boyle was to be paid his blackmail from the proceeds of the crime.

Both Forker and James Whitla denied the accusation. Boyle was identified by Wesley Sloss as the man who had taken Willie Whitla from his classroom. Accordingly, Boyle's statement regarding the blackmail was not believed.

Mrs. Boyle was released from prison after serving ten years. Her husband spent the remainder of his life in confinement, dying at Riverside Penitentiary on January 23, 1920.

Suggested Readings

"Clamor for Man the Boyles Accuse," New York Times, May 10, 1909, p. 1.

Cohen, Bruce J. Crime in America. Itasca, Ill.: Peacock, 1970.

"Get Kidnappers, Recover Money," New York Times, March 24, 1909, p. 1.

Hibbert, Christopher. Roots of Evil: A Social History of Crime and Punishment. New York: Funk & Wagnalls, 1968.

"Kidnap Schoolboy and Demand $10,000," New York Times, March 19, 1909, p. 1.

"Kidnapped Boy Is Recovered," New York Times, March 23, 1909, p. 3.

"Kidnapper Boyle Moved to Mercer," New York Times, March 27, 1909, p. 2.

"Kidnapper Suspect Held in Cleveland," New York Times, March 20, 1909, p. 1.

"Kidnapping Victims: Tragic Aftermaths," Saturday Evening Post, April, 1976.

Messick, Hank, and Burt Goldblatt. Kidnapping: The Illustrated History. New York: Dial Press, 1974.

Slattery, William, ed. Abduction: Fiction Before Fact. New York: Grove Press, 1974.

Smith, Edward H. "The Whitla Kidnaping," in his Mysteries of the Missing. New York: Lincoln MacVeagh-The Dial Press, 1927.

Sutherland, Edwin H. On Analyzing Crime. Edited by Karl Schussler. Chicago: University of Chicago Press, 1972.

"To Arrest Woman as Kidnapper," New York Times, March 21, 1909, p. 1.

Winslow, Robert W. Crime in a Free Society. Encino, Calif.: Dickenson, 1968.

THE LOS ANGELES TIMES BOMBING (1910)

At seven minutes past one on the morning of Saturday, October 1, 1910, a dynamite blast aided by a gas leak demolished the four-story Los Angeles Times building at First Street and Broadway, Los Angeles, California. Twenty-one persons were killed.

Los Angeles was then an open-shop city, with businessmen and industrialists determined to oppose trade unions and closed shops. General Harrison Gray Otis, publisher of the Times, was particularly set against all aspects of unionism. Through his efforts Los Angeles was a place where hours were long, pay was low, and workers were virtually helpless to protest.

In 1910 Labor decided to unionize the city. Strikes and riots broke out and in October matters came to a head when the Times building was dynamited. General Otis blamed the unions for the act and the city hired detective William J. Burns to investigate the explosion. Burns and his organization collected evidence sufficient to charge three men with the crime: Ortie McManigal, James B. MacNamara and his brother, John J. MacNamara. McManigal and James B. MacNamara were members of the Structural Iron Workers Union and John J. MacNamara was the Union's secretary-treasurer.

The three men were arrested in the Middle West, kidnaped and illegally transported to California. There McManigal confessed that, under orders from John J. MacNamara, he and James B. MacNamara had dynamited the Times building.

Samuel Gompers, president of the American Federation of Labor, was convinced that the MacNamaras were not guilty and that the explosion was sabotage designed to generate anti-union sentiment. He hired Clarence Darrow, a prominent criminal lawyer, to defend the MacNamara brothers.

Darrow conducted extensive investigations, seeking to discredit McManigal's confession. No evidence of the MacNamara's innocence was to be found, and eventually the two brothers confessed to the crime. Lincoln Steffens, journalist and reformer, entered the case on the side of the MacNamaras, negotiating an out-of-court settlement. The defendants changed their pleas from "not guilty" to "guilty," and were sentenced to prison, John to serve fifteen years and his brother receiving a life sentence.

In January, 1912, Darrow was indicted on two counts of attempted jury bribing. His defense was conducted in part by Earl Rogers, a flamboyant and highly successful criminal lawyer. Darrow made his own final summation to the jury, over Rogers' protests. The jury, out only 34 minutes, returned with a "not guilty" verdict. Three months later a second trial, covering the other count in the indictment, was held. Darrow was defended by Jerry Giesler, a legal associate of Earl Rogers. This case resulted in a hung jury, following which the second indictment was dismissed.

Suggested Readings

Cohen, Alfred, and Joe Chisholm. Take the Witness! New York: Stokes, 1934.

Cook, Roy. Leaders of Labor. Philadelphia: Lippincott, 1966.

Crandall, Allen. The Man from Kinsman. Sterling, Colo.: Published by the author, 1965.

Darrow, Clarence. The Story of My Life. New York: Scribner's, 1932.

Giesler, Jerry. The Jerry Giesler Story, as told to Pete Martin. New York: Simon & Schuster, 1960.

Gurko, Miriam. Clarence Darrow. New York: Crowell, 1965.
Harrison, Charles Yale. Clarence Darrow. New York: Jonathan Cape and Harrison Smith, 1931.
Heaps, Willard Allison. Riots, U.S.A., 1765-1970. New York: Seabury Press, 1970.
Hynd, Alan. In Pursuit: The Cases of William J. Burns. Camden, N.J.: Nelson, 1968.
Iman, Raymond S., and Thomas W. Koch. Labor in American Society. Chicago: Scott, Foresman, 1965.
Johnsen, Julia E., comp. The Closed Shop. New York: Wilson, 1942.
Lens, Sidney. Unions and What They Do. New York: Putnam, 1968.
Litwack, Leon F., ed. The American Labor Movement. Englewood Cliffs, N.J.: Prentice-Hall, 1962.
Meltzer, Milton. Bread--and Roses: The Struggle of American Labor. New York: Knopf, 1967.
Rayback, Joseph G. A History of American Labor. New York: The Free Press, 1959.
St. Johns, Adele Rogers. Final Verdict. Garden City, N.Y.: Doubleday, 1962.
Steffens, Lincoln. The Autobiography of Lincoln Steffens. New York: Harcourt, Brace, 1931.
Stone, Irving. Clarence Darrow for the Defense. Garden City, N.Y.: Doubleday, 1941.
Weinberg, Arthur, ed. Attorney for the Damned. New York: Simon & Schuster, 1957.

THE DOROTHY ARNOLD DISAPPEARANCE (1910)

One of the unexplained disappearances of the 20th century is that of Dorothy Arnold who, on the afternoon of Monday, December 12, 1910, vanished from Fifth Avenue, New York City, and was never seen again.

Dorothy Arnold was the 25-year-old daughter of a wealthy importer of perfumes. She was the niece of Associate Justice Rufus Wheeler Peckham of the United States Supreme Court and her family was socially prominent in Washington, New York and Philadelphia. A Bryn Mawr graduate, she was a participant in the activities of what was then called the "younger set." She was considered "active, cheerful, intelligent, and talented."

On the day of her disappearance Dorothy Arnold left her father's home shortly before noon, ostensibly to shop for an evening dress. Her mother offered to accompany her but she demurred, saying she would telephone if she found a gown to her liking. The subsequent investigation of her movements that day disclosed that she had, about noon, purchased a box of candy at a store on Fifty-Ninth Street, charging the purchase to her father's account. At half past one she bought a volume of fiction at Brentano's book shop at Twenty-Seventh Street and Fifth Avenue. This, like the candy, she charged to her father's account. At about two o'clock she encountered a girl friend and the friend's mother on the street. They chatted briefly and parted. So far as is known, Dorothy Arnold was never seen alive again.

When the girl did not return home for dinner her parents, though irritated at what they considered thoughtlessness in not telephoning, were not alarmed. However, at midnight, when she had still not arrived, her father communicated with several of her friends by telephone. None of them could furnish any information concerning her whereabouts and Arnold contacted his attorney, John S. Keith. A search was begun but it was not until January 26 that the girl's disappearance was mentioned in the newspapers.

Private detectives had been engaged to locate the girl. An examination of her room revealed no preparations to leave for an extended period. She had taken no clothes except those she was wearing. She had left her checkbook behind and had less than thirty dollars in her purse. Her bank account showed no large or unexplained recent withdrawals. None of her valuable jewelry was missing.

Various leads were investigated but they invariably led nowhere. A man "of a well-known family in another city" with whom she had had correspondence and who "had considered himself engaged to marry her" was investigated. It was found that, three months before Dorothy Arnold's disappearance, she and this man had both been in Boston at the same time, but the police were unable to establish any connection between this fact and the girl's sudden absence.

Several theories concerning Dorothy Arnold's disappearance were offered. Kidnaping was suggested as an explanation, as were death in an automobile accident and secret burial, amnesia, elopement, and arrest and imprisonment under another name for some crime such as shoplifting. These were all considered and they, as well as other theories, were checked through and came to nothing.

Eventually Dorothy Arnold's father declared his belief in her death. When he passed away in 1922 he left no provision for her in his will, stating "I am satisfied that she is dead."

In 1916 a thief, arrested in Providence, Rhode Island, told the police that he had helped bury the girl's body in the cellar of a house near West Point. Although the cellar was carefully dug up, no trace of any body was found there.

Suggested Readings

"Arnold Girl Gone Now Fifty-Five Days," New York Times, February 5, 1911, p. 9.

"Arnolds Tell All in Search for Girl," New York Times, January 30, 1911, p. 1.

Churchill, Allen. "The Girl Who Never Came Back," American Heritage Magazine, August, 1960.

________. They Never Came Back. Garden City, N.Y.: Doubleday, 1960.

Cohen, Bruce J. Crime in America. Itasca, Ill.: Peacock, 1970.

"Deny Miss Arnold Went Away by Boat," New York Times, February 1, 1911.
"Dorothy Arnold May Be Home Today," New York Times, January 28, 1911, p. 1.
"Follow New Clue to Miss Arnold," New York Times, January 27, 1911, p. 1.
Hibbert, Christopher. Roots of Evil: A Social History of Crime and Punishment. New York: Funk & Wagnalls, 1968.
"Kidnapping Victims: Tragic Aftermaths," Saturday Evening Post, April, 1976.
Messick, Hank, and Burt Goldblatt. Kidnapping: The Illustrated History. New York: Dial Press, 1974.
"Mrs. Arnold May Be Seeking Girl Abroad," New York Times, January 31, 1911, p. 1.
"Niece of Peckham Strangely Missing," New York Times, January 26, 1911, p. 1.
"Police Drop Work on the Arnold Case," New York Times, February 7, 1911, p. 7.
Rinehart, J. C. "Tracing People Lost, Stolen and Strayed Away," American City, October, 1929.
Slattery, William, ed. Abduction: Fiction Before Fact. New York: Grove Press, 1974.
Smith, Edward H. "Dorothy Arnold," in his Mysteries of the Missing. New York: Lincoln MacVeagh-The Dial Press, 1927.
Sutherland, Edwin H. On Analyzing Crime. Edited by Karl Schussler. Chicago: University of Chicago Press, 1972.
Winslow, Robert W. Crime in a Free Society. Encino, Calif.: Dickenson, 1968.

THE TRIANGLE SHIRTWAIST COMPANY FIRE (1911)

On Saturday, March 25, 1911, one of the most disastrous fires ever to occur in the United States swept over the premises of the Triangle Shirtwaist Company in New York City. One hundred and forty-six persons lost their lives and many others were injured.

The Triangle Shirtwaist Company occupied the three top floors of the Asch Building, a ten-story loft at the intersection of Greene Street and Washington Place in lower Manhattan. At 4:45 in the afternoon, immediately after the day's work had been completed, fire was discovered in a rag bin. It was presumed that the fire was caused by a carelessly dropped cigarette or lighted match.

Despite efforts to extinguish the blaze it spread rapidly, the wooden cutting tables, piled high with cloth, being highly flammable. The fire hose in the hall proved to be rotten and the valve on the wheel was rusted shut and could not be turned. A sheet of flame poured out of the windows of the ninth and tenth floors of the building.

About 600 employees of the Triangle Shirtwaist Company, mostly girls, were on the premises when the fire started. A dozen or so escaped from the eighth floor by the fire escape. Others found the stairway exit door locked and were unable to open it. Two hundred girls made their escape down the unlighted stairs. On the ninth floor there was a mad rush for the elevators. These made several trips, each time carrying 25 or 30 girls. Some, driven by the unbearable heat, flung themselves down the shafts. Nineteen bodies were found above one elevator on the Greene Street side of the Asch Building. On the Washington Place side an elevator gave way under the weight of the girls who had flung themselves on top of it. Some girls slid down the elevator cables and bodies were found wedged between the cars and the walls of the elevator shafts.

The fire trucks which sped to the scene were unable to raise their extension ladders higher than the sixth story of the building. Men and women jumped from windows, dying when they struck the street. At least fifty persons were killed in this manner.

As a result of the fire the Department of Buildings belatedly condemned the Asch Building as unsafe. It was generally realized that fire and safety regulations were largely ignored in the needle-trade sweatshops, and the Triangle Shirtwaist fire was a shocking example of the result of such noncompliance.

Max Blanck and Isaac Harris, owners of the factory, were indicted, the charge being manslaughter in the first and second degree in connection with the death of Margaret Schwartz, one of the employees who had died in the fire. They were defended by the noted attorney Max D. Steuer and the prosecution was handled by Assistant District Attorneys Bostwick and Rubin. Justice Thomas T. C. Crain was the presiding judge.

The trial was characterized by perjured testimony on both sides. The jury brought in a highly unpopular verdict of "not guilty." No other manslaughter indictments were ever brought against either Blanck or Harris.

The Triangle Shirtwaist fire resulted in the adoption of a stringent building code, revision of labor laws, and organization of labor in the needle trades.

Suggested Readings

"Acquittal in the Triangle Case," Outlook, June 6, 1912.

"Criticisms of the Work of the Factory Investigating Commission," Survey, June 13, 1914.

Epstein, Melech. Jewish Labor in the U. S. A. New York: Ktav Publishing House, 1969.

Haywood, Charles F. General Alarm. New York: Dodd, Mead, 1967.

Hurwitz, Maximilian. The Workmen's Circle: Its History, Ideals, Organization, and Institutions. New York: Workmen's Circle, 1936.

Limpus, Lowell M. History of the New York Fire Department. New York: Dutton, 1940.

"Lives at $75," Literary Digest, March 28, 1914.

Lorwin, Lewis L. The Women's Garment Workers. New York:

AMS Press, 1924.
McFarlane, A. E. "Triangle Fire: The Story of a Rotten Risk," Collier's, May 17, 1913.
Morris, John V. "A New Age: Motors and Two Platoons: Triangle Shirtwaist Fire," in his Fires and Firefighters. Boston: Little, Brown, 1953.
Morris, Richard B. "The Triangle Fire Case," in his Fair Trial. New York: Knopf, 1952.
"147 Dead, Nobody Guilty," Literary Digest, January 6, 1912.
Steuer, Aron. Max D. Steuer, Trial Lawyer. New York: Random House, 1950.
Stolberg, Benjamin. Tailor's Progress. Garden City, N.Y.: Doubleday, 1944.
"Triangle Fire, Three Years After," Survey, April 11, 1914.
Waldman, Louis. Labor Lawyer. New York: Dutton, 1945.
________. "Sweatshop Firetrap," in Kartman, Ben, and Leonard Brown, eds. Disaster! New York: Pellegrini & Cudahy, 1948.

THE ZACHARIAH WALKER LYNCHING (1911)

In 1911 the steel town of Coatesville, Pennsylvania (population about 11,000) was the scene of one of the most brutal lynchings in American history. The victim was Zachariah Walker, an uneducated Negro wagon driver who allegedly shot and killed Edgar Rice, a policeman employed by the Worth Brothers Company, an iron and steel manufacturing corporation.

On the evening of Saturday, August 12, 1911, Walker, having become semi-intoxicated from drinking gin, put a pistol in his pocket and started to walk from his home to Coatesville. He entered a small patch of woods. Shortly afterwards Rice, hearing two shots, went to the woods to investigate. On the way he encountered a fleeing man who told Rice that a Negro had attempted to hold him up and had fired two pistol shots but without hitting him.

Rice entered the woods. Another shot was heard, followed by two more. The policeman staggered from the woods and in a few minutes dropped dead. There were two bullet holes in his body.

Edgar Rice, although a policeman, was well liked in Coatesville. Chief of Police Charles E. Umsted, equally popular, instructed the members of his seven-man force to get Rice's killer. "I don't care how you get him but get him," he said.

Various clues pointed to Walker as the killer and on the morning of Sunday, August 13, he was found hiding in a barn. The posses organized to search for him closed in. Walker left the barn, climbed a cherry tree and attempted to commit suicide by shooting himself. The attempt was unsuccessful; he succeeded only in inflicting himself with a bullet wound in the jaw. He was taken to the local hospital where the bullet was removed. He was then strapped to a bed and his right leg was secured by a chain to the foot of the bed.

Tension mounted and feelings ran high in Coatesville. Questioned by Chief Umsted, Walker stated that when in the woods and "feeling pretty good," he had "fired his pistol in the air just to whoop things up," and this had frightened the man who ran from him. He had made no attempt to rob anyone. When Rice had encountered him shortly thereafter he had killed Rice in self-defense. "The policeman," he said, "came over and placed me under arrest. I knew ... I would serve time for carrying a weapon and I resisted him. Rice ... made a lunge at me with his club, then dropped it and reached for his revolver.... I had my gun out first and fired two shots into him. Rice shot at me once and missed."

That night a crowd gathered around the hospital. Talk of "lynching that nigger" spread. The crowd grew larger. Miss Lena Gray Townsend, superintendent of the hospital, became nervous and attempted to call Umsted, but was unable to contact him. Eventually he was located and informed of the situation, which he dismissed as "a lot of hot air," and refused to call out the police force, although ordered to do so by Richard D. Gibney, chairman of the Coatesville Police Committee.

A masked man armed with a pistol appeared suddenly at the hospital and provoked the crowd into action. He was joined by other masked men who broke into the hospital and dragged Walker, his chain and a portion of the bedstead outside. One man yelled, "Burn him! Burn him!"

Chief Umsted, when contacted a second time, went to the hospital, talked to Miss Townsend and some others, and then went home. The mob dragged Walker to a nearby farm belonging to Mrs. Sarah Jane Newlin. There a pyre of logs and fence rails was quickly put together, on the top of which the half-unconscious Negro was thrown. The wood was set afire and when Walker attempted to escape he was pushed back into the flames. Twenty minutes later he was dead.

Chief Umsted and police officer Stanley Hower were indicted for involuntary manslaughter and twelve other men were indicted for murder in varying degrees. Seven of the cases came to trial and all seven defendants were acquitted. The other cases were dropped.

Suggested Readings

Ames, Jessie (Daniel). The Changing Character of Lynching, 1931-1941. With a Discussion of Recent Developments in This Field. Atlanta: Commission on Interracial Cooperation, 1942.

Archer, Jules. Riot: A History of Mob Action in the United States. New York: Hawthorn Books, 1974.

"Blot on Civilization: Coatesville Lynching," Outlook, August 26, 1911.

Caughey, John Walton. Their Majesties, The Mob. Chicago: University of Chicago Press, 1960.

"Coatesville Lynchers Free," Literary Digest, May 18, 1912.
Cutler, James Elbert. Lynch-Law: An Investigation into the History of Lynching in the United States. New York: Negro University Press, 1969.
Goldman, Eric F. "Summer Sunday," American Heritage Magazine, June, 1964.
Heaps, Willard Allison. Riots, U.S.A., 1765-1970. New York: Seabury Press, 1970.
"The Lessons of Coatesville," Nation, August 31, 1911.
National Association for the Advancement of Colored People. Thirty Years of Lynching in the United States, 1889-1918. New York: Arno Press and the New York Times, 1969.
Nock, Albert Jay. "What We All Stand For: The Coatesville Lynching," American Magazine, February, 1913.
Raper, Arthur F. The Tragedy of Lynching. Chapel Hill, N.C.: University of North Carolina Press, 1933.
Shay, Frank. Judge Lynch: His First Hundred Years. New York: I. Washburn, 1938.
Wells-Barnett, Ida B. On Lynching: Southern Horrors. New York: Arno Press and the New York Times, 1969. (Originally written 1892.)
White, Walter Francis. Rope and Faggot. New York: Arno Press, 1969.

THE "TITANIC" DISASTER (1912)

Shortly before midnight on Sunday, April 14, 1912, the White Star luxury liner "Titanic" struck an iceberg in the North Atlantic. She was on her maiden voyage from Southampton to New York and three hours after striking the iceberg she sank.

The "Titanic," thought to be unsinkable, was the largest and most modern vessel of her time. Many prominent Americans were aboard. Her first class passengers included Colonel John Jacob Astor, Benjamin Guggenheim, F. D. Millet, Isidor Straus, Archibald W. Butt and William T. Stead, all wealthy persons. In addition there were 724 steerage passengers. J. Bruce Ismay, president and managing director of the White Star Lines, was making the trip, as was naval architect Thomas Andrews, builder of the vessel. Captain Edward J. Smith was in command.

During the day wireless warnings of icebergs in the vicinity had been received. The liner "Californian," thirty miles away, was blocked by an ice field and shut down her engines. Her radio operator sent out an "all ships" warning which was not received on the "Titanic," her operator being busy transmitting commercial messages by way of the Cape Race relay station in Newfoundland. A previous warning message, though received, had not been delivered to Captain Smith.

The iceberg with which the liner collided tore a gash 300 feet long in her starboard side beneath the water line. Water rushed into six of the sixteen compartments, and Andrews realized that the ship was doomed.

The "Titanic" carried sixteen lifeboats. The passengers, minimizing the danger, were reluctant to enter them and several were lowered to the water only partially filled. John George Phillips, wireless operator, sent out a distress call. Answers were received from the "Frankfort" and the "Mt. Temple," neither of which was in a position to render effective help. Rockets fired from the "Titanic" were seen but ignored by the crew of the "Californian." At 12:36 a message from the "Carpathia" was received. This vessel, then 58 miles away, was coming to the rescue.

The "Titanic" sank at 2:27 A.M. Of the 2,223 persons aboard, 1,517 were lost. The survivors were saved by the "Carpathia," which arrived a few hours later.

Suggested Readings

Baldwin, Hanson W. Sea Fights and Shipwrecks. New York: Museum Press, 1956.

Bancroft, Caroline. The Unsinkable Mrs. Brown. Boulder, Colo.: Johnson Publishing Co., 1963.

Beesley, Lawrence. The Loss of the Titanic. Port Arthur, Tex.: 7 C's Press, 1973.

Bullock, Shan F. A Titanic Hero: Thomas Andrews, Shipbuilder. Port Arthur, Tex.: 7 C's Press, 1973.

Gracie, Archibald. The Truth About the Titanic. Port Arthur, Tex.: 7 C's Press, 1973.

Hill, George Roy, and John Whedon. "A Night to Remember" (television drama), in The Writers Guild of America Presents the Prize Plays of Television and Radio--1956. Foreword by Clifton Fadiman. New York: Random House, 1957.

Hoehling, Adolph A. Great Ship Disasters. New York: Cowles, 1971.

Lord, Walter. A Night to Remember. New York: Holt, 1955.

Marcus, Geoffrey. The Maiden Voyage. New York: Viking Press, 1969.

Mielke, Otto. Disaster at Sea. New York: Fleet, 1958.

Mizner, Wilson. "You're Dead!" (short story), in Grayson, Charles, comp. Stories for Men. Boston: Little, Brown, 1936.

Paine, Ralph D. Lost Ships and Lonely Seas. New York: Century, 1921.

Vandenberg, Philipp. "The Mummy on the Titanic," in his The Curse of the Pharaohs. Translated by Thomas Weyr. Philadelphia: Lippincott, 1975.

Wallechinsky, David, and Irving Wallace. "The Unsinkable 'Titanic'," in their The People's Almanac. Garden City, N.Y.: Doubleday, 1975.

THE BECKER-ROSENTHAL MURDER CASE (1912)

The political corruption which existed in New York City from the days of the infamous Tweed Ring in the latter part of the 19th century continued for several decades. In 1912 public morals were at a low ebb, with Tammany Hall, the local Democratic political machine, taking graft and milking the city treasury of large sums of money.

The New York police department was also graft-ridden. Brothel proprietors and gamblers were permitted to operate their unlawful enterprises provided they shared their illicit profits with police officials. One such was Police Lieutenant Charles Becker, the trusted personal assistant of Police Commissioner Rhinelander Waldo. Becker directed the "strongarm squads" whose functions included the stamping out of gambling in the city, and was also a partner of the gambler Herman Rosenthal in an establishment located on West Forty-Fifth Street.

Rosenthal and Becker fell out. Becker had the establishment raided and then assigned policemen to remain there to see that it did not reopen. Rosenthal then talked to Herbert Bayard Swope, a reporter on the New York World. Swope published an exposé in his paper, his article implicating Becker as a secret partner in the gambling establishment and a sharer in the profits generated by it.

Shortly before two o'clock on the morning of Tuesday, July 16, 1912, Herman Rosenthal was shot to death by four men just outside the Metropole Hotel on Forty-Third Street, near Broadway. He had been summoned from the hotel dining room by Harry Vallon, a friend of Bridgey Webber, another gambler and proprietor of a poker club on Forty-Second Street. The four men escaped in a 1909 model gray Packard touring car bearing license plate No. 41313 N.Y. The Packard was pursued by three policemen in a taxi but was not intercepted.

District Attorney Charles Seymour Whitman had long been an enemy of organized crime and political corruption and he announced through the newspapers that he suspected Becker of being implicated in Rosenthal's murder. Becker and Whitman met at the station house where the dead gambler's body had been taken, and it was here that Charles Gallagher, an unemployed singer who had witnessed the shooting, gave the police the correct license number of the getaway Packard. He was immediately booked as a material witness and locked in a cell, whereupon Whitman ordered him released. The Packard was traced through its license number to Louis Libby, who had purchased it in order to enter the car rental business. His partner in this venture was William Shapiro, who had driven the car on the morning of the murder. Libby and Shapiro were formally accused of the murder of Herman Rosenthal. Later, in a statement they made to District Attorney Whitman, they indicated that the car had been ordered by Bald Jack Rose, whose real name was Jacob Rosenzweig. Rose, a gambler and collector of shakedown money for Becker, had not been in the car when it made its getaway following the shooting. Libby and Shapiro were exonerated.

Police Commissioner Waldo relieved Becker of his command of the "strongarm squad" until his name should be cleared of all suspicion. Rose turned himself in to Assistant Police Commissioner George Dougherty. He gave an account of his movements during the night of July 15-/16, denying any knowledge of Rosenthal's death until after it had occurred. He was, however, confined to the Tombs on a technical holding charge after further interrogation by Whitman.

Whitman ordered the arrest of Bridgey Webber and gamblers Sam Paul and Jack Sullivan (Jacob Reich). Shapiro, having been guaranteed immunity from a murder charge by Whitman, implicated others, including Rose, Harry Vallon, and Sam Schepps, the last two also gamblers. These men were all in the Packard prior to the murder, as was Dago Frank (Frank Cirofici), a young gunman who worked for Big Jack Zelig, an East Side gangster. Also in the car were Lefty Louie (Louis Rosenberg), Gyp the Blood (Harry Horowitz) and Whitey Lewis (Jacob Seidenschmer), all gunmen employed by Zelig.

Webber and Paul were arraigned. Sullivan was held as a material witness. Harry Vallon was also arraigned.

Whitman was not satisfied to convict "three small-time gamblers for killing a fourth," and offered immunity to help him locate and convict "the big fish--the police, for example." George Fredericks (or Hendricks), a Harlem saloon keeper, informed the police where they might find Dago Frank. Frank was found and arrested, and the saloon keeper was shot dead in his place of business by two men shortly thereafter.

Big Jack Zelig, Becker's employee, was shot to death on a streetcar by a hired killer named Philip "Red Phil" Davidson. Found guilty of second degree murder, Davidson was sentenced to twenty years' imprisonment. Who hired him to kill Zelig or why he was hired has never been satisfactorily determined.

On July 29, 1912, Police Lieutenant Charles Becker was indicted for the murder of Herman Rosenthal. Bald Jack Rose testified that Becker had promised Jack Zelig, then a prisoner in the Tombs, his freedom and $200 for Rosenthal's murder, and that Zelig, upon release, had hired Dago Frank, Lefty Louie, Gyp the Blood, and Whitey Lewis (the "Four Gunmen") to kill Rosenthal in public.

Judge John W. Goff presided at Becker's trial. John F. McIntyre acted as Becker's chief counsel and the prosecution was headed by Whitman. Goff, like the District Attorney, was a bitter enemy of all grafting officials and his attitude became plain as the trial progressed.

Becker, in a lengthy trial, was found guilty of murder in the first degree. On October 30, 1912, Judge Goff sentenced him to death.

The Four Gunmen, defended by attorney Charles A. Wahle, were tried together, Judge Goff presiding and Whitman handling the prosecution. The four were found guilty of murder and sentenced to death at Sing Sing Prison. Their conviction was referred to the Court of Appeals and was sustained. They died in the electric chair on April 13, 1914.

Becker's attorneys appealed the death sentence handed down by Judge Goff. Represented by attorney John W. Hart, the Court of Appeals reversed Becker's conviction and ordered a new trial on the grounds that, among other things, Goff's open hostility to the defendant and his counsel had made a fair trial impossible.

Becker's second trial was held before Judge Samuel Seabury, with Whitman again heading the prosecution and W. Bourke Cockran, assisted by Martin Manton, acting as defense counsel. The proceedings began on May 2, 1914 and, like the first trial, resulted in a conviction and the death sentence. The sentence was appealed; this time to the lower court's judgment was affirmed.

An appeal for pardon was made to Whitman who, in November, 1914, had been elected Governor of New York State. Whitman declined to pardon the convicted officer, though he did grant a stay until the matter could be considered by the New York County Supreme Court. Justice John Ford of the latter court denied the motion for a new trial.

On July 30, 1915, Police Lieutenant Charles Becker died in the electric chair at Sing Sing Prison.

Suggested Readings

Asbury, Herbert. The Gangs of New York. New York: Knopf, 1927.

______. Sucker's Progress. New York: Dodd, Mead, 1938.

Crane, Milton. The Sins of New York. New York: Boni & Gaer, 1947.

Katcher, Leo. The Big Bankroll. New York: Harper, 1958.

Kunstler, William H. "Murder for Hire: The State of New York versus Charles Becker," in his First Degree. New York: Oceana Publications, 1960.

Logan, Andy. Against the Evidence: The Becker-Rosenthal Affair. New York: McCall, 1970.

Lynch, Denis T. "Boss" Tweed: The Story of a Grim Generation. New York: Boni & Liveright, 1927.

Pink, Louis. Gaynor--The Tammany Mayor Who Swallowed the Tiger. New York: International Press, 1931.

Root, Jonathan. One Night in July: The True Story of the Rosenthal-Becker Murder Case. New York: Coward-McCann, 1961.

Smith, Mortimer. William Jay Gaynor. Chicago: H. Regnery & Co., 1951.

Steffens, Lincoln. The Autobiography of Lincoln Steffens. New York: Harcourt, Brace, 1931.

Stoddard, Theodore. Master of Manhattan. New York: Longmans Green, 1931.

Sullivan, Mark. Our Times, The United States, 1900-1925. New York: Scribner's, 1928-1935.

Symons, Julian. A Pictorial History of Crime. New York: Crown Publishers, 1966.

Tebbel, John. The Life and Good Times of William Randolph Hearst. New York: Dutton, 1952.

Thompson, Craig. Gang Rule in New York. New York: Dial

Press, 1940.
Tully, Andrew. Era of Elegance. New York: Funk & Wagnalls, 1947.
Werner, M. R. It Happened in New York. New York: Coward-McCann, 1957.
________. Tammany Hall. Garden City, N.Y.: Doubleday, 1928.

THE PUJO MONEY TRUST INVESTIGATION (1912-1913)

With the publication of Henry Demarest Lloyd's Wealth Against Commonwealth in 1894, the charge that a minute group of financiers and capitalists controlled American industry, credit, and transportation gained increasing acceptance. Following the Civil War the great railroads, industrial firms, commercial banks, insurance companies, and other business enterprises had mushroomed from small beginnings into giant business organizations. By such devices as mergers, voting trusts, and interlocking directorates, control had been concentrated in the hands of a few money masters who maximized their own positions of power and gave little or no consideration to the public good.

The charges made by Lloyd in his book were amplified and substantiated during the first decade of the 20th century by such "muckrakers" as Lincoln Steffens, Ida M. Tarbell, Upton Sinclair, and Ray Stannard Baker. Such revolts of the workers as the Pullman Strike, the Haymarket Riot, and the strikes of railroad and steel workers emphasized the injustice of a situation which permitted a few to enjoy virtually unlimited wealth and power while thousands of others suffered abject poverty.

In February, 1912, the House of Representatives directed the Committee on Banking and Currency to conduct an investigation into the practices of business management in the United States. This investigation was to determine whether or not the control exercised by a few men over financial affairs constituted a "money trust." The ultimate objective of the investigation was to determine what, if any, legislation regarding the situation might be required.

A fact-finding sub-committee, headed by Arsène Paulin Pujo, Chairman of the Committee on Banking and Currency, conducted an intensive investigation in 1912/13. Samuel Untermyer, a prominent attorney, acted as legal counsel for the investigators. This subcommittee, in a report to Congress in 1913, revealed that the charges of monopoly made by the muckrakers were essentially true, at least so far as the control of wealth and industrial power were concerned.

The Pujo Report stated specifically that officers of four New York financial institutions were closely allied and held a total of 341 directorships in banks, insurance companies, steamship companies, public utilities and trading corporations having aggregate resources of $22,245,000,000. Specific business organizations were mentioned by name, as were those officers who held multi-directorships. It gave facts and figures but did not make any recommendations.

The information disclosed by the Pujo Committee proved valuable in the 1912 Presidential campaign and helped strengthen the position of the reformers in President Woodrow Wilson's first administration. Much legislation was passed by Congress and also by the individual states as a result of the information unearthed by the investigation.

Opinions on the situation differed widely. Those who favored the concentration of wealth argued that a few large concerns administered by men of vision and ability who had, by their outstanding qualifications, risen to the top of the corporate ladder, could operate more efficiently than could a larger number of small firms with few resources and mediocre management. Others pointed out that a few men exercising great financial power and responsible to no one for their actions would manipulate production, eliminate business rivals, and use the industries they controlled for private gain regardless of the welfare of the public.

Suggested Readings

Berle, Adolph A., and Gardiner C. Means. Modern Corporation and Private Property. New York: Harcourt, Brace, 1968.

Commager, Henry Steele, ed. "The Concentration of Control of Money and Credit," (Doc. No. 388) in his Documents of American History, 8th edition. New York: Appleton, 1968.

________. "The Concentration of Wealth," (Doc. No. 407) in his Documents of American History, 8th edition. New York: Appleton, 1968.

Corey, Lewis. The House of Morgan. New York: AMS Press, 1930.

Dos Passos, John. "The House of Morgan," in his Nineteen-Nineteen. New York: Random House, 1931.

Flynn, John T. Men of Wealth. New York: Simon & Schuster, 1941.

King, Willford I. Wealth and Income of the People of the United States. New York: Johnson Reprint, 1969.

LaFollette, Robert M. LaFollette's Autobiography: A Personal Narrative of Political Experiences. Madison, Wis.: University of Wisconsin Press, 1960.

Lloyd, Henry Demarest. Wealth Against Commonwealth. New York: Harper, 1894.

Moody, John. The Truth About the Trusts. Westport, Conn.: Greenwood Press, 1968.

Reiger, C. C. The Era of the Muckrakers. Chapel Hill, N.C.: University of North Carolina Press, 1932.

Report of the Committee Appointed Pursuant to House Resolutions 429 and 504 to Investigate the Concentration and Control of Money and Credit ("Pujo Committee"), 1913.

THE TAMPICO INCIDENT (1914)

The "Tampico Incident" was the outgrowth of the political situation in Mexico prior to World War I. Porfirio Diaz had, as a benevolent despot, ruled the country for over forty years. A few persons had become exceedingly rich by exploiting Mexico's vast natural resources, but the great mass of people were extremely poor.

In 1910 Francisco Madero organized a revolt, drove Diaz out of the country, and became president. Madero, in turn, was overthrown by General Victoriano Huerta, who became provisional president. In February, 1913, Madero was arrested and shot while allegedly attempting to escape. It is generally thought that Huerta was responsible for his death.

The leading European nations recognized Huerta as President of Mexico but President Woodrow Wilson of the United States refused to so recognize him, on the grounds that his government was founded on violence, contrary to the Mexican constitution.

Under these circumstances Huerta had little liking for America or Americans. This dislike was made quite clear when, on April 9, 1914, a paymaster and several sailors from the U. S. S. "Dolphin," while picking up supplies at the dock in Tampico, Mexico, were arrested by a detachment of soldiers of Huerta's army. This was one of a series of incidents indicating "the irritation and contempt" felt by Huerta for the United States.

Admiral Charles Mayo considered the Tampico Incident extremely serious, in that permitting it to go unnoticed would set a very bad precedent. Accordingly, he demanded an apology from the Huerta government and a 21-gun salute to the American flag. The apology was made but the salute was not given. President Wilson preferred to ignore the affair after the apology but felt constrained to support Admiral Mayo.

On April 20, 1914, Wilson addressed Congress, requesting authority to "use the armed forces of the United States in such ways and to such an extent as may be necessary to obtain from General Huerta and his adherents the fullest recognition of the rights and dignity of the United States, even amidst the distressing conditions now unhappily obtaining in Mexico." Both House and Senate gave the President the necessary authority, and on April 22 Admiral Fletcher landed marines at Vera Cruz, seized the customs house and prevented the landing of munitions from a German ship.

This action not only brought the United States and Mexico to the brink of war but also threatened to unite all of Mexico behind Huerta.

At this point the "A. B. C. Powers" (Argentina, Brazil, and Chile) made a proposal of mediation which was accepted. Huerta resigned his provisional presidency on July 15, 1914, and was driven from Mexico. The American troops were soon withdrawn. A conference of the A. B. C. Powers, the United States and several lesser Latin-American countries decided to recognize General Venustiano Carranza as President of Mexico in 1915.

Suggested Readings

Baker, Ray Stannard. Woodrow Wilson, Life and Letters. Westport, Conn.: Greenwood Press, 1939.

_______, and W. E. Dodd, eds. The Public Papers of Woodrow Wilson. New York: Kraus Reprint, 1927.

Beals, Carleton. Mexican Maze. Westport, Conn.: Greenwood Press, 1971.

Callahan, James M. American Foreign Policy in Mexican Relations. New York: Cooper Square, 1932.

Commager, Henry Steele, ed. "Mediation Protocol of the A. B. C. Conference," (Doc. No. 396) in his Documents of American History, 8th edition. New York: Appleton, 1968.

_______. "The Tampico Incident," (Doc. No. 395) in his Documents of American History, 8th edition. New York: Appleton, 1968.

Curti, Merle E. Bryan and World Peace. New York: Octagon Books, 1969.

Gruening, Ernest H. Mexico and Its Heritage. Westport, Conn.: Greenwood Press, 1968.

Lingley, Charles Ramsdell, and Allen Richard Foley. Since the Civil War. New York: Appleton, 1935.

Munro, Dana C. The United States and the Caribbean Area. New York: Johnson Reprint, 1966.

Perkins, Dexter. The United States and the Caribbean. Cambridge, Mass.: Harvard University Press, 1966.

Rippy, James Fred. The United States and Mexico. New York: Crofts, 1931.

_______, José Vasconcelos and Guy Stevens. Mexico: American Foreign Policies Abroad. New York: Crofts, 1928.

Robertson, William S. Hispanic-American Relations With the United States. Edited by David Kinley. Millwood, N.Y.: Kraus Reprint, 1923.

Stuart, Graham H., and James L. Tigner. Latin America and the United States. Englewood Cliffs, N.J.: Prentice-Hall, 1974.

THE DISAPPEARANCE OF AMBROSE BIERCE (1914)

Ambrose Bierce, regarded by some as "the foremost master of the short story after Edgar Allan Poe," was born in Meigs County, Ohio, in 1842. He served in the Civil War, rising from private to the brevet rank of major. Following his military service he became a professional writer. In London, where he lived for a time, his early short stories, criticisms, and sketches, which were acid, ironical, and cutting, made his name well known.

Returning to America, he settled in San Francisco where he continued to write, contributing many witty and caustic items to the "Prattle" column of the San Francisco Examiner and becoming known as "the dean of Western writers." Later, in Washington, D.C., he continued producing satirical short stories, poems, essays, and miscellaneous literary material. He also found time to manage Western ranches and mining properties and to attack political corruption.

In 1913 Mexico was experiencing one of its many revolutions. Francisco Madero, that country's president, had been arrested and shot on the orders of Victoriano Huerta, who had deposed him and become a virtual dictator. Bierce was living in Washington when General Frederick Funston was sent with a detachment of troops to patrol the border along the Rio Grande. Rebels Venustiano Carranza and Pancho Villa were active in Northern Mexico and an invasion of that country by the United States was considered a distinct possibility.

Historians do not agree on the motives which prompted Bierce to depart for Mexico. It is generally thought that the man, then in his seventies, "not as spry as he used to be; old, long written-out and ready for sleep," was reminded of the old military days of the Civil War. He may have felt an irresistable urge to participate once again in exciting events "south of the border." It has been suggested that he was going to Mexico as a correspondent. Others said that he planned to join Carranza's Constitutionalists in the capacity of a military adviser.

After a visit to his California home Bierce went south. In the fall of 1913 he is said to have crossed the border into Mexico and in January, 1914, Carrie Christianson, his secretary, received a letter from him mailed from Chihuahua.

Nothing further was heard from him and in September, 1914, Mrs. H. D. Cowden, his daughter, appealed to the State Department in Washington. The State Department instructed the American chargé d'affaires in Mexico to make an investigation and gave General Funston similar instructions. These actions disclosed no information concerning the fate or whereabouts of the missing man. It was conjectured that he might be with one of the independent rebel forces in the Mexican mountains and thus out of touch with the main body.

Rumors began to circulate. It was considered possible that Bierce had joined Pancho Villa as a military aide and that he had been killed in action at Chihuahua. Other rumors were bruited about. A friend stated that, prior to going to Mexico, Bierce had told him that he was seeking "a soldier's grave." Other friends maintained that such an action was foreign to Bierce's character; that, as a fatalist, he would take what came but "would not go out and seek a conclusion."

So it went. Bierce was reported to be in England, a major on the staff of Lord Horatio Hubert Kitchener. The British War Office stated that Bierce's name did not appear on any of its records. He was reported to have been executed by a military firing squad some time in 1918. In 1919 George F. Weeks, an old friend of the missing writer, while visiting in Mexico City, encountered a native soldier who claimed to have served with Bierce on Villa's staff. The soldier stated that Bierce had been executed in 1915 for alleged gun-running. Partial confirmation of this last report was obtained by U. H. Wilkins, a reporter on the staff of the San Francisco Bulletin.

Other rumors concerning the fate of one of America's famous men of letters appeared. He had been ambushed and shot by Villa, who thought Bierce was about to betray him. He died of disease or battle wounds. He survived the military campaigns of Mexico and died of old age under another name, declining to reveal his true identity for reasons known only to him.

All this is conjecture. Today Ambrose Bierce is undoubtedly dead, but the details of his life after early 1914 are still an unsolved mystery.

Suggested Readings

Bierce, Ambrose. The Collected Writings of Ambrose Bierce. New York: Citadel Press, 1946.

________. The Letters of Ambrose Bierce. Edited by Bertha Pope. Staten Island, N.Y.: Gordian Press, 1967.

Boynton, Percy H. More Contemporary Americans. East Saint Clair Shores, Mich.: Scholarly Press, 1927.

Brooks, Van Wyck. Emerson and Others. New York: Octagon Books, 1973.

De Castro, Adolphe. "Ambrose Bierce as He Really Was," The American Parade, October, 1926.

Fatout, Paul. Ambrose Bierce: The Devil's Lexicographer. Norman, Okla.: University of Oklahoma Press, 1951.

Gaer, Joseph, ed. Ambrose Gwinnett Bierce: A Bibliography and Biographical Data. New York: Burt Franklin, 1935.

Grattan, C. Hartley. Bitter Bierce, A Mystery of American Letters. New York: Cooper Square, 1929.

Grenander, Mary Elizabeth. Ambrose Bierce. New York: Twayne, 1971.

McWilliams, Carey. Ambrose Bierce: A Biography. Hamden, Conn.: Shoe String Press, 1967.

Morrill, Sibley S. Ambrose Bierce, F. A. Mitchell-Hedges and the Crystal Skull. San Francisco: Cadleon Press, 1972.

Smith, Edward H. "The Ambrose Bierce Irony," in his Mysteries of the Missing. New York: Lincoln MacVeagh-The Dial Press, 1927.

Smith, Paul Jordan. On Strange Altars. Philadelphia: Richard West, 1924.

Starrett, Vincent. Bibliography of Ambrose Bierce. New York: Centaur Press, 1928.

Sterling, George. In the Midst of Life. New York: Modern Library, 1927.

________. "Shadow Maker," American Mercury, September, 1925.

Wiggins, Robert A. Ambrose Bierce. Minneapolis: University of Minnesota Press, 1964.

Wilson, Edmund. Patriotic Gore: Studies of the Literature of the Civil War. New York: Oxford University Press, 1962.

THE SINKING OF THE "LUSITANIA" (1915)

The "Lusitania," a British passenger liner owned by the Cunard Line, was torpedoed and sunk by a German submarine on May 7, 1915, ten miles off Kinsdale Head, Ireland. Some 1,152 persons were lost, 124 of them being American citizens.

Casualties included Elbert Hubbard and Justus Miles Forman, authors; Alfred G. Vanderbilt, sportsman; and Charles Frohman, theatrical producer.

While war was raging in Europe, America was then neutral. The "Lusitania" was unarmed, but the Germans declared that she was carrying contraband of war. They declared further that Americans had been warned that they took passage on English vessels at their peril, this warning, signed "Imperial German Embassy," having appeared in American newspapers on the day the "Lusitania" sailed from New York.

The incident of the sinking inflamed public feeling in America and generated a strong sentiment to declare war at once. President Woodrow Wilson, however, requested reparations from Germany. That country, nevertheless, refused to accept responsibility for the deaths of the Americans aboard the liner but did agree not to sink further ships without first giving warning.

Suggested Readings

Baldwin, Hanson W. Sea Fights and Shipwrecks. New York: Museum Press, 1956.

Bemis, Samuel F. A Diplomatic History of the United States. New York: Cooper Square, 1950.

________, ed. American Secretaries of State and Their Diplomacy. New York: Cooper Square, 1927-1929.

Bryan, William Jennings, and Mary B. Bryan. The Memoirs of William Jennings Bryan. Philadelphia: Winston, 1925.

Coletta, Paolo E. William Jennings Bryan. Lincoln, Neb.: University of Nebraska Press, 1969.

Commager, Henry Steele, ed. "The First 'Lusitania' Note," (Doc. No. 405) in his Documents of American History, 8th edition. New York: Appleton, 1968.

Corbett, Edmund V., ed. "Torpedoing of the Lusitania" in his Great True Stories of Tragedy and Disaster. New York: Archer House, 1963.

Grattan, C. Hartley. Why We Fought. Indianapolis: Bobbs-Merrill, 1969.

Hoehling, Adolph A. Great Ship Disasters. New York: Cowles, 1971.

House, Edward M. The Intimate Papers of Colonel House. Edited by Charles Seymour. Saint Clair Shores, Mich.: Scholarly Press, 1971.

Lingley, Charles Ramsdell, and Allen Richard Foley. Since the Civil War. New York: Appleton, 1935.

Mielke, Otto. Disaster at Sea. New York: Fleet, 1958.

Paine, Ralph D. Lost Ships and Lonely Seas. New York: Century, 1921.
Seymour, Charles. American Diplomacy During the World War. Hamden, Conn.: Shoe String Press, 1973.
Simpson, Colin. The Lusitania. Boston: Little, Brown, 1972.
Wallechinsky, David, and Irving Wallace. "The 'Lusitania'," in their The People's Almanac. Garden City, N.Y.: Doubleday, 1975.
Werner, M. R. Bryan. New York: Harcourt, Brace, 1929.

THE DISAPPEARANCE OF JIMMY GLASS (1915)

In May, 1915, Charles L. Glass and his family vacationed on the Frazer farm near the small town of Greeley, Pennsylvania. On the morning of May 11 Jimmy Glass, his four-year-old son, strayed into a nearby field to watch William Losky, a farmhand who was busy plowing. Jimmy was last seen by his father who was sunning himself on the farmhouse porch.

The father went inside for a glass of water. Returning to the porch he met his wife, who had just come back from the post office. She inquired as to Jimmy's whereabouts and the child was then found to be missing. Losky, when questioned, said only that he had observed the boy crawling through the fence and had not seen him since.

A search for the missing child was immediately inaugurated. Abduction was suspected and reports that two men and a woman had been seen in an automobile with a crying child were made but never substantiated. A bloodhound was used in an effort to trail the boy through the field but without success. Neighbors and friends formed a searching party which explored the area thoroughly but found nothing. This was followed by a second search made by state police officers; like the first, this search disclosed no trace of young Jimmy Glass.

Descriptions of the boy were circulated. His picture was published in newspapers and shown on motion picture screens. Later appeals were made by radio, and pictures of him were displayed on post office bulletin boards throughout the country.

Gypsies were suspected of kidnaping the boy. Gypsy camps were investigated and traveling bands of gypsies were traced, in one case as far as Puerto Rico. The missing Jimmy Glass was not found.

In Jersey City an egg bearing the inscription, "Help! James Glass held captive in Richmond, Va." was found in a grocery store. The egg was traced and the appeal for help was found to be the practical joke of a fifteen-year-old boy.

One ransom note demanding that $5,000 be placed in a milk bottle which was to be left at a designated spot was received. The bottle, filled with stage money, was picked up by a shoe-shine stand proprietor who turned out to have nothing to do with the case. The writer or writers of the note were never caught.

Charles and Mrs. Glass traveled to Norman, Oklahoma, having been advised that their missing son was there. The boy they saw turned out to be someone else's child.

In December, 1923, Otto Winckler was hunting rabbits near Greeley. He found a small human skull, a pair of children's shoes, and a few buttons. The boy's father and Captain Rooney of the Jersey City police were summoned. The father identified the shoes as those Jimmy Glass had been wearing on the day of his disappearance and the structure of the skull revealed the cause of the cowlick which had characterized his appearance.

It was surmised that the boy had fallen into a small pond and drowned, "only to have his bones cast up again by the droughty fall eight years later." Captain Rooney also inferred that Jimmy Glass may have been kidnaped, perhaps killed, with the body then being deposited in the pond.

The mystery still remains unsolved.

Suggested Readings

"Bones May Be Clue to 'Jimmy' Glass," New York Times, December 2, 1923, p. 20.

Churchill, Allen. They Never Came Back. Garden City, N.Y.: Doubleday, 1960.

Cohen, Bruce J. Crime in America. Itasca, Ill.: Peacock, 1970.

"Father Identifies Bones of Glass Boy, Missing Nine Years," New York Times, December 3, 1923, p. 1.

"Fear Boy Is Kidnapped," New York Times, May 16, 1915, Part III, p. 4.

Hibbert, Christopher. Roots of Evil: A Social History of Crime and Punishment. New York: Funk & Wagnalls, 1968.

"Kidnapping Victims: Tragic Aftermaths," Saturday Evening Post, April, 1976.

"Lost Boy Turns Up in Far Oklahoma," New York Times, August 9, 1915, p. 4.

Messick, Hank, and Burt Goldblatt. Kidnapping: The Illustrated History. New York: Dial Press, 1974.

"On Trail of Missing Boy," New York Times, July 13, 1915, p. 6.

"Pray For Boy's Return," New York Times, May 17, 1915, p. 5.

Rinehart, J. C. "Tracing People, Lost, Stolen and Strayed Away," American City, October, 1929.

"Seek Fingerprints on Glass Boy's Toys," New York Times, August 13, 1915, p. 18.

Slattery, William, ed. Abduction: Fiction Before Fact. New York: Grove Press, 1974.

Smith, Edward H. "The Return of Jimmy Glass," in his Mysteries of the Missing. New York: Lincoln MacVeagh-The Dial Press, 1927.

Sutherland, Edwin H. On Analyzing Crime. Edited by Karl Schussler. Chicago: University of Chicago Press, 1972.

Winslow, Robert W. Crime in a Free Society. Encino, Calif.: Dickenson, 1968.

THE FORD PEACE MISSION (1915-1916)

In 1915 and for some time before, Henry Ford, the automobile manufacturer, had been extremely outspoken on the subject of the European War. On one occasion he stated, "I hate war, because war is murder, desolation, and destruction." His pronouncements against armed conflict were such that James Couzens, vice president and treasurer of the Ford Motor Company, felt called upon to resign both these positions, though staying on as a director.

Working for world peace were, among others, Rosika Schwimmer, an Hungarian author, lecturer, and woman suffragist, and Louis P. Lochner, secretary of the International Federation of Students. In November, 1915, these two met with Ford and, showing him evidence that apparently both neutrals and belligerents were receptive to mediation, got him to agree to finance a campaign which would work in "continuous mediation" for a peace acceptable to all concerned. Ford's wife Clara and such associates as Dean Marquis and William Livingston opposed the scheme but he resolved to support it. Later, Clara Ford was won over.

On November 21, 1915, a meeting was held in New York City. It was attended by Ford, Jane Addams of Chicago's Hull House, Dean George W. Kirchwey of Columbia University, Paul Kellogg of the Survey, Louis P. Lochner and Rosika Schwimmer. Here it was decided to send an official mediating commission or private group to Europe. Lochner and Ford were to go to Washington to receive President Woodrow Wilson's official sanction, and Ford agreed to charter an ocean-going vessel to transport the delegates. The Scandinavian-American liner "Oscar II" was chartered that evening.

President Wilson refused to appoint an official commission, even though Ford offered to finance it. Ford then stated to reporters that he intended to "assemble a group of the biggest and most influential peace advocates in the country, who can get away, on this ship," and that he would accompany them to Europe. The reporters regarded the idea as impractical and the published news stories concerning it were satirical.

Ford and Rosika Schwimmer sent out telegrams to various pacifists asking them to make the trip. When the "Oscar II" sailed from Hoboken on December 4, 1915, a large crowd, including William Jennings Bryan and Thomas Alva Edison, gathered to see it off. Before it left, Berton Braley the poet and Marian Rubicam were married in the saloon. Someone handed Bryan a cage containing two squirrels.

Over 150 pacifists and students were in the Ford party. On December 9 a dispute broke out between various members of the group concerning the issue of America's preparedness for war. This dispute was reported in the press but before the "Oscar II" had reached Norway the entire delegation pledged itself to "peace by continuing mediation."

Henry Ford contracted a severe cold and took to his bed. On the freezing morning of December 18 the ship docked at Oslo and Ford, having collapsed, went directly to his hotel to rest. He never appeared in public in Norway. Jenkin Lloyd Jones and other members of the peace mission made a poor impression when speaking before an audience at the University of Christiania.

Ford continued ill. On December 23 he left the group and started home. It is assumed that he began to realize privately that the peace mission had been badly managed. According to John Dos Passos, "the reporters had kidded him so that he got cold feet."

The mission carried on in Ford's absence. It went to Stockholm but, without Ford, financing became precarious and "disagreements began to divide those in charge. Uncertainty developed as to Ford's own wishes."

From Stockholm the group moved on to Denmark and from there to Holland. Some members, disagreeing with others, withdrew and returned home. In January, 1916, having completed its task, the peace party returned to the United States, the students on the "Noordam" and the delegates on the "Rotterdam." T. N. Pockman, writing in the New York Tribune, said "the comedy of errors is over. During its two months' run the show has aroused more lively interest, cynical amusement, and sheer pity than possibly any other in history."

Suggested Readings

Bainbridge, John, and Russell Maloney. "The Innocent Voyage" (in series: "Where Are They Now?"), New Yorker, March 9, 1940.

Beales, Arthur C. The History of Peace. New York: Garland, 1931.

Bolton, Sarah K. Lives of Poor Boys Who Became Famous. New York: Crowell, 1939.

Burlingame, Roger. Henry Ford. New York: Knopf, 1954.

Caldwell, Cyril Cassidy. Henry Ford. New York: Messner, 1947.

Clancy, Louise B. The Believer: The Life of Mrs. Henry Ford. New York: Coward-McCann, 1960.

Curti, Merle E. Bryan and World Peace. New York: Octagon Books, 1969.

_______. Peace or War: The American Struggle, 1635-1936. Boston: J. S. Canner, 1936.

Dos Passos, John. "Tin Lizzie," in his The Big Money. New York: Random House, 1936.

Flynn, John T. Men of Wealth. New York: Simon & Schuster, 1941.

Ford, Henry, and Samuel Crowther. My Life and Work. Garden City, N.Y.: Doubleday, 1922.

Gilbert, Miriam. Henry Ford, Maker of the Model T. Boston: Houghton Mifflin, 1962.

Hagedorn, Hermann. Americans: A Book of Lives. New York: John Day, 1946.

Hershey, Burnet. The Odyssey of Henry Ford and the Great Peace Ship. New York: Taplinger Publishing Co., 1967.

Marquis, Samuel S. Henry Ford: An Interpretation. Boston:

Little, Brown, 1923.
Millis, Walter. The Road to War. New York: Howard Fertig, 1970.
Nevins, Allan, with the collaboration of Frank Ernest Hill. Ford: Expansion and Challenge. New York: Scribner's, 1957.
Neyhart, Louise Albright. Henry Ford, Engineer. Boston: Houghton Mifflin, 1950.
Simonds, William A. Henry Ford: His Life, His Work, His Genius. Indianapolis: Bobbs-Merrill, 1943.
Sullivan, Mark. Our Times, The United States, 1900-1925. New York: Scribner's, 1928-1935.
Sward, Keith. The Legend of Henry Ford. New York: Holt, Rinehart, 1948.
Werner, M. R. Bryan. New York: Harcourt, Brace, 1929.

THE PERSHING PUNITIVE EXPEDITION (1916)

In the year 1916, diplomatic relations between Mexico and the United States were extremely strained. Mexico had had a series of presidents, including Benito Juarez, Sebastián Lerdo de Tejada, Porfirio Diaz, Francisco Indalecio Madero, Diaz (who was reelected), Victoriano Huerta, and Venustiano Carranza. Carranza had taken power in 1914 and Pancho Villa, cattle thief, bandit, revolutionary leader and one-time associate of Carranza in the revolution against Huerta, immediately instigated a revolt against his former associate, as did the revolutionist Emiliano Zapata.

The situation was further complicated by the interventions of foreign governments seeking to protect the interests of their nationalists in Mexico. As early as 1913 President Woodrow Wilson, in a message to Congress, had reviewed the Mexican situation. In 1915 six Latin American powers met in Washington at Wilson's suggestion, to confer. They decided to recognize Carranza as the leader of a de facto government. The rebel leaders, with the exception of Pancho Villa, laid down their arms.

Villa, on March 9, 1916, crossed the American border into New Mexico. With a contingent of followers he raided the town of Columbus, partly burning it and killing sixteen persons.

The situation was such that the United States did not wish war with Mexico, yet such an invasion as Villa's could not be permitted to go unheeded. It was also thought that a repetition of this raid could arouse American feeling to the point where war would be inevitable. Consequently, the American authorities sought some forceful move short of actual warfare.

In this situation General John J. Pershing was sent in command of a punitive expedition to Mexico. The purpose of the expedition was to prevent further attacks from Mexico on the southern border of the United States. President Carranza, while not actually consenting to the expedition, tolerated it for a time but eventually protested the presence of American troops on Mexican soil, and they were withdrawn. The expedition was partially successful. It did prevent further raids but did not result in Villa's capture.

Historians are generally agreed that Pershing handled an extremely delicate situation with diplomacy and tact. He and his men were sufficiently equipped to have penetrated as far south as Mexico City. Further, there was the possibility that his activities might have irritated the Mexicans to the point of their declaring war, something to be avoided if at all possible. Such a war would have nullified the whole purpose of the expedition, would have instigated new raids, and would have compelled American intervention. The skill and tact displayed by Pershing in the situation led eventually to his appointment as commander-in-chief of the American Expeditionary Forces in World War I and as Chief of Staff of the United States army.

Suggested Readings

Baker, Ray Stannard. Woodrow Wilson, Life and Letters. Westport, Conn.: Greenwood Press, 1939.

________, and W. E. Dodd, eds. The Public Papers of Woodrow Wilson. New York: Kraus Reprint, 1927.

Braddy, Haldeen. Pershing's Mission in Mexico. El Paso: Texas Western University Press, 1966.

Callahan, James M. American Foreign Policy in Mexican Relations. New York: Cooper Square, 1932.

Clendennen, Clarence C. The United States and Pancho Villa. Ithaca, N.Y.: Cornell University Press, 1961.

Commager, Henry Steele, ed. "Wilson's Mobile Address," (Doc. No. 394) in his Documents of American History, 8th edition. New York: Appleton, 1968.

________. "Wilson's Special Message on Mexican Relations," (Doc. No. 393) in his Documents of American History, 8th edition. New York: Appleton, 1968.

Gruening, Ernest H. Mexico and Its Heritage. Westport, Conn.: Greenwood Press, 1968.

Link, Arthur S. Wilson: Confusion and Crisis. Princeton, N.J.: Princeton University Press, 1964.

________. Wilson the Diplomatist: A Look at His Major Foreign Policies. Baltimore: Johns Hopkins University Press, 1957.

Pinchon, Edgcum. Viva Villa. New York: Arno Press, 1970.

Rippy, James Fred. The United States and Mexico. New York: Crofts, 1931.

________, José Vasconcelos and Guy Stevens. Mexico: American Foreign Policies Abroad. New York: Crofts, 1928.

Robertson, William S. Hispanic-American Relations With the United States. Edited by David Kinley. Millwood, N.Y.: Kraus Reprint, 1923.

Tannenbaum, Frank. Mexico: The Struggle for Peace and Bread. New York: Knopf, 1950.

Tompkins, Colonel Frank. Chasing Villa: The Story Behind the Story of Pershing's Expedition Into Mexico. Harrisburg, Pa.: Stackpole Co., 1935.

THE MOONEY-BILLINGS CASE (1916-1939)

Nine people were killed and forty wounded when a bomb was thrown into a Preparedness Day parade in San Francisco on July 22, 1916. A bomb squad, organized immediately afterwards, arrested labor agitator Thomas J. Mooney, Mooney's wife Rena, machinist Warren K. Billings and jitney driver Israel Weinberg. These four were charged with the crime.

Indictments for murder were returned by the grand jury, but later the charges against Weinberg and Rena Mooney were dropped. Thomas Mooney and Warren K. Billings were tried and convicted of first degree murder. Mooney was sentenced to death and Billings to life imprisonment. The case attracted worldwide attention and doubts arose as to the justice of the proceedings and the convictions.

In September, 1917, President Woodrow Wilson made a formal request for an investigation by the Mediation Commission. The following January the Commission recommended that Wilson intercede with the Governor of California on behalf of the convicted prisoners, as considerable doubt as to their guilt existed. Wilson wrote to the Governor on three occasions, and Mooney's sentence was commuted to life imprisonment. Efforts by various interested groups to reopen the case or to secure pardons for Mooney and Billings were unsuccessful.

A sub-committee of the Wickersham Committee on Lawless Enforcement of Law investigated the matter in 1931 and reported its findings. Professor Z. Chaffee, W. H. Pollack and C. S. Stern, members of the sub-committee, did not publish their report along with the others published by the Wickersham Committee, feeling that their function was to inquire into general principles rather than specific cases. However, because of the great demand for publication, the findings of these three men were privately printed when Congress failed to provide for publication. This privately printed report indicated that the sub-committee found that the investigation and trial of the two accused bombers had been characterized by "flagrant violations of the statutory laws of California by both the police and the prosecution, a deliberate attempt to arouse public prejudice [against Mooney and Billings]," and coaching of witnesses by the prosecution in a manner which made their testimony "a vouching for perjured testimony."

After 23 years of demonstrations and repeated court appeals, Mooney and Billings were pardoned and released by Governor C. L. Olson of California in 1939.

Suggested Readings

Adamic, Louis. Dynamite: The Story of Class Violence in America. Gloucester, Mass.: Peter Smith, 1959.

Borchard, E. M. Convicting the Innocent. New York: Da Dapo Press, 1974.

Boyer, Richard D., and Herbert M. Morais. Labor's Untold

Story. New York: Cameron Associates, 1955.
Chaplin, Ralph. Wobbly: The Rough and Tumble Story of an American Radical. Chicago: University of Chicago Press, 1948.
Commager, Henry Steele, ed. "The Mooney-Billings Case," (Doc. No. 471) in his Documents of American History, 8th edition. New York: Appleton, 1968.
Cross, Ira B. A History of the Labor Movement in California. Berkeley, Calif.: University of California Press, 1935.
Frost, Richard H. The Mooney Case. Stanford, Calif.: Stanford University Press, 1968.
Gentry, Curt. Frame-up: The Incredible Case of Tom Mooney and Warren Billings. New York: Norton, 1967.
Hays, Arthur Garfield. Trial by Prejudice. New York: Covici, Friede, 1933.
Hopkins, Edward J. What Happened in the Mooney Case. New York: Brewer, Warren, 1932.
Hunt, Henry T. The Case of Thomas J. Mooney and Warren K. Billings: Abstract and Analysis of Record Before Governor Young of California. New York: National Mooney-Billings Committee, 1929.
Karson, Marc. American Labor Unions and Politics, 1900-1918. Boston: Beacon Press, 1965.
Knight, Robert Edward Lee. Industrial Relations in the San Francisco Bay Area, 1900-1918. Berkeley, Calif.: University of California Press, 1960.
Lamparski, Richard. "Warren K. Billings," in his Whatever Became Of...? (Second Series). New York: Crown Publishers, 1968.
Older, Fremont. My Own Story. New York: Macmillan, 1926.
Slobodek, Mitchell. A Selective Bibliography of California Labor History. Los Angeles: Institute of Industrial Relations, University of California Press, 1964.
Wells, Evelyn. Fremont Older. New York: Appleton, 1936.

THE CREATION OF BOYS TOWN (1917)

Boys Town, a home for indigent boys, is the brainchild of Father Edward Joseph Flanagan, an Irish-American Catholic priest. It is the outgrowth of the Workingman's Hotel, founded by him in 1914.

Father Flanagan, having been ordained, served as assistant pastor at St. Peter's Church, Omaha, Nebraska. While there he established the Workingman's Hotel as a temporary home for destitute men. Becoming concerned with the problem of rehabilitating underprivileged youths, he established the Home for Homeless Boys at Omaha. When the Home opened in 1917 five boys were admitted. Father Flanagan's belief that "there is no such thing as a bad boy" was widely quoted and his Home grew rapidly.

In due course the philanthropic priest acquired a parcel of land in Douglas County, ten miles west of Omaha. Here he established Boys Town with an initial investment of ninety dollars borrowed from well-wishers. Boys Town is open to boys of all religions and races. It is governed by a mayor and six commissioners elected by the boys from among themselves. Most of the inmates are from eleven to seventeen years of age. The institution, which is supported entirely by voluntary contributions, today accommodates about 900 youths.

Boys Town occupies about 1,000 acres. Its facilities include a gymnasium, an assembly hall, a library, dormitories and school buildings. In 1936 it was incorporated as a village.

From its inception Boys Town has had as its aim "the sympathetic guidance and re-education of underprivileged youths representing all races, creeds, and colors." In 1938 it was the subject of a popular motion picture called "Boys Town."

Suggested Readings

"Boys Town Buries Father Flanagan," Life, May 31, 1948.
"Father Flanagan," Newsweek, May 24, 1948.
Flanagan, Edward Joseph, as told to Ford McCoy. Understanding Your Boy. New York: Rinehart, 1950.
Johnsen, John Reuben. Representative Nebraskans. Lincoln, Neb.: Johnsen Publishing Co., 1954.
"No Bad Boys," America, May 29, 1948.
Oursler, Fulton, and Will Oursler. Father Flanagan of Boys Town. Garden City, N.Y.: Doubleday, 1949.
Rotten, Elisabeth. Children's Communities. Paris: 1949.
Schultz, Gladys Denny. "Boy-Handling Tips from Boys Town," Better Homes and Gardens, March, 1940.
Staudacher, Rosemarian V. Children Welcome. New York: Farrar, Strauss, 1963.
Wegner, George. Father Flanagan und seine Jungenstadt. Vienna: 1957.

THE ZIMMERMANN TELEGRAM (1917)

In early 1917 diplomatic relations between the United States and Germany were extremely strained. Germany was engaged in intensive submarine warfare against England and other members of the Allies, not including the United States, which did not declare war on Germany until April 6. By March, 1917, every important event pointed toward war with Germany. On March 1 the American administration published a remarkable message, known as the "Zimmermann Telegram." This had been sent to the German minister in Mexico by Arthur Zimmermann, the German Secretary of State for Foreign Affairs.

The Zimmermann Telegram instructed the German minister to propose an alliance between Germany and Mexico in the event that the United States entered the European War. Should the German forces emerge victorious from the War, Mexico would receive Arizona, New Mexico, and Texas. The Telegram provided further that General Venustiano Carranza, the Mexican president, was to be requested to ask Japan to leave the Allies and join the Mexico-German combination. Last of all, the German minister was to advise Carranza of Germany's belief that intensive and ruthless submarine warfare would shortly mean the defeat of England.

The Zimmermann Telegram and the plot it outlined were disclosed by the British Intelligence Service. The message had been sent by three different routes, all of which were intercepted by the British. British wireless operators caught one transmission sent from Nauen, Germany to Sayville, Long Island, for forwarding to Mexico. A second transmission was made by way of the pro-German Swedish foreign office, and a third was sent to Count Johann-Heinrich von Bernstorff, German ambassador at Washington, by way of the American State Department. Besides catching these three separate sendings, the British purchased a copy from some "approachable" person in Mexico.

These messages were in code but, unknown to the Germans, the British Intelligence Service had succeeded in breaking the code some time before.

The Zimmermann Telegram caused great anti-German feeling, particularly in the Southwest United States, which had been promised to Mexico provided that country accepted Germany's proposition and the German powers won the war.

Suggested Readings

Commager, Henry Steele, ed. "The Zimmermann Note," (Doc. No. 417) in his Documents of American History, 8th edition. New York: Appleton, 1968.

Dos Passos, John. Mr. Wilson's War. Garden City, N.Y.: Doubleday, 1962.

Dulles, Allen W. The Craft of Intelligence. New York: Harper, 1963.

Garraty, John A. Woodrow Wilson: A Great Life in Brief. New York: Knopf, 1956.

Grattan, C. Hartley. Why We Fought. Indianapolis: Bobbs-Merrill, 1969.

Hendrick, Burton J. "The Zimmermann Telegram to Mexico," World's Work, November, 1925.

________, ed. The Life and Letters of Walter Hines Page. Saint Clair Shores, Mich.: Scholarly Press, 1971.

House, Edward M. The Intimate Papers of Colonel House. Edited by Charles Seymour. Saint Clair Shores, Mich.: Scholarly Press, 1971.

Lingley, Charles Ramsdell, and Allen Richard Foley. Since the Civil War. New York: Appleton, 1935.

Tuchman, Barbara W. The Zimmermann Telegram. New York: Macmillan, 1966.

THE FOURTEEN POINTS (1918)

President Woodrow Wilson, in an address to Congress on January 8, 1918, set forth the Fourteen Points that were his framework for a peace settlement of World War I. He covered war aims and peace terms in what came to be known as the "Fourteen Points," which were later taken as the basis for peace negotiations after the war ended on November 11, 1918. The Points were based on a report prepared for the President by the Inquiry, a commission organized by Colonel Edward M. House for the purpose of studying Allied and American policy.

Wilson, in explaining the attitude of the United States, said that "the processes of peace, when they are begun, shall involve ... no secret understandings of any kind." He advocated the freedom of the seas in both peace and war and the removal of economic barriers between nations. He also advocated the reduction of armaments, the impartial adjustment of colonial claims, and the evacuation of territories occupied by Germany, such as Belgium, France, Russia, and the Balkan States. Other points covered were the freeing of all French territory and the righting of the "wrong done to France by Prussia in 1871 in the matter of Alsace-Lorrain [wresting provinces from France by Germany] which has unsettled the peace of the world for nearly fifty years."

He proposed, as one of his Fourteen Points, "a readjustment of the frontiers of Italy ... along clearly recognizable lines of nationality," and an opportunity for "the people of Austria-Hungary ... to be accorded the freest opportunity of autonomous development."

"Rumania, Serbia, and Montenegro," Wilson stated, "should be evacuated, occupied territories restored, Serbia accorded free and secure access to the sea ... and the relations of the several Balkan States to one another be determined by friendly counsel...." plus certain other provisions. The peoples subject to Turkish rule, he indicated, should be given the opportunity to develop along lines chosen by themselves. He also advocated the establishment of a Polish state which "should include the territories inhabited by indisputably Polish populations," and the establishment of a general association of nations to guarantee the safety of large and small states alike.

Both Germany and Austria replied to Wilson's address, but not in a manner to make possible a cessation of hostilities. Colonel House, Frank Cobb, and Walter Lippmann prepared a semi-official interpretation of the Fourteen Points which was designed to clarify various aspects which could be interpreted in more than one way. In October, 1918, the German government entered into negotiations with President Wilson. Germany sought an armistice on the basis of the Fourteen Points. Wilson, though willing to conclude an armistice on this basis, was not able to bind the Allies to any such agreement. Eventually the Allies accepted the Fourteen Points as a basis for an armistice, but with a reservation on the second Point, which concerned the freedom of the seas.

Suggested Readings

Andrist, Ralph K., ed. in charge. The American Heritage History of the 20's and 30's. New York: American Heritage, 1970.
Bailey, Thomas Andrew. Woodrow Wilson and the Lost Peace. New York: Quadrangle/The New York Times Co., 1963.
Baker, Ray Stannard. Woodrow Wilson and the World Settlement. Gloucester, Mass.: Peter Smith, 1958.
_______. Woodrow Wilson, Life and Letters. Westport, Conn.: Greenwood Press, 1939.
_______, and W. E. Dodd, eds. The Public Papers of Woodrow Wilson. New York: Kraus Reprint, 1927.
Braeman, John, ed. Wilson. Englewood Cliffs, N.J.: Prentice-Hall, 1972.
Commager, Henry Steele, ed. "The Fourteen Points," (Doc. No. 423) in his Documents of American History, 8th edition. New York: Appleton, 1968.
Dos Passos, John. "Meester Veelson," in his Nineteen-Nineteen. New York: Random House, 1931.
_______. Mr. Wilson's War. Garden City, N.Y.: Doubleday, 1962.
Garraty, John A. Henry Cabot Lodge: A Biography. New York: Knopf, 1953.
_______. Woodrow Wilson: A Great Life in Brief. New York: Knopf, 1956.
Hansen, Harry. "The Forgotten Men of Versailles," in Leighton, Isabel, ed. The Aspirin Age, 1919-1941. New York: Simon & Schuster, 1949.
House, Edward M. The Intimate Papers of Colonel House. Edited by Charles Seymour. Saint Clair Shores, Mich.: Scholarly Press, 1971.
Link, Arthur S. Wilson: Confusion and Crisis. Princeton, N.J.: Princeton University Press, 1964.
_______. Wilson the Diplomatist: A Look at His Major Foreign Policies. Baltimore: Johns Hopkins University Press, 1957.
Sann, Paul. The Lawless Decade. New York: Crown Publishers, 1957.
Seymour, Charles. American Diplomacy During the World War. Hamden, Conn.: Shoe String Press, 1973.
Smith, Gene. When the Cheering Stopped: The Last Years of Woodrow Wilson. New York: Morrow, 1964.
Wilson, Woodrow. The Case for the League of Nations. Edited by H. Foley. Port Washington, N.Y.: Kennikat Press, 1967.

THE FORD-TRIBUNE LIBEL SUIT (1919)

On June 23, 1916, the Chicago Tribune published an editorial titled "Ford Is an Anarchist." The Ford mentioned was Henry Ford, the American automobile manufacturer. The occasion for the diatribe was an erroneous report that employees of the Ford

Motor Company who, as members of the National Guard, were mustered out to serve with General John J. Pershing in his 1916 Punitive Expedition, forfeited their positions with the Company. The editorial declared in part that "if Ford allows this rule of his shop to stand, he will reveal himself not merely as an ignorant idealist, but as an anarchist enemy of the nation which protects him and his wealth."

At the urging of his attorney Alfred Lucking, Ford decided to sue the Chicago Tribune for libel. On September 7, 1916, he filed suit for $1,000,000 in the Federal District Court of Northern Illinois. His attorneys, feeling that an impartial trial was not to be had in a geographical area served by the Tribune, requested a change of venue. This was granted and the case was eventually heard in Mt. Clemens, Michigan, a small town previously known only for its mineral springs beneficial to rheumatic patients. This was in early May, 1919.

Extensive preparations were made for the trial. Carpenters remodeled the small courtroom to accommodate batteries of lawyers, stenographers, and reporters. Telegraph instruments were installed to permit direct wire communication with leading cities. The hotels were filled to capacity. Both the Ford and the Tribune forces set up headquarters, the former in the Colonial and Medea Hotels, and the latter, led by the newspaper's editor and publisher Colonel Robert R. McCormic, in the Park Hotel.

The six attorneys representing Ford were headed by Alfred Lucking and Alfred Murphy. The Tribune's legal staff contained no fewer than nine members, and included Weymouth Kirtland and Elliott G. Stevenson. Judge James G. Tucker presided on the bench.

The trial began in the crowded courtroom on the afternoon of Monday, May 12, 1919. Both sides agreed to a struck jury, which was completed on the third day. Its membership, all males, consisted of ten farmers, one retired farmer, and one road-builder. One juror, the owner of a Ford automobile, remarked, "That would not prejudice me against Mr. Ford."

Judge Tucker ruled that "all issues appurtenant to the case, rather than just those pertinent to it, ought to be threshed out." This permitted counsel for the Tribune to exercise great latitude in cross-examination. Testimony bearing upon personal views and character was to be admitted as evidence, and this, as it turned out, had the effect of putting the plaintiff--in this case Henry Ford --on trial. This became very clear when Kirtland, in his opening statement, "dragged in appeals to Americanism, patriotism, and national vigilance."

A total of 120 witnesses testified in the case, the star witness being Ford himself. He was called to the stand on July 14 and, for eight days, was subjected to a merciless inquisition by Stevenson. Ford's ignorance in many fields of knowledge was shown, the Tribune attorney contending that a man as ignorant as Ford was shown to be in fields other than engineering and manufacturing was completely unqualified to guide or even speak intelligently in politico-social fields. It was in this cross-examination that the witness said that he believed Benedict Arnold to be a writer, that the War of 1812 was "about aggression," and that history was bunk.

The trial lasted twelve weeks. On August 14 it went to the jury and, after ten hours of deliberation, the jury found the Chicago Tribune guilty of libel. The Tribune was fined six cents.

On July 30, 1941, Colonel McCormick wrote Ford, saying, in part, "It occurs to me on this, our birthday" (both men were born on July 30), "to write you and say I regret the editorial we published about you so many years ago. I only wonder why the idea never occurred to me before. It was the product of the war psychology which is bringing us so many similar expressions today."

Suggested Readings

Bolton, Sarah K. Lives of Poor Boys Who Became Famous. New York: Crowell, 1939.

Burlingame, Roger. Henry Ford. New York: Knopf, 1954.

Caldwell, Cyril Cassidy. Henry Ford. New York: Messner, 1947.

Clancy, Louise B. The Believer: The Life of Mrs. Henry Ford. New York: Coward-McCann, 1960.

Dos Passos, John. "Tin Lizzie," in his The Big Money. New York: Random House, 1936.

Flynn, John T. Men of Wealth. New York: Simon & Schuster, 1941.

Ford, Henry, and Samuel Crowther. My Life and Work. Garden City, N.Y.: Doubleday, 1922.

Gilbert, Miriam. Henry Ford, Maker of the Model T. Boston: Houghton Mifflin, 1962.

Glasscock, C. B. The Gasoline Age. Indianapolis: Bobbs-Merrill, 1937.

Hagedorn, Hermann. Americans: A Book of Lives. New York: John Day, 1946.

"Henry Ford at Bay," Forum, LXII, 1919.

Marquis, Samuel S. Henry Ford: An Interpretation. Boston: Little, Brown, 1923.

Nevins, Allan, with the collaboration of Frank Ernest Hill. Ford: Expansion and Challenge. New York: Scribner's, 1957.

Neyhart, Louise Albright. Henry Ford, Engineer. Boston: Houghton Mifflin, 1950.

Simonds, William A. Henry Ford: His Life, His Work, His Genius. Indianapolis: Bobbs-Merrill, 1943.

Sullivan, Mark. Our Times, The United States, 1900-1925. New York: Scribner's, 1928-1935.

Sward, Keith. The Legend of Henry Ford. New York: Holt, Rinehart, 1948.

THE BOSTON POLICE STRIKE (1919)

In 1919 the Boston police were badly underpaid. Salaries were based on an annual minimum of $1,100 and members of the force were required to provide their own uniforms. Their efforts to secure salary increases were unsuccessful.

As a result of this situation the Boston police formed a union and affiliated with the American Federation of Labor, although this had been forbidden by Police Commissioner Curtis. Curtis immediately brought charges against nineteen members of the police union, found them guilty of disobedience, and summarily suspended them.

The police threatened to strike. A committee appointed by Mayor Peters to arbitrate the dispute was unable to resolve it, and on September 9, 1919, 1,117 members of the 1,544-man Boston police force went out on strike.

Without adequate police restraint looters, rioters and other lawless persons ran riot. The Mayor requested state troops. A volunteer police force attempted to maintain order. On the following day Governor Calvin Coolidge of Massachusetts called out the State Guard.

The inexperienced guardsmen were goaded into firing into a South Boston mob and two people were killed. The rioting continued and other persons were killed and wounded. The Central Labor Union considered calling a general strike in favor of the police officers but, because of public opinion, which was against the police, decided against such a move. Police Commissioner Curtis, with the concurrence of Governor Coolidge, discharged the nineteen men whom he had previously suspended. He then recruited a new police force.

Samuel Gompers, president of the American Federation of Labor, wired Coolidge that Curtis' action was both autocratic and unwarranted. Coolidge's reply has gone down in history: "There is no right to strike against the public safety by anyone, anywhere, any time."

By his firm action in calling in the militia Coolidge successfully suppressed the strike, became a national hero, and won reelection overwhelmingly. His action made him a national political figure and helped boost him into the 1920 vice-presidential nomination. The Harding-Coolidge ticket won easily and when President Warren G. Harding died at the Palace Hotel, San Francisco, in 1923, Coolidge became the 30th President. He was elected to a second term in 1924.

Suggested Readings

Abels, Jules. _In the Time of Silent Cal_. New York: Putnam, 1969.

Allen, Frederick Lewis. _Big Change: America Transforms Itself, 1900-1950_. New York: Harper, 1952.

_______. _Only Yesterday_. New York: Harper, 1931.

Andrist, Ralph K., ed. in charge. _The American Heritage History of the 20's and 30's_. New York: American Heritage, 1970.

Coolidge, Calvin. _The Autobiography of Calvin Coolidge_. New York: Cosmopolitan Book Corp., 1929.

Fuess, Claude M. _Calvin Coolidge, Man from Vermont_. Boston:

Little, Brown, 1940.
Gompers, Samuel. Seventy Years of Life and Labour. Clifton, N.J.: Augustus M. Kelley, 1966.
Green, Horace. The Life of Calvin Coolidge. New York: Duffield, 1924.
McCoy, Donald R. Calvin Coolidge: The Quiet President. New York: Macmillan, 1967.
Moran, Philip R., ed. Calvin Coolidge, 1872-1933. Chronology--Documents--Bibliographical Aids. Dobbs Ferry, N.Y.: Oceana Publications, 1970.
Quint, Howard F., and R. H. Ferrell, eds. Calvin Coolidge: The Talkative President. Amherst, Mass.: Amherst, 1964.
Russell, Francis. "The Strike That Made a President," American Heritage Magazine, October, 1963.
Sann, Paul. The Lawless Decade. New York: Crown Publishers, 1957.
Stern, Gerald E., ed. Gompers. Englewood Cliffs, N.J.: Prentice-Hall, 1971.
Stone, Irving. "Calvin Coolidge: A Study in Inertia," in Leighton, Isabel, ed. The Aspirin Age, 1919-1941. New York: Simon & Schuster, 1949.
Washburn, R. M. Calvin Coolidge: His First Biography. Boston: Small, Maynard, 1923.
White, William Allen. A Puritan in Babylon: The Story of Calvin Coolidge. New York: Macmillan, 1938.

THE WESLEY EVEREST LYNCHING (1919)

Following World War I, soldiers returning from overseas to their homes on the Pacific Coast found that the lumber industry was dominated by ten monopoly groups. These groups, with a mere 1,802 holders, controlled 1,209,800,000,000 square feet of standing timber. During the war these lumber tycoons had realized enormous profits from the government contracts they had negotiated and from the high prices for which wood products were selling. Finding that the Industrial Workers of the World had infiltrated their logging camps, the monopolists organized the Employers Association and the Legion of Loyal Loggers with the object of overcoming the I.W.W., whom they characterized as "Reds" and radicals.

Wesley Everest, a logger and ex-soldier, became an active member of the lumberjacks' local. On Armistice Day, 1919, in Centralia, Washington, rumors circulated that the local I.W.W. hall was to be raided. Warren O. Grimm, a former army officer, was chosen by the local businessmen to lead the forces of the Citizen's Protective League against the I.W.W.

The Armistice Day parade passed the I.W.W. hall without incident. Then, on the way back, the parade halted outside the hall and paraders broke the door down. Shots were fired and Grimm and another ex-soldier were hit. Everest, in the hall, shot out the cartridges in his rifle and fled. The paraders, who had become a mob, and had brought along a rope with which they had intended to hang Britt Smith, the I. W. W. secretary, pursued Everest. He ran for the river and started to wade across. Then, trapped, he offered to submit to arrest by proper authority. The mob rushed him and, with his revolver, he shot and killed Dale Hubbard, an ex-soldier and nephew of a prominent Centralia lumberman.

Everest was captured, beaten, taken to jail, and confined in a cell. That night the city lights were turned off and a mob broke into the jail, seized the prisoner, and took him out to the Chehalis River bridge. There he was mutilated, hanged, and shot.

At the coroner's inquest it was determined that Everest had broken out of jail and committed suicide by hanging and self-inflicted gun shots. Sixteen loggers, members of the I. W. W., were tried, sentenced to jail and died in Walla Walla penitentiary.

Suggested Readings

Ames, Jessie (Daniel). The Changing Character of Lynching, 1931-1941, With a Discussion of Recent Developments in This Field. Atlanta: Commission on Interracial Cooperation, 1942.

Archer, Jules. Riot: A History of Mob Action in the United States. New York: Hawthorn Books, 1974.

Caughey, John Walton. Their Majesties, The Mob. Chicago: University of Chicago Press, 1960.

Cook, Roy. Leaders of Labor. Philadelphia: Lippincott, 1966.

Cutler, James Elbert. Lynch-Law: An Investigation into the History of Lynching in the United States. New York: Negro University Press, 1969.

Dos Passos, John. "Paul Bunyan," in his Nineteen-Nineteen. New York: Random House, 1931.

Heaps, Willard Allison. Riots, U. S. A., 1765-1970. New York: Seabury Press, 1970.

Iman, Raymond S., and Thomas W. Koch. Labor in American Society. Chicago: Scott, Foresman, 1965.

Johnsen, Julia E., comp. The Closed Shop. New York: Wilson, 1942.

Johnson, Donald. The Challenge to American Freedoms. Lexington, Ky.: University of Kentucky Press, 1963.

Lens, Sidney. Unions and What They Do. New York: Putnam, 1968.

Litwack, Leon F., ed. The American Labor Movement. Englewood Cliffs, N. J.: Prentice-Hall, 1962.

Main, Quincy. History of the American Legion. Houston: Abbotsford Publishing Co., 1975.

Mandelbaum, Seymour J. The Social Setting of Intolerance. Chicago: Scott, Foresman, 1964.

Meltzer, Milton. Bread--and Roses: The Struggle of American Labor. New York: Knopf, 1967.

Murray, Robert K. The Red Scare: A Study in National Hysteria, 1919-1920. Minneapolis: University of Minnesota Press, 1955.

National Association for the Advancement of Colored People. Thirty Years of Lynching in the United States, 1889-1918. New York: Arno Press and the New York Times, 1969.
Preston, William, Jr. Aliens and Dissenters: Federal Suppression of Radicals, 1903-1933. Cambridge, Mass.: Harvard University Press, 1963.
Raper, Arthur F. The Tragedy of Lynching. Chapel Hill, N.C.: University of North Carolina Press, 1933.
Shay, Frank. Judge Lynch: His First Hundred Years. New York: I. Washburn, 1938.
Werstein, Irving. Pie in the Sky: An American Struggle, the Wobblies and Their Times. New York: Delacorte, 1969.
White, Walter Francis. Rope and Faggot. New York: Arno Press, 1969.

THE RED SCARE (1919-1920)

The "Red Scare" came to a head on the afternoon of Friday, January 2, 1920, when agents of the Department of Justice conducted raids on the headquarters of various Communist and allegedly Communist organizations in major cities throughout the country. Homes, clubs, pool halls and meeting halls were broken into. Property was destroyed and those arrested were jailed and held incommunicado, denied legal counsel and interrogated at length. Those who could prove American citizenship were released; aliens were held for longer periods before being given their liberty. Members of the Communist Party and Communist Labor Party were detained in jail, from where the Department of Justice hoped to deport them.

Nearly 5,000 persons were arrested in the next two days, with another 1,000 being picked up in the following two weeks. These arrests were made with complete disregard for due process of law. They were conducted with the complete approval of President Woodrow Wilson's attorney general, Alexander Mitchell Palmer, and were known as the "Palmer Raids."

Since before the turn of the century capital and labor had been working at cross purposes. Capital's exploitation of the working classes had resulted in the formation of trade unions. These, in order to enforce their demands for better working conditions for their members, had resorted to strikes. During World War I public feeling, once the United States was an active participant in the hostilities, was that any effort by the workers to hamper or interfere with the war effort was unpatriotic, unAmerican and Communistic.

American radicals, banding together in such organizations as the Industrial Workers of the World, founded in Chicago in 1905, became prime targets for persecution on the grounds that they were in sympathy with Germany and other enemies of America. This feeling, prevalent during the war years, did not cease with the Armistice of November, 1918. Conscientious objectors, sentenced to prison "while our boys were over there making the world safe for democracy," remained in jail, the Wilson administration refusing them amnesty.

Patriotic societies, such as the American Civil Liberties Union, abounded. These advocated "one hundred percent Americanism" and propagandized through fraternal orders and schools.

Wartime hysteria had not died down. The Boston police strike of 1919 was condemned by the newspapers as "Bolshevick." A widespread strike in the steel industry was similary categorized. Several public officials had had crude homemade bombs thrown at them or sent to them by mail. Race riots between blacks and whites and between war veterans and non-veterans broke out.

Public clamor for governmental action to arrest and deport the radicals became intense. In August, 1919, Attorney General Palmer responded by establishing the General Intelligence Division of his department. This was headed by G. Edgar Hoover. An anti-radical agency, it collected information on so-called un-American organizations and leaders of such organizations. On November 7, 1919, Hoover's agents conducted a raid on the headquarters and branches of the Union of Russian Workers, a labor society. Other similar raids were made on suspected radicals elsewhere. On December 21, 249 men were deported on the army transport "Buford." Then, on January 2, the greatest of these raids, as described above, was made.

Following this the "Red Scare" experienced a decline. Sociologists, however, have stated that "perhaps it never really ended." Anti-radical training and seminars characterized U.S. Army instruction in the 1920's, and again in World War II. The restrictions on immigration grew tighter. Loyalty oaths, an "American plan" for labor unions and textbook censorship became characteristics of American society. In June, 1920, federal judge George W. Anderson, in the case of Colyer v. Skeffington, "found justice department methods to have been brutal and unjust, and its raids sordid and disgraceful." Attorney General Palmer, who is thought to have condoned the raids because of presidential ambitions, failed to receive the Democratic nomination in 1920.

Suggested Readings

Allen, Frederick Lewis. Only Yesterday. New York: Harper, 1931.

Andrist, Ralph K., ed. in charge. The American Heritage History of the 20's and 30's. New York: American Heritage, 1970.

Blum, John M. Woodrow Wilson and the Politics of Morality. Translated by Patsy Southgate. Boston: Little, Brown, 1956.

Chafee, Zechariah, Jr. Free Speech in the United States. Cambridge, Mass.: Harvard University Press, 1941.

Coben, Stanley. A. Mitchell Palmer: Politician. New York: Columbia University Press, 1963.

Demaris, Ovid. The Director. New York: Harper's Magazine Press, 1975.

Draper, Theodore. The Roots of American Communism. New

York: Viking Press, 1957.
Garraty, John A. Woodrow Wilson: A Great Life in Brief. New York: Knopf, 1956.
Heaps, Willard Allison. Riots, U. S. A., 1765-1970. New York: Seabury Press, 1970.
Johnson, Donald. The Challenge to American Freedoms. Lexington, Ky.: University of Kentucky Press, 1963.
Mandelbaum, Seymour J. The Social Setting of Intolerance. Chicago: Scott, Foresman, 1964.
Murray, Robert K. The Red Scare: A Study in National Hysteria, 1919-1920. Minneapolis: University of Minnesota Press, 1955.
Post, Louis F. The Deportations Delirium of 1920. Chicago: Charles H. Kerr, 1923.
Preston, William, Jr. Aliens and Dissenters: Federal Suppression of Radicals, 1903-1933. Cambridge, Mass.: Harvard University Press, 1963.
Sann, Paul. The Lawless Decade. New York: Crown Publishers, 1957.

THE PONZI POSTAL EXCHANGE SWINDLE (1919-1920)

Charles Ponzi was an Italian swindler and petty thief who, in 1919 and 1920, persuaded gullible investors that he had literally "invented money." After a checkered career as a gambler, waiter, card sharp, bank clerk, and convict, he discovered that postal reply coupons, purchased in Spain for one cent each, American money, could be redeemed in the United States for ten cents. This situation existed because postal reply coupons were redeemed at a rate fixed by treaty which did not reflect the actual rate of exchange.

Ponzi opened an office in Boston and on his first day in business collected $250 from his friends. Shortly afterwards he paid back $375. His slogan, "50% profit on your money in 45 days; double it in six months," spread rapidly and by the spring of 1920 the Securities Exchange Company which he had established was accepting investments at the rate of $250,000 per day.

Ponzi's operation became a matter of national notice. No one seemed to realize that postal exchange coupons were redeemed in stamps, not cash, and that Ponzi would have to have hundreds of agents in Europe purchasing coupons for him in order to handle the large sums invested in his scheme.

In the meantime Ponzi had purchased a large estate and was spending lavishly on automobiles, vintage wines, swimming pools, expensive clothes, and similar accoutrements of great wealth. His sun, however, was setting. The postal authorities had started an investigation. Their inquiry showed that, although during the previous six months something less than a million dollar's worth of postal reply coupons had been issued, Ponzi had apparently acquired ten million dollars dealing in them. Richard Grozier, publisher of the Boston Post, sent reporters to question the financier. He stated that his postal coupon idea was merely a front for certain high level financial transactions which he did not divulge.

Eventually it was realized that the Italian entrepreneur was using one of the oldest of swindles: taking the money invested for himself and repaying the early investors with money deposited by more recent ones. When, on August 11, 1920, Ponzi was identified as a former bank clerk who had served a three-year prison term for forgery, United States Federal agents entered his office, impounded its contents, and arrested him. He was indicted by the Federal Government, pleaded guilty, and was sentenced to prison. After serving three and a half years of a five-year sentence he was released, indicted by the State of Massachusetts for grand larceny, tried, convicted, and sent to prison for seven years. Released in 1934, he was deported to his native Italy, having never taken out United States citizenship. From Italy he traveled to Rio de Janeiro where he died penniless in 1949.

Suggested Readings

Deeson, A. F. L. Great Swindlers. New York: Drake Publishers, 1972.
Dunn, Donald Harley. Ponzi! The Boston Swindler. New York: McGraw-Hill, 1975.
Hynd, Alan. "The Pied Piper of Boston," in his Murder, Mayhem and Mystery. New York: A. S. Barnes & Company, 1958.
MacDougall, Curtis D. Hoaxes. New York: Macmillan, 1940.
McMasters, D. H. "Ponzi Press Agent Tells Inside Story of Scoop," Editor and Publisher, July 5, 1952.
"One of the Slickest of Them All," Newsweek, April 1, 1957.
Russell, Francis. "Bubble, Bubble--No Toil, No Trouble," American Heritage Magazine, February, 1973.
Sann, Paul. The Lawless Decade. New York: Crown Publishers, 1957.
"Take My Money," Time, January 31, 1949.
Thomas, H., and D. L. Thomas. "Ponzi's Monumental Swindle," Reader's Digest, April, 1952.

THE BLACK SOX SCANDAL (1919-1921)

Although eight members of the 1919 Chicago White Sox baseball team were involved in deliberately losing the world series of that year to the Cincinnati Reds, the affair is remembered as the "Black Sox scandal." In 1919 the Chicago team was one of the best in the history of the sport. Its roster of players included such outstanding stars as Eddie Collins, Ray Schalk, Eddie Cicotte, Lefty Williams, Urban Faber, Dick Kerr, Buck Weaver, Chick Gandil, Shoeless Joe Jackson, Swede Risberg, Happy Felsh, Jimmy Collins and Fred McMullen.

The Sox's opponents in the series were the Cincinnati Reds, a good team but completely outclassed by the Chicago players. When the two met at Cincinnati for the opening game on October 1, 1919, the White Sox were 1 to 5 favorites in the early betting odds. Later the odds switched to even money.

Eight of the players on the White Sox team had been bribed by a group of gamblers to lose the series to the Cincinnati team. Star players made deliberately clumsy misplays and rumors that the games had been "fixed" spread before the series was over. It was whispered that Arnold Rothstein, a New York gambler, had engineered the rigging of the games, although this has never been proved. It is known that Abe Attell, gambler and former world's featherweight boxing champion, was implicated, and that he had declared to Chick Gandil and Eddie Cicotte that Rothstein was backing him.

The Cincinnati Reds won the series in the eighth of the nine-game series.

The following year Cicotte disclosed his knowledge of the swindle. Charles A. Comiskey, owner of the Chicago club, immediately suspended Cicotte and the seven other players involved. These were Risberg, McMullen, Williams, Gandil, Weaver, Jackson and Felsh.

On September 28, 1920, these eight players were indicted by the Grand Jury of Cook County, Illinois, on a charge of fraud. Tried in 1921, they were acquitted, but the damage done to professional baseball was such that some club owners realized that the three-man National Baseball Commission was ineffectual. They negotiated with Judge Kenesaw Mountain Landis, a federal jurist, who agreed to administer the game as National Commissioner. This position he retained until his death in 1944. The eight players who had been involved in the Black Sox scandal never again played major league baseball.

Suggested Readings

Allen, Lee. The American League Story. New York: Hill & Wang, 1962.

________. The National League Story. New York: Hill & Wang, 1965.

Asinof, Eliot. Eight Men Out: The Black Sox and the 1919 World Series. New York: Holt, Rinehart, 1963.

Durant, John. Highlights of the World Series. New York: Hastings House, 1971.

________. The Story of Baseball in Words and Pictures. New York: Hastings House, 1973.

Einstein, Charles. The Fireside Book of Baseball. New York: Simon & Schuster, 1956.

Katcher, Leo. The Big Bankroll. New York: Harper, 1958.

Lieb, Frederick G. "Big Blotch on Baseball's Escutcheon," in his The Story of the World Series. New York: Putnam, 1949.

Meany, Thomas. Baseball's Greatest Teams. New York: Barnes, 1949.

Powers, James. Baseball Personalities. New York: Field, 1949.
Sann, Paul. The Lawless Decade. New York: Crown Publishers, 1957.
Seymour, Harold. Baseball: The Early Years. New York: Oxford University Press, 1960.
Smith, Ira L., and H. Allen Smith. Three Men on Third. Garden City, N.Y.: Doubleday, 1951.
Smith, Robert M. Heroes of Baseball. Cleveland: World, 1952.
Thompson, Lewis, and Charles Boswell. "Say It Ain't So, Joe!," American Heritage Magazine, June, 1960.
Verral, Charles S. Mighty Men of Baseball. New York: Aladdin Books, 1955.
Voigt, David Quentin. American Baseball: From Gentlemen's Sport to the Commissioner System. Norman, Okla.: University of Oklahoma Press, 1966.

THE NOBLE EXPERIMENT (PROHIBITION) (1919-1933)

Prohibition, called by President Herbert Hoover "The Noble Experiment," is defined as "a form of sumptuary legislation which attempts to abolish the manufacture and sale of alcoholic liquors." Various states, beginning with Maine in 1846, passed laws prohibiting the sale of liquor. By 1915 only three states, Nevada, New Jersey, and Pennsylvania, had neither statewide prohibition nor some form of local option.

When the United States entered World War I in 1917, Congress enacted a law which forbade the manufacture of all intoxicants except beer and wine. A Presidential proclamation in December of the same year reduced the legal alcoholic content of beer to 2-3/4% by weight, effective January 21, 1918. Shortly afterwards Congress passed a resolution to amend the Constitution of the United States (the Eighteenth Amendment), which went into effect January 16, 1920, providing for national prohibition. This Amendment was finally ratified on January 29, 1919, by all the states except Connecticut and Rhode Island.

The National Prohibition Act, popularly known as the Volstead Act after its sponsor, Senator Andrew J. Volstead of Montana, was passed on October 28, 1919, over the veto of President Woodrow Wilson. The purpose of this Act was to enforce prohibition of the manufacture, sale, and transportation of intoxicating liquors under the Eighteenth Amendment. Wayne B. Wheeler, an American lawyer and general counsel of the Anti-Saloon League of America, was instrumental in securing the passage of both the Eighteenth Amendment and the Volstead Act, claiming authorship of the latter.

In June, 1920, the Supreme Court unanimously upheld the constitutionality of both the Amendment and the Act. In 1925 laws were passed by Congress making the Prohibition Unit a federal bureau administered by the Secretary of the Treasury. This was transferred to the Department of Justice in July, 1930.

Enforcement of the prohibition laws proved impossible. The American people, feeling that their personal rights were being interfered with unjustly, patronized bootleggers and speakeasies. Gangsters, such as Alphonse Capone, George "Bugs" Moran, Dutch Schultz, Waxey Gordon and the Genna brothers, dominated the illegal liquor market, manufacturing gin, bootleg whiskey, and beer, or smuggling it in from Canada and the West Indies. People manufactured "home brew" and "bathtub gin." Legal industrial alcohol was diverted into illegal channels and the poisonous substances added to make such industrial alcohol unpotable were partially removed by chemists using denaturing processes.

In 1927 opposition to prohibition became organized and in that year the Association Against the Prohibition Amendment was founded. Various states repealed or altered their prohibition laws.

In 1929 Herbert Hoover, the newly elected president, who was considered a "dry" by the electorate, was not convinced that prohibition's far-reaching purpose was being achieved and he had promised that a study of the enforcement problem would be made by a governmental commission. In that year such a commission was appointed. George Woodward Wickersham, a New York lawyer and former attorney general of the United States, was designated to head it. His eleven-man commission, officially known as "The National Commission on Law Observance and Enforcement," and popularly called "The Wickersham Commission," made its report to President Hoover on January 19, 1931, after nineteen months of investigation. The report declared that three factors stood in the way of enforcing prohibition. These were: 1) great profits were being made in the illicit liquor traffic; 2) the public, in general, did not regard violation of the Eighteenth Amendment as criminal; 3) the states regarded enforcement as a task which properly should be performed by the federal government alone.

The members of the Wickersham Commission were by no means unanimous in their recommendations. Two favored outright repeal of the Eighteenth Amendment, five urged further trial of it, and the other four recommended modification.

In 1932 both the Republican and the Democratic parties, in their nominating conventions, indicated their favoring repeal of the Eighteenth Amendment. In February, 1933, Congress voted to submit the Twenty-First Amendment repealing the Eighteenth, to the State Conventions. When Utah voted in favor of the repeal Amendment on December 5, 1933, becoming the 36th state to do so, the dry law was at an end.

Suggested Readings

Allen, Frederick Lewis. Only Yesterday. New York: Harper, 1931.

Andrist, Ralph K., ed. in charge. The American Heritage History of the 20's and 30's. New York: American Heritage, 1970.

Asbury, Herbert. The Great Illusion: An Informal History of Prohibition. Garden City, N.Y.: Doubleday, 1950.

________. "The Noble Experiment of Izzy and Moe," in Leighton,

Isabel, ed. The Aspirin Age, 1919-1941. New York: Simon & Schuster, 1949.
Cherrington, Ernest H. Evolution of Prohibition in the United States of America. Montclair, N.J.: Patterson Smith 1969.
Coffey, Thomas M. The Long Thirst. New York: Norton, 1975.
Commager, Henry Steele, ed. "The Constitution of the United States," (Doc. No. 87) in his Documents of American History, 8th edition. New York: Appleton, 1968.
________. "National Prohibition Cases," (Doc. No. 433) in his Documents of American History, 8th edition. New York: Appleton, 1968.
________. "The Volstead Act," (Doc. No. 432) in his Documents of American History, 8th edition. New York: Appleton, 1968.
________. "W.C.T.U. Declaration of Principles," (Doc. No. 357) in his Documents of American History, 8th edition. New York: Appleton, 1968.
Dobyns, Fletcher. The Amazing Story of Repeal. Chicago: Willett, Clark, 1940.
Jenkins, Alan. The Twenties. New York: Universe Books, 1974.
Krout, John Allen. The Origins of Prohibition. New York: Knopf, 1925.
Lyle, John H. The Dry and Lawless Years. Englewood Cliffs, N.J.: Prentice-Hall, 1960.
Merz, Charles. The Dry Decade. New York: Doubleday, 1931.
Odegard, Peter H. Pressure Politics, The Story of the Anti-Saloon League. New York: Columbia University Press, 1928.
Ostrander, Gilman M. The Prohibition Movement in California. Berkeley, Calif.: University of California Press, 1957.
Pasley, Fred. Al Capone: The Biography of a Self-Made Man. Plainview, N.Y.: Books for Libraries, 1931.
Sann, Paul. The Lawless Decade. New York: Crown Publishers, 1957.
Sellers, James B. The Prohibition Movement in Alabama. Chapel Hill, N.C.: University of North Carolina Press, 1943.
Sinclair, Andrew. Era of Excess: A Social History of the Prohibition Movement. New York: Harper, 1964.
Whitener, David Jay. Prohibition in North Carolina. Chapel Hill, N.C.: University of North Carolina Press, 1945.

THE ELWELL MURDER CASE (1920)

Joseph Bowne Elwell, a former hardware salesman who had turned bridge-whist expert and become a nationally recognized authority on the game, was mysteriously murdered on June 11, 1920. His wife, Helen Darby Elwell, had recently requested a divorce, charging that he had deserted her for a younger woman. He lived alone in a three-story brownstone house on West 70th Street, New York City.

At 8:35 on the morning of June 11 Mrs. Marie Larsen, his housekeeper, came to his home. She found Elwell sitting in a chair in the parlor. He was dressed in pajamas and the toupee he customarily wore was not on his head, nor were his artificial teeth in his mouth. He was alive but unconscious, and a large bullet hole was in his forehead. Blood was dripping from the wound and onto a letter in his lap. Other letters, apparently received in the morning's mail, were on the floor beside him.

Mrs. Larsen notified the police, who rushed Elwell to Bellevue Hospital where he died two hours later without regaining consciousness. It was first assumed that Elwell was a suicide, but the police found no weapon near his chair. They did, however, find a .45 caliber cartridge on the floor and the fatal bullet, when pried from the plaster where it had lodged, was also a .45. It was established at the autopsy that the bullet had been fired at close range.

A search of the dead man's home disclosed that robbery had not been the motive for the killing, as $400 in cash and jewelry valued at $7,000 were found in the bedroom. Also found were 55 toupees in a closet and a lavishly appointed boudoir.

When the police checked Elwell's movements prior to his death they established the fact that, on the previous evening, he had dined at the Ritz-Carlton Hotel with Mr. and Mrs. Walter Lewisohn and Mrs. Lewisohn's sister, Miss Viola Kraus. Miss Kraus had recently been divorced from Victor Von Schlegel, a business executive. Von Schlegel and his companion, called by the tabloids "The Lady in Black," walked by.

After dinner Elwell and his companions attended the "Midnight Frolic" at the New Amsterdam Theater. Following the performance he decided to go home alone and the other three, plus Octavio Figueroa, whom they had met at the theater, departed in a taxi. This was at 1:30 on the morning of June 11.

Further police investigation disclosed that Lewisohn had telephoned Elwell's home about a projected trip to Long Beach and had been advised of the tragedy. They learned that Miss Kraus had also telephoned Elwell at 2:30 on the morning of June 11 to express the hope that he was not angry for "having to go home alone," and had talked to him at that time. She said she had the impression, from the way he spoke, that he had a visitor. She had no idea who the visitor might be.

Von Schlegel, when questioned, stated that he had gone directly home from the theater the night before, and that when "The Lady in Black" heard of Elwell's death she had immediately left for her home in Minnesota.

Elwell's wife, who had not seen her husband for several years, was obviously not involved in the matter. His neighbors, when interviewed by the police, shed a new light on his affairs. They testified that he was in the habit of entertaining women--"hundreds of women"--who would arrive at his home in the morning and leave late in the afternoon. Mrs. Larsen, the housekeeper, said that every woman Elwell brought into his home was known to her only as "Miss Wilson."

Other facts came out as the investigation continued. Elwell had once been driven away from Louisville, Kentucky, by the family of a girl he had seduced. Several similar scandals and evidence of clandestine affairs and love triangles came to light. The taxi driver who picked up the party on the street outside the New Amsterdam Theater stated that he had carried three passengers, not four as others claimed. These he identified as Mr. and Mrs. Lewisohn and Octavio Figueroa.

Joseph Elwell's love life became front page news in the tabloids. His secretary, his chauffeur and others indicated that he had been intimate with "at least fifty women, most of them married."

Other clues were found, but they led nowhere. From the circumstantial evidence available the police concluded that Elwell had been shot by a man who knew him well, as the .45 with which the murder was committed was too heavy a weapon for a woman to handle. Elwell was extremely vain and would not permit a woman to see him sans teeth and toupee. The fact that the shot was fired at close range indicated that Elwell did not suspect his murderer of homicidal intentions and permitted him to come as close as he did before firing.

Joseph Bowne Elwell's murderer was never discovered.

Suggested Readings

Abrahamsen, David, M.D. The Murdering Mind. New York: Harper & Row, 1973.

Collins, Ted, ed. "Joseph B. Elwell," in his New York Murders. New York: Duell, Sloan, 1944.

"The Elwell Case," New Republic, July 21, 1920.

Goren, C. H. "The Elwell Murder Case," McCall's, October, 1962.

"J. B. Elwell, Whist Expert and Race Horse Owner, Slain," New York Times, June 12, 1920, p. 1.

Pearson, Edmund. "Mr. Elwell," in his Masterpieces of Murder. Boston: Little, Brown, 1963.

Reinhardt, James Melvin. The Psychology of Strange Killers. Springfield, Ill.: Thomas, 1962.

Sann, Paul. The Lawless Decade. New York: Crown Publishers, 1957.

Sterling, Hank, pseud. "The 55 Toupees Murder," in his Ten Perfect Crimes. New York: Stravon Publishers, 1954.

Symons, Julian. A Pictorial History of Crime. New York: Crown Publishers, 1966.

Van Dine, S. S., pseud. The Benson Murder Case (fiction). New York: Scribner's, 1926.

Woollcott, Alexander. "Five Classic Crimes," in his Long, Long Ago. New York: Viking Press, 1943.

THE RAGGED STRANGER MURDER (1920)

Shortly after ten o'clock on the evening of June 21, 1920, Carl Wanderer, a Chicago butcher and former army lieutenant, was walking home from a motion picture theater with his pregnant wife Ruth. A neighbor, sitting on his front porch, saw them pass and also saw an unkempt, poorly dressed man apparently following them.

Very shortly thereafter shots rang out from the vestibule of the house where the Wanderers lived with Ruth's mother. James Wilson, a resident of an apartment on the ground floor of the Wanderer home, opened his front door to see Ruth lying on the floor and Carl Wanderer beating a man's head on the floor and shouting, "You've killed her!"

Both Ruth and the man had been shot, Ruth with two bullets and the man, Edward Joseph Ryan, who was to become known as the "Ragged Stranger," with three. Ruth died a few minutes later in her home, Ryan at the Ravenswood Hospital. Neither made a dying statement.

Carl Wanderer's story was that, upon entering the vestibule, someone had attempted to rob him, that he had shot the would-be robber with the .45 army Colt he always carried, and that the robber, also armed, had shot and killed his wife. Altogether ten shots were fired, none of which had hit Wanderer.

At first the widower was lauded by the Chicago newspapers for his defense of his wife against an armed robber, and was dubbed the "Husband Hero." He was interviewed by Ben Hecht, then a reporter on the staff of the Chicago Daily News. Hecht felt that Wanderer's story did not ring true, and that he was entirely too cheerful and dramatic for one who had just lost his wife and unborn child under such tragic circumstances. Hecht's story in the Daily News reflected his suspicions.

Police Sergeant Michael Grady of the Chicago Homicide Squad had been assigned to the Wanderer case. He had taken possession of the two revolvers found in the Wanderer vestibule. Both were .45 caliber Colts. One, Wanderer's, was stamped "U.S. Army." The other, without such stamping, bore the factory serial number C2282. The bullets which had killed Ruth Wanderer had been discharged from Ryan's revolver and those which had killed Ryan came from Wanderer's army Colt.

Grady's suspicions of foul play, first aroused by Hecht's newspaper article, were augmented when he learned that Wanderer had grown tired of married life and had told his father that he "felt trapped and wished he was back in the army." The police sergeant traced the ownership of Ryan's revolver to Fred Wanderer, a cousin of Carl. Fred Wanderer stated to Grady that Carl had borrowed the weapon from him on the evening of June 21, not saying why he wanted it.

Sergeant Grady sent for Wanderer. At the police station the latter was examined and at last broke down and confessed. He had enjoyed his adventurous service as an army officer and did not wish to continue "a confining life with a wife who thought of nothing but having a big family and saving pennies." He saw her immediate death as the only way out of his dilemma.

Wanderer's plan was to recruit a derelict from Chicago's skid row, offer him a job, and arrange a rendezvous at which Ruth would be present. He would then shoot the two of them, one with each of the two revolvers, and tell a story of a resisted hold-up attempt which had fatal consequences.

Carl Wanderer was tried for the murder of his wife. He repudiated his confession but was found guilty and sentenced to serve 25 years confinement in Joliet. Tried again, this time for the murder of Ryan, the "Ragged Stranger," he was convicted and sentenced to death. He was hanged in 1921.

Suggested Readings

Bromberg, Walter. Mold of Murder: A Psychiatric Study of Homicide. New York: Grune & Stratton, 1961.

Guttmacher, M. S. The Mind of the Murderer. Freeport, N.Y.: Books for Libraries, 1960.

Hecht, Ben. A Child of the Century. New York: Simon & Schuster, 1954.

_______. Gaily, Gaily. Garden City, N.Y.: Doubleday, 1963.

Hynd, Alan. "The Riddle of the Ragged Stranger," in his Murder, Mayhem and Mystery. New York: A. S. Barnes, 1958.

Jesse, F. Tennyson. Murder and Its Motives. London: Harrap, 1952.

Reinhardt, James Melvin. The Psychology of Strange Killers. Springfield, Ill.: Thomas, 1962.

"Shot at Husband's Side," New York Times, June 23, 1920, p. 6.

"Wanderer Denies Murder," New York Times, October, 21, 1920, p. 16.

Woollcott, Alexander. "Five Classic Crimes," in his Long, Long Ago. New York: Viking Press, 1943.

THE WALL STREET EXPLOSION (1920)

The financial center of the United States lies at the intersection of Wall and Broad Streets in New York City. In 1920 the Sub-Treasury Building and the United States Assay Office occupied the north side of Wall Street. The offices of J. P. Morgan & Company, international bankers, occupied the southeast corner. On the southwest corner was the excavation for the cellar of the yet-to-be-built annex of the New York Stock Exchange. Next to the excavation, on Broad Street, stood the Exchange.

This geographical concentration of financial power exemplified the combination of government and financial influence which was anathema to the radical element of the day. Shortly before noon on Thursday, September 16, 1920, a TNT bomb of tremendous power exploded in front of the Assay Office and across the street from the Morgan building. Thirty people were killed outright and hundreds were injured. The interior of the Morgan offices was wrecked and windows for blocks around were broken. In the House of Morgan the chief clerk was killed and dozens were injured, some requiring hospitalization. J. P. Morgan was abroad at the time and so was not injured. One Morgan partner had his hand cut by flying glass but the others, being in conference on the other side of the building, were not hurt.

The remains of a horse-drawn wagon and the carcass of a horse were found in the middle of Wall Street. The immediate evidence showed that the bomb had, in all probability, been brought to Wall and Broad Streets in the wagon where it had been detonated by a time-fuse mechanism while the driver made his escape.

The investigation, extending over several months, disclosed little. Slugs which composed part of the bomb and which had been driven into surrounding buildings by the force of the explosion were found to be sash weights cut in two. The shoes of the dead horse were traced to the blacksmith who had nailed them to the horse's hoofs. The blacksmith was unable to furnish any clues concerning the man who had brought the animal to him, other than to say that he "looked Sicilian." Other clues were carefully run down but led nowhere. Suspected radicals were rounded up, questioned at length, and released for want of evidence.

One clue, leading nowhere like the others, consisted of five sheets of paper found in a mailbox a few blocks from the scene of the explosion. These sheets were crudely lettered with the following misspelled message:

Remember

We will not tolerate

any longer

Free the politi Cal

Prisoners or it will be

Sure death to all oF you

American Anarchists

Fighters

As of this date the identity of the perpetrators of the Wall Street Explosion still remains unknown.

Suggested Readings

Allen, Frederick Lewis. Only Yesterday. New York: Harper, 1931.

Andrist, Ralph K., ed. in charge. The American Heritage History

of the 20's and 30's. New York: American Heritage, 1970.
"Bomb Death List Increased to 36," New York Times, September 30, 1920, p. 9.
"Experts to Ignite TNT in Bomb Test," New York Times, September 28, 1920, p. 6.
"Explosives Found to be Missing in Hunt for Bomb Plot Clue; Fischer Starts for New York," New York Times, September 20, 1920, p. 1.
"Flynn Encouraged in His Bomb Hunt," New York Times, September 23, 1920, p. 1.
Hynd, Alan. "The Great Wall Street Explosion," in Kartman, Ben, and Leonard Brown, eds. Disaster! New York: Pellegrini & Cudahy, 1952.
Sutherland, Sidney. "The Mystery of the Wall Street Explosion," Liberty, April 26, 1930.
"Wall Street Explosion Kills 30, Injures 300; Morgan Office Hit, Bomb Pieces Found; Toronto Fugitive Sent Warnings Here," New York Times, September 17, 1920, p. 1.
"Wall Street Massacre," Independent & Weekly Review, October 2, 1920.
"Wall Street's Bomb Mystery," Literary Digest, October 2, 1920.

THE SACCO-VANZETTI CASE (1920-1927)

The Sacco-Vanzetti case, one of the most famous in American jurisprudence, involved the prosecution of Nicola Sacco, an Italian shoe worker, and Bartolomeo Vanzetti, an Italian fish peddler. The two men were accused of the murders of Frederick A. Parmenter, paymaster of a shoe factory, and Alessandro Beradelli, a guard. These murders occurred during a payroll robbery in South Braintree, Massachusetts on April 15, 1920.

Sacco and Vanzetti were arrested on May 5. The arrest was made following an investigation which disclosed that they were two of four men who had rented an automobile which allegedly was used in the robbery. When taken into custody both men, avowed anarchists, were armed with pistols. Vanzetti also had in his possession several 12-gauge shotgun shells. A 12-guage shotgun had been used in an unsuccessful payroll robbery at Bridgewater, Massachusetts on December 24, 1919. Vanzetti was tried for the Bridgewater crime, found guilty, and sentenced to prison.

Subsequently both Sacco and Vanzetti were indicted for the South Braintree murders and brought to trial. Judge Webster Thayer presided and on July 14, 1921, the two Italians were found guilty. They were sentenced to death.

The jury's verdict was received with anger by radicals throughout the world, the radicals contending that the two anarchists had been tried on trumped-up evidence and convicted for their political beliefs.

A motion for a new trial made by the defense in 1922 was denied. The two men remained in Dedham Jail while their attorneys sought to free them. On November 18, 1925, Celestino Madeiros, a fellow prisoner, confessed to Sacco that he had participated in the South Braintree crime and that neither Sacco nor Vanzetti was in any way involved.

On October 24, 1926, Judge Thayer denied the petition submitted to him based on the Madeiros confession and on other evidence. The Massachusetts Supreme Judicial Court refused an appeal on technical grounds.

The case was carried to Governor Alvan Tufts Fuller of Massachusetts. The governor made a personal investigation of the matter. He appointed an advisory committee consisting of Abbott Lawrence Lowell, president of Harvard University, Samuel Wesley Stratton, president of Massachusetts Institute of Technology, and Robert Grant, a judge, to consider the case. Felix Frankfurter, then a professor of law at Harvard University, exonerated the two defendants in a public review of the trial. Governor Fuller, on August 3, published his decision that the defendants had received a fair trial and were guilty. He indicated further that the advisory committee concurred in his conclusions.

A series of stays of execution was granted. Public feeling flared high and protest demonstrations were held in many American and foreign cities. On August 23, 1927, Sacco, Vanzetti, and Madeiros were executed in the electric chair at the Massachusetts State Prison, Charlestown. Just prior to dying Vanzetti declared his innocence.

Suggested Readings

Allen, Frederick Lewis. Only Yesterday. New York: Harper, 1931.

Andrist, Ralph K., ed. in charge. The American Heritage History of the 20's and 30's. New York: American Heritage, 1970.

Aymar, Brandt, and Edward Sagarin. "Nicola Sacco and Bartolomeo Vanzetti," in their A Pictorial History of the World's Great Trials. New York: Crown Publishers, 1967.

Busch, Francis X. Prisoners at the Bar. Indianapolis: Bobbs-Merrill, 1952.

Churchill, Allen. The Year the World Went Mad. New York: Crowell, 1960.

Commager, Henry Steele, ed. "Bartolomeo Vanzetti's Last Statement in Court," in his Documents of American History, 8th edition. New York: Appleton, 1968.

Ehrmann, Herbert B. The Case That Will Not Die: Commonwealth vs. Sacco and Vanzetti. Boston: Little, Brown, 1969.

_______. The Untried Case: The Sacco-Vanzetti Case and the Morelli Gang. New York: Vanguard, 1933.

Felix, David. Protest: Sacco-Vanzetti and the Intellectuals. Bloomington, Ind.: Indiana University Press, 1965.

Frankfurter, Felix. "The Case of Sacco and Vanzetti," Atlantic Monthly, March, 1927.

_______. The Case of Sacco and Vanzetti: A Critical Analysis for Lawyers and Laymen. Boston: Little, Brown, 1927.

Grossman, James. "The Sacco-Vanzetti Case Reconsidered," Commentary, January, 1962.
Jenkins, Alan. The Twenties. New York: Universe Books, 1974.
Joughin, Louis, and Edmund M. Morgan. The Legacy of Sacco and Vanzetti. New York: Harcourt, Brace, 1948.
Musmanno, Michael A. After Twelve Years. New York: Knopf, 1939.
Russell, Francis. "Sacco Guilty, Vanzetti Innocent?," American Heritage Magazine, June, 1962.
_____. Tragedy at Dedham. New York: McGraw-Hill, 1962.
Sacco, Nicola, and Bartolomeo Vanzetti. The Letters of Sacco and Vanzetti. Edited by Marion D. Frankfurter and Gardner Jackson. New York: Viking, 1928.
The Sacco-Vanzetti Case. Transcript of the Record of the Trial of Nicola Sacco and Bartolomeo Vanzetti in the Courts of Massachusetts and Subsequent Proceedings, 1920-27. 6 vols. New York: Holt, 1928.
Sann, Paul. The Lawless Decade. New York: Crown Publishers, 1957.
Stark, Louis. "A Case That Rocked the World," in Baldwin, Hanson W., and Shepard Stone, eds. We Saw It Happen. New York: Simon & Schuster, 1939.
Stong, Phil. "The Last Days of Sacco and Vanzetti," in Leighton, Isabel, ed. The Aspirin Age, 1919-1941. New York: Simon & Schuster, 1949.
Symons, Julian. A Pictorial History of Crime. New York: Crown Publishers, 1966.
Weeks, Robert P. Commonwealth vs. Sacco and Vanzetti. Englewood Cliffs, N.J.: Prentice-Hall, 1958.

MAH-JONGG, THE CROSSWORD PUZZLE BOOK AND OTHER CRAZES OF THE TWENTIES (1920-1929)

During the 1920's a number of crazes swept the people of the United States. One of these was an adaptation of the old Chinese game of Mah-Jongg which was extremely popular in 1923 and 1924.

Shortly after World War I, Joseph P. Babcock, the Soochow representative of the Standard Oil Company, had become interested in the game. He codified and simplified the rules for the use of occidentals and the game was introduced into English-speaking communities in Shanghai by the White brothers. Here its popularity was such that W. A. Hammond, a San Francisco businessman, decided to import Mah-Jongg sets and merchandise them, which he did in the year 1922.

Hammond conducted a nationwide advertising campaign and by 1923 the game had become so popular that American manufacturers were producing sets. Before the craze had faded away in 1924 a Mah-Jongg League of America had been organized, playing rules had been established, and Chinese costumes, designed to be worn while the game was being played, were on the market.

Mah-Jongg was originally played with 108 cards, each bearing the name of a Chinese hero. Later the cards were replaced with ivory tablets on which the heroes' names were engraved. As played in America and Europe, tiles of ivory or plastic and bamboo were used, with a complete set containing 144 tiles. These are divided into six suits: 36 each of bamboos, circles and characters, 16 winds, 12 honors, and 8 seasons, although the seasons are often omitted. The tiles in each suit are numbered, and each tile is matched by three other identical tiles. For example, there are four number one characters, four number two characters, and so on up to nine.

Mah-Jongg is usually played by four people, designated as north, south, east, and west wind. The tiles are mixed and then each player builds a wall seventeen tiles long and two tiles high, all tiles being placed face downward. The walls are brought together to form a square from which each player removes tiles until the east wind player has fourteen and each of the other players has thirteen tiles.

Each player in turn draws a tile from the wall, adds it to his hand, and discards a tile he does not need. The object of the game is to acquire four sets of three identical tiles each and one set of two similar tiles. A set may also consist of any numerical sequence of the same suit.

The winning player is the one who first assembles his four groups and final pair. When the game is played for money each losing player pays the winner a sum previously agreed on.

The Mah-Jongg craze came and went, as did such later popular pastimes as jigsaw puzzles and hula hoops. The game, however, is still played in many homes and social clubs in the United States, Europe, and elsewhere.

Another of the crazes which swept the United States in the 1920's was a unique publication called the *Cross-Word Puzzle Book*, a volume which launched two youthful publishers on a successful business career.

Early in January, 1924, Richard Lee Simon became aware of the crossword puzzles appearing in the *New York World* when he was asked if a book of such puzzles was available. Simon, who was about to enter the publishing business, investigated and found that no such book was to be had. He learned that the idea of the crossword puzzle, an American invention, was a direct descendant of the ancient word square. The first crossword puzzle appeared in the Sunday supplement of the *New York World* in December, 1913. It was constructed by Arthur Winn, a member of the newspaper's staff. Over the next decade the crossword puzzle became a regular feature in the *World*. Franklin Pierce Adams mentioned it frequently in his column, "The Conning Tower," which appeared regularly in the same newspaper.

Simon and his partner, Max Lincoln Schuster, decided to bring out a crossword puzzle book. This, with a convenient pencil attached, first appeared in April, 1924. The puzzles were prepared by Prosper Buranelli, F. Gregory Hartswick, and Margaret Petherbridge, puzzle editors of the *World*.

The Cross-Word Puzzle Book was an overnight best-seller and within a year had sold several hundred thousand copies. Other publishers entered the field and virtually every newspaper in the country published its daily crossword, with the solution appearing in the following edition. Sales of dictionaries and Roget's Thesaurus boomed. Winston Churchill is said to have remarked, concerning the craze, "I should think that ninety per cent of the people believe that there was but one Roman emperor and that his name was Nero."

The crossword puzzle craze subsided somewhat in 1925 but many people, having acquired the "bug," furnished a market for the puzzle magazines which are published today. Newspapers continue to carry the crossword and such puzzles are by no means a thing of the past. Paperback books of such brain-teasers are available at newsstands everywhere, and many variations of the original crossword are now published. These include such special puzzles as anacrostics, cryptograms, cross sums, laddergrams, solicross, and subject crosswords, such as Bible crosswords, television crosswords and historical crosswords.

Following the success of the original 1924 edition, Simon and Schuster brought out many subsequent books of similar nature. Sales of the series totaled several million copies.

A number of question and answer books, typified by the popular Ask Me Another, followed the crossword puzzle craze. This fad, though popular, was short-lived and gave over to the game of contract bridge which evolved from the earlier games of auction bridge and bridge-whist. Contract bridge made heroes of such authorities as Ely Culbertson, Sidney S. Lenz and P. Hal Sims.

Other crazes and fads came and went. Certain songs, notably "Yes, We Have No Bananas" and "Sonny Boy," were briefly popular. Early in the decade Emile Coué, a French exponent of auto-suggestion, developed and introduced a system of psychotherapy known as Couéism. Here the patients repeated the phrase, "day by day in every way I'm getting better and better." Dances, such as the charleston, varsity drag and black bottom, were transiently popular. The discovery of the tomb of the Egyptian pharaoh Tutankhamen near Luxor by the English Egyptologists Howard Carter and George E. S. M. Herbert, Earl of Carnarvon, influenced ladies' clothing fashions and hair styling which suddenly acquired an Egyptian motif. These and other crazes and fads caused the third decade of the twentieth century to become known as the "roaring twenties" and the "era of wonderful nonsense."

Suggested Readings

Allen, Frederick Lewis. Big Change: America Transforms Itself, 1900-1950. New York: Harper, 1952.

______. Only Yesterday. New York: Harper, 1931.

Andrist, Ralph K., ed. in charge. The American Heritage History of the 20's and 30's. New York: American Heritage, 1970.

Carter, Howard, and A. C. Mace. The Tomb of Tut-ank-Amen. New York: Cooper Square, 1954.

Churchill, Allen. The Year the World Went Mad. New York: Crowell, 1960.
Coué, Emile. Self-Mastery Through Conscious Autosuggestion. London: Allen, 1922.
Cross Word Puzzle Dictionary of the English Language. Philadelphia: Winston, 1959.
Jenkins, Alan. The Twenties. New York: Universe Books, 1974.
Morris, Lloyd R. Not So Long Ago. New York: Random House, 1949.
Sann, Paul. The Lawless Decade. New York: Crown Publishers, 1957.
Strauser, Kitty, and Lucille Evans. Mah-Jongg, Anyone? A Manual of Modern Play. Rutland, Vt.: Tuttle, 1964.
Walker, Stanley. The Night Club Era. New York: Blue Ribbon Books, 1933.

THE JASPER COUNTY MURDER FARM (1921)

Peonage is defined as "the practice of holding persons in servitude or partial slavery, as to work off debt." It is illegal in America because it involves involunatry servitude, something specifically forbidden by the Constitution (Article XIII, December 18, 1865). In spite of its being prohibited, however, peonage has existed in certain rural areas of the United States since Civil War days. Southern Negroes in particular were victimized in this fashion, being required to work without compensation to pay off obligations that, somehow, became never-ending.

It was customary for white plantation owners to visit local courts, pay the fines imposed on Negroes for various infractions of the law, and then require the Negroes to work out the fines on the owners' lands.

One such owner was John S. Williams, a 54-year-old white man who operated a 2,000-acre plantation in Jasper County, Georgia. On February 18, 1921, Williams was interviewed by federal agents George W. Brown and A. J. Wismer who were investigating a report that peonage was being practiced on the Williams plantation. The investigators were particularly interested in one Gus Chapman, a Negro laborer who had fled from the plantation. Clyde Manning, Williams' 27-year-old Negro foreman, stated that neither he nor his employer had ever caught Chapman. Williams said that Chapman had been caught by someone else and returned to the plantation. The federal agents departed, saying that they had found no conditions which would lead to prosecution.

A month later a white boy named Cash found the drowned body of a Negro in a stream in neighboring Newton County. He summoned help and that day two other Negro bodies were reclaimed from the river. It was found that the victims had been murdered by being weighted, bound, and then thrown alive from a bridge into the river.

The investigation which followed led to the Williams plantation, and Clyde Manning was arrested on suspicion of murder. Promised immunity, he confessed that he and John Williams had murdered no fewer than eleven men, and he conducted law enforcement officers to the graves of the various victims. Williams, though claiming to be the subject of a frame-up by a neighbor, was arrested. The story of his peonage activities then came out. Once a Negro was brought to the Williams place he found it virtually impossible to escape. His debt was never repaid and he moved from a position of peonage to one of slavery. Negroes who did attempt to run away and who were recaptured were flogged, punished in other ways, and sometimes murdered. Manning, in fear of Williams, felt that he had to carry out the latter's orders or suffer dire consequences.

John Williams' sons, Huland, Marvin, and LeRoy, operated plantations some five miles away. Claude Freeman, a Negro, acted as foreman for the sons, on whose plantations peonage was practiced as it was on the plantation of their father.

A gruesome story of whippings and murders gradually unfolded as the investigation progressed. It became evident that, by the spring of 1921, John S. Williams and his sons had all committed murder and that peonage was practiced on all plantations owned by them. Williams Senior stated that he took men from jails, paid their fines, and only worked them until he had been repaid. However, he never said what the wage rate was and no record of any Negro who had been so "rescued" from jail by Williams leaving his plantation could be found. The blacks were afraid to leave, were in mortal terror of Williams, and feared for their lives.

Following the visit by federal agents Brown and Wismer, Williams Senior told his foreman Clyde Manning that it would be necessary "to do away with some of those boys." Apparently he reasoned that by "doing away" with some of the Negroes, he could show, by their absence, that they had not been held on the plantation against their wills. Manning demurred but assisted with the "doing away," as he felt that Williams would kill him if he did not.

Within a week the killings began. Together Williams and Manning murdered at least eleven Negroes by shooting and drowning. It was one of the drowned bodies found by the boy Cash that led to the investigation and subsequent arrests of John S. Williams and Clyde Manning. Williams' sons had left the county.

Williams' trial opened in Covington, Georgia, on April 5, 1921. Manning was tried two months later, having testified against Williams in the latter's trial. Williams was found guilty and sentenced to life imprisonment. Manning was also found guilty and received a similar sentence but his case was successfully appealed on a technicality. Tried again, he was found guilty and sentenced to life imprisonment. Williams' sons were never tried.

Manning died on a Georgia chain gang. Williams was killed while attempting to prevent a jail break on a Georgia prison farm where he had earned the position of trusty.

Suggested Readings

Barry, Richard. "Slavery in the South Today," Cosmopolitan Magazine, March, 1907.
Conrad, David Eugene. The Forgotten Farmers. Urbana, Ill.: University of Illinois Press, 1965.
Daniel, Pete. Shadow of Slavery: Peonage in the South, 1901-1969. Urbana, Ill.: University of Illinois Press, 1972.
_______. "We Are Going to Do Away With Those Boys," American Heritage Magazine, April, 1972.
Dorsey, Hugh M. The Negro in Georgia. Atlanta: Bowers, 1921.
"The Fruits of Peonage," New Republic, April 20, 1921.
"Georgia's Death Farm," Literary Digest, April 16, 1921.
Hart, Albert Bushnell. "Peonage and the Public," Survey, April 9, 1921.
"Peonage, A Mere Symptom," Review of Reviews, June, 1921.
"Peonage in Georgia," Independent, April 16, 1921.
Trial Transcript, "Georgia vs. John S. Williams," Newton Superior Court, March term, 1921.
Waskow, Arthur I. From Race Riot to Sit-in. Garden City, N.Y.: Doubleday, 1966.

THE VAROTTA KIDNAP MURDER (1921)

One afternoon in the late fall of 1920 Adolfo Varotta, the eldest son of Salvatore Varotta, an Italian-American truck driver, was badly injured in a collision incurred while riding with his father in the latter's truck. The boy was taken to Bellevue Hospital where it was determined that extensive surgical attention was needed.

Varotta was a poor man and his plight came to the attention of a wealthy woman who agreed to underwrite the cost of the boy's hospitalization. This woman also visited Varotta at his New York tenement home in the Italian district. Neighbors, seeing the chauffeur-driven limousine in which the woman made her visits, imagined that the boy's father had found a wealthy patron and envied him what they considered to be his good fortune.

The neighbors' impression that Salvatore Varotta had become rich was amplified when he brought suit for $50,000 against the driver of the truck which had collided with his, causing the accident in which his son Adolfo had been injured. This impression was magnified further when the father purchased a ramshackle second-hand car for $150, hoping to earn some money with it. The purchase of the car led the neighbors to believe that Varotta's lawsuit had been settled in his favor, although it had not. The other driver was as poor as Varotta and could not have paid any such judgment.

The false rumor of Varotta's new-found prosperity resulted in the kidnaping for ransom of Giuseppe ("Joe") Varotta, the five-year-old brother of Adolfo. On May 24, 1921, young Joe left his home and went to a nearby candy store to spend a penny a passer-by had given him. He was never seen alive again, except by the kidnapers.

Salvatore Varotta, on returning home from work, inquired for his son Joe, who was not to be found. After a short wait the father made a search for the boy. This proved unsuccessful and the police were notified. The officer to whom Varotta spoke took the matter lightly, saying that the boy would certainly come back.

Young Joe Varotta had not returned the following day, and the postman brought his father a letter. This, written in Italian, stated that the writer was a member of the Black Hand, a powerful secret society; that the society had taken the boy and demanded $2,500 for his release; and that if the police were notified the boy would be killed and further vengeance would be taken on the surviving members of his family. Joe's father was to obtain the money immediately and have it ready for further instructions.

Salvatore Varotta, unable to pay the sum demanded, reported the matter to Police Captain Archibald McNeill, who placed the case in the hands of Detective Sergeant Michael Fiaschetti, head of the New York Italian Squad. Fiaschetti was convinced that the kidnaping was the work of amateurs rather than of professional criminals. He suspected friends or neighbors who knew something but not enough of Varotta's affairs.

Fiaschetti's first move was to install Mrs. Rae Nicoletti, an Italian-speaking policewoman, in Varotta's home. She was to play the part of a cousin of Mrs. Varotta. Mrs. Nicoletti noticed that the Varotta house was under constant observation from across the street. Then a neighbor, Santo Cusamano, visited Varotta and urged him to pay the ransom. Varotta protested that he could raise not more than $500, which would be advanced by the "cousin." Other neighbors, including Mrs. Mary Pogano and Antonio Marino and his wife, also visited the Varottas and urged them to pay the kidnapers.

Cusamáno returned with the news that the Black Handers would accept $500. Mrs. Nicoletti reported these visits to Fiaschetti.

On June 1 a stranger came to the Varotta home and stated that he represented the kidnapers and would return the following night to collect the money. Varotta obtained $500 in marked money from Fiaschetti who planted another detective, John Pellegrino, in the Varotta home disguised as a plumber.

On the following night two men arrived. One was the emissary of the Black Hand who had called on June 1. The other, whose name turned out to be John Melchione, was unknown to Varotta.

The money was paid, following which Pellegrino stepped from his place of concealment, arrested and handcuffed the emissary, whose name was later found to be Roberto Raffaelo. Other detectives, who had concealed themselves in the neighborhood, arrested Marino, his stepson James Ruggieri, Cusamano, Melchione, and five others. These men were lodged in jail.

Joe Varotta was not found until July 11, when John Derahica, a Polish laborer, discovered the body of a small boy in the Hudson

River near Piermont, New York. The body was identified as that of the missing boy, who had been strangled and thrown into the river.

Certain facts concerning the kidnap-murder came out at the trial. Neighbors of Varotta had assumed him to have become, by their standards, wealthy, and had conceived the plot to abduct his son and hold him for ransom. Roberto Raffaelo had been recruited by James Ruggieri to act as a go-between and he, in turn, had enlisted his friend John Melchione, neither of these men being known to the Varottas. Each had been promised $500 for his participation in the plot.

After the ten men had been arrested another confederate had taken young Joe Varotta to the Hudson River, strangled and drowned him, as it was feared that he could, if alive, identify his kidnapers. He had been seized in the candy store, gagged, placed in a barrel, and taken to a hiding place.

The identities of the confederate or confederates were never discovered, the known participants refusing to talk. All were tried and convicted. Cusamano, Marino, and Ruggieri were sentenced to death but these sentences were commuted to life imprisonment. Melchione went insane and was sent to Matteawan for life. Raffaelo turned state's evidence, for which his original death sentence was commuted to imprisonment.

Suggested Readings

Cohen, Bruce J. Crime in America. Itasca, Ill.: Peacock, 1970.

"Hastens Murder Trial," New York Times, July 2, 1921.

Hibbert, Christopher. Roots of Evil: A Social History of Crime and Punishment. New York: Funk & Wagnalls, 1968.

"Kidnapping Victims: Tragic Aftermaths," Saturday Evening Post, April, 1976.

Lewis, Norman. Honored Society: A Searching Look at the Mafia. New York: Putnam, 1964.

Messick, Hank, and Burt Goldblatt. Kidnapping: The Illustrated History. New York: Dial Press, 1974.

Rinehart, J. C. "Tracing People, Lost, Stolen and Strayed Away," American City, October, 1929.

"Seven Indicted as Murderers," New York Times, July 1, 1921, p. 24.

Slattery, William, ed. Abduction: Fiction Before Fact. New York: Grove Press, 1974.

Smith, Edward H. "The Fates and Joe Varotta," in his Mysteries of the Missing. New York: Lincoln MacVeagh-The Dial Press, 1927.

Sondern, Frederick, Jr. Brotherhood of Evil: The Mafia. New York: Farrar & Rinehart, 1959.

Sutherland, Edwin H. On Analyzing Crime. Edited by Karl Schussler. Chicago: University of Chicago Press, 1972.

"Three More Taken in Verotta Case," New York Times, July 8, 1921, p. 6.

"Varotta Kidnapper Guilty of Murder," New York Times, August

20, 1921, p. 1.
"Varotta Kidnapper Sentenced to Die," New York Times, August 23, 1921, p. 17.
Winslow, Robert W. Crime in a Free Society. Encino, Calif.: Dickenson, 1968.

THE MOVIE SCANDALS AND THE HAYS OFFICE (1921-1923)

By the year 1920 the production of motion pictures for public viewing had become a multi-million dollar industry. Each week over 35 million customers attended theaters where films were shown. Pictures ranged in type from westerns through the custard pie comedies of Mack Sennett and Hal Roach to dramas with such provocative titles as "Adam's Rib," "Don't Change Your Husband," "Why Change Your Wife?," "Forbidden Fruit," "The Golden Bed," and "For Better or Worse." Many stars enjoyed wide audience acceptance and lavish salaries. In an atmosphere combining glamor and extremely hard work many of these public figures became involved in scandals which did great harm to the industry.

On September 5, 1921, Roscoe "Fatty" Arbuckle, an extremely popular comedian of his day, became involved in a sordid affair at the St. Francis Hotel in San Francisco. Virginia Rappe, a 23-year-old actress, was injured following sexual intercourse with the 350-pound comedian. The cause of her death four days later was peritonitis, following a rupture of the bladder.

Arbuckle was charged with manslaughter and stood trial three times, the first two resulting in hung juries and the third in an acquittal. The comedian's career was finished and the studio discarded his unreleased pictures, thereby incurring a great financial loss.

The dead body of bachelor William Desmond Taylor, chief director of Famous Players-Lasky Studios, was found by Henry Peavy on the morning of February 2, 1922, in the study of his Hollywood bungalow. He had been shot in the heart by an unknown assailant.

The murder investigation implicated Mary Miles Minter, Mary Pickford and Mabel Normand, all highly paid motion picture favorites. There was no evidence, however, that any of these luminaries had killed the director, but newspaper headlines carried the story for weeks. To this day the identity of Taylor's killer has never been determined, nor has a motive for the murder been found.

Other prominent motion picture celebrities succumbed to drugs, alcohol, or both. Wallace Reid, the extremely popular actor in such epics as "The Birth of a Nation," "Forever," and a series of pictures concerned with automobile racing, died on January 18, 1923, a victim of narcotics. Barbara LaMarr, "one of the real beauties of the screen in the twenties," became a drug addict and committed suicide. Olive Thomas, another actress, had died of poison in 1920. Art Acord, a western cowboy star, became an alcoholic and committed suicide.

The motion picture executives realized that the industry required policing in terms of decency, both as the term pertained to the personal lives of the performers and to the content of the pictures in which they appeared. In 1922 Will H. Hays, an American lawyer, politician, Presbyterian elder and former postmaster-general, was hired to head the Motion Picture Producers and Distributors of America. This organization, popularly known as the "Hays Office," was to police the industry and pass on the suitability of films produced for public display. Morals clauses were inserted in all contracts between studios and the actors and actresses they employed.

Hays presided as president of his Office until 1945.

Suggested Readings

Bardeché, Maurice, and Robert Brasillach. The History of Motion Pictures. New York: Norton, 1938.
Blum, Daniel C. A Pictorial History of the Silent Screen. New York: Putnam, 1953.
Blumer, Herbert. Movies, Delinquency and Crime. New York: Macmillan, 1933.
Carmen, Ira Harris. Movies, Censorship and the Law. Ann Arbor, Mich.: University of Michigan Press, 1966.
DeMille, William C. Hollywood Saga. New York: Dutton, 1939.
Fenin, George N., and William K. Everson. The Western: From Silents to Cinerama. New York: Orion Press, 1962.
Griffith, Richard, and Arthur Mayer. "The Big Clean-Up," in their The Movies. New York: Bonanza Books, 1957.
Hays, Will H. Memoirs. Garden City, N.Y.: Doubleday, 1955.
Hynd, Alan. "Murder in Hollywood," American Mercury, November, 1949.
Jacobs, Lewis. The Rise of the American Film. New York: Harcourt, Brace, 1939.
Jennings, Gary. The Movie Book. New York: Dial Press, 1963.
Moley, Raymond. The Hays Office. Indianapolis: Bobbs-Merrill, 1945.
Morris, Lloyd R. Not So Long Ago. New York: Random House, 1949.
Pickard, Roy. Companion to the Movies from 1903 to the Present Day. New York: Hippocrene Books, 1972.
Randall, Richard S. Censorship of the Movies. Madison, Wis.: University of Wisconsin Press, 1968.
Sann, Paul. The Lawless Decade. New York: Crown Publishers, 1957.
Schillaci, Anthony. Movies and Morals. Notre Dame, Ind.: Fides Publishers, 1968.
Schumach, Murray. Face on the Cutting-Room Floor: The Story of Movie and Television Censorship. New York: Morrow, 1964.
Sennett, Mack, as told to Cameron Shipp. King of Comedy.

Garden City, N. Y.: Doubleday, 1954.
Treadwell, Bill. Fifty Years of American Comedy. Hicksville, N. Y.: Exposition Press, 1951.
Vizzard, Jack. See No Evil: Life Inside a Hollywood Censor. New York: Simon & Schuster, 1970.
Wagenknecht, Edward. The Movies in the Age of Innocence. Norman, Okla.: University of Oklahoma Press, 1962.
Zukor, Adolph, with Dale Kramer. The Public Is Never Wrong. New York: Putnam, 1953.

THE MILLION DOLLAR GATE AND THE LONG COUNT (1921 and 1927)

Prizefighting has been engaged in by man since the days of the ancient Greeks. Many famous matches have been staged and one of the best-remembered modern encounters was the one between the American Jack Dempsey, then world's heavyweight champion, and Georges Carpentier, the French contender. The match was held at Boyle's Thirty Acres in Jersey City, New Jersey, on July 2, 1921. It was for the heavyweight championship of the world. Under the management of promoter Tex Rickard it produced boxing's first "million dollar gate."

The fight went four rounds, of which Carpentier won the second, and Dempsey won the other three. In the fourth round Dempsey floored Carpentier twice, the first for a count of nine and the second for a ten-count knockout.

The betting odds favored Dempsey 3 to 1. Carpentier broke his hand in the second round. The bout was seen by more than 90,000 persons, with more than 81,000 paid admissions. Gate receipts totaled $1,565,000. Of this sum Dempsey received $300,000. Carpentier received $200,000 and Rickard and his partners, as promoters, made a profit of approximately $500,000 after expenses, which included the shares going to the fighters. Carpentier was never heavyweight champion but did hold the light-heavyweight championship, winning it in 1920 and losing it in 1922.

Dempsey defended his title six times, having won it from Jess Willard on July 4, 1919 at Toledo, Ohio, when he scored a three-round knockout. As champion he scored knockouts over contenders Billy Miske, Bill Brennan, Carpentier and Luis Angel Firpo. He won a decision over Tommy Gibbons in a fifteen-round match and also won by a knockout in a non-title "elimination" fight with Jack Sharkey.

On September 23, 1926, he lost the heavyweight boxing title to Gene Tunney by decision in a ten-round match. Dempsey, having defeated Sharkey as mentioned above, fought Tunney for the title a second time. This match, staged at Soldier's Field, Chicago, on September 22, 1927, resulted in a victory for Tunney. It was then that the famous "long count" occurred, something still argued about by boxing fans. In the seventh round Dempsey delivered a quick succession of blows, knocking Tunney to the canvas. The rules by which the bout was fought required that in the event of a knockdown the other boxer was to go immediately to a neutral corner. Dave Barry, the referee, started his count over the fallen Tunney and then, realizing that Dempsey was standing over his victim, ordered him to a corner. He then took up the count over Tunney, giving the latter the advantage of an additional number of seconds above the customary ten in which to recouperate. The official timekeeper indicated that four additional seconds had been allowed but some reporters put the figure at seven.

Tunney waited until the last possible moment before rising to continue the fight. He was awarded the victory on points, winning seven of the ten rounds. He retained the title, retiring undefeated in July, 1928.

Suggested Readings

Andrist, Ralph K., ed. in charge. The American Heritage History of the 20's and 30's. New York: American Heritage, 1970.

Carpenter, Harry. Boxing: A Pictorial History. Chicago: Regnery, 1975.

________. Masters of Boxing. New York: Barnes, 1964.

Dempsey, Jack, in collaboration with Myron M. Stearns. Round by Round: An Autobiography. New York: McGraw-Hill, 1940.

Durant, John. Heavyweight Champions. New York: Hastings House, 1967.

Fleischer, Nathaniel S. Jack Dempsey. New Rochelle, N.Y.: Arlington, 1972.

Grombach, John V. The Saga of Sock: A Complete Story of Boxing. New York: Barnes, 1949.

Heimer, Melvin Leighton. Long Count. New York: Atheneum, 1969.

Heller, Pete. "In This Corner---!" New York: Simon & Schuster, 1973.

Inglis, William. "Dempsey the Dynamiter" and "Tunney, Captain of Fistic Industry," in his Champions Off Guard. New York: Vanguard Press, 1932.

Jenkins, Alan. The Twenties. New York: Universe Books, 1974.

Johnson, Alexander. Ten and Out. New York: I. Washburn, 1947.

Kearns, Jack "Doc," with Oscar Fraley. The Million Dollar Gate. New York: Macmillan, 1966.

Lardner, Rex. Legendary Champions. New York: American Heritage Press, 1972.

McCallum, Dennis. The World Heavyweight Boxing Championship. Radnor, Pa.: Chilton, 1974.

Odd, Gilbert. Boxing: The Great Champions. London: Hamlyn, 1974.

Rainbolt, Richard. Boxing's Heavyweight Champions. Minneapolis: Lerner, 1975.

Samuels, Charles. Majestic Rube: The Life and Gaudy Times of Tex Rickard. New York: McGraw-Hill, 1957.

Sann, Paul. The Lawless Decade. New York: Crown Publishers, 1957.

Schoor, Gene, with Henry Gilfond. The Jack Dempsey Story. New York: Messner, 1954.

Tunney, Gene. "My Fights With Jack Dempsey," in Leighton, Isabel, ed. The Aspirin Age, 1919-1941. New York: Simon & Schuster, 1949.

THE HALL-MILLS MURDER MYSTERY (1922 and 1926)

One of the still unsolved murder mysteries of the 20th century began on the morning of Saturday, September 16, 1922, when the bodies of the Reverend Edward Wheeler Hall and Mrs. Eleanor Mills were found under a crabapple tree on the old Phillips Farm off DeRussey's Lane near New Brunswick, New Jersey. Death had been caused by shooting, and Mrs. Mills' throat had been slashed. The bodies had been arranged by someone who placed them side by side under the tree, Hall's arm under Mrs. Mills' head. Their faces were covered, his by a panama hat, hers by a scarf. Hall's wallet and gold watch were missing. Incriminating letters from Mrs. Mills to Hall were strewn over the dead bodies and one of Hall's calling cards was propped against his shoe.

Hall had been the 41-year-old rector of fashionable St. John's Church and Mrs. Mills had been a choir singer in the same church. Both were married. The bodies were discovered by Raymond Schneider and his girl friend Pearl Bahmer who were out for a stroll. Clifford Hayes, a friend of Schneider, was later arrested and charged with the double murder but no motive for his alleged action could ever be established.

Four years later, in 1926, the case was reopened when Arthur S. Riehl filed a petition for annulment of his marriage to Louise Geist who had, at the time of the murders, been a maid in the Hall home. Riehl charged that his wife had been bribed to withhold pertinent information during the 1922 investigation. The divorce suit came to the attention of Philip Payne, managing editor of the New York Mirror, a tabloid newspaper. Payne located the calling card which had been propped against Hall's shoe and charged, in a news story published in the Mirror on July 17, 1926, that it held the fingerprint of Willie Stevens, Mrs. Hall's brother.

Willie Stevens, his brother Henry, and Mrs. Frances S. Hall were arrested, as was Henry Carpender, Mrs. Hall's cousin. The latter was held for separate trial.

The trial of Mrs. Hall and the Stevens brothers was one of the most sensational events of the decade. It was presided over by Justice Charles W. Parker of the New Jersey Supreme Court, with Judge Frank L. Cleary of the Somerset, New Jersey, court sitting. The special prosecutor was State Senator Alexander Simpson, and Robert M. McCarter and Clarence E. Case acted as defense counsel.

More than two hundred witnesses testified. Many of these said that they had been bribed to withhold testimony at the original investigation. Amorous letters written by Hall to Mrs. Mills were introduced as evidence. Mrs. Hall stated under oath that she had no knowledge of her husband's affair with the choir singer. All defendants pleaded "not guilty."

The state produced an eye-witness to the murder. This was Mrs. Jane Gibson, a pig-farmer who became known as the "Pig Woman." She was extremely ill and was carried into court from the hospital on a stretcher and gave her testimony from a bed which had been provided for her use.

Mrs. Gibson testified that, on the night of September 15, 1922, she had ridden her mule Jenny to DeRussey's Lane in pursuit of persons in a wagon who she believed were stealing corn from her cornfield. She said that she saw Willie and Henry Stevens and Mrs. Hall in the lane and heard the sound of shots.

On cross-examination Defense Counsel Case brought out a number of discrepancies in the Pig Woman's testimony. At the preliminary hearing she had identified Henry Carpender as being present in DeRussey's Lane. At the trial she identified, not Carpender, but Henry Stevens. Her testimony was thoroughly discredited.

The Stevens brothers, Carpender, and Mrs. Hall all testified for the defense and stood up well under the prosecutor's cross-examination.

After 29 days of courtroom dramatics the jury brought in the verdict of "not guilty." Henry Carpender was not tried. Mrs. Hall and her brothers brought a libel suit against the New York Mirror which was settled out of court for a reputed $50,000.

The Hall-Mills murder mystery still remains unsolved.

Suggested Readings

Andrist, Ralph K., ed. in charge. The American Heritage History of the 20's and 30's. New York: American Heritage, 1970.

Bliven, Bruce. "The Hall-Mills Case," New Republic, December 1, 1926.

Busch, Francis X. They Escaped the Hangman. Indianapolis: Bobbs-Merrill, 1953.

Jenkins, Alan. The Twenties. New York: Universe Books, 1974.

Kunstler, William M. The Minister and the Choir Singer: The Hall-Mills Murder Case. New York: Morrow, 1964.

________. "Murder for Revenge: The State of New Jersey versus Frances Hall, Henry Stevens and William Stevens," in his First Degree. New York: Oceana Publications, 1960.

Morris, Richard B. "The Clergyman, The Choir Singer and the Pigwoman," in his Fair Trial. New York: Knopf, 1952.

Pearson, Edmund. "Five Hours in Court: At the Hall-Mills Trial," Outlook, December 15, 1926.

Roberts, Carl E. B. The New World of Crime. London: Noble, 1933.

Runyon, Damon. Trials and Other Tribulations. Philadelphia: Lippincott, 1947.

Sann, Paul. The Lawless Decade. New York: Crown Publishers, 1957.

Wilson, Colin, and Patricia Pitman. Encyclopedia of Murder.

New York: Putnam, 1962.
Woollcott, Alexander. "Five Classic Crimes," in his Long, Long Ago. New York: Viking Press, 1943.

THE TEAPOT DOME SCANDAL (1923)

A number of political scandals occurred during the administration of President Warren G. Harding, the most notorious of these being the one popularly called "Teapot Dome."

Proven naval oil reserves in Wyoming (Teapot Dome) and California (Elk Hills) were transferred to the jurisdiction of the Department of the Interior by Presidential executive order in 1921, at the urging of Edwin Denby, Secretary of the Navy, and Albert B. Fall, Secretary of the Interior.

In 1922 Fall leased the Teapot Dome reserves to Harry F. Sinclair, president of the Mammoth Oil Company and also leased the Elk Hills reserves to Edward L. Doheney of the Pan American Oil Company. These leases were negotiated secretly and without competitive bidding.

In 1923 Senator Tom Walsh of Montana, following investigation by the Public Lands Committee of the Senate, disclosed that Fall had been bribed to negotiate these leases with at least $100,000 from Doheney and had also received $300,000 from Sinclair. Fall resigned his cabinet position that same year.

In 1924 Fall, Doheny, and Sinclair were indicted for conspiracy, and Fall was also indicted for accepting bribes. In 1926 Fall and Doheny were acquitted of the conspiracy charge but Fall, in 1929, was tried and found guilty of accepting bribes, and was imprisoned. In 1931 he appealed his case, but the conviction was upheld and a subsequent appeal to the Supreme Court was refused by that Court.

Suggested Readings

Adams, Samuel Hopkins. The Incredible Era: The Life and Times of Warren Gamaliel Harding. Boston: Houghton Mifflin, 1939.
________. "The Timely Death of President Harding," in Leighton, Isabel, ed. The Aspirin Age, 1919-1941. New York: Simon & Schuster, 1949.
Allen, Frederick Lewis. Only Yesterday. New York: Harper, 1931.
Andrist, Ralph K., ed. in charge. The American Heritage History of the 20's and 30's. New York: American Heritage, 1970.
Bates, J. Leonard. The Origins of Teapot Dome. Urbana, Ill.: University of Illinois Press, 1963.
________. "The Teapot Dome Scandal and the Election of 1924," American Historical Review, January, 1955.
Beard, Charles A., and Mary R. Beard. The Rise of American Civilization. Gloucester, Mass.: Peter Smith, 1966.

Bliven, Bruce. "Tempest Over Teapot," American Heritage Magazine, August, 1965.
Commager, Henry Steele, ed. "Teapot Dome," (Doc. No. 452) in his Documents of American History, 8th edition. New York: Appleton, 1968.
Jenkins, Alan. The Twenties. New York: Universe Books, 1974.
McCoy, Donald R. Calvin Coolidge: The Quiet President. New York: Macmillan, 1967.
Noggle, Burl. Teapot Dome: Oil and Politics in the 1920's. Baton Rouge, La.: Louisiana State University Press, 1962.
Ravage, Marcus E. The Story of Teapot Dome. New York: Burt Franklin, 1947.
Russell, Francis. The Shadow of Blooming Grove: Warren G. Harding and His Times. New York: McGraw-Hill, 1968.
Sann, Paul. The Lawless Decade. New York: Crown Publishers, 1957.
Sinclair, Andrew. The Available Man: The Life Behind the Masks of Warren Gamaliel Harding. New York: Macmillan, 1965.
Werner, M. R., and John Starr. Teapot Dome. New York: Viking Press, 1959.
White, William Allen. Masks in a Pageant. Saint Clair Shores, Mich.: Scholarly Press, 1930.
________. A Puritan in Babylon: The Story of Calvin Coolidge. New York: Macmillan, 1938.

THE "BROADWAY BUTTERFLY" MURDER (1923)

Dorothy Keenan, also known as "Dot King" and the "Broadway Butterfly," was the 27-year-old mistress of a "Mr. Marshall." She was murdered in her New York apartment at 144 West 57th Street in March, 1923. Who killed her and why she was killed have never been determined.

Dot King, as she preferred to be called, was the second eldest of four children. She had been married at eighteen and divorced shortly before her death. She and her husband had separated less than two years after their marriage and never saw each other again. Her father had cut her out of his will and expelled her from his home.

As a professional dress model and later, as a nightclub hostess, Dot King met many men, and was not adverse to accepting presents from them in return for her favors. In 1921 she met "Mr. Marshall" in a nightclub and became his mistress. He established her in the West 57th Street apartment. When he visited her he was always accompanied by a man named Wilson, who "Mr. Marshall" said was his secretary.

"Mr. Marshall" was frequently out of town on business and Dot King met a small-time criminal named Albert Guimares, who also used the name "Al Morris." "Mr. Marshall" maintained the apartment, paid the bills, and gave Dot King expensive presents. Guimares, on the other hand, often abused her, in spite of which she gave him presents of jewelry and tailormade clothes.

On March 16, 1923, Ella Bradford, Dot King's Negro maid, arrived at the apartment around noon. She found her employer lying dead in her bed. The maid notified the police, who found a bottle which had once contained chloroform and also a wad of cotton in the bed. The apartment had been searched and jewels and a number of expensive gowns were missing.

Dot King's death was first considered suicide or, possibly, accidental, as the inhaling of chloroform had become a substitute for narcotics and alcohol when these were unobtainable. It was reasoned that the girl was not alone when she died, and that her companion had stolen her belongings.

The medical examiner declared, after the autopsy, that Dot King had been murdered. This was substantiated by scratches on her face and the position of her left arm which was twisted behind her back. The chloroform bottle, when found, was not corked. Also, the serial number on the label of the bottle had been scratched off.

A further search of the apartment disclosed a man's small black pocket comb which was found in the bed and which had presumably fallen from the assailant's pocket. More cotton and a man's umbrella were found in an urn. Other items, including a pair of pajamas, a woman's handbag, and a pair of men's kid gloves were discovered under the bed, as though they had been shoved out of sight. These, as clues, led nowhere.

The girl's mother testified that she had had jewelry valued at $20,000 in the apartment and that she had urged her daughter to keep it in a safer place. The mother knew of "Mr. Marshall" and Guimares and told the police of them. She said that her daughter had broken off her relationship with Guimares some three months before and that he had threatened her. When located, Guimares stated that he had seen the girl following the breakup but denied quarreling with her. The police were unable to hold him as a suspect but did arrest him for carrying a concealed weapon.

"Mr. Marshall," through his lawyer, offered to tell his story to Acting District Attorney Ferdinand Pecora. When interviewed he stated that he had, with Wilson, gone to Dot King's apartment at 11:30 on Wednesday, that Wilson had left, and that he remained until 1:55 A.M. when he took a taxi to his hotel where he arrived about 2:10. His arrival time at the hotel, as stated, was checked and found true. The district attorney then issued a statement exonerating "Mr. Marshall" and Wilson from any connection with the murder.

Guimares was released on bail. "Mr. Marshall" was identified as John Kearsley Mitchell, a Philadelphia financier and millionaire.

The police discovered that a number of men besides Mitchell and Guimares had keys to Dot King's apartment, and that she had one hanging in the elevator "just in case." It would have been entirely possible for anyone to enter and leave the building undetected by using the stairs.

District Attorney Pecora received a letter concerning a blackmail attempt against Mitchell. The blackmailer, the letter stated, hoped to enlist Dot King in the extortion plot.

Other tips were received. These involved Mitchell and Guimares and other men as well. Many of the letters received were anonymous. The police investigated these leads but were never able to connect anyone with the crime. The identity of whoever murdered the "Broadway Butterfly" is still unknown.

Suggested Readings

Abrahamsen, David, M.D. The Murdering Mind. New York: Harper & Row, 1973.

"America's Murder Record," Outlook, July 11, 1923.

Collins, Ted. "Dot King," in his New York Murders. New York: Duell, Sloan, 1944.

"Comb is Chief Clue to Man Who Killed and Robbed Model," New York Times, March 17, 1923, p. 1.

"Find Model Dead, Her Jewels Gone," New York Times, March 16, 1923, p. 3.

"Our Appalling Murder Record," American City, January, 1923.

Reinhardt, James Melvin. The Psychology of Strange Killers. Springfield, Ill.: Thomas, 1962.

Sann, Paul. The Lawless Decade. New York: Crown Publishers, 1957.

"Says Model's Slayer Wanted His Letters," New York Times, March 18, 1923, p. 1.

Sterling, Hank, pseud. "Diamonds, Pretty Pink Toes and Homicide," in his Ten Perfect Crimes. New York: Stravon Publishers, 1954.

THE DAWES PLAN (1924)

Following World War I the matter of reparations to be paid the Allied Powers by Germany for damages to civilian population and property came up for consideration. In May, 1921, the amount was fixed by an Allied reparations commission at 100 million pounds per year plus 26 per cent of Germany's exports. Germany protested that she could not make these payments, and France declared her intention of holding the Ruhr Valley, occupied by her, until the German payments were made.

The resulting disagreement led to the commission appointing two committees to make recommendations for settling the question. General Charles Gates Dawes was president of one of these committees and the reparations plan made by this committee took its name--the "Dawes Plan"--from him.

The Dawes Plan stipulated that Germany was to pay the sum of 100 million gold marks the first year, 1,220 million the second year, a like sum the third year, 1,750 million the fourth year, and 2,500 million each year thereafter. The Plan also made recommendations concerning fiscal control of the German currency and the organization of an international control of reparations.

On April 11, 1924, the report of the committee was adopted by the reparations committee and the governments concurred in the plan on August 30. It went into effect shortly after being accepted by the German Reichstag.

The Dawes Plan, however, did not settle the problem of reparations. It stated neither the total amount to be paid by Germany nor the number of years the payments were to be made. In 1929 it was superseded by the Young Plan, named after Owen D. Young who had been associated with the original Dawes Plan and who, as chairman of a committee of experts representing Belgium, France, Great Britain, Japan, and Germany, met with them in Geneva to revise the Dawes Plan.

The Young Plan became effective in 1930. It was intended to reduce the annual payments to sums that Germany could pay, and to establish the total number of payments to be made. In 1931 a moratorium on payments was declared and since then no further payments have been received.

Suggested Readings

Baruch, Bernard M. The Making of the Reparation and Economic Sections of the Treaty. New York: Howard Fertig, 1970.
Brandes, Joseph. Herbert Hoover and Economic Diplomacy: Department of Commerce Policy, 1921-1928. Pittsburgh: University of Pittsburgh Press, 1962.
Dawes, Charles G. A Journal of Reparations. London: Macmillan, 1939.
Duroselle, Jean-Baptiste. From Wilson to Roosevelt: Foreign Policy of the United States, 1913-1945. Cambridge, Mass.: Harvard University Press, 1963.
Feis, Herbert. The Dipolomacy of the Dollar, 1919-1932. Baltimore: Johns Hopkins University Press, 1950.
Hoover, Herbert Clark, and Hugh Gibson. The Problems of Lasting Peace. Garden City, N.Y.: Doubleday, 1943.
Keynes, John Maynard. The Economic Consequences of the Peace. New York: Harper & Row, 1971.
________. A Revision of the Treaty. Plainview, N.Y.: Books for Libraries, 1973.
Leach, Paul R. That Man Dawes. Chicago: University of Chicago Press, 1953.
Moulton, Harold G., and Leo Pasvolsky. War Debts and World Prosperity. Washington, D.C.: Brookings Institution, 1932.
Perkins, Dexter. Charles Evans Hughes and American Democratic Statesmanship. Boston: Little, Brown, 1956.
Pusey, Merlo John. Charles Evans Hughes. New York: Macmillan, 1951.
Soule, George. Prosperity Decade: From War to Depression, 1917-1929. New York: Holt, Rinehart, 1947.
Szladits, Lola L. Owen D. Young, Book Collector. New York: New York Public Library, 1974.
Williams, Benjamin H. Economic Foreign Policy in the United States. New York: Macmillan, 1929.

THE PAUL WHITEMAN SYMPHONIC JAZZ CONCERT (1924)

Until the five-piece Original Dixieland Jazz Band, under the leadership of cornetist Nick La Rocca, opened at Reisenweber's Cabaret in New York City on January 26, 1917, jazz was considered the music of the honky-tonk, the brothel, and the cheap dance hall. Suddenly it became "respectable" and the "upper classes," having previously limited their dancing largely to the sedate waltz, took to such other dances as the turkey trot, the bunny hug, the maxixe and, eventually, the fox trot, the varsity drag, the rhumba and the twist.

Following the success of the La Rocca orchestra, other such orchestras, playing music intended for dancers, were organized and led by such pioneers as Art Hickman, Jan Garber, Vincent Lopez and Meyer Davis. These orchestras played in the ballrooms of the better hotels and at private society parties.

One of the orchestra leaders who helped make jazz "respectable" was Paul Whiteman, a former symphony violinist, who organized his first musical aggregation in San Francisco in 1919. His idea was to play scored arrangements of light classics with a rhythm to which people could dance.

On February 12, 1924, Whiteman presented his symphonic jazz concert at Aeolian Hall in New York. As the concert took place on Lincoln's birthday, it was also known as the "Emancipation Proclamation of Jazz." Such musical luminaries as Walter Damrosch, Mischa Elman, Leopold Godowsky, Jascha Heifetz, Fritz Kreisler, John McCormack, Sergei Rachmaninoff, John Philip Sousa, Moriz Rosenthal, Leopold Stokowski, and Igor Stravinsky were present.

Whiteman's purpose in presenting his concert was, in his own words, "to show these skeptical people the advance which had been made in popular music from the day of discordant early jazz to the melodious form of the present."

The opening selection was an intentionally "corny" rendition of "Livery Stable Blues." This was intended to show just how crude such music had been "before Whiteman." The program included 26 selections by such accepted composers as Sir Edward Elgar, Rudolf Friml, and Edward MacDowell, "adapted to dance music." Victor Herbert composed a special suite for the occasion. The highlight of the evening was George Gershwin's "Rhapsody in Blue," arranged for piano and orchestra by Ferde Grofé, and with the composer at the piano. The extremely difficult clarinet glissando opening this selection was played by Ross Gorman.

Whiteman estimated that he lost $7,000 on the concert but recouped this sum many times over from the publicity it generated. He became known as the "King of Jazz" and advanced the cause of this type of music immeasurably. Other large dance orchestras were organized and found bookings easier to get than before the Whiteman concert. Also, the intelligentsia began to consider jazz seriously.

Paul Whiteman and the orchestras he directed remained dominant in the "sweet jazz" field until his retirement as a conductor in the 1940's. In 1943 he became musical director for a large broadcasting company. He died in 1967.

Suggested Readings

Bakeless, Katherine. Story-Lives of American Composers. Philadelphia: Lippincott, 1953.
Berger, Melvin. Masters of Modern Music. New York: Lothrop, 1970.
Ewen, David. A Journey to Greatness: The Life and Music of George Gershwin. New York: Holt, 1956.
_______. Men of Popular Music. New York: Prentice-Hall, 1949.
_______. The Story of George Gershwin. New York: Holt, 1943.
_______, ed. and comp. American Composers Today. New York: Wilson, 1949.
Gilman, Lawrence. "Music," New York Tribune, February 13, 1924.
Goldberg, Isaac. George Gershwin: A Study in American Music. New York: Simon & Schuster, 1931.
Jenkins, Alan. The Twenties. New York: Universe Books, 1974.
Keepnews, Orrin, and Bill Grauer, Jr. A Pictorial History of Jazz. New York: Crown Publishers, 1971.
Stearns, Marshall W. The Story of Jazz. New York: Oxford University Press, 1956.
Ulanov, Barry. The History of Jazz in America. New York: Da Capo Press, 1972.
Vedey, Julian, pseud. Band Leaders. London: Rockliff, 1950.
Whiteman, Paul, and Mary Margaret McBride. Jazz. New York: J. H. Sears, 1926.

THE LOEB-LEOPOLD MURDER CASE (1924)

In the spring of 1924 Richard A. Loeb, eighteen, and Nathan F. Leopold, Jr., nineteen, kidnaped and murdered fourteen-year-old Bobby Franks, the son of a neighbor. Loeb and Leopold were brilliant students, members of wealthy Chicago families, and had attempted to commit the "perfect crime" for reasons of ego satisfaction rather than for pecuniary gain.

The kidnap victim, chosen at random, was lured into an automobile and murdered. His body was stuffed into a rural culvert. Ransom demands were made to the parents of the victim, but the plot failed as Leopold had dropped his glasses near the dead body of the Franks boy and these glasses, found by the authorities, were traced to him.

A charge of kidnaping and murder was brought against the two boys, and their families engaged Clarence Darrow, a prominent criminal lawyer, to defend them.

The trial was held in the summer of 1924 in Chicago, with Judge John R. Caverly hearing the case without a jury. Darrow pleaded his clients guilty and on August 22 began an impassioned twelve-hour plea spread over three days, in which he put forth the hypothesis that his clients, as "victims of wealth," had deliberately planned a cold-blooded murder. This murder, he stated, "was the senseless, terrible act of immature and diseased brains, which were the result of conditions of heredity and environment over which [his clients] had no control."

Three weeks later the court reconvened to hear Judge Caverly's decision. Loeb and Leopold were sentenced to life imprisonment despite the fact that the public had been clamoring for the death sentence. Loeb was killed by a fellow prisoner at Joliet in 1936. Leopold was released from prison in 1963 and died in 1971.

Suggested Readings

Andrist, Ralph K., ed. in charge. The American Heritage History of the 20's and 30's. New York: American Heritage, 1970.

Busch, Francis X. Prisoners at the Bar. Indianapolis: Bobbs-Merrill, 1952.

Crandall, Allen. The Man from Kinsman. Sterling, Colo.: Published by the author, 1965.

Darrow, Clarence. The Story of My Life. New York: Scribner's, 1932.

Gurko, Miriam. Clarence Darrow. New York: Crowell, 1965.

Harrison, Charles Yale. Clarence Darrow. New York: Jonathan Cape and Harrison Smith, 1931.

Higdon, Hal. The Crime of the Century. New York: Putnam, 1975.

Hynd, Alan. "The Case of the Curious Cubs," in his Murder, Mayhem and Mystery. New York: A. S. Barnes, 1958.

"'Intellectual' Murder in Chicago," Literary Digest, July 5, 1924.

Jenkins, Alan. The Twenties. New York: Universe Books, 1974.

Kunstler, William H. "Murder for Thrills: The State of Illinois versus Richard A. Loeb and Nathan F. Leopold, Jr.," in his First Degree. New York: Oceana Publications, 1960.

Leopold, Nathan F., Jr. Life Plus 99 Years. Garden City, N.Y.: Doubleday, 1958.

Messick, Hank, and Burt Goldblatt. Kidnapping: The Illustrated History. New York: Dial Press, 1974.

Sann, Paul. The Lawless Decade. New York: Crown Publishers, 1957.

Stone, Irving. Clarence Darrow for the Defense. Garden City, N.Y.: Doubleday, 1941.

Symons, Julian. A Pictorial History of Crime. New York: Crown Publishers, 1966.

Weinberg, Arthur, ed. Attorney for the Damned. New York: Simon & Schuster, 1957.

GITLOW VS. PEOPLE OF NEW YORK (1925)

The Supreme Court case, Gitlow vs. People of New York, decided in 1925, concerned the matter of freedom of speech. This freedom is guaranteed by the First Amendment to the Constitution of the United States and by state constitutions, but the question of just how far that right may be used is "one of the most perplexing that the courts have to face." In the case of U.S. vs. Schenck, ruled on by the Supreme Court in 1919, Justice Oliver Wendell Holmes stated, "the most stringent protection of free speech would not protect a man in falsely shouting fire in a theater and causing a panic. It does not even protect a man from an injunction against uttering words that may have all the effect of force."

The case of Gitlow vs. People of New York involved the validity of the New York Criminal Anarchy Act of 1902, and was appealed from the Supreme Court of the State of New York.

Benjamin Gitlow and three others had been indicted, tried, and found guilty on two counts of statutory criminal anarchy. Specifically, he and his associates had, in violation of the Anarchy Act of 1902, "advocated, advised, and taught the duty, necessity, and propriety of overthrowing and overturning organized government by force, violence, and unlawful means, by certain writings therein set forth, entitled The Left Wing Manifesto ..." and, on a second count, "had printed, published, and knowingly circulated and distributed a certain paper called The Revolutionary Age, containing the writings set forth in the first count, advocating, advising, and teaching the doctrine that organized government should be overthrown by force, violence, and unlawful means...."

The defendants argued that as "there was no evidence of any concrete result flowing from the publication of the Manifesto, or of circumstances showing the likelihood of such result...," the Anarchy Act contravened the due process clause of the Fourteenth Amendment, including the "liberty of speech and of the press." The question to be decided by the Supreme Court was whether or not the statute deprived Gitlow "of his liberty of expression, in violation of the due process clause of the Fourteenth Amendment."

The Court found that the Manifesto did in fact advocate "mass strikes and revolutionary mass action" to "progressively foment industrial disturbances" and "overthrow and destroy organized parliamentary government." It held that the Fourteenth Amendment "does not confer an absolute right to speak or publish without responsibility whatever one may choose..." and that a state "may punish those who so abuse this freedom by utterances inimical to the public welfare...." It held further that the Fourteenth Amendment "does not protect publications or teachings which tend to subvert or imperil the government ..." or "publications prompting the overthrow of the government by force."

Justice Edward T. Sanford, rendering the Court's decision, upheld the Anarchy Act as constitutional on the grounds that the state, in taking measures it considers necessary to protect the public peace and safety, is not acting in an arbitrary manner.

Justices Oliver Wendell Holmes and Louis D. Brandeis dissented, with Holmes reading the dissenting opinion. These two jurists held that only publication of an idea was alleged and that the state was in no danger as a result of such publication. Holmes referred to the 1919 Supreme Court decision in the case of U. S. vs. Schenck and observed that if the Manifesto, as alleged, was an incitement rather than a theory, so is every idea an incitement.

Suggested Readings

Bent, Silas. Justice Oliver Wendell Holmes: A Biography. New York: AMS Press, 1969.

Bowen, Catherine Drinker. Yankee from Olympus: Justice Holmes and His Family. Boston: Little, Brown, 1944.

Carson, Hampton L. The History of the Supreme Court. New York: Burt Franklin, 1972.

Chafee, Zechariah, Jr. Free Speech in the United States. Cambridge, Mass.: Harvard University Press, 1941.

Commager, Henry Steele, ed. "Abrams v. United States," (Doc. No. 427) in his Documents of American History, 8th edition. New York: Appleton, 1968.

_______. "The Constitution of the United States," (Doc. No. 87) in his Documents of American History, 8th edition. New York: Appleton, 1968.

_______. "Gitlow v. People of New York," (Doc. No. 456) in his Documents of American History, 8th edition. New York: Appleton, 1968.

_______. "Schenck v. United States," (Doc. No. 426) in his Documents of American History, 8th edition. New York: Appleton, 1968.

Frankfurter, Felix. Mr. Justice Holmes and the Supreme Court. New York: Atheneum, 1965.

James, Leonard F. The Supreme Court in American Life. Chicago: Scott, Foresman, 1964.

Johnson, Gerald White. The Supreme Court. New York: Morrow, 1962.

McCloskey, Robert G. The American Supreme Court. Chicago: University of Chicago Press, 1960.

Warren, Charles. The Supreme Court in United States History. Boston: Little, Brown, 1923.

THE DEATH OF FLOYD COLLINS (1925)

In February, 1925, a young Kentuckian named Floyd Collins was exploring an underground cave at Sand Cave, a few miles from Mammoth Cave. An experienced spelunker, he was hoping to find a tourist attraction to compete with the other limestone formations in the area. A loosened boulder some 125 feet from the surface pinned him in the passage he was exploring. He never got out alive.

Tragic as it was, news of the episode might never have been heard other than locally had not W. B. Miller, a reporter on the staff of the Louisville Courier-Journal, become aware of Collins' plight. He crawled down the slippery underground passageway and interviewed the trapped man. He described the efforts of the rescue party in vivid journalistic prose which overnight became page 1 material in virtually every newspaper in the country.

The dramatic situation attracted great crowds of people to the cave's mouth. A "city" of over a hundred tents sprang up and state troops and barbed wire barriers were used to restrain the curious. The rescue parties worked in shifts around the clock, trying to free Collins without causing a cave-in. They were hampered by the small size of the passage in which the Kentuckian was imprisoned and were forced to pass back loosened dirt and rocks hand by hand.

The rescue efforts were unavailing. On February 26, 1925, Floyd Collins died, still a prisoner in his cave. The following day the New York Times announced in screaming headlines:

> FIND FLOYD COLLINS DEAD IN CAVE
> TRAP ON 18TH DAY; LIFELESS AT LEAST
> 24 HOURS; FOOT MUST BE AMPUTATED
> TO GET BODY OUT

In terms of tragedies, Collins' death was a minor incident. However, it had great dramatic news value and gripped the imaginations of millions of readers. A composer of folk music wrote the song, "The Death of Floyd Collins," which achieved a degree of popularity in country music circles.

Suggested Readings

Allen, Frederick Lewis. Only Yesterday. New York: Harper, 1931.

Andrist, Ralph K., ed. in charge. The American Heritage History of the 20's and 30's. New York: American Heritage, 1970.

Bent, Silas. Ballyhoo, The Voice of the Press. New York: Boni & Liveright, 1927.

Burman, B. L. "Kentucky's Crazy Cave War," Collier's, June 6, 1953.

"Collins Dead in Cave; Rock Holds Body Fast," New York Daily Mirror, February 27, 1925, p. 1.

"Discovery Down Deep," Newsweek, January 9, 1956.

"Find Floyd Collins Dead in Cave Trap on 18th Day; Lifeless at Least 24 Hours; Foot Must Be Amputated To Get Body Out," New York Times, February 27, 1925, p. 1.

"The Floyd Collins Tragedy," Literary Digest, February 28, 1925.

Miller, W. B. "Our Fight to Save Floyd Collins," Reader's Digest, April, 1960.

Russell, O. D. "Floyd Collins in the Sand Cave," American Mercury, November, 1937.

Sann, Paul. The Lawless Decade. New York: Crown Publishers, 1957.

THE SCOPES "MONKEY TRIAL" (1925)

In the year 1925 the sovereign state of Tennessee passed a law which prohibited the teaching of any theory that denied the story of Genesis in the Bible and taught instead that man had descended from a lower order of animals. The American Civil Liberties Union decided to finance a test case to determine this anti-evolution law's constitutionality.

John Thomas Scopes, a biology instructor in the Rhea County High School at Dayton, Tennessee, agreed, at the urging of George Rappelyea, a mining engineer, to teach the forbidden doctrine and to permit himself to be arrested and tried.

William Jennings Bryan, a fundamentalist, lawyer, and politician, offered his services as a member of the prosecution. Clarence Darrow, the foremost criminal lawyer of his day, immediately volunteered his services, without charge, for the defense. Both offers were accepted. Darrow was assisted by Arthur Garfield Hays and Dudley Field Malone, both prominent attorneys.

The trial was one of the most sensational of the decade. The presiding judge, John T. Raulston, and the Tennessee citizens were strongly prejudiced against Darrow and favored Bryan. When the trial started on July 10, 1925, it became obvious that the issue was not the guilt or innocence of Scopes but the anti-evolution law itself. Essentially, the defense wished to "stress the importance of removing legal restrictions on the freedom to think and teach."

The highlight of the trial came on July 20 when Bryan accepted the invitation of the defense to take the stand and testify as an expert on the Bible. Through the swelteringly hot July afternoon Darrow attacked Bryan's religious beliefs, insisting on precise answers to each question. Eventually Bryan, after an exceedingly uncomfortable session in the witness chair, contradicted himself and Judge Raulston adjourned the court for the day. Bryan, despite his wish, did not have the opportunity to cross-examine Darrow, who requested the judge to submit the case to the jury with instructions to find Scopes guilty.

Scopes was convicted and fined a nominal $100. The conviction was reversed on technical grounds by the Tennessee Supreme Court.

Suggested Readings

Allen, Frederick Lewis. Only Yesterday. New York: Harper, 1931.

Allen, Leslie C., ed. Bryan and Darrow at Dayton: The Record and Documents of the "Bible-Evolution Trial." New York: Arthur Lee, 1925.

Aymar, Brandt, and Edward Sagarin. "John Thomas Scopes," in their A Pictorial History of the World's Great Trials. New York: Crown Publishers, 1967.

Coletta, Paolo E. William Jennings Bryan. Lincoln, Neb.: University of Nebraska Press, 1969.

Crandall, Allen. The Man from Kinsman. Sterling, Colo.: Published by the author, 1965.
Darrow, Clarence. The Story of My Life. New York: Scribner's, 1932.
Ginger, Ray. Six Days or Forever? Tennessee v. John Thomas Scopes. Boston: Beacon Press, 1958.
Grebstein, Sheldon N., ed. Monkey Trial. Boston: Houghton Mifflin, 1960.
Gurko, Miriam. Clarence Darrow. New York: Crowell, 1965.
Haldeman-Julius, Anna Marcet. Clarence Darrow's Two Great Trials. Girard, Kans.: Haldeman-Julius, 1927.
Harrison, Charles Yale. Clarence Darrow. New York: Jonathan Cape and Harrison Smith, 1931.
Hays, Arthur Garfield. Let Freedom Ring. New York: Liveright, 1937.
_______. Trial by Prejudice. New York: Covici, Friede, 1933.
Jenkins, Alan. The Twenties. New York: Universe Books, 1974.
Lawrence, Jerome, and Robert E. Lee. Inherit the Wind (play). New York: Random House, 1955.
_______. "Inherit the Wind" (play condensation), in Mantle, Burns, ed. The Best Plays of 1954-1955. New York: Dodd, Mead, 1955.
Levine, Lawrence W. Defender of the Faith: William Jennings Bryan: The Last Decade, 1915-25. New York: Oxford University Press, 1965.
Michelson, Charles. "Darrow vs. Bryan," in Hutchens, John K., and George Oppenheimer. The Best in the World. New York: Viking Press, 1973.
Sann, Paul. The Lawless Decade. New York: Crown Publishers, 1957.
Scopes, John Thomas, and James Pressley. Center of the Storm: Memoirs of John T. Scopes. New York: Holt, Rinehart, 1967.
Stone, Irving. Clarence Darrow for the Defense. New York: Doubleday, 1941.
Tompkins, Jerry R., ed. D-Days at Dayton. Baton Rouge, La.: Louisiana State University Press, 1965.
Weinberg, Arthur, ed. Attorney for the Damned. New York: Simon & Schuster, 1957.
Werner, M. R. Bryan. New York: Harcourt, Brace, 1929.
The World's Most Famous Court Trial: Tennessee Evolution Case. Cincinnati: National Book Co., 1925.

THE COURTMARTIAL OF BILLY MITCHELL (1925)

William (Billy) Mitchell was a career soldier. He had enlisted in the army as a private in 1898 and had been promoted through the ranks to brigadier general, a rank seldom attained by one who is not a graduate of the military academy at West Point or some other "approved" academy.

In World War I Mitchell, as a combat aviator, had gained practical experience with military aircraft. He became convinced that the future of national safety and defense lay in the speedy development of an adequate air force rather than in the outmoded battleships favored by the general staff and high-ranking army and navy officers.

Mitchell and his colleagues, in 1921, demonstrated the effectiveness of aerial bombing by sinking several obsolete vessels, including four former German warships and the reputedly unsinkable German battleship "Ostfriedsland" off the coast of Virginia. This demonstration, though successful, alienated many old-time army and navy officers who had not been trained in air warfare.

For his struggle with what he considered the "hidebound brass," Mitchell had been demoted from brigadier general to colonel, and had been transferred to San Antonio, Texas, where he held the insignificant post of air officer with the Eighth Corps.

On September 3, 1925, the heavy dirigible "Shenandoah," while on a trip from Lakehurst, New Jersey to St. Paul, Minnesota, was torn to pieces in a thunderstorm over Ava, Ohio. Lieutenant Commander Zachary Landsdowne, captain of the dirigible, and 13 members of his crew of 28 were killed. Lansdowne's wife immediately charged "politics" as responsible for her husband's death. A board of inquiry was appointed to investigate the tragedy. Mitchell, speaking to a group of reporters in San Antonio, accused his superiors of "incompetency, criminal negligence, and almost treasonable administration by the War and Navy Departments." This interview, published in the New York Times, carried the subhead, "Expects Arrest Monday."

Mitchell was formally charged under the 96th Article of War, which prohibits "all disorders and neglects to the prejudice of good order and military discipline [and] all conduct of a nature to bring discredit upon the military service..." and states that these "shall be taken cognizance of by a ... courtmartial and punished by the discretion of such court." Essentially, the prosecution's case was based upon what Mitchell had said, and his defense, in turn, based its case on the truth of his statements.

The Mitchell courtmartial opened on October 28, 1925, in Washington, D.C. The court, headed by Major General Charles P. Summerall, consisted of five other major generals, including Douglas MacArthur, six brigadier generals, and three officers of lesser rank as law members and judges advocate.

Fiorello La Guardia, then a congressman and a former flyer, referred to the court as "a pack of beribboned dog robbers of the general staff."

Mitchell was defended by Representative Frank R. Reid of Illinois. Before the trial got under way Generals Summerall, F. W. Sladen and Albert J. Bowley were ousted on legal grounds, and Major General Robert L. Howze replaced Summerall as president of the court.

Eight specific charges were made against the defendant, each one based on particular remarks and accusations made by him against the navy and war departments, including his tirade after the "Shenandoah" tragedy. Reid countered by granting that Mitchell had indeed made the statements as charged, but held that the law of libel, a punishable offense, did not apply. His client, he maintained, had attacked a system only, and that it was impossible to libel the war department, which was an intangible body. He pointed out that President Calvin Coolidge, on June 3, 1925, had endorsed the right of officers to express their views before their fellow citizens.

The trial went on, with Reid interjecting various technicalities into the hearings. On November 9 the defense read into the record a total of 66 accusations made by Mitchell and announced it would prove the truth and validity of each. Many witnesses were called by both sides, to be asked and to answer involved technical and military questions. Major Carl Spaatz testified that most of the air service was "either obsolescent or obsolete." Major H. H. Arnold attacked the use of out-of-date planes and produced a long list of casualties directly attributable to the use of such aircraft.

Other witnesses included Mrs. Zachary Lansdowne, who testified that she had been asked by Captain Paul Foley of the Naval Board of Inquiry to tell a false story on the witness stand. The prosecution's attempt to have her testimony stricken from the record was overruled.

The star witness for the defense was retired Admiral William S. Sims. In 1920 he had submitted a lengthy report charging that serious errors had been made by the U. S. Navy Department in the management of naval operations during World War I. Shattering navy tradition with every word, this graduate of the naval academy at Annapolis bluntly stated that "any invading fleet could be destroyed by a properly organized land-based air force." He remarked that most admirals were uneducated, and that the battleship, as a weapon of war, was obsolete and had been replaced by the aircraft carrier.

After the prosecution had completed its case on December 17, Mitchell, speaking for himself, took only a few minutes. After three hours of deliberation the court brought in its verdict. Mitchell was found guilty of the charge and of all specifications. He was sentenced to be suspended from rank, command, and duty, with forfeiture of all pay and allowances for five years.

Billy Mitchell resigned from the army in 1926. In 1945 the Senate voted to confer on him posthumously the Congressional Medal of Honor and, at the same time, promoted him to the rank of major general. Mitchell Field, the airport at Milwaukee, Wisconsin, has been named after him.

Suggested Readings

Andrist, Ralph K., ed. in charge. The American Heritage History of the 20's and 30's. New York: American Heritage, 1970.

Aymar, Brandt, and Edward Sagarin. "General Billy Mitchell," in their A Pictorial History of the World's Great Trials. New York: Crown Publishers, 1967.

Burlingame, Roger. General Billy Mitchell: Champion of Air Defense. New York: McGraw-Hill, 1952.

Davis, Burke. The Billy Mitchell Affair. New York: Random House, 1967.
DiMona, Joseph. "The Court-Martial of Billy Mitchell," in his Great Court-Martial Cases. New York: Grosset & Dunlap, 1972. Also in Rubenstein, Richard E., ed. Great Courtroom Battles. Chicago: Playboy Press, 1973.
Duke, Neville Frederick, ed. The Saga of Flight. New York: John Day, 1961.
Hyde, Margaret Oldroyd. Flight Today and Tomorrow. New York: Whittlesey House, 1962.
Levine, Isaac Don. Mitchell: Pioneer of Air Power. New York: Duell, Sloan, 1958.
Lewellen, John Bryan, and Irwin Shapiro. The Story of Flight. New York: Golden Press, 1959.
Maizlish, I. L. Wonderful Wings: The Story of Aviation. Evanston, Ill.: Row, 1941.
Miller, Francis Trevelyan. The World in the Air: The Story of Flying in Pictures. New York: Putnam, 1930.
Mitchell, Ruth. My Brother Bill: The Life of General "Billy" Mitchell. New York: Harcourt, Brace, 1953.
Mitchell, William. Unpublished Diary. Library of Congress.
New York Times, October 29 to December 19, 1925.
Sherrill, Robert. Military Justice Is to Justice as Military Music Is to Music. New York: Harper, 1969.
Wallechinsky, David, and Irving Wallace. "William 'Billy' Mitchell," in their The People's Almanac. Garden City, N.Y.: Doubleday, 1975.
Whitehouse, Arch. Billy Mitchell: America's Eagle of Air Power. New York: Putnam, 1962.

THE SWEET TRIALS (1925-1926)

During and following World War I, industrialism in the city of Detroit, Michigan drew thousands of workers from the South to that city. An acute housing shortage resulted and many persons, the Negroes in particular, found it extremely difficult to obtain adequate places in which to live. When Negroes attempted to move into white neighborhoods they often met with violent resistance.

In September, 1925, Dr. Ossian Sweet, a Negro physician, purchased a house in Detroit and he, with his wife, baby, and two brothers, moved into it. This house was in an all-white neighborhood. A crowd gathered in the street before the house, milling around until almost 3:00 A.M., but no untoward incidents occurred at that time. The Sweets, fearing trouble from their white neighbors, sat in darkness inside their home.

The following day a large, menacing crowd gathered outside the Sweet home. Dr. Sweet's brother Otis, a dentist, together with three friends, arrived by taxi. Otis Sweet and his friends were stoned by the crowd. Windows were broken and shots were fired. A group of policemen entered the home and arrested the occupants. A white man had been killed during the shooting and the Negroes were held for murder.

The National Association for the Advancement of Colored People entered the case and arranged with the criminal lawyers Clarence Darrow and Arthur Garfield Hays to conduct the defense.

The first trial commenced in October, 1925. Eleven defendants were charged. Judge Frank Murphy declared a mistrial when, after 46 hours, the jury failed to reach a verdict.

Five months later Henry Sweet, Dr. Ossian Sweet's brother, a college student, was tried separately. He had, in the first trial, admitted firing a revolver but the defense had shown that the bullet which had killed the white man had not been shot from Henry Sweet's weapon.

Darrow emphasized the point that the issue at hand was not murder but racial prejudice. He convinced the jury that mob violence did not make right. The jury found Henry Sweet not guilty and charges against the ten other defendants were dropped. The Sweet family, however, did not reoccupy their new home. Between the first and second trials someone had attempted to burn it down.

Suggested Readings

Crandall, Allen. The Man from Kinsman. Sterling, Colo.: Published by the author, 1965.

Darrow, Clarence. The Story of My Life. New York: Scribner's, 1932.

Gurko, Miriam. Clarence Darrow. New York: Crowell, 1965.

Haldeman-Julius, Anna Narcet. Clarence Darrow's Two Great Trials. Girard, Kans.: Haldeman-Julius, 1927.

Harrison, Charles Yale. Clarence Darrow. New York: Jonathan Cape and Harrison Smith, 1931.

Hays, Arthur Garfield. Let Freedom Ring. New York: Liveright, 1937.

________. Trial by Prejudice. New York: Covici, Friede, 1933.

Myrdal, Gunnar. An American Dilemma: The Negro Problem and American Democracy. New York: Harper, 1962.

Schwartzmann, Ruth, and Joseph Stein. The Law of Personal Liberties. New York: Oceana Publications, 1955.

Stone, Irving. Clarence Darrow for the Defense. Garden City, N.Y.: Doubleday, 1941.

Weinberg, Arthur, ed. Attorney for the Damned. New York: Simon & Schuster, 1957.

THE FLORIDA LAND BOOM (1925-1926)

In 1920 Miami, Florida had a population of only 30,000. By 1925, according to the state census, it had grown to 75,000, not counting the hordes of visitors who swarmed in from the North. This huge increase in population was brought about by one of the largest land booms in the history of the United States.

Certain Florida entrepreneurs saw a means of becoming wealthy by exploiting the unique characteristics of their state. The climate was excellent. It was convenient to the northern states with their concentrated populations and cold winters. The automobile had made mobility comparatively easy, and the national prosperity encouraged the belief that great wealth could be achieved quickly by means of speculation in Florida real estate.

Such land developers as George Edgar Merrick, Carl G. Fisher, D. P. Davis and Joseph W. Young launched building projects on a giant scale, advertising them extensively. Merrick, operating out of Miami on the East Coast, called his suburban development "Coral Gables," and hired William Jennings Bryan to lecture on the beauties of Florida. Fisher promoted Miami Beach and is said to have made nearly $40,000,000 selling lots. Davis sold three million dollars worth of land at Tampa on the west coast, and Young built Hollywood-by-the-Sea.

By 1925 the general public was purchasing "anything, anywhere, just so it was in Florida." Stories of the prodigious profits to be made in Florida real estate were legion. Prices skyrocketed and, according to one typical account, a lot in the business center of Miami which had originally sold for $800 was resold in 1924 for $150,000. A piece of land near Miami, bought for $25 in 1896, was disposed of in 1925 for $150,000.

Many of the stories of gigantic speculative profits were true but, unfortunately, these profits were mostly on paper. Lots were bought with the idea of reselling at a large profit. "Binders" covering a small percentage of the purchase price were in common use and purchasers hoped to unload before further payments became due. By mid-1925 property speculation reached an all-time high, with an estimated 2,000 real estate offices and 25,000 agents operating feverishly in the Miami area alone.

By January of 1926 it began to be apparent that new buyers were fewer in number than they had been in the Fall of the previous year, and that many of the purchasers were extremely anxious to dispose of their land to other buyers. Many holders of binders were defaulting on their payments. In some cases land which had been sold to a series of purchasers had come back to the original owner, laden with unpaid taxes and assessments.

On the morning of September 18, 1926, the second of two tropic hurricanes hit Miami, causing tremendous damage. It swept across Florida, leaving destruction in its wake and virtually ending the state's land boom.

By 1927, according to Homer V. Vanderblue, most of the Miami real estate offices had gone out of business. D. P. Davis' promotion was in bankruptcy and many Florida cities were experiencing difficulties in collecting their taxes.

In 1929 the Mediterranean fruit fly destroyed much of Florida's citrus crop and in 1930 Lake Worth, Miami, Sanford and West Palm Beach, as well as 21 other cities in the state, had gone into default on principal or interest on their bonds.

Suggested Readings

Allen, Frederick Lewis. Only Yesterday. New York: Harper, 1931.

Andrist, Ralph K., ed. in charge. The American Heritage History of the 20's and 30's. New York: American Heritage, 1970.

Fisher, Jane (Watts). Fabulous Hoosier: A Story of American Achievement. New York: McBride, 1947.

"Florida Boom Examined," Literary Digest, May 9, 1925.

Isman, Felix. "Forida's Land Boom," Saturday Evening Post, August 22, 1925.

Jenkins, Alan. The Twenties. New York: Universe Books, 1974.

Miller, J. L. "In the Land of the Realtor," Outlook, January 13, 1926.

Roberts, Kenneth L. "Florida Fever," Saturday Evening Post, December 5, 1925.

Sann, Paul. The Lawless Decade. New York: Crown Publishers, 1957.

Shelby, Gertrude Mathews. "Florida Frenzy," Harper's Magazine, January, 1926.

Smiley, Nixon. Yesterday's Florida. Miami: Seemann, 1974.

Weidling, Philip J., and August Burghard. Checkered Sunshine: The Story of Fort Lauderdale, 1793-1955. Gainesville: University of Florida Press, 1966.

THE SNYDER-GRAY SASH WEIGHT MURDER (1927)

A number of sordid and sensational murders were committed in the United States during the 1920's. One of these, which made newspaper headlines from coast to coast, was committed at Queen's Village, New York. The murder victim was art editor Albert Snyder. He was killed by his 32-year-old wife Ruth and her 34-year-old paramour Henry Judd Gray, a corset salesman. Their object was money; Gray had recently been insured for $50,000.

The murder first came to light when, at two o'clock on a Sunday morning in March, 1927, Ruth Snyder, bound and gagged, pounded at the bedroom door of her nine-year-old daughter Lorraine. The daughter removed the gag and her mother screamed, "Get help!" A neighbor was awakened. He found the battered body of Albert Snyder in bed. It was later determined that Snyder had been bludgeoned to death with a sash weight.

Gray and Ruth Snyder had been lovers from the time they first met in the summer of 1925, although Gray was married and had a daughter not quite twelve years old. Eventually the two decided to kill Ruth's husband. Several attempts to poison him were made but were unsuccessful. Then the lovers beat him to death as he lay asleep. Ruth stated to the authorities that burglars had entered her home, gagged and tied her, killed her husband and stolen a number of articles. These, however, were found cached in various places in the house.

That night Ruth Snyder confessed and implicated Gray. The latter was arrested in Syracuse.

The two were tried together for first degree murder, the trial being held in the Queens County courthouse at Long Island City, Judge Townsend Scudder presiding. The prosecution was handled by Assistant District Attorney Charles W. Froessel. Ruth Snyder was defended by attorney Edgar F. Hazleton and Gray was represented by attorney Samuel L. Miller and others.

The trial was one of the most sensational of the decade. It was reported in the newspapers by such luminaries as David Belasco, Billy Sunday, Mary Roberts Rinehart, David Wark Griffith, Peggy Hopkins Joyce, Will Durant, Aimee Semple McPherson and Dr. John Roach Stratton.

Each defendant charged the other with being the instigator of the homicide. The two were found guilty by the jury, which deliberated only ninety-eight minutes. They were sentenced to death and were executed in the electric chair at Sing Sing Prison on January 12, 1928. Although photographers were barred from the death chamber, one enterprising newsman took a picture of Ruth Snyder with a concealed camera strapped to his leg.

Suggested Readings

Abrahamsen, David, M.D. The Murdering Mind. New York: Harper & Row, 1973.

Andrist, Ralph K., ed. in charge. The American Heritage History of the 20's and 30's. New York: American Heritage, 1970.

Bromberg, Walter. Mold of Murder: A Psychiatric Study of Homicide. New York: Grune & Stratton, 1961.

Churchill, Allen. The Year the World Went Mad. New York: Crowell, 1960.

Guttmacher, M. S. The Mind of the Murderer. Freeport, N.Y.: Books for Libraries, 1960.

Jenkins, Alan. The Twenties. New York: Universe Books, 1974.

Jesse, F. Tennyson. Murder and Its Motives. London: Harrap, 1952.

Kunstler, William H. "Murder at a Premium: The People of the State of New York versus Ruth Snyder and Henry Judd Gray," in his First Degree. New York: Oceana Publications, 1960.

Lawes, Warden Lewis E. Meet the Murderer! New York: Harper, 1940.

Lester, David, and Gene Lester. Crime of Passion: Murder and the Murderer. Chicago: Nelson Hall, 1975.

Reinhardt, James Melvin. The Psychology of Strange Killers. Springfield, Ill.: Thomas, 1962.

Sann, Paul. The Lawless Decade. New York: Crown Publishers, 1957.

"The Snyder Murder Mystery," Outlook, May 18, 1927.

Sparrow, Gerald. Women Who Murder. New York: Abelard, 1970.

Symons, Julian. A Pictorial History of Crime. New York: Crown Publishers, 1966.

Wallechinsky, David, and Irving Wallace. "The Snyder-Gray Case," in their The People's Almanac. Garden City, N.Y.: Doubleday, 1975.
Wilson, Colin, and Patricia Pitman. Encyclopedia of Murder. New York: Putnam, 1962.
Woollcott, Alexander. "Five Classic Crimes," in his Long, Long Ago. New York: Viking Press, 1943.

THE LINDBERGH FLIGHT (1927)

In 1919 Raymond Orteig, a French-born restaurateur and proprietor of the Lafayette and Brevoort Hotels in New York City, offered a prize of $25,000 for the first non-stop New York to Paris flight. On May 19, 1927, three pilots were waiting for clearing weather that they might take off on such a flight. These three were Clarence Chamberlin, Lieutenant-Commander Richard Evelyn Byrd, and Charles Augustus Lindbergh.

Lindbergh decided that, in spite of a drizzle, he had a fair chance for good weather over the ocean, and planned to start the following morning. Receiving further weather forecasts, he had his single-engine propeller-driven monoplane, "The Spirit of St. Louis," moved to Roosevelt Field. At 7:40 on the morning of Friday, May 20, 1927, he took off, arriving safely at Le Bourget Airfield outside Paris on the morning of May 21, 33-1/2 hours later.

Lindbergh's plane was stocked with 451 gallons of gasoline and 20 gallons of oil. He had no lights, radio, heat, de-icing equipment, or automatic pilot. The plane made the trip without incident.

The "Lone Eagle," as Lindbergh came to be called, was greeted at the French airport by cheering crowds. President Calvin Coolidge sent a navy cruiser to bring him and his plane back to the United States. In Washington the President made a long and impressive address of welcome and congratulation. Millions of New Yorkers turned out to welcome him when he flew there from Washington, accompanied by an escort of 23 army planes. Mayor James J. Walker complimented him at city hall, declaring, "New York City is yours. I don't give it to you; you won it!" That city spent $71,000 welcoming him and the ticker tape parade up Broadway generated some 1,800 tons of paper.

Lindbergh became a national idol. The former stunt flyer and airmail pilot proved unassuming and modest. Schools, streets, and restaurants were named after him. He was commissioned a colonel, awarded the Congressional Medal of Honor and the Distinguished Flying Cross. Until his death in 1974 he remained a national hero.

Suggested Readings

Allen, Frederick Lewis. Only Yesterday. New York: Harper, 1931.
Andrist, Ralph K., ed. in charge. The American Heritage History of the 20's and 30's. New York: American Heritage, 1970.
Churchill, Allen. The Year the World Went Mad. New York: Crowell, 1960.
Davis, Kenneth Sydney. The Hero: Charles A. Lindbergh and the American Dream. Garden City, N.Y.: Doubleday, 1959.
De Leeuw, Adele Louise. Lindbergh: Lone Eagle. Philadelphia: Westminster Press, 1949.
Duke, Neville Frederick, ed. The Saga of Flight. New York: John Day, 1961.
Hyde, Margaret Oldroyd. Flight Today and Tomorrow. New York: Whittlesey House, 1962.
Jenkins, Alan. The Twenties. New York: Universe Books, 1974.
Lardner, John. "The Lindbergh Legends," in Leighton, Isabel, ed. The Aspirin Age, 1919-1941. New York: Simon & Schuster, 1949.
Lewellen, John Bryan, and Irwin Shapiro. The Story of Flight. New York: Golden Press, 1959.
Lindbergh, Charles A. The Spirit of St. Louis. New York: Scribner, 1953.
________. "We." New York: Putnam, 1927.
"Lindbergh Does It! To Paris in 33-1/2 Hours; Flies 1,000 Miles Through Snow and Sleet; Cheering French Carry Him Off Field," New York Times, May 22, 1927, p. 1.
Maizlish, I. L. Wonderful Wings: A Story of Aviation. Evanston, Ill.: Row, 1941.
Miller, Francis Trevelyan. The World in the Air: The Story of Flying in Pictures. New York: Putnam, 1930.
O'Brien, Patrick Joseph. The Lindberghs: The Story of a Distinguished Family. Philadelphia: International Press, 1935.
Ross, Walters. The Last Hero: Charles A. Lindbergh. New York: Manor Books, 1974.
Sann, Paul. The Lawless Decade. New York: Crown Publishers, 1957.
Wise, William. Charles A. Lindbergh, Aviation Pioneer. New York: Putnam, 1970.

THE SINKING OF THE "S-4" (1927)

On the evening of December 17, 1927, the newly-commissioned naval submarine "S-4" was making a trial run off Wood End, near Provincetown, Massachusetts. The "Paulding," a navy destroyer, had made the trip from Boston to search for rum runners. The "S-4," emerging from the water after a dive, broke surface under the destroyer's port bow, colliding violently with it.

The "Paulding" stopped and a boat was lowered. The "S-4" went down.

A surfboat was launched from the shore and after several hours of sweeping with a grapnel, the "S-4" was snagged but shortly thereafter the hook gave way. The grappling was resumed with better equipment obtained from the "Bushnell," mother ship of the "S-4," which had been summoned.

At 10:45 the grapnel caught the "S-4" again, thus providing a guide for divers from the salvage vessel "Falcon" to descend to the sunken submarine.

Reaching the "S-4," a diver heard signals--six raps coming from the torpedo room. This indicated that six men were alive in that part of the submarine. No similar evidence was forthcoming from the conning tower, engine room, or control room.

The "Falcon" had brought Admiral Brumby to Provincetown and he, as ranking officer, took charge of the rescue operation. The admiral decided to try to blow the ballast tanks and float up the undamaged stern of the "S-4," a decision which later proved to be the wrong one, as two compartments were flooded and a ballast tank had ruptured. Fred Michels, a member of the rescue party, had been brought to the surface unconscious and was removed to Boston on the "Falcon" in order to save his life. The people of Provincetown, thinking that the naval authorities were abandoning the rescue efforts, felt extremely bitter towards the admiral and his staff.

The navy, declining the proffered assistance of a civilian wrecking concern whose equipment was in nearby Boston, elected to send its own equipment from more distant New York. Because of bad weather it took three days for the navy equipment to reach Provincetown.

The six crew members of the sunken "S-4" signaled by tapping on the hull in Morse code that they were still alive but that the air was bad and getting worse. An attempt to hook up a compartment salvage air line failed and bad weather prevented further diving.

On March 3, 1928, three months after she sank, the "S-4" was brought to the surface. Thirty-four members of her crew had died. The sequence of events comprising the tragedy was reconstructed. Following the collision with the "Paulding" the "S-4" had gone down bow first, hit the bottom hard, and leveled off on even keel. Lieutenant Fitch, a submarine officer, and five torpedo men were forward; Lieutenant Commander Jones, another officer, was in the control room, from which most of the machinery, still functioning, could be operated. When water suddenly came pouring in, an attempt was made to close the forward ventilation valves; the bulkhead valve would not close.

It was found that a green baize curtain which had been draped before the door of the captain's stateroom had become entangled with the valve, rendering it useless. Some men drowned immediately. Six were driven by the rush of water from the control room, with its compressed air and controls, to the torpedo room where they died lingering deaths by suffocation.

Suggested Readings

Beach, S. "Down to the Sea Without Safety Devices: The 'Paulding'-'S-4' Collision," Independent, January 7, 1928.
Ellsberg, Commander Edward. "Inside the 'S-4'," Collier's, June 16, 1928.
_______. Men Under the Sea. New York: Dodd, Mead, 1939.
_______. On the Bottom. New York: Dodd, Mead, 1929.
_______. "Safety for Our Submarines," World's Work, March, 1928.
Fredericks, Pierce G. "About: The 'S-4'," New York Times Magazine, December 15, 1957, pp. 20-21.
Green, F. C. "Brains and Braid in the 'S-4' Debate," Independent, January 14, 1928.
"Heroic Deeds of the 'S-4' Salvage Divers," Literary Digest, January 14, 1928.
"The Loss of the 'S-4'," World's Work, February, 1928.
Mielke, Otto. Disaster at Sea. New York: Fleet, 1958.
O'Donnell, J. "How Men Act in a Sunken Submarine," Collier's, August 7, 1926.
"The Risks Men Face in Subs," Popular Mechanics, March, 1928.
Rowland, J. T. "The Why of the 'S-4' Disaster," Scientific American, March, 1928.
Stephens, Edward. Submarines: The Story of Underwater Craft from the Diving Bell of 300 B.C. to Nuclear-powered Ships. New York: Golden Press, 1962.
"Submarine Death Traps: 'S-4' Disaster," Literary Digest, December 31, 1927.
Terrett, C. "What Is the Truth About the 'S-4'?," Outlook, January 11, 1928.
Vorse, Mary Heaton. "The 'S-4' Has Been Sunk!" in Kartman, Ben, and Leonard Brown, eds. Disaster! New York: Pellegrini & Cudahy, 1948.
_______. Time and the Town. New York: Dial Press, 1942.
"Whitewash or Soapsuds? The 'S-4' Disaster Investigation," Independent, March 3, 1928.
"Who Is to Blame for Our Submarine Disasters?," Literary Digest, January 7, 1928.
Zim, Herbert Spencer. Submarines, the Story of Undersea Boats. New York: Harcourt, Brace, 1942.

THE ARNOLD ROTHSTEIN MURDER (1928)

One of the most successful--and notorious--"sure-thing" gamblers of the early 20th century was Arnold Rothstein. He was fatally shot on November 4, 1928, and died two days later. When asked by the police who shot him, he replied, "I won't tell you." The identity of his murderer has never been established.

Six weeks previous to the shooting Rothstein had participated in an all-night poker game at the Park Central Hotel, Broadway and

56th Street, New York City. Other players in the game were "Titanic" Thompson, "Nigger Nate" Raymond, George "Hump" McManus, Jimmy Meehan, the Boston brothers, Sammy and Meyer, and Willie McCabe, the last a member of gangster Alphonse Capone's organization.

The poker game started at ten in the evening and lasted until nine the following morning. The stakes were extremely high, and one by one the players dropped out until only Rothstein and Raymond remained. When the game ended Rothstein had lost $320,000. When asked to pay up he stated that he did not carry such sums around with him and that "the boys would have to wait a day or so."

On the evening of November 4 Rothstein was at Lindy's restaurant on Broadway between 49th and 50th Streets when he received a call. He went to the telephone to answer and was heard to shout, "I won't pay them a cent!" He then handed Jimmy Meehan a .38 caliber revolver, saying that he had to go over to the Park Central "to see McManus," and asked Meehan to keep the revolver for him.

Half an hour after he left the restaurant Rothstein stumbled down the back stairs of the Park Central Hotel. He had been shot in the stomach. Lawrence Fallon, the house detective, and a timekeeper named Paddy were sitting in the watchman's shack at the foot of the stairs. Fallon asked, "Who shot you?" Rothstein replied, "Never mind. Get me a taxi."

Rothstein refused to tell the police anything concerning his assailant. A new .38 caliber revolver with one shot fired from it was dropped from a window of the Park Central Hotel. It was recovered by Al Bender, a taxi driver. Its ownership has never been established.

It was determined that the poker game mentioned above had been played in room 309 of the hotel, and that the room had been occupied by George "Hump" McManus, who had registered under the name of "George Richards." McManus had disappeared but there was a hole in the window screen which could have been caused by a revolver being thrown through it.

One by one the gamblers who had been present at the poker game were interrogated by the police and released for want of evidence. McManus was located six weeks later, questioned, and dismissed.

The newspapers charged laxity on the part of the police. They stated that Rothstein's private papers, which could incriminate important New York politicians, had not been acquired and no effort to guard them had been made. Mayor James J. Walker ordered Police Commissioner Warren to either resign or produce Rothstein's murderer. The district attorney promised immunity to anyone involved in the case except the actual killer if he would turn state's evidence. It was rumored that Ruth Keyes, a "free lance model," had been used by persons unknown to lure Rothstein to his death. She was located, questioned, and released.

When Rothstein's records were examined they disclosed nothing incriminating regarding any politicians or underworld characters. McManus, together with John Doe and Richard Roe, was indicted, tried a year later, and released on a directed verdict of acquittal.

Suggested Readings

Adamic, Louis. "Racketeer," New Republic, January 7, 1931.

Asbury, Herbert. The Gangs of New York. New York: Knopf, 1927.

_______. Sucker's Progress. New York: Dodd, Mead, 1938.

Clarke, Donald Henderson. In the Reign of Rothstein. New York: Vanguard Press, 1929.

Collins, Ted, ed. "Arnold Rothstein," in his New York Murders. New York: Duell, Sloan, 1944.

Crouse, Russel. Murder Won't Out. New York: Doubleday, 1932.

Fowler, Gene. Beau James. New York: Viking Press, 1949.

_______. The Great Mouthpiece. New York: Covici, Friede, 1931.

Katcher, Leo. The Big Bankroll. New York: Harper, 1958.

Merz, Charles. The Dry Decade. New York: Doubleday, 1931.

Rothstein, Carolyn (Green). Now I'll Tell. New York: Vanguard Press, 1934.

"Rothstein, Gambler, Mysteriously Shot; Refuses to Talk," New York Times, November 5, 1928, p. 1.

Sterling, Hank, pseud. "The $320,000 Welsh ... and a Bullet in the Gut," in his Ten Perfect Crimes. New York: Stravon Publishers, 1954.

Sullivan, Edward Dean. Chicago Surrenders. New York: Vanguard Press, 1930.

Symons, Julian. A Pictorial History of Crime. New York: Crown Publishers, 1966.

Walker, Stanley. The Night Club Era. New York: Blue Ribbon Books, 1933.

THE WRONG-WAY RUN (1929)

One of the outstanding post-season college football games is the Rose Bowl Game, played at Pasadena, California, on New Year's Day. In this classic the champion of the Pacific Coast Conference is pitted against the outstanding non-Pacific Coast college team in the country.

On January 1, 1929, the University of California Golden Bears played against the Golden Tornado of Georgia Tech. A star of the California team was halfback Benny Lom, extremely light for a football player but one of the best gridiron men of his day. California's center was captain-elect Roy Riegels, a large man and, compared to Lom, a slow runner.

California had the ball and was driving towards the Georgia goal line. On the fourth down Georgia stopped the drive on its own 36-yard line. Lom threw a forward pass which he failed to complete and the ball came into the possession of the Georgia team.

On the next play Georgia fumbled and Riegels scooped up the ball. As running with a recovered fumble was then authorized by football rules, Riegels set off for the goal line. Unfortunately, however, he headed, not for the goal Georgia was defending, but for the other goal.

Lom, realizing that Riegels was running the wrong way, took off after him. The California rooters screamed, "No, Roy, No!," but Riegels, intent on making a touchdown, failed to heed them. Lom, shouting at his teammate to stop, pursued him a total of 63 yards. He made a desperate flying tackle on the one-yard line, bringing Riegels and football to the ground half a foot from the goal line.

Deep in the end zone, Lom attempted to punt out of danger. However, his kick was blocked. Georgia Tech made a safety, the first score of the game, and led, 2-0.

At the conclusion of the game the final score saw Georgia the winner, 8-7. Riegels' 63-yard wrong-way run was the deciding factor in the contest. Football fans still talk about this one.

Suggested Readings

Danzig, Allison. The History of American Football. Englewood Cliffs, N.J.: Prentice-Hall, 1956.

Durant, John, and Les Fetter. Highlights of College Football. New York: Hastings House, 1970.

Kaye, Ivan. Good Clean Violence. Philadelphia: Lippincott, 1973.

Liss, Howard. Great Moments in Football. New York: Cowles, 1970.

McCallum, John D. College Football, U.S.A., 1869-1971. Greenwich, Conn.: Hall & Fame, 1971.

Otto, J. R. Football. Mankato, Minn.: Creative Educational Society, 1961.

"Riegels's 60-yard Run Toward Wrong Goal Helps Georgia Tech Win on Coast, 8-7," New York Times, January 2, 1929, p. 22.

Russell, Fred. Big Bowl Football. New York: Ronald Press, 1963.

Samuelsen, Rube. The Rose Bowl Game. Garden City, N.Y.: Doubleday, 1951.

Schulberg, Budd. "Memories of Post Rose Bowls," TV Guide, December 28, 1974.

Treat, Roger L. The Official Encyclopedia of Football. New York: Barnes, 1967.

Walsh, Christy, and G. C. Whittle, eds. Intercollegiate Football. New York: Doubleday, 1934.

Ward, Gene, comp. Football Wit and Humor. New York: Grosset & Dunlap, 1970.

Weyland, Alexander M. American Football: Its History and Development. New York: Appleton, 1926.

______. The Saga of American Football. New York: Macmillan, 1955.

THE ST. VALENTINE'S DAY MASSACRE (1929)

During prohibition days gangsterism flourished in Chicago probably more than in any other geographical area. Fantastic sums of money were to be made by dealing in illegal whisky, gin, and beer. Gangs of hoodlums, attracted by the enormous profits of bootlegging, often resorted to armed warfare in their struggle to monopolize the business.

One of the Chicago gangs was headed by Alphonse Capone and another was guided by George "Bugs" Moran. Moran's gang, operating on the north side of the city, repeatedly highjacked shipments of liquor consigned to Capone, bombed saloons that were purchasing his beer, and attempted, often successfully, to murder his followers.

Early in 1929 Capone visited Florida, thus providing himself with an unshakable alibi for what was to follow in the multiple murders known as the "St. Valentine's Day Massacre."

Moran had purchased a shipment of highjacked whisky and had arranged to have it delivered at a former garage, used by him as a warehouse. This building, located at 2122 North Clark Street, was ostensibly the home of the S. M. C. Cartage Company. The delivery was to be made at 10:30 on the morning of St. Valentine's Day, February 14, 1929. Seven members of Moran's gang were waiting at the warehouse to help unload the shipment.

Five men drove up to the Clark Street establishment in a Cadillac. They entered the warehouse and witnesses later testified that three of the men were wearing policemen's uniforms. Moran, together with Ted Newberry and Willie Marks, members of his gang, arrived, saw the Cadillac, assumed that a police raid was in progress, and hurriedly left.

A few minutes after the men entered the warehouse, machine gun and shotgun shots rang out. The men reappeared, the first two with their hands raised, the other three, wearing police uniforms, holding pistols to their backs. The men climbed into the Cadillac and drove away.

Later it was found that Pete Gusenberg, James Clark, alias Kasheller, Adam Heyer, Al Weinshank, Reinhardt H. Schwimmer, and Johnny May, all followers of Moran, had been lined up against a wall, gunned to death, and had died instantly. Gusenberg's brother Frank survived a short time but died without identifying the killers.

Rewards totaling $100,000 were offered for the apprehension of the murderers. The notoriously corrupt Chicago police were suspect. Eventually the Cadillac used in the murder was located and traced to a "James Morton" of Los Angeles. "Machine Gun" Jack McGurn, a Capone gangster, was suspected of being implicated in the massacre but the authorities were never able to prove this.

Two Thompson sub-machineguns were eventually located in the possession of Fred "Killer" Burke, a St. Louis gangster. Major Calvin H. Goddard, a ballistics expert, was able to show that these weapons were among those used in the St. Valentine's Day Massacre.

Burke was captured in April, tried for another murder in Michigan, convicted, and sentenced to life imprisonment in that state.

The authorities tried for many years to determine the identities of the other killers, but Burke is the only man who can be said with any certainty to have had a hand in the affair. The warehouse on North Clark Street where the Massacre took place was razed in 1967.

Suggested Readings

Allen, Frederick Lewis. Only Yesterday. New York: Harper, 1931.

Andrist, Ralph K., ed. in charge. The American Heritage History of the 20's and 30's. New York: American Heritage, 1970.

Asbury, Herbert. Gem of the Prairie: An Informal History of the Chicago Underworld. New York: Knopf, 1940.

_______. The Great Illusion: An Informal History of Prohibition. Garden City, N.Y.: Doubleday, 1950.

_______. "The St. Valentine's Day Massacre," '47 The Magazine of the Year, September, 1947.

Burns, Walter Noble. The One-Way Ride. Indianapolis: Bobbs-Merrill, 1954.

Churchill, Allen. A Pictorial History of American Crime. New York: Holt, Rinehart, 1964.

Cooper, Courtney Ryley. Ten Thousand Public Enemies. New York: Blue Ribbon Books, 1935.

Dedmon, Emmett. Fabulous Chicago. New York: Random House, 1953.

DeLacey, Charles. "The True Inside on Chicago's Notorious St. Valentine's Day Massacre," True Detective Mysteries, March-April, 1931.

Gunther, John. "The High Cost of Hoodlums," Harper's Monthly Magazine, October, 1929.

Hynd, Alan. The Giant Killers. New York: Robert M. McBride, 1945.

Jenkins, Alan. The Twenties. New York: Universe Books, 1974.

Kefauver, Estes. Crime in America. New York: Doubleday, 1951.

Kobler, John. Capone. New York: Putnam's, 1971.

Lewis, Lloyd, and Henry Justin Smith. Chicago: The History of Its Population. Englewood Cliffs, N.J.: Prentice-Hall, 1960.

Lyle, John H. The Dry and Lawless Years. Englewood Cliffs, N.J.: Prentice-Hall, 1960.

Merz, Charles. The Dry Decade. New York: Doubleday, 1931.

Ness, Eliot, with Oscar Fraley. The Untouchables. New York: Messner, 1957.

Pasley, Fred. Al Capone: The Biography of a Self-Made Man. Plainview, N.Y.: Books for Libraries, 1931.

Redston, George, and Kendell F. Crossen. The Conspiracy of

Death. Indianapolis: Bobbs-Merrill, 1965.
Sann, Paul. The Lawless Decade. New York: Crown Publishers, 1957.
Sullivan, Edward Dean. Chicago Surrenders. New York: Vanguard Press, 1930.
________. Rattling the Cup on Chicago Crime. New York: Vanguard Press, 1930.
Symons, Julian. A Pictorial History of Crime. New York: Crown Publishers, 1966.
Waller, Irle. Chicago Uncensored. New York: Exposition Press, 1965.

THE WALL STREET CRASH (1929)

Following the economic depression of 1921 one of the greatest bull markets in history gripped the American public. "Coolidge Prosperity" characterized much of that President's administration (1923-1929). Economists have attributed the soaring prices on the stock markets to several causes: the orgy of over-speculation coupled with margin accounts, excessive diversion of credits into stock market loans, thus encouraging speculation, the over-expansion of industry, disregard by the investing public of signs that stocks were "dangerously high" in price and were "overbought," and a general feeling that stock prices would continue to rise in price for an indefinite period.

In 1919 sales of stocks on the New York Stock Exchange totaled 315,983,000 shares. Ten years later sales had increased to 1,125,000,000 shares and in January, 1929, new securities had been issued in the amount of over a billion dollars.

By March 4, 1928, the stock market had started upward in a highly sensational manner. Such market favorites as General Motors, Radio, United States Steel, and Montgomery Ward were climbing feverishly in price. The volume of shares traded on the New York Exchange each day increased enormously.

The Federal Reserve authorities attempted to stem the flood of speculation by raising the rediscount rate. Financial analysts such as Roger Babson advised caution and expressed the opinion that a price decline was imminent. Babson and other harbingers of gloom were looked at askance and the boom went on, though punctuated with some sudden drops and immediate recoveries. Between September 19 and October 4, 1929, prices had declined badly. This was optimistically regarded as a "technical readjustment," and people rushed to purchase stocks at the comparatively low prices to which they had fallen.

On October 24 prices collapsed on both the New York Stock and the Curb Exchanges and over 19,000,000 shares changed hands. Six major banking institutions, including J. P. Morgan & Company and the Chase National Bank, organized a pool to steady the market, each providing $40,000,000. Richard Whitney, president of the Stock Exchange, handled their purchases on the Exchange floor, which helped temporarily to shore up the sagging market to some degree. On October 28 losses on the two exchanges exceeded $10,000,000,000, with some issues falling from 100 to 500 points.

On "Black Tuesday," October 29, 1929, the great crash came. Huge blocks of stock were thrown on the market and very few buyers appeared. By the end of the day's trading 16,410,030 shares had been sold on the New York Exchange and the New York Times average had fallen nearly forty points. The high-speed ticker which reported prices from the Exchange floor fell far behind the trading.

On October 30 John D. Rockefeller announced that he and his son were buying "sound common stocks" for investment. They hoped that this announcement would restore public confidence in the market but were unable to stem the continued decline.

The New York Exchange was closed on Friday and Saturday, November 1 and 2, in order to permit brokers to catch up on their paperwork. A short session on Thursday, October 31, saw a minor rally in prices, but during the following week, in a series of short sessions, the market fell further, reaching the lowest level of prices for the year 1929 on November 13.

The Wall Street Crash of 1929 saw the end of the great Bull Market of the 1920's and ushered in the Great Depression of the 1930's.

Suggested Readings

Allen, Frederick Lewis. Big Change: America Transforms Itself, 1900-1950. New York: Harper, 1952.

________. Only Yesterday. New York: Harper, 1931.

________. Since Yesterday: The Nineteen-Thirties in America, September 3, 1929-September 3, 1939. New York: Harper, 1940.

Andrist, Ralph K., ed. in charge. The American Heritage History of the 20's and 30's. New York: American Heritage, 1970.

Arnold, Thurman. "The Crash and What It Meant," in Leighton, Isabel, ed. The Aspirin Age, 1919-1941. New York: Simon & Schuster, 1949.

Galbraith, John Kenneth. "The Great Wall Street Crash," Harper's Magazine, October, 1954. Reprinted in Knowles, Horace, ed. Gentlemen, Scholars and Scoundrels. New York: Harper, 1959.

Jenkins, Alan. The Twenties. New York: Universe Books, 1974.

Leuchtenburg, William E. The Perils of Prosperity. Chicago: University of Chicago Press, 1958.

Lingley, Charles Ramsdell, and Allen Richard Foley. Since the Civil War. New York: Appleton, 1935.

Mitchell, Broadus, Depression Decade, 1929-1941. New York: Holt, Rinehart, 1947.

Pecora, Ferdinand. Wall Street Under Oath: The Story of Our Modern Money Changers. New York: Simon & Schuster, 1939.

Robbins, Lionel. The Great Depression. New York: Macmillan, 1934.

Sann, Paul. The Lawless Decade. New York: Crown Publishers, 1957.
Warren, Harris G. Herbert Hoover and the Great Depression. New York: Oxford University Press, 1959.
Wilson, Edmund. The American Jitters: A Year of the Slump. Plainview, N.Y.: Books for Libraries, 1932.

THE DISAPPEARANCE OF JUDGE CRATER (1930)

One of the great unsolved mysteries of all time is the disappearance of Judge Joseph Force Crater who was last seen stepping into a taxi on West 45th Street, New York City, at 9:15 P.M. on Wednesday, August 6, 1930, and who vanished into complete oblivion.

Judge Crater was appointed a justice of the Supreme Court of the State of New York by Franklin Delano Roosevelt, then governor of that state, in April, 1930. The appointment was made to fill out the unexpired term of Justice Proskauer who had retired, and which would terminate in December. Crater, an astute politician, a law professor at Fordham University and New York University Law School, and a brilliant attorney, had been selected by Roosevelt who had rejected the men nominated by the Tammany and anti-Tammany wings of the Democratic organization. He had held the office of president of the Cayuga Democratic Club, a Tammany organization, for fifteen years. When it was established that City Magistrate George F. Ewald had bought his post from Martin J. Healy, a leader of the Club, for $12,000, Judge Crater, though innocent, was thought to be implicated. This became one of the matters considered following Crater's mysterious disappearance.

On August 3, 1930, the Judge, who had been vacationing at Belgrade Lakes, Maine, where he had a summer cottage, was driven to the railroad station by Fred Kahler, his chauffeur. He had received a telephone call summoning him to New York and had departed at once, although he had previously indicated that he planned to remain at the cottage an additional two weeks.

Crater arrived at his Fifth Avenue New York apartment at about ten o'clock on the morning of August 4. That day he visited his chambers at the courthouse, handled some routine matters, and wrote a personal letter to his niece.

On August 5 he had dinner at Billy Haas' restaurant on West 45th Street with William Klein, an attorney, and a Broadway chorus girl. He told Klein that he planned to return to Maine for a three-week vacation.

On the following day, Wednesday, August 6, Crater appeared in his chambers. He wrote two bearer checks, one for $3,000, drawn on the Chase National Bank, and the other, in the amount of $2,100, on the Empire Trust Company. He asked Joseph A. Mara, his court attendant, to cash these two checks. Mara later testified that he had done so and turned the money over to Judge Crater.

Crater and Mara then took a taxi to Crater's apartment, taking with them a briefcase and four portfolios. Crater told Mara at that time that he planned to go for a swim in Westchester that afternoon.

Instead of going to Westchester, however, Crater went to Times Square, where he purchased a ticket to a current play from Joseph Grainsky, a ticket broker. At 9:15 that night he was seen entering the taxi on West 45th Street from which he disappeared completely from the sight of his family and friends.

His absence was first discovered on August 7 when his chauffeur, who had come from Belgrade Lakes to drive him back from New York, failed to find him in his apartment. After waiting a day the chauffeur telephoned Crater's wife who, in turn, contacted some of the judge's New York friends. A quiet search was made, and by the end of the month the news of Crater's disappearance had leaked out.

Tips and rumors appeared by the dozen. These were laboriously checked out. They led nowhere. One man reported picking up a hitch-hiker who stated that he was Judge Crater. When questioned he said that the name meant nothing to him at the time, but when he was shown a picture of the judge he identified it as a picture of the man he had picked up on August 28 at Port Chester, New York. Helen Murray, a pharmacist, said that Crater had entered her brother's drug store at Phillipsburg, Pennsylvania, on August 8, two days after his disappearance. He was also reported to have been seen at the Showplace, a Lynbrook, Long Island, roadhouse.

The grand jury which investigated the case handed down a report in which complete failure was admitted.

When Mrs. Crater visited her husband's New York apartment she found in a bureau drawer a large envelope which had not been there when the police searched the place looking for clues. In the envelope was $6,690 in currency, the deed to the Maine home, the lease on the apartment, a long list of debtors, and a short message: "Am very weary. Love, Joe."

To this day it is not known whether Judge Crater planned to disappear, was kidnaped and murdered, or whether he is still alive. On September 12 his family received a note from Detroit demanding $20,000 ransom and stating that Crater had "something wrong with his head." A second note, demanding only $10,000, was received on September 16. These notes were dismissed as hoaxes.

Judge Joseph Force Crater was later declared legally dead but his fate remains an unsolved mystery.

Suggested Readings

Alexander, J. "What Happened to Judge Crater?," Saturday Evening Post, September 10, 1960.

Churchill, Allen. They Never Came Back. Garden City, N.Y.: Doubleday, 1960.

Clapp, Jane. The Vanishing Point. Metuchen, N.J.: Scarecrow Press, 1961.

Crater, Ella, with Oscar Fraley. Empty Robe. Garden City, N.Y.: Doubleday, 1961.
"Hollywood Herring Crosses Trail of Lost Jurist," Newsweek, September 12, 1936.
Lee, H. "The Mystery of the Vanished Judge," Coronet, February, 1951.
Manning, G. "The Most Tantalizing Disappearance of Our Time," Collier's, July 29, 1950.
Meehan, T. "Case No. 13595," New York Times Magazine, August 7, 1960, pp. 27-28.
Sterling, Hank, pseud. "Vanished from the Face of the Earth," in his Ten Perfect Crimes. New York: Stravon Publishers, 1954.
"Weird Clue in the Crater Mystery," Life, November 16, 1959.

THE GREAT DEPRESSION (1930-1939)

Following the stock market crash of October, 1929, the American nation entered the longest and darkest economic depression in its history. This, known as the "Great Depression," lasted until the late 1930's.

The gross national product, which amounted to \$104,400,000,000 in 1929, shrank to \$74,200,000,000 in 1933. Industrial production declined by half. Unemployment was widespread, with one out of every four persons in the labor force without a job in the year 1933. In that year the national income had fallen to a low of \$40,200,000,000, compared to the 1929 figure of \$87,800,000,000, with farmers suffering the greatest loss of income.

The depression became worldwide. President Herbert Hoover hoped that voluntary cooperation between capital and labor would maintain production and payrolls. Unfortunately it did not. Hoover established the Reconstruction Finance Corporation (R.F.C.), a loan agency designed to assist large business organizations such as railroads and banks. This later became one of the major agencies of President Franklin Delano Roosevelt's New Deal.

President Hoover was able to have Congress authorize new funds to deter the foreclosure of farm mortgages and the Home Loan Bank Act, designed to prevent the foreclosure of mortgages on homes, was also passed.

The relief issue proved a sore point. Hoover felt that unemployment was a problem to be assumed and solved by local resources; Congress wished the federal government to battle the depression by giving direct relief and spending great sums for public works. Eventually Hoover capitulated and signed the Emergency Relief and Construction Act, which provided \$300,000,000 for local relief loans and \$1,500,000,000 for public works which were to be self-liquidating. In spite of these efforts, however, the depression became worse.

When President Roosevelt entered the White House on March 4, 1933, the country's banking system had virtually collapsed and the economic system was approaching stagnation. It was then that Roosevelt ordered a bank holiday. Banks were closed until their financial structures could be examined. Those which were found to be solvent were permitted to reopen; others stayed closed.

Men and women stood ready and willing to work but plants stayed idle and jobs were nonexistent. Bread lines formed. Soup kitchens fed hungry people. Unemployed men sold apples on street corners. Shanty towns known as "Hoovervilles" were established at the outskirts of many cities and towns. Men and boys took to the road in search of work, but many of them found none. Cotton was selling at five cents a pound and wheat was going at 35 cents per bushel. Farmers, unable to regain the cost of raising their crops, destroyed them, as when, in August, 1932, dairy farmers in Iowa poured milk on the ground rather than send it to market where it would be sold at a loss.

Sergeant Walter W. Waters led a group of World War I veterans to Washington in a bonus march. There they demanded the immediate payment of a bonus due in 1945. Waters' men were dispersed by American soldiers commanded by General Douglas MacArthur, following the orders of President Hoover.

Utopian schemes to end poverty were proposed by such men as Dr. Francis Everett Townsend and Huey P. Long. These, though accepted by many persons as the way out of their personal financial difficulties, were economically unsound and unworkable. Father Charles E. Coughlin, the "radio priest," advocated the nationalization of banks, utilities and natural resources.

Eventually the American people regained some sense of optimism. Signs of recovery appeared in 1937, but this was followed by another sharp recession, almost as bad as that of 1929. Conditions had again improved somewhat by mid-1938 and recovered still more in 1939 when defense spending in preparation for World War II created jobs in plenty for the first time in a decade.

Suggested Readings

Allen, Frederick Lewis. Big Change: America Transforms Itself, 1900-1950. New York: Harper, 1952.

________. Since Yesterday: The Nineteen-Thirties in America, September 3, 1929-September 3, 1939. New York: Harper, 1940.

Andrist, Ralph K., ed. in charge. The American Heritage History of the 20's and 30's. New York: American Heritage, 1970.

Bernstein, Irving. The Lean Years: A History of the American Worker, 1920-1933. Boston: Houghton Mifflin, 1960.

________. The Turbulent Years: A History of the American Worker, 1933-1941. Boston: Houghton Mifflin, 1970.

Bird, Caroline. The Invisible Scar. New York: David McKay, 1936.

Goldston, Robert C. The Great Depression: The United States in the Thirties. Indianapolis: Bobbs-Merrill, 1968.

Leuchtenburg, William E. Franklin D. Roosevelt and the New Deal, 1932-1940. New York: Harper, 1963.
Lindop, Edmund. Turbulent America: The Turbulent Thirties. New York: Watts, 1970.
Meltzer, Milton. Brother, Can You Spare a Dime? The Great Depression, 1929-1933. New York: Knopf, 1969.
Mitchell, Broadus. Depression Decade, 1929-1941. New York: Holt, Rinehart, 1947.
Phillips, Cabell. New York "Times" Chronicle of American Life: From the Crash to the Blitz, 1929-1939. New York: Macmillan, 1969.
Robbins, Lionel. The Great Depression. New York: Macmillan, 1934.
Romasco, Albert U. The Poverty of Abundance: Hoover, the Nation, the Depression. New York: Oxford University Press, 1965.
Rublowsky, John. After the Crash: America in the Great Depression. New York: Crowell, 1970.
Sann, Paul. The Lawless Decade. New York: Crown Publishers, 1957.
Schlesinger, Arthur M., Jr. The Age of Roosevelt. Boston: Houghton Mifflin, 1957.
Shannon, David A. The Great Depression. Englewood Cliffs, N.J.: Prentice-Hall, 1960.
Sperling, John C. Great Depressions: 1837-1844, 1893-1898, 1929-1939. Chicago: Scott, Foresmen, 1966.
Steinbeck, John. The Grapes of Wrath (fiction). New York: Viking Press, 1939.
Terkel, Studs. Hard Times: An Oral History of the Great Depression in America. New York: Pantheon Books, 1970.
Warren, Harris G. Herbert Hoover and the Great Depression. New York: Oxford University Press, 1959.
Wecter, Dixon. The Age of the Great Depression, 1929-1941. New York: Macmillan, 1948.
Werstein, Irving. A Nation Fights Back: The Depression and Its Aftermath. New York: Messner, 1962.
Wilson, Edmund. The American Jitters: A Year of the Slump. Plainview, N.Y.: Books for Libraries, 1932.

FATHER DIVINE'S PEACE MISSION (1930-1965)

The American Negro religious cult leader known as "Father Divine" was born George Baker near Savannah, Georgia, about 1882. Information concerning his early years is fragmentary, contradictory, and unbelievable. It has been established that for a number of years he was an itinerant preacher in several states in the American South.

In 1915 Baker appeared in Harlem where, under the name of Major M. J. Divine, he later founded the Peace Mission Movement. At this time he substituted the prefix "Father" for "Major," dropping the initials "M. J." No record of military service performed by him has ever been found, and the "Father" is self-bestowed, as was any ordainment he may have claimed. In spite of this, many of his followers regarded him as the personification of God.

Over the years Baker won thousands of devoted adherents, both white and colored. These followers lived in residences known as "heavens" and took vows of sexual abstinence, unselfishness, and frugality. Most were employed, turning their earnings over to their mentor, who in turn provided them with food and other necessities. Members of the Peace Mission took names such as "Sunbeam Willing" and "Quiet Devotion," and were known as "angels."

Although "Father Divine" claimed a membership of from two to twenty million in his organization, it has been estimated that the membership did not at any time exceed 500,000. During the depression of the 1930's, many "angels" operated cheap hotels, low-cost restaurants, dry cleaning and barber shops, and other business enterprises. These were all run on a semi-charitable basis.

George Baker--or "Father Divine"--with his saying of "Peace! It's wonderful!" died in 1965. His followers felt that he did much to assist and rehabilitate the poor in New York, and later in Philadelphia, to which city he moved his headquarters in 1942. His critics, on the other hand, feel that he was demagogic and self-seeking, deliberately exploiting those "angels" who became members of the Peace Mission. At his death his assets were estimated at more than ten million dollars.

Suggested Readings

Bach, Marcus Louis. They Have Found a Faith. Indianapolis: Bobbs-Merrill, 1946.
Clark, Glenn. Fishers of Men. Boston: Little, Brown, 1928.
Harris, Sara. Father Divine, Holy Husband. Garden City, N.Y.: Doubleday, 1953.
"Life with 'God'," Newsweek, December 4, 1950.
McKelway, St. Clair. True Tales from the Annals of Crime and Reality. New York: Random House, 1951.
"Made in Heaven," Time, August 19, 1946.
"Mortal Mystery," Newsweek, June 10, 1957.
"The New Mrs. Divine," Life, August 19, 1946.
"Peace and A-a-a-men," Newsweek, October 26, 1953.
"Prophet and a Divine Meet," Life, September 28, 1953.
"Swiss Heaven," Time, October 14, 1946.
The Unofficial Observer, pseud. American Messiahs. New York: Da Capo Press, 1975.
Wallechinsky, David, and Irving Wallace. "Father Divine," in their The People's Almanac. Garden City, N.Y.: Doubleday, 1975.

THE STARR FAITHFULL MURDER (1931)

Starr Faithfull, nee Wyman, was the 25-year-old stepdaughter of Stanley E. Faithfull, an Anglo-American chemist. She was an alcoholic, irresponsible in her many affairs with men, and drifting aimlessly and bewilderedly through life. Her dead body was found on the sands of Long Beach, Long Island, on the morning of Monday, June 8, 1931. Medical evidence indicated that she had probably been murdered by drowning. If murdered, her killer has never been discovered. Some evidence exists that she may have committed suicide.

Newspaper pictures show Starr Faithfull to have been an extremely beautiful girl. The product of a broken home, she exhibited some evidence of mental disorder. Her two diaries, discovered after her death, described in detail some fourteen years of esoteric sex experiences, yet she professed to be afraid of men. She had a long affair with an "elderly friend of the family, a man high up in New England politics." In 1927 this "friend" obtained a release of claim from her family in return for $20,000 paid them for medical expenses "on account of Starr's illness."

The girl was in the habit of attending, uninvited, shipboard farewell parties where, in prohibition America, she was able to obtain free drinks. Whenever the Cunard Line steamer "Franconia" was in port she visited it. Here she met and fell in love with Dr. George Jameson-Carr, the ship's surgeon. On May 29, 1931, she deliberately remained aboard when the ship sailed and it was necessary to put her ashore on a passing tug.

On the afternoon of Thursday, June 4, she attended a party being given by the actress Miriam Hopkins, from which she returned home drunk. That evening she attended another party, coming home at two o'clock the following morning. The last time her family saw her alive was at 9:30 the same morning--Friday, June 5--when she left the house again.

That night, when she failed to return home, her stepfather checked with the missing persons bureau and the local hospitals without result. Her body, badly bruised and clothed only in a dress, was found on the morning of Monday, June 8, by a beachcomber. Stanley Faithfull made the identification and declared that his stepdaughter had been murdered.

The autopsy revealed that the girl had had food about three hours before she died, had consumed no alcohol for approximately twelve hours, had recently been sexually intimate with a man, and that sand was present in her lungs and trachea. She was known to have been a strong swimmer and the presence of sand in her body suggested that she had been forcibly drowned in shallow water near the shore.

Starr Faithfull's two diaries were discovered by the police although her stepfather claimed to have destroyed them. The men mentioned in the diaries were interviewed by the authorities but none of them could be connected to the murder. Dr. Alexander G. Gettler, New York City's toxicologist, announced that a further autopsy of the corpse revealed the presence of sufficient veronal in the liver to convince him that the girl had been asleep when she entered the water.

The dead girl's movements after she left home on the morning of June 5 were traced. She had had a marcel at a beauty shop and had apparently gone from there to the Cunard Line pier where she attended a stateroom party on the "Mauretania." She was seen leaving in the company of a young ship's officer. She and the officer were driven to her home in a taxi but she did not enter the house. That afternoon she was seen at the Cunard Line pier again in the company of the same officer. The taxi driver, Murray Edelman, drove her home again. Edelman stated that she was extremely drunk. Her mother said that she did not come home at that time and the taxi driver did not see her actually enter the building.

At about 4:20 that afternoon she was seen at the entrance to the Channin Building at 42nd Street and Lexington Avenue. She was then in a drunken stupor. Two men assisted in placing her in a taxi in which she rode to a drugstore in Flushing, Long Island. There she purchased two bottles of whiskey, and she and the taxi driver, Sy Brockman, drank part of one of them. Later he drove her to a speakeasy where the trail ended.

One theory is that Starr Faithfull went from the speakeasy to the "Mauretania" where, after making love with her officer friend, she attempted to swim ashore from the boat which, by that time, was some seven miles off Long Beach. It is also thought that she may have been thrown from the vessel and drowned, her body washing ashore. However, the tides at that part of Long Island are such that this seems improbable.

Dr. George Jameson-Carr of the "Franconia" reported from England that he had received three letters from the girl written on May 30, June 2, and June 4. The June 4 letter said, in part, "I am going ... to end my worthless, disorderly bore of an existence --before I ruin everyone else's life as well.... I hate everything so ... life is horrible."

The Faithfull family, when shown the letters, declared them to be forgeries written by the murderer to deceive the police. One set of handwriting experts declared the letters to be genuine; another pronounced them spurious.

Whether Starr Faithfull was murdered or died by her own hand has never been definitely established. The case remains an unsolved mystery.

Suggested Readings

Abrahamsen, David, M.D. The Murdering Mind. New York: Harper & Row, 1973.

"Body of Missing Girl Found at Long Beach," New York Times, June 9, 1931, p. 1.

Collins, Ted, ed. "Starr Faithfull," in his New York Murders. New York: Duell, Sloan, 1949.

Crouse, Russel. Murder Won't Out. New York: Doubleday, 1932.

"The Faithfull Murder Case," Outlook, July 1, 1931.

"Grand Jury Hears 15 Today in Murder of Starr Faithfull," New York Times, June 12, 1931, p. 1.
Jenkins, Alan. The Twenties. New York: Universe Books, 1974.
Markey, Morris. "The Mysterious Death of Starr Faithfull," in Leighton, Isabel, ed. The Aspirin Age, 1919-1941. New York: Simon & Schuster, 1949.
Packer, Peter. Love Thieves (fiction). New York: Holt, 1962.
"Politician Accused of Aiding in Murder of Miss Faithfull," New York Times, June 10, 1931, p. 1.
Reinhardt, James Melvin. The Psychology of Strange Killers. Springfield, Ill.: Thomas, 1962.
"Sister Seeks Clues in Faithfull Diary," New York Times, June 14, 1931, p. 1.
Sterling, Hank, pseud. "The Only Clues--Sex and Alcohol," in his Ten Perfect Crimes. New York: Stravon Publishers, 1954.
Symons, Julian. A Pictorial History of Crime. New York: Crown Publishers, 1966.

THE WINNIE RUTH JUDD TRUNK MURDERS (1931)

The murder of two women by Winnie Ruth Judd, on or about October 16, 1931, became a gruesome spectacular which captured the attention of the world.

The murderess was a 26-year-old nurse. Her victims were Agnes Anne LeRoi and Hedvig Samuelson, both approximately her own age. Agnes was an X-ray technician and the Samuelson girl was a teacher. The three were apparently the best of friends and shared a Phoenix, Arizona apartment until Winnie moved out early in October in order to be nearer the Grunow Clinic where she was employed.

On the evening of October 17 Winnie arranged to ship two trunks from her former apartment at 2929 North Second Street, Phoenix, to Los Angeles. The next day she went by train to the California city, the trunks accompanying her. At Los Angeles a baggage handler noticed blood leaking from the larger of the trunks and notified his supervisor. Winnie appeared at the baggage office but did not take possession of the trunks and the incident was reported to the police. Lieutenant Frank Ryan opened the baggage and discovered the dismembered bodies of two women. These were determined to be the bodies of Agnes Anne LeRoi and Hedvig Samuelson.

The Phoenix police were notified and an order to arrest Winnie Ruth Judd on suspicion of murder was issued. Jason McKinnell, Winnie's brother, told the police that his sister had confessed to killing the two women and wanted him to "pick up the two trunks and throw them in the ocean." Dr. William C. Judd, the husband from whom Winnie was separated, testified that he believed his wife to be mentally deranged, that she had killed the two girls "and was perfectly justified in doing so."

A gigantic search was started. On October 23 Winnie surrendered to the authorities at Edward Mallory's undertaking parlor. She had been shot in the left hand. One of her attorneys, former Judge Louis P. Russell, later stated that Winnie had shot Agnes Anne LeRoi after Agnes had killed Hedvig Samuelson and threatened to kill her and that she had been shot in the hand while trying to defend herself.

Winnie was extradited from California to Arizona. On November 3 she was charged with murdering her two former roommates. Her trial began on January 19, 1932. The prosecution was handled by Acting County Attorney Lloyd Andrews and two deputy attorneys. Justice Howard C. Speakman presided. She was tried on a charge of murdering the LeRoi girl but not with causing the death of Hedvig Samuelson.

The trial lasted 21 days, and she was found guilty of murder in the first degree and sentenced to hang. She was transferred to the Arizona State Prison at Florence.

Her legal counsel, Samuel Franks, Paul Schenck and Joseph Zaversack, were replaced by Edward J. Flanigan and O. V. Willson, with others being engaged later. Briefs for a rehearing were filed on September 1, these alleging judicial error by the Maricopa County Supreme Court. The higher court affirmed the conviction.

Winnie implicated John J. Halloran, a Phoenix lumber dealer, as an accomplice. Halloran was indicted by the grand jury, which also petitioned the State Board of Prisons and Paroles to commute Winnie's death sentence to life imprisonment. This was denied by the Board. Halloran was tried and the Court dismissed the complaint.

In April, 1933, Winnie Ruth Judd was the subject of a sanity trial. Attorney Willson represented her. She was declared insane and her death sentence was amended to one of confinement at the Arizona State Hospital for the Insane. Transferred to the Hospital, she escaped on October 24, 1939, but was found and returned. This was the first of seven escapes from custody, the last one being made on October 8, 1962. Using the name of Susan Leigh Clark, she obtained employment as housekeeper in a Central California town.

In 1969 she was identified and extradited to Arizona. Her sentence was commuted to the time spent in the Hospital as though served in prison, a total of 29 years, 154 days, not counting escape time. She then returned to California a free woman.

Suggested Readings

Abrahamsen, David, M.D. The Murdering Mind. New York: Harper & Row, 1973.

Bromberg, Walter. Mold of Murder: A Psychiatric Study of Homicide. New York: Grune & Stratton, 1961.

Dobkins, J. Dwight, and Robert J. Hendricks. Winnie Ruth Judd: The Trunk Murders. New York: Grosset & Dunlap, 1973.

"Indicts Halloran in Judd Murders," New York Times, December 31, 1932, p. 32.

"Judd Appeal Is Filed," New York Times, April 8, 1932, p. 30.
"Mrs. Judd Sentenced to Die on Scaffold May 11," New York Times, February 25, 1932, p. 44.
Reinhardt, James Melvin. The Psychology of Strange Killers. Springfield, Ill.: Thomas, 1962.
"Ruth Judd Files Plea," New York Times, December 28, 1932, p. 7.
"Ruth Judd to Hang for Arizona Slaying," New York Times, December 13, 1932, p. 15.
Sparrow, Gerald. Women Who Murder. New York: Abelard, 1970.
Still, Larry. Limits of Sanity. New York: St. Martin's Press, 1973.

THE SCOTTSBORO BOYS (1931-1937)

One of the most sensational and long-drawn-out 20th century series of jury trials involved the nine black "Scottsboro Boys" accused of the rape of two white women on March 25, 1931.

The nine were unemployed vagrants who, searching for work in the midst of the depression, had hitched rides on a freight train in Alabama. A fight with some white boys, also riding on the train, had resulted in the whites being thrown off by the Negroes. At Paint Rock, Alabama, the Negroes were arrested by a sheriff's posse and were then accused of rape by Victoria Price and Ruby Bates, white prostitutes who had been on the train with the white boys.

The nine Negroes--Haywood Patterson, Roy Wright, Andy Wright, Olen Montgomery, Ozie Powell, Willie Roberson, Eugene Williams, Charlie Weems, and Clarence Norris--were taken to Scottsboro, the county seat. There, in a trial characterized by racial prejudice and a threatening mob, they were quickly convicted on the rape charge. All, with the exception of Roy Wright, were sentenced to death. At this time the youngest of the Scottsboro Boys was thirteen, the eldest nineteen. One was virtually blind, one suffering from venereal disease, and all were undernourished and unwell.

On review the case came before the United States Supreme Court, which overturned the convictions on the ground that there had been no effective appointment of counsel for the defendants.

Samuel Liebowitz, a prominent New York criminal lawyer, was approached by William Patterson, national secretary of International Labor Defense, an organization with Communist connections, and asked to defend the Scottsboro Boys at their second trial. Although Liebowitz did not agree with the social views of the I.L.D., he agreed to represent the Boys on humanitarian grounds and without fee.

The second trial, which began on March 28, 1933, was held at Decatur, Alabama, Judge James E. Horton presiding. Again the nine Negro defendants were found guilty in a trial that differed little from that of 1931, even though Ruby Bates testified that everything she had said in the first trial had been false. Judge Horton overturned the verdict, stating that he could find no evidence of guilt in the trial record. For this he was overwhelmingly defeated in the next election.

Judge William Callahan, an ignorant, bigoted Southerner, presided at the third trial, which opened on November 27, 1934. Again the atmosphere in the courtroom was one of race hatred and again the Scottsboro Boys were found guilty. This trial, like the first, was reviewed by the Supreme Court, which ruled that systematic exclusion of Negroes from juries, which had been shown to be the case in the other trials, was grounds for reversal.

The fourth trial, starting on January 19, 1936, again found Judge Callahan on the bench, and this time Negroes were called for jury duty but were not selected to serve. Callahan showed the same prejudice he had displayed at the previous trial. In a three-day session Haywood Patterson, who was tried alone, was found guilty and sentenced to 75 years' confinement.

Prosecutor Thomas Knight attempted to make a behind-the-scenes deal with defense counsel Liebowitz which, upon being submitted to Judge Callahan, was indignantly refused. The trials resumed on July 13, 1937, Judge Callahan again presiding. Clarence Norris, Andy Wright, and Charlie Weems were again convicted, each in a separate hearing. Ozie Powell was convicted of stabbing Edgar Blalock, a deputy sheriff, and sentenced to twenty years' confinement.

At this point Thomas S. Lawson, assistant attorney-general, announced that the charges of rape against five of the defendants, namely Montgomery, Roberson, Williams, Roy Wright, and Powell, were dropped, although Powell would be required to serve his sentence for stabbing Blalock.

Eventually the five remaining Scottsboro Boys were given "justice." All were paroled except Haywood Patterson, who escaped from the Kilby Prison Farm and has never been found. Andy Wright violated his parole and was returned to prison, finally being released in 1950. Since then some of the Boys have dropped from sight and others have died. It is believed that Ruby Bates and Victoria Price have also passed away.

Suggested Readings

American Civil Liberties Union. Report on the Scottsboro, Alabama, Case. New York: 1931.

Aymar, Brandt, and Edward Sagarin. "Scottsboro Boys," in their A Pictorial History of the World's Great Trials. New York: Crown Publishers, 1967.

Belfrage, Sally. "The Scottsboro Boys Today," Fact, December, 1966.

Breckenridge, Sophonisba P. Social Work and the Courts. Chicago: University of Chicago Press, 1934.

Carter, Dan T. "A Reasonable Doubt," American Heritage Magazine, October, 1968.

________. Scottsboro: A Tragedy of the American South. Baton Rouge, La.: Louisiana State University Press, 1969.
Chalmers, Allan Knight. They Shall Be Free. Garden City, N.Y.: Doubleday, 1951.
Endore, Guy. The Crime at Scottsboro. Hollywood, Calif.: Hollywood Scottsboro Committee, 1938.
Four Free, Five in Prison--on the Same Evidence; What the Nation's Press Says About the Scottsboro Case. New York: Scottsboro Defense Committee, 1937.
Fraenkel, Osmond K. Ozie Powell (and others), Petitioners vs. the State of Alabama.... New York: Court Press, 1933.
Patterson, Haywood, petitioner. Transcript of Record. Supreme Court of the United States.... Washington, D.C.: Judd & Detweiler, Inc., 1937.
________, with Earl Conrad. Scottsboro Boy. Garden City, N.Y.: Doubleday, 1950.
Record, Wilson. Race and Radicalism. Ithaca, N.Y.: Cornell University Press, 1965.
Reynolds, Quentin. Courtroom. New York: Farrar, Strauss, 1950.
Scottsboro: A Record of a Broken Promise. New York: Scottsboro Defense Committee, 1939.
Scottsboro: The Shame of America. The True Story and the True Meaning of the Famous Case. New York: Scottsboro Defense Committee, 1936.
Symons, Julian. A Pictorial History of Crime. New York: Crown Publishers, 1966.
Tindall, George Brown. The Emergence of the New South. Baton Rouge, La.: Louisiana State University Press, 1967.

THE MASSIE CASE (1932)

The Massie Case involved a sensational murder trial, the last of many in which Clarence Darrow acted as defense counsel. The locale was Honolulu and the defendants were Naval Lieutenant Thomas H. Massie, Mrs. Granville Fortescue, his mother-in-law, and two sailors, members of Massie's command. The charge was murder, the victim being Joseph Kahahawai, an Hawaiian athlete and football player.

Lieutenant Massie and his wife Thalia had attended a party one evening in the spring of 1932. Following an argument with her husband Mrs. Massie left the party and started home. On the way she was attacked by five men: two Hawaiians, one Chinese, and two Japanese. These beat her severely and broke her jaw in two places. On the following day she identified four of the men and they were arrested and brought to trial. The jury, consisting primarily of natives, disagreed and the men were released on bail until a new trial could be held.

Believing that the four men would not be punished by the law for the assault on his wife, Lieutenant Massie decided to handle the matter himself. He located one of the Japanese and beat him until he secured a confession. Massie's attorney, however, indicated that a confession so obtained would not be admissible in court. The Lieutenant, with the assistance of Mrs. Fortescue and two sailors, Albert O. Jones and Edward J. Lord, kidnaped Kahahawai and one of them shot him. The four were arrested and charged with murder.

Clarence Darrow was brought into the case as defense counsel. Racial tensions mounted and hostility flared up. In his defense Darrow emphasized the emotional considerations involved, stating that the case showed "the effect of sorrow and mishap on human minds."

Following a two-day deliberation the jury found the accused guilty of manslaughter and recommended leniency. The judge sentenced the defendants to serve ten years in prison. The Governor of Hawaii, however, commuted the sentence to one hour. The four served their sentence sitting in the governor's palace and were then discharged. Prior to his death in 1966 Jones confessed that he was the one who did the shooting.

Clarence Darrow then persuaded the Island authorities to drop the case against the remaining natives who had assaulted Thalia Massie and the affair was officially terminated.

Suggested Readings

Andrist, Ralph K., ed. in charge. The American Heritage History of the 20's and 30's. New York: American Heritage, 1970.
Crandall, Allen. The Man from Kinsman. Sterling, Colo.: Published by the author, 1965.
Darrow, Clarence. The Story of My Life. New York: Scribner's, 1932.
Fortescue, Mrs. Granville. "The Honolulu Martyrdom," Liberty, July 30, August 6, August 30, 1932.
Gurko, Miriam. Clarence Darrow. New York: Crowell, 1965.
Harrison, Charles Yale. Clarence Darrow. New York: Jonathan Cape and Harrison Smith, 1931.
Martin, John Barlow. "Murder on His Conscience," Saturday Evening Post, April 2, April 9, April 16, April 23, 1955.
"A Sailor Confesses to Old Hawaii Killing," Life, October 7, 1966.
Stone, Irving. Clarence Darrow for the Defense. Garden City, N.Y.: Doubleday, 1941.
Van Slingerland, Peter. Something Terrible Has Happened. New York: Harper, 1966.
Weinberg, Arthur, ed. Attorney for the Damned. New York: Simon & Schuster, 1957.
Wright, Theon. Rape in Paradise. New York: Hawthorn Books, 1966.

THE LINDBERGH KIDNAPING (1932)

What has been referred to as the "crime of the century" involved the kidnaping of Charles Augustus Lindbergh, Jr., the 20-month-old son of Charles Lindbergh, the famous "Lone Eagle." The tragic event occurred on Tuesday, March 1, 1932, some time between nine and ten o'clock in the evening.

The child had been removed from his crib at the home of his parents near Hopewell, New Jersey. A homemade wooden ladder was discovered outside the bedroom window. An envelope containing a hand-written letter demanding $50,000 ransom was found on the windowsill of the child's room.

Because of Lindbergh's prominence the kidnaping precipitated a situation not unlike a three-ring circus. The Lindbergh home was overrun with reporters and state policemen. The Lindbergh servants were questioned, as were those of Dwight Morrow, Mrs. Lindbergh's father. Lindbergh, together with Colonel Norman H. Schwartzkopf and others, enlisted the aid of Mickey Rosner, who claimed to have connections with the underworld. Ransom notes containing interlocking circles, one red and one blue, and punctured by three holes, were received. A clairvoyant was consulted.

Gaston Bullock Means, a government investigator and confidence man, convinced Mrs. Evalyn Walsh McLean of Washington, D.C. that he could, because of his familiarity with the underworld, contact the kidnapers. He mulcted Mrs. McLean of $104,000 and was subsequently tried, convicted of extortion, and sentenced to Leavenworth for fifteen years. He died in prison.

Eventually Dr. John Francis Condon, a Bronx school teacher, volunteered to act as intermediary between the Lindberghs and the kidnaper or kidnapers. He advertised in the Home News, was answered by a letter containing the red and blue circles and the three holes and indicating that he would be satisfactory as a go-between.

Condon, calling himself "Jafsie," acknowledged receipt of the letter by inserting an advertisement in the New York American. He was then instructed to rendezvous with the kidnaper at Woodlawn Cemetery in the Bronx. Here he met a man who called himself "John" and who spoke with a German accent. "John" told Condon that the kidnaped boy was being held on a boat somewhere by a six-member gang.

Other people entered the scene. Three men claimed to have been contacted by the kidnapers and one, Commodore John Hughes Curtis, had been elected to act as intermediary and was to deal with a man named Sam who demanded $25,000 ransom. This story was later found to be a fabrication.

More letters and more newspaper advertisements followed. Then, on April 12, Condon paid "John" $50,000 in bills at St. Raymond's Cemetery in the Bronx. For this he received a note indicating that little Charles was on a boat, the "Nelly," off Martha's Vineyard, Massachusetts. Lindbergh flew over the area for two days but found nothing.

On May 12 William Allen, a colored truck driver, discovered the corpse of a small child in the woods near the Lindbergh home. The corpse was identified by Lindbergh and Betty Gow, the child's nurse, as that of the kidnaped infant. Other persons, however, including Dr. Philip Van Ingen, the baby's physician, doubted the identification.

The ransom money paid "John" by Condon consisted largely of gold certificates and the serial numbers of these bills had been recorded. Lists of these numbers had been widely distributed and a few of the ransom bills had turned up but could not be traced.

The country went off the gold standard in 1933 and as of May 1 of that year it became illegal for American citizens to retain metallic gold coins or gold certificates in their possession. A man calling himself J. J. Faulkner exchanged $2,990 in ransom gold certificates for legal tender at the Federal Reserve Bank, New York City. He was never found.

In September, 1934, a man bought gasoline at a Bronx service station with a gold certificate which proved to be one of the ransom bills. The station attendant, suspicious of his customer's furtive manner, made a note of the automobile license number. This was traced to Bruno Richard Hauptmann, a German-American carpenter living in the Bronx.

Hauptmann's home was searched. Approximately $15,000 of ransom money was found hidden in the garage back of his home. The ladder used by the kidnaper was found to be constructed from wood removed from the floor of Hauptmann's attic. Condon's telephone number was found written on a wall in the German's home.

Hauptmann denied any connection with the kidnaping. He stated that the money found in his garage had been part of the belongings of Isadore Fisch, a friend who had asked him to keep them for him when he took a trip to Germany. Fisch had died abroad and Hauptmann, to whom Fisch had owed money had, according to Hauptmann's statement, hidden the bills and passed them surreptitiously because they were illegal gold certificates.

The trial of Bruno Richard Hauptmann in Flemington, New Jersey, which started in January, 1935, was one of the most sensational of the decade. His defense counsel, Edward J. Reilly, hired by the *New York Daily Mirror*, did a particularly poor job of defending his client. The trial was covered by radio, with Gabriel Heatter as commentator. Newspapers headlined each new development. Newsreel cameras took pictures of the proceedings.

Hauptmann was found guilty of kidnaping and murder and was sentenced to die in the electric chair. He was executed on April 4, 1936.

Suggested Readings

Andrist, Ralph K., ed. in charge. *The American Heritage History of the 20's and 30's.* New York: American Heritage, 1970.

Aymar, Brandt, and Edward Sagarin. "Bruno Richard Hauptmann," in their *A Pictorial History of the World's Great Trials.* New York: Crown Publishers, 1967.

Bruno Richard Hauptmann, Petitioner vs. State of New Jersey, Respondent. On Certiorari. Respondent's brief.... Trenton, N.J.: MacRellish & Quigley Co., 1935.

Busch, Francis X. Prisoners at the Bar. Indianapolis: Bobbs-Merrill, 1952.

Condon, Dr. John F. Jafsie Tells All. New York: Jonathan Lee, 1936.

Haldeman-Julius, Anna Marcet. The Lindbergh-Hauptmann Kidnap-Murder Case. Girard, Kans.: Haldeman-Julius, n.d.

Haring, John Vreeland. The Hand of Hauptmann: The Handwriting Expert Tells the Story of the Lindbergh Case. Plainfield, N.J.: Hamer Publishing Co., 1937.

Hynd, Alan. "Why the Lindbergh Case Was Never Solved," in his Murder, Mayhem and Mystery. New York: A. S. Barnes, 1958.

"Kidnapping Victims: Tragic Aftermaths," Saturday Evening Post, April, 1976.

Kunstler, William H. "Murder in Flight: The State of New Jersey versus Bruno Richard Hauptmann," in his First Degree. New York: Oceana Publications, 1960.

Lardner, John. "The Lindbergh Legends," in Leighton, Isabel, ed. The Aspirin Age, 1919-1941. New York: Simon & Schuster, 1949.

Messick, Hank, and Burt Goldblatt. Kidnapping: The Illustrated History. New York: Dial Press, 1974.

O'Brien, Patrick Joseph. The Lindberghs: The Story of a Distinguished Family. Philadelphia: International Press, 1935.

Pease, Frank. The "Hole" in the Hauptmann Case? New York: Published by the author, 1936.

Reynolds, Quentin. Courtroom. New York: Farrar, Straus, 1950.

Ross, Walters. The Last Hero: Charles A. Lindbergh. New York: Manor Books, 1974.

Schoenfeld, Dr. Dudley D. Crime and the Criminal: A Psychiatric Study of the Lindbergh Case. New York: Covici, Friede, 1936.

The State of New Jersey, Defendant in Error, vs. Bruno Richard Hauptmann, Plaintiff in Error. Somerville, N.J.: Somerset Press, 1935.

Symons, Julian. A Pictorial History of Crime. New York: Crown Publishers, 1966.

Waller, George. Kidnap: The Story of the Lindbergh Case. New York: Dial Press, 1961.

Wendel, Paul H. The Lindbergh-Hauptmann Aftermath. Brooklyn, N.Y.: Loft Publishing Co., 1940.

Whipple, Sidney B. The Lindbergh Crime. New York: Blue Ribbon Books, 1935.

________. The Trial of Bruno Richard Hauptmann, Edited With a History of the Case. New York: Doubleday, 1937.

Woollcott, Alexander. "Five Classic Crimes," in his Long, Long Ago. New York: Viking Press, 1943.

THE BONUS ARMY (1932)

The "Bonus Army," also known as the "Bonus Expeditionary Force," consisted of some 15,000 unemployed veterans of World War I who descended on Washington, D.C., in 1932, and camped there, demanding immediate payment of war bonuses. On July 28 federal troops commanded by General Douglas MacArthur and acting under the orders of President Herbert Hoover drove them out of the city.

The matter of bonuses for former members of the armed forces had occupied the attention of such organizations as the American Legion as early as 1919. At its Minneapolis meeting that year the Legion considered favorably what it called "adjusted compensations" for war veterans. President Warren G. Harding, in 1922, vetoed a bonus bill which had passed both houses of Congress, and a similar bill was vetoed by President Calvin Coolidge the following year. In 1924 a bill calling for adjusted compensation certificates to mature in 1945 was passed over Coolidge's veto. In February, 1931, Congress passed, this time over Hoover's veto, the Bonus Loan Act. This amended the World War Adjusted Compensation Act in that it authorized an increase in the loan basis of adjusted silver certificates and reduced the interest charge.

The depression brought a demand for full and immediate payment of these certificates and the Bonus Army, led by Sergeant Walter W. Waters of World War I, marched on Washington to impress this demand. A bill in the Senate, providing for the immediate cash payment of the certificates at face value, was defeated by Hoover and the Bonus Army was expelled from Washington.

General MacArthur described the Bonus Army as "a bad-looking mob animated by the spirit of revolution." Will Rogers, the humorist, took the opposite view, writing, "just think what 15,000 clubwomen would have done in Washington even if they wasn't hungry." Fiorello La Guardia, then congressman from New York, wired President Hoover, "Soup is cheaper than tear bombs and bread is better than bullets in maintaining law and order in these times of depression, unemployment, and hunger."

Suggested Readings

Andrist, Ralph K., ed. in charge. The American Heritage History of the 20's and 30's. New York: American Heritage, 1970.

Bakke, E. Wright. Citizens Without Work. New Haven, Conn.: Yale University Press, 1940.

______. The Unemployed Worker. New Haven, Conn.: Yale University Press, 1940.

Commager, Henry Steele, ed. "Roosevelt's Veto of the Soldiers' Bonus Bill," (Doc. No. 496) in his Documents of American History, 8th edition. New York: Appleton, 1968.

Hatch, Alden. Franklin D. Roosevelt, An Informal Biography. New York: Holt, 1947.

Hunt, Frazier. The Untold Story of Douglas MacArthur. New York: Devin-Adair, 1955.
Inshaw, David. Herbert Hoover, American Quaker. New York: Farrar, Straus, 1950.
Leuchtenburg, William E. Franklin D. Roosevelt and the New Deal, 1932-1940. New York: Harper, 1963.
Lyons, Eugene. Herbert Hoover: A Biography. Garden City, N.Y.: Doubleday, 1964.
MacArthur, General Douglas. Reminiscences. New York: McGraw-Hill, 1964.
Main, Quincy. History of the American Legion. Houston: Abbotsford Publishing Co., 1975.
Rice, Arnold S., ed. Herbert Hoover, 1874-1964. Chronology, Documents, Bibliographical Aids. Dobbs Ferry, N.Y.: Oceana Publications, 1971.
Schnittkind, Henry Thomas. Franklin Delano Roosevelt. New York: Putnam, 1962.
Seldes, Gilbert. The Years of the Locust: America, 1929-1932. New York: Da Capo Press, 1973.
Warren, Harris G. Herbert Hoover and the Great Depression. New York: Oxford University Press, 1959.
Waters, W. W., and William B. E. F. White. The Whole Story of the Bonus Army. New York: Arno Press, 1933.
Whitney, Courtney. MacArthur: His Rendezvous With History. New York: Knopf, 1956.

THE DECLINE AND FALL OF SAMUEL INSULL (1932-1934)

Samuel Insull, an Anglo-American business magnate, between 1892 and 1932 had built up the utilities in the American Middle West until he had monopolistic control over electric and gas companies, traction companies, and coal mines in 23 states and in Canada. His holding company, Middle West Utilities, dominated by him through the financial device of the voting trust, gave him control over one-twelfth of the power output of America. From his headquarters in Chicago he manipulated a complicated, financially unsound maze of interlocking financial corporations, "intertwined in a tangle no bookkeeper has ever been able to unravel."

Eventually Insull's shaky, over-expanded financial empire collapsed and he was forced to step down. He was president of eleven companies, chairman of 65, and held directorships in 85 others. On June 6, 1932, he resigned from all of them.

Indicted for fraud, Insull fled with his wife to Canada, and from there to Paris, then in turn to Italy and to Athens, Greece. He was expelled from Greece as an undesirable alien and, on March 15, 1934, dressed as a woman, left Athens in order to avoid deportation to the United States. He was apprehended and returned to Chicago.

Insull's trial attracted national attention. It was brought out that he had "made an error of some ten million dollars in accounting, but that it had been an honest error," and the jury found him not guilty.

He retired from business and, until his death in 1938, led a quiet life "spending the pension of $21,000 a year that the directors of his old companies had dutifully restored to him."

Suggested Readings

Bonbright, James C. The Holding Company: Its Public Significance and Its Regulation. Clifton, N.J.: Augustus M. Kelley, 1932.

"The Crash of the Insull Dynasty," Christian Century, June 15, 1932.

Danielian, N. D. "From Insull to Injury: A Study in Financial Jugglery," Atlantic Monthly, April, 1933.

Dawson, M. "Insull on Trial," Nation, November 28, 1934.

Dos Passos, John. "Power Superpower," in his The Big Money. New York: Random House, 1936.

Flynn, John T. "Up and Down With Sam Insull," Collier's, December 10, December 17, December 24, 1932.

Forbes, B. C. "Samuel Insull," in his Men Who Are Making America, 4th edition. New York: B. C. Forbes Publishing Co., 1917.

Holbrook, Stewart Hall. The Age of the Moguls. Garden City, N.Y.: Doubleday, 1953.

McDonald, Forrest. Insull. Chicago: University of Chicago Press, 1962.

________. "Samuel Insull and the Movement for State Utility Regulatory Commissions," Business History Review, Autumn, 1958.

Ramsay, M. L. Pyramids of Power: The Story of Roosevelt, Insull and the Utility Wars. Indianapolis: Bobbs-Merrill, 1937.

Sann, Paul. The Lawless Decade. New York: Crown Publishers, 1957.

Stuart, William H. The Twenty Incredible Years as "Heard and Seen by William H. Stuart." Chicago: M. A. Donohue, 1935.

Wooddy, Carl H. The Case of Frank L. Smith: A Study of Representative Government. New York: Morrow, 1974.

THE DUST BOWL DEVASTATION (1932-1935)

An area comprising approximately 96,000,000 acres in the southern part of the Great Plains region of the United States and including parts of Colorado, Kansas, New Mexico, Oklahoma, and Texas is popularly known as the "Dust Bowl." In the early 1930's a combination of factors which had been building up for several decades culminated in heavy damage to the area from wind erosion.

The soil of the Dust Bowl area is largely loessal (wind-deposited), and an additional large quantity is outwash carried down from the Rocky Mountains by rain water. These soils were, in their native state, largely held in place by the hardy grasses which covered the ground, and thus withstood the combination of long recurrent droughts and intermittent heavy rains which are characteristic of the region.

Prior to World War I homesteaders settled in the Dust Bowl area. They planted the land to wheat and row crops. Because of the practicability of large-scale farming, the availability of modern agricultural machinery, and the high price of wheat between 1914 and the middle 1920's, the acreage devoted to that grain was vastly expanded.

Those farmers who did not use their land for agricultural purposes devoted it to the raising of cattle. Both types of land use left the soil exposed to the danger of wind erosion. The natural protective grass was plowed under by the wheat farmer, and the extensive cattle herds trampled this grass into the ground and grazed it short.

The dangers of erosion were augmented by a number of crop failures from 1930 to 1935. A series of severe droughts was experienced, beginning in the early 1930's, and the soil began to be blown away by the winds. In some cases the clay, silt, and organic matter in the soil was carried hundreds of miles away. The heavier components, such as sand, accumulated in drifts against barns, fences, and houses. "In many places three to four inches of topsoil were blown away, and sand and silt dunes four to ten feet in height were formed."

It was estimated that in one area near the center of the Dust Bowl eighty per cent of the land suffered from wind erosion during the drought years. Of this, at least forty per cent was severely eroded.

The combination of crop failure, loss of fertile soil, drought, and the depressed state of the economy in the 1930's meant financial ruin for many Dust Bowl farmers. Many of these migrated west to California with their families, hoping to secure employment there. Others, remaining behind, found it necessary to accept government relief.

In 1935 federal and state governments inaugurated intensive programs to rehabilitate the Dust Bowl and prevent the recurrence of the extensive soil erosion which had wrought such havoc. Large areas have been reclaimed by the seeding of huge tracts in grass, rotating wheat, sorghum, and fallow, and employing the techniques of contour plowing, terracing, and strip planting. Long shelter belts, consisting of trees which break the force of the wind, were planted. Government agronomists estimate that by 1940 the area subject to serious wind damage had been reduced by half, and that since that year it has been further substantially reduced.

Suggested Readings

Andrist, Ralph K., ed. in charge. The American Heritage History of the 20's and 30's. New York: American Heritage, 1970.

Bagnold, R. A. The Physics of Blown Sand and Desert. New York: Halsted Press, 1965.

Bennett, Hugh Hammond. Elements of Soil Conservation. New York: McGraw-Hill, 1955.
Brink, Wellington. Big Hugh: The Father of Soil Conservation. New York: Macmillan, 1951.
Chorn, H. F. "Dust Storms in the S. W. Plains Area" (U. S.), Monthly Weather Review, August, 1936.
"Facts About Winter Dust Storms," U. S. Department of Agriculture Leaflet No. 394, 1955.
Johnson, Vance. Heaven's Tableland: The Dust Bowl Story. New York: Farrar, Straus, 1947.
Parkinson, G. R. "Dust Storms Over the Great Plains," Bulletin of the American Meteorological Society, May, 1956.
Reeves, George S. A Man from South Dakota. New York: Dutton, 1950.
Sears, Paul B. Deserts on the March. Norman, Okla.: University of Oklahoma Press, 1947.
Steinbeck, John. The Grapes of Wrath (fiction). New York: Viking Press, 1939.
Svobida, Lawrence. An Empire of Dust. Caldwell, Idaho: Caxton Printers, 1940.
Tinkle, Lon. The Story of Oklahoma. New York: Random House, 1962.
Warn, G. F. "Some Dust Storm Conditions on the Southern High Plains," Bulletin of the American Meterological Society, 1952.

THE RECONSTRUCTION FINANCE CORPORATION (1932-1953)

The Reconstruction Finance Corporation Act, signed by President Herbert Hoover on January 22, 1932, authorized the formation of the Reconstruction Finance Corporation, or R. F. C. , as it was commonly called. The independent agency created by the Act was "to provide emergency financing facilities for financial institutions; to aid in financing agriculture, commerce, and industry; to purchase preferred stock, capital notes or debentures of banks, trust companies, and insurance companies; and to make loans and allocations of its funds as prescribed by law." Capital stock of $500,000,000 was to be entirely subscribed by the federal government.

In 1948 Congress reduced the capital stock to $100,000,000 and retired the excess. The R. F. C. was also authorized to issue to the United States treasury its own obligations, such as notes, bonds, and debentures, in the amount of its outstanding loans that it might borrow funds with which to perform its functions.

Originally classed as an emergency agency, the R. F. C. , in 1939, was grouped with certain other government units, a consolidation which became the Federal Loan Agency. In 1942 it was transferred to the Department of Commerce and three years later reverted to the Federal Loan Agency. When the latter was abolished in 1947 the R. F. C. assumed its functions.

The R. F. C. was charged with a number of additional responsibilities, one of which was to act as fiscal agent for the disbursement and collection of the loans of the Commodity Credit Corporation of the Department of Agriculture. It also made loans to various federal agencies, banks, trust companies, industrial enterprises, state and local governments, and livestock credit corporations, among others. Huge sums were disbursed by the R. F. C. for the purchase of securities offered by the Public Works Administration, other government agencies, and private business corporations.

In 1951 and 1952 congressional investigations disclosed evidence of corruption and fraud by R. F. C. officials. In July, 1953, the Agency was abolished by congressional action and it was replaced by the Small Business Administration.

Suggested Readings

Andrist, Ralph K., ed. in charge. The American Heritage History of the 20's and 30's. New York: American Heritage, 1970.

Bakke, E. Wright. Citizens Without Work. New Haven, Conn.: Yale University Press, 1940.

_______. The Unemployed Worker. New Haven, Conn.: Yale University Press, 1940.

Bernstein, Irving. The Lean Years: A History of the American Worker, 1920-1933. Boston: Houghton Mifflin, 1960.

_______. The Turbulent Years: A History of the American Worker, 1933-1941. Boston: Houghton Mifflin, 1970.

Goldston, Robert C. The Great Depression: The United States in the Thirties. Indianapolis: Bobbs-Merrill, 1968.

Hatch, Alden. Franklin D. Roosevelt, An Informal Biography. New York: Holt, 1947.

Kimmel, Lewis. Federal Budget and Fiscal Policy, 1789-1958. Washington, D. C.: Brookings Institution, 1959.

Leuchtenburg, William E. Franklin D. Roosevelt and the New Deal, 1932-1940. New York: Harper, 1963.

Moley, Raymond, with the assistance of Elliot A. Rosen. The First New Deal. New York: Harcourt, Brace, 1966.

Romasco, Albert U. The Poverty of Abundance: Hoover, the Nation, the Depression. New York: Oxford University Press, 1965.

Schlesinger, Arthur M., Jr. "The First Hundred Days of the New Deal," in Leighton, Isabel, ed. The Aspirin Age, 1919-1941. New York: Simon & Schuster, 1949.

Schnittkind, Henry Thomas. Franklin Delano Roosevelt. New York: Putnam, 1962.

Seldes, Gilbert. The Years of the Locust: America, 1929-1932. New York: Da Capo Press, 1973.

Shannon, David A. The Great Depression. Englewood Cliffs, N. J.: Prentice-Hall, 1960.

Warren, Harris G. Herbert Hoover and the Great Depression. New York: Oxford University Press, 1959.

Wecter, Dixon. The Age of the Great Depression, 1929-1941. New York: Macmillan, 1948.

THE ULYSSES CASE (1933)

James Joyce, an Irish writer, published his novel Ulysses in France in the year 1922. This book earned international fame for the writer, as well as some degree of notoriety. Basing his story on the Odyssey of the Greek poet Homer, he describes a day in the lives of certain middle class people living in Dublin, Ireland, in the spring of 1904. His principal characters are Leopold Bloom, an Irish Jew, and Stephen Dedalus, an Irish schoolmaster and writer. Joyce's main themes are Bloom's symbolic search for a son and Dedalus' "growing sense of dedication as a writer."

Ulysses is characterized by the technique of "interior monologue" as a means of character portrayal, showing the innermost thoughts of Bloom, Dedalus, and those with whom they come in contact. Certain passages of the book were considered obscene and pornographic and when copies of the book were imported to the United States the customs officials confiscated them under the provisions of the Tariff Act of 1890.

This Act authorized the confiscation and destruction of imported books which were of an "obscene" character. On October 10, 1929, Senator Bronson M. Cutting of New Mexico had, in a vitriolic speech, attacked this provision of the Act.

Following the seizure of the imported volumes suit was brought. The Supreme Court, before which the case was tried in 1933, denied the government's motion for a decree of forfeiture and destruction.

In rendering its decision the Court held that Ulysses was not pornographic because it was not "written for the purpose of exploiting obscenity." The Court held that Joyce had "attempted ... with astonishing success to show how the screen of consciousness with its ever-shifting kaleidoscopic impressions carries ... not only what is in the focus of each man's observation of the actual things about him, but also in a penumbral zone residua of past impressions, some recent and some drawn up by association from the domain of the subconscious. He shows how each of these impressions affects the life and behavior of the character which he is describing."

Continuing, the Court pointed out that "had Joyce not made an honest attempt to develop his peculiar technique, the result would be unfaithful to this technique and so would be artistically inexcusable." The use of words criticized as "dirty" the Court held to be "old Saxon words known to almost all men ... and to many women." They were "words as would be naturally and habitually used by the types of folk whose life, physical and mental, Joyce [was] seeking to describe."

Ulysses was held to be "a sincere and honest book ... the criticisms of which [were] entirely disposed of by its rationale."

In conclusion, the Court stated, "Ulysses may, therefore, be admitted into the United States."

Suggested Readings

Adams, Robert Martin. Surface and Symbol: The Consistency of James Joyce's Ulysses. New York: Oxford University Press, 1962.

Anderson, Chester Grant. James Joyce and His World. New York: Viking Press, 1967.

Brennan, Joseph. Three Philosophical Novelists: James Joyce, André Gide, Thomas Mann. New York: Macmillan, 1964.

Budgen, Frank Spencer Curtis. James Joyce and the Making of Ulysses. Bloomington, Ind.: Indiana University Press, 1960.

Colum, Mary, and Padriac Colum. Our Friend James Joyce. Garden City, N.Y.: Doubleday, 1958.

Commager, Henry Steele, ed. "Ginzburg v. United States," (Doc. No. 670) in his Documents of American History, 8th edition. New York: Appleton, 1968.

________. "United States v. One Book Called 'Ulysses'," (Doc. No. 487) in his Documents of American History, 8th edition. New York: Appleton, 1968.

Curran, C. P. James Joyce Remembered. New York: Oxford University Press, 1968.

Edel, Leon. James Joyce, The Last Journey. New York: Gotham Book Mart, 1947.

Elmann, Richard. James Joyce. New York: Oxford University Press, 1959.

Ernst, M. L., and A. Lindley. The Censor Marches On. New York: Da Capo Press, 1972.

________, and W. Seagle. To the Pure: A Study of Obscenity and the Censor. New York: Da Capo Press, 1970.

Freund, Gisèle, and V. B. Carieton. James Joyce in Paris: His Final Years. New York: Harcourt, Brace, 1965.

Givens, Sean, ed. James Joyce: Two Decades of Criticism. New York: Vanguard Press, 1948.

Goldberg, Samuel Louis. The Classical Temper: A Study of James Joyce's Ulysses. New York: Barnes & Noble, 1961.

________. James Joyce. New York: Grove Press, 1962.

Gorman, Herbert. James Joyce. New York: Farrar & Rinehart, 1939.

Hodgart, Matthew, and Mabel P. Worthington. Song in the Works of James Joyce. New York: Columbia University Press, 1959.

Jenkins, Alan. The Twenties. New York: Universe Books, 1974.

Jones, William Powell. James Joyce and the Common Reader. Norman, Okla.: University of Oklahoma Press, 1955.

Joyce, James. Ulysses. New York: Modern Library, 1946. (First published in 1922.)

Joyce, Stanislaus. My Brother's Keeper. New York: Viking Press, 1958.

Kenner, Hugh. Dublin's Joyce. Bloomington, Ind.: Indiana University Press, 1956.

Magalaner, Marvin. Joyce: The Man, The Work, The Reputation. New York: New York University Press, 1956.
_______. Time of Apprenticeship: The Fiction of James Joyce. New York: Abelard-Schuman, 1959.
Morris, William E., and Clifford A. Nault, Jr., eds. Portrait of an Artist. New York: Odyssey Press, 1962.
Morse, Josiah Mitchell. The Sympathetic Alien. New York: New York University Press, 1959.
Noon, William T. Joyce and Aquinas. New Haven, Conn.: Yale University Press, 1957.
Pound, Ezra Loomis. Pound/Joyce: The Letters of Ezra Pound to James Joyce, With Pound's Essays on Joyce. New York: New Directions, 1967.
Schutte, William M. Joyce and Shakespeare: A Study in the Meaning of Ulysses. New Haven, Conn.: Yale University Press, 1957.
Slocum, John J., and Herbert Cahoon. A Bibliography of James Joyce, 1882-1941. New Haven, Conn.: Yale University Press, 1953.
Strong, Leonard A. G. The Sacred River: An Approach to James Joyce. New York: Pellegrini & Cudahy, 1951.
Sultan, Stanley. The Argument of Ulysses. Columbus, Ohio: Ohio State University Press, 1964.
Thornton, Weldon. Allusions in Ulysses. Chapel Hill, N.C.: University of North Carolina Press, 1968.
Tindall, William York. James Joyce: His Way of Interpreting the Modern World. New York: Scribner's, 1950.
_______. A Reader's Guide to James Joyce. New York: Noonday Press, 1959.

THE TENNESSEE VALLEY AUTHORITY (1933)

The Tennessee Valley Authority, commonly known as the T.V.A., is one of forty-odd wholly government-owned corporations. Under the sponsorship of Senator George W. Norris it was created by the Tennessee Valley Act of May 18, 1933. It is, in some respects, strikingly different from any other national government agency. It employs over 24,000 people and, unlike other agencies which are created to perform a single major function, is charged with the responsibility of administering a whole array of diversified tasks. It is authorized to control floods, improve navigation, and generate and sell electric power.

The overall object of creating the T.V.A. was to improve the economic and social conditions of the Tennessee Valley region. These aims were achieved by activities in reforestation, industrial and community development, test-demonstration farming, and the establishment of recreational facilities.

Flood control involved excess water both in rivers and on land. The construction of reservoirs and dams has eliminated floods in the Tennessee Valley. Nine dams were built on the

Tennessee River as well as a series of dams on the upper tributaries. These dams make it possible to maintain a constant water level which, in turn, permits the transportation of tons of freight by water.

Electric current was produced by additional steam generators constructed for that purpose. Most but not all of this electricity is sold to distributors rather than direct to consumers.

Soil conservation was effected by the manufacture and distribution of fertilizers which enabled the eroded land to produce vegetation which would hold soil on the land areas draining into the reservoirs. Over 150,000,000 trees were planted, with the double object of producing timber and holding the topsoil.

Other activities of the T.V.A. include the development of a program for retrieving the mineral wealth of the region, the stocking of rivers and reservoirs with game fish, control of the malaria mosquito, and the establishment of libraries and park areas.

Suggested Readings

Chase, Stuart. Rich Land, Poor Land. New York: AMS Press, 1969.

Clapp, Gordon R. The T.V.A.: An Approach to the Development of a Region. Chicago: University of Chicago Press, 1955.

Commager, Henry Steele, ed. "The Tennessee Valley Act," (Doc. No. 479) in his Documents of American History, 8th edition. New York: Appleton, 1968.

Dillon, Marye. Wendell Willkie. Philadelphia: Lippincott, 1952.

Duffus, R. L. The Valley and Its People. New York: Knopf, 1944.

Finer, Herman. The T.V.A.: Lessons for International Application. Montreal: International Labor Office, 1944.

Hall, Ford P., Pressly S. Sikes, and others. American National Government: Law and Practice. New York: Harper, 1949.

Hatch, Alden. Franklin D. Roosevelt, An Informal Biography. New York: Holt, 1947.

Leuchtenburg, William E. Franklin D. Roosevelt and the New Deal, 1932-1940. New York: Harper, 1963.

Lilienthal, David E. T.V.A.: Democracy on the March. New York: Harper, 1953.

Lord, Russell. Behold Our Land. New York: Da Capo Press, 1974.

Lowitt, Richard. George W. Norris: The Persistence of a Progressive. Urbana, Ill.: University of Illinois Press, 1971.

Norris, George W. Fighting Liberal: The Autobiography of George W. Norris. New York: Macmillan, 1945.

Peare, Catherine O. The FDR Story. New York: Crowell, 1962.

Pritchett, C. Herman. The Tennessee Valley Authority. Chapel Hill, N.C.: University of North Carolina Press, 1943.

Riesch, Anna Lou. Conservation Under Franklin D. Roosevelt (unpublished Ph.D. dissertation, University of Wisconsin, 1952).

Roosevelt, James, and Sidney Shalett. Affectionately, F.D.R.: A Son's Story of a Lonely Man. London: Harrap, 1960.

Satterfield, M. H. "T. V. A. --State-Local Relationships," American Political Science Review, XL, 1946.
Schnittkind, Henry Thomas. Franklin Delano Roosevelt. New York: Putnam, 1962.
Selznick, Philip. T. V. A. and the Grass Roots. New York: Harper, 1953.
Spero, Sterling D. Government Jobs. Philadelphia: Lippincott, 1945.
Weingast, David Elliott. Franklin D. Roosevelt, Man of Destiny. New York: Messner, 1952.

THE UNION STATION MASSACRE (1933)

One of the most spectacular attempts to rescue a convicted criminal from the hands of the law occurred at the Union Station, Kansas City, Missouri, on the morning of Saturday, June 17, 1933. It was then that outlaws Verne Miller, Arthur "Pretty Boy" Floyd and Adam Richetti shot and killed Frank "Jelly" Nash, an Oklahoma police chief and two Kansas City law officials. In addition a guard and two unarmed F. B. I. agents were wounded, one mortally.

Miller was a former sheriff and World War I soldier turned gangster. Floyd was a bank robber and murderer, as was Richetti. The three were attempting to free Nash from F. B. I. agents who had captured him following his escape from Leavenworth prison and were returning him to custody.

Frank "Jelly" Nash had had a checkered career as a criminal. In 1913 he had shot and killed an accomplice in a petty theft. For this he was tried, convicted and sentenced to serve life in the Oklahoma State Penitentiary. He was paroled in 1918 but in 1920 was back in prison, having been convicted of safecracking. In 1922 he was discharged and promptly became a member of a gang of bank robbers headed by Al Spencer. Spencer was killed following an attempt to rob a train. Nash, who was a participant in the robbery, made his escape. The Federal Bureau of Investigation entered the case when it was learned that the robbers had stolen a pouch of registered mail, thus involving the federal authorities.

After various escapades Nash was captured in El Paso, Texas. He was sent to the federal penitentiary at Leavenworth, from which he later escaped. After participating in a series of bank robberies he was captured again, this time in Hot Springs, Arkansas, by F. B. I. agents Joe Lackey, Frank Smith and police chief Otto Reed of McAlester, Oklahoma. News of the capture was relayed to Verne Miller in Kansas City. Miller was Nash's friend and had participated in bank robberies with him.

Miller and Pretty Boy Floyd, with the assistance of Adam Richetti, arranged to attempt to rescue Nash from his captors when they arrived by train at the Kansas City Union Station on his way to Leavenworth. Floyd and Richetti, strangers to Miller, were enlisted by Johnny Lazia, a local racketeer to whom they "owed a favor." Lazia was a member of the political machine dominated by Tom Pendergast, who controlled Kansas City's vice, crime and rackets.

On the morning of June 17 the Missouri Pacific train pulled into the Union Station on schedule. Miller, Floyd and Richetti were sitting in a parked car awaiting the train's arrival. Nash and his three guards, Lackey, Smith and Reed, left the train. They were met by F. B. I. agents Ray Caffrey and Reed Vetterli and by Kansas City police detectives W. J. "Red" Grooms and Frank Hermanson.

Nash and his escort proceeded to Caffrey's two-door sedan. The prisoner was instructed to enter and sit behind the wheel until the others could enter. He was then to slide over to the right.

At this moment Miller, Floyd and Richetti, all heavily armed, stepped from between the parked cars and opened fire. Nash was killed, as were Reed and detectives Grooms and Hermanson. Lackey was hit three times and Vetterli received a flesh wound in the arm. Caffrey died on the way to the hospital. Only Frank Smith was unhurt. None of the three bandits had been wounded. Within seconds they had reentered their car and raced away.

Though badly hurt, Joe Lackey eventually recovered. In due course Adam Richetti, Verne Miller and Pretty Boy Floyd were found to have been the ones who participated in the massacre. A year later Miller was killed by gangster associates near Detroit. Floyd was killed in October, 1934, on a farm near East Liverpool, Ohio, while attempting to drive to a sanctuary in the Oklahoma Cookston Hills, near his home. Richetti died in the gas chamber at Missouri State Penitentiary on October 7, 1938, having been tried and convicted on a charge of murdering Frank Hermanson.

The Union Station Massacre resulted in Congress passing new legislation authorizing the arming of F. B. I. agents and giving them authority to make arrests, both of which were previously forbidden.

Suggested Readings

Clayton, Merle. Union Station Massacre: The Shootout That Started the FBI's War on Crime. Indianapolis: Bobbs-Merrill, 1975.

Dorsett, Lyle W. The Pendergast Machine. New York: Oxford University Press, 1968.

Elman, Robert. Fired in Anger. Garden City, N. Y.: Doubleday, 1968.

"Five Slain in Battle by Gang to Free Oklahoma Bandit," New York Times, June 18, 1933, p. 1.

"Floyd Shoots Way Out of Iowa Trap," New York Times, October 12, 1934, p. 1.

Milligan, Maurice Morton. The Inside Story of the Pendergast Machine by the Man Who Smashed It. New York: Scribner's, 1948.

"Pretty Boy Floyd Slain as He Flees by Federal Men," New York Times, October 23, 1934, p. 1.

Reddig, William M. Tom's Town: Kansas City and the Pendergast Legend. Philadelphia: Lippincott, 1947.
Sanders, Bruce. "Slaughter at Union Station," in his Murder in Big Cities. New York: Roy Publishers, 1962.
Steckmesser, Kent Ladd. "Oklahoma Robin Hood," American West, January, 1970.

THE TUFVERSON DISAPPEARANCE (1933)

Agnes Colonia Tufverson was, at 43, a spinster and a highly successful New York attorney. In the summer of 1933 she vacationed in Europe and on the boat train between London and Southampton met Captain Ivan Ivanovich Poderjay, a charming Yugoslavian army officer. She fell madly in love and returned to New York with him, an engaged woman. She was not aware that her fiancé was a confidence man with a checkered past, or that he was already married. On December 4, 1933, the two were "united in holy wedlock" at the Little Church Around the Corner in New York City.

The newly wedded couple booked passage on the steamship "Hamburg," planning to spend their honeymoon in Europe. The "Hamburg" was scheduled to sail on December 20. On that day the new bride called her two sisters in Detroit to say goodbye and to introduce them to her husband by telephone. On January 2, 1934, the sisters received a cable signed "Agnes" and reading "Going to India via France." They never heard from their sister again.

When after three months Agnes failed to appear at her office, the police were notified. The police conducted an investigation and learned that, following his marriage, Poderjay had moved into his wife's New York apartment and that he had brought with him a steamer trunk and some other luggage. The investigation also confirmed that on December 19 Agnes had had a conversation with Paul Shapiro, proprietor of a cleaning and dyeing establishment.

On December 20 Eva, Agnes' Negro maid, had helped the newlyweds pack for their honeymoon. A discrepancy arose when Agnes' sisters disclosed that in their telephone conversation Poderjay had said that he intended to live in Europe permanently, but that Agnes had expressed her intention of returning to New York not later than the following April. The police learned also that Poderjay had, on December 20, made several purchases, one of which was ten dollars' worth of razor blades.

Clement Price drove Agnes and Poderjay and their luggage to the Hamburg-American Line pier in his taxi on the afternoon of December 20. They returned to the apartment late that night without their luggage. Eva, who was still cleaning up, was, with the exception of Poderjay, the last person to see Agnes alive. When Eva returned to the apartment on December 22 she found Poderjay there alone. He told her that his wife had gone to Philadelphia, but the police later learned that he had told other tenants in the building that she had gone to Europe ahead of him. Shortly afterwards Poderjay left the apartment, taking with him a trunk and six handbags. He boarded the White Star steamer "Olympic," on which he had booked a single passage, and sailed for Europe.

The New York police investigation disclosed further facts. Agnes Tufverson had withdrawn all her money from the bank where it had been deposited and had given Poderjay a $5,000 draft on a London bank. Her securities were not to be found. In checking with Scotland Yard in London, the New York authorities learned that Poderjay had been charged with swindling two Englishwomen and an English musician. These people refused to prosecute, hoping that they might eventually get their money back. He was also found to be a deserter from the French Foreign Legion.

Eventually Poderjay was located in Vienna, living with a French woman named Suzanne Marguerite Ferrand. At the request of the New York police the Viennese authorities investigated Poderjay's past and brought out details of a career which included many unsavory affairs with women, active service in the Austrian and Yugoslavian armies and the French Foreign Legion, and a brief post as a banker. He had bigamously married a Madam Zhika in 1926 and had "divorced" her in 1933, shortly before his marriage to Agnes Tufverson.

The Viennese police took Poderjay into custody. He claimed to be completely ignorant of Agnes' whereabouts. His story was that he had married her "in name only" as a favor to her, she not wishing to return to the United States still a 43-year-old spinster. He had, he said, married Suzanne Marguerite Ferrand in London and had quarreled at the New York dock with Agnes because she wished to travel with him as his wife, although he told her that he already had a wife. Accused of committing bigamy, he disagreed. His marriage to Agnes, as a civil ceremony and mere act of kindness, did not count; to a Yugoslav Catholic only a Catholic religious ceremony was valid. He told conflicting stories concerning the sources of his funds, but apparently these were furnished him by the various women with whom he had become involved. He remained in the custody of the Viennese police pending further investigation.

In New York the authorities continued their own investigation. Feeling that the trunks used by Agnes and Poderjay might furnish useful information, they checked into the luggage which had been removed from the apartment. Eva, the maid, accounted for three trunks which had been taken to the "Hamburg" on December 20, and stated further that when she last saw Poderjay in the apartment on December 22, he had with him another trunk she had never seen before. The police then learned that Poderjay had purchased a trunk from Lipkin Brothers, New York luggage dealers, on December 21. Further investigation disclosed that Poderjay had had his luggage carted to the White Star pier, and that the luggage included one trunk and six handbags. White Star's records showed that the "Olympic" carried three trunks belonging to Poderjay when he left New York on December 22. Two "Olympic" stewards recalled Poderjay's having one trunk with him in his cabin. "It was open most of the time," they said, "and we saw it through customs."

Two trunks containing articles which had belonged to Agnes Tufverson were found in Poderjay's Vienna apartment. Also found there were letters written by Suzanne Marguerite Ferrand to Poderjay, urging him to get money "no matter how ... to enable us to live as we may desire." It was apparent that Poderjay had married Agnes for her money.

The Viennese police and New York authorities concluded that Agnes Tufverson was dead. Four trunks had been taken from the New York apartment and only three were accounted for, those which had gone aboard the "Olympic." The police felt that the fourth trunk, containing Agnes' dismembered body, had been sent to Poderjay's outside stateroom and that he had dropped it over the side of the ship in midocean. How he got the trunk aboard undetected was never explained, provided he did get it aboard.

As no corpus delicti was ever found, it was not possible to charge Poderjay with murder. He was extradited to the United States, tried for and convicted of bigamy, sentenced to and served five years in Auburn Prison, and then deported. In his native country he dropped from sight.

Suggested Readings

Churchill, Allen. They Never Came Back. Garden City, N.Y.: Doubleday, 1960.

Cohen, Bruce J. Crime in America. Itasca, Ill.: Peacock, 1970.

Gribble, Leonard. They Had a Way with Women. New York: Roy Publishers, 1967.

Hibbert, Christopher. Roots of Evil: A Social History of Crime and Punishment. New York: Funk & Wagnalls, 1968.

"His 'Second Wife' Missing Here, Captain Is Arrested in Vienna," New York Times, June 14, 1934, p. 1.

"Mystery Stirs Police in Three Cities," Newsweek, June 23, 1934.

"Poderjay Is Held on Formal Charges as Murder Suspect," New York Times, June 23, 1934, p. 1.

"Poderjay Luggage Held No Secrets, Stewards Assert," New York Times, June 21, 1934, p. 1.

"Poderjay Mystery Centered on Trunk," New York Times, June 17, 1934, p. 1.

"Poderjay Trunk Hunted in Europe," New York Times, June 18, 1934, p. 4.

"Poderjay's Bride Is Believed Dead by Vienna Police," New York Times, June 19, 1934, p. 1.

"Police Seek Trunk in Tufverson Case," New York Times, June 15, 1934, p. 1.

"Police Widen Hunt in Tufverson Case," New York Times, June 16, 1934, p. 1.

"Red Stains in Tufverson Trunk Are Analyzed by Police in Vienna," New York Times, June 20, 1934, p. 1.

Rinehart, J. C. "Tracing People, Lost, Stolen and Strayed Away," American City, October, 1929.

Sterling, Hank, pseud. "The Absent Corpus Delicti and the Bigamous Captain," in his Ten Perfect Crimes. New York:

Stravon Publishers, 1954.
Sutherland, Edwin H. On Analyzing Crime. Edited by Karl Schussler. Chicago: University of Chicago Press, 1972.
Winslow, Robert W. Crime in a Free Society. Encino, Calif.: Dickenson, 1968.

THE NATIONAL INDUSTRIAL RECOVERY ACT (1933-1936)

The National Industrial Recovery Act, passed by Congress on June 16, 1933, was part of President Franklin D. Roosevelt's New Deal program to restore economic stability following the Wall Street Crash of October, 1929, and the subsequent depression.

By 1933 industrial production and commercial activity were in the doldrums. Unemployment was rampant and widespread. Many businessmen found themselves unable to avoid bankruptcy, and stringent corrective action was needed.

The N. I. R. A. called for the drafting by the various trade associations of codes which would represent standards of fair competition within each industry, would regulate wages and working hours, eliminate child labor, and be enforceable at law. It also dealt with the rights of employees to form unions for bargaining purposes and prohibited employers from requiring their employees to join or not to join any particular labor union or organization. In addition, it authorized the President to inaugurate a system of public works projects.

Following the passage of the Act, Roosevelt established the National Recovery Administration (N. R. A.) and appointed General Hugh S. Johnson administrator. The N. R. A. was symbolized by a blue eagle which was to appear on merchandise produced under the appropriate code.

Johnson, during the following eighteen months, set up codes for approximately 98% of American industry. Unemployment, as a result, showed some decline and industrial production made substantial gains. A blanket code, called the President's Reemployment Program, was prepared. This provided for a forty-hour work-week, a minimum wage and the virtual elimination of children under sixteen from the work force.

Various subsidiary boards were established to assist in the operation of the N. R. A., one of which was the important National Labor Board. This organization, headed by Senator Robert F. Wagner of New York, was concerned with the settlement of industrial disputes. This Board was replaced, about a year later, by the National Labor Relations Board, which was in no way connected with the N. R. A.

Eventually complaints were received in Washington that many industrialists were deliberately violating the N. R. A. codes. It became increasingly difficult for employers to maintain "the lowest schedule of prices on which higher wages and increasing employment can be maintained" because, as payrolls went up, retail prices also rose. Various prominent men, including Paul Block, William

Randolph Hearst, Alfred E. Smith, Gerard Swope, Henry I. Harriman, and Ogden Mills, became increasingly critical of the entire New Deal program, declaring it socialistic, bureaucratic, and un-American. This led to a sweeping reorganization of the Act in September, 1934.

Finally, in May, 1935, the constitutionality of the National Recovery Act and other aspects of the New Deal were passed on by the United States Supreme Court. Two cases, the Hot Oil Case and the Schecter Poultry Case, were decided, and in both of them the N. R. A. was declared unconstitutional. The Farm Moratorium Case, heard in the same month, found the Frazier-Lemke Farm Bankruptcy Act, a part of the program of agricultural relief, also unconstitutional. In general, the Court held that Congress could not legally delegate its legislative authority and that regulation of intrastate commerce by federal authorities was illegal.

As a result of this, President Roosevelt terminated the life of the National Recovery Act on January 1, 1936. Later, some of its more important features were reenacted.

Suggested Readings

Allen, Frederick Lewis. Since Yesterday: The Nineteen-Thirties in America, September 3, 1929-September 3, 1939. New York: Harper, 1940.

Andrist, Ralph K., ed. in charge. The American Heritage History of the 20's and 30's. New York: American Heritage, 1970.

"A Balance Sheet to the New Deal," New Republic, June 10, 1936.

Beard, Charles A., and George H. Smith. The Future Comes. Westport, Conn.: Greenwood Press, 1972.

Bernstein, Irving. The Turbulent Years: A History of the American Worker, 1933-1941. Boston: Houghton Mifflin, 1970.

Brown, Douglas V., et al. The Economics of the Recovery Program. New York: Da Capo Press, 1971.

Commager, Henry Steele, ed. "Cotton Textile Code," (Doc. No. 485) in his Documents of American History, 8th edition. New York: Appleton, 1968.

________. "The National Recovery Act," (Doc. No. 484) in his Documents of American History, 8th edition. New York: Appleton, 1968.

________. "Schecter Poultry Corp. v. United States," (Doc. No. 486) in his Documents of American History, 8th edition. New York: Appleton, 1968.

Dearing, Charles L., et al. The ABC of the NRA. Dubuque, Ia.: William C. Brown, 1934.

Goldston, Robert C. The Great Depression: The United States in the Thirties. Indianapolis: Bobbs-Merrill, 1968.

Hatch, Alden. Franklin D. Roosevelt: An Informal Biography. New York: Holt, 1947.

Ickes, Harold L. Back to Work. New York: Da Capo Press, 1935.

Johnson, Hugh S. The Blue Eagle from Egg to Earth. New York: Doubleday, 1935.

Leuchtenberg, William E. Franklin D. Roosevelt and the New Deal, 1932-1940. New York: Harper, 1963.
Lingley, Charles Ramsdell, and Allen Richard Foley. Since the Civil War. New York: Appleton, 1935.
Lyon, Levere H., et al. The National Recovery Administration. New York: Da Capo Press, 1972.
Moley, Raymond, with the assistance of Elliot A. Rosen. The First New Deal. New York: Harcourt, Brace, 1966.
Peare, Catherine O. The FDR Story. New York: Crowell, 1962.
Schlesinger, Arthur M., Jr. The Age of Roosevelt. Boston: Houghton Mifflin, 1957.
Schnittkind, Henry Thomas. Franklin Delano Roosevelt. New York: Putnam, 1962.
Seldes, Gilbert. The Years of the Locust: America, 1929-1932. New York: Da Capo Press, 1973.
Shannon, David A. The Great Depression. Englewood Cliffs, N.J.: Prentice-Hall, 1960.
Wecter, Dixon. The Age of the Great Depression, 1929-1941. New York: Macmillan, 1948.

THE CIVILIAN CONSERVATION CORPS (1933-1942)

The purpose of the Civilian Conservation Corps, or C.C.C., set up by the federal government in April, 1933, was twofold: to establish a program for the conservation of such natural resources as soil, timber, and water, and to provide work and training for young unmarried men who were unemployed.

The Bill known as the C.C.C. Reforestation Act, establishing the agency, was one of four signed on the last day of March, 1933. The agency was originally known as Emergency Conservation Work but in June, 1937, when Congress extended its period of operation, the popular name, Civilian Conservation Corps, was made official. The agency became part of the Federal Security Agency in 1939 and in June, 1942, Congress voted to abolish the C.C.C. within one year. About six months later it was liquidated by Presidential Order.

During its nine and a quarter-year existence the Civilian Conservation Corps furnished employment for some three million unmarried male citizens of the United States between the ages of 18 and 25. These enrollees were lodged in camps set up in various parts of the country. Under the supervision of army officers the men were set to work on forest-clearing and planting, road-building, and flood control projects. Each man received, in addition to board, a cash payment of $30 monthly, of which at least $25 was sent home to dependents.

While the primary purpose of the Civilian Conservation Corps was to provide employment and training for those enrolled in the program, it added somewhere between one and a half and two billion dollars to the value of the national domain at a cost of under three billion dollars. Some four thousand fire observation towers were constructed, three billion trees were planted, miles of telephone lines were installed, and many forest trails were constructed. It was also found that the program improved the health and morale of the young men assigned to it. Educational courses were made available to the participants, thus benefiting young men deprived of further schooling elsewhere, and some writers have referred to the C.C.C. as the "University of the Woods."

Suggested Readings

Allen, Frederick Lewis. Since Yesterday: The Nineteen-Thirties in America, September 3, 1929-September 3, 1939. New York: Harper, 1940.

Andrist, Ralph K., ed. in charge. The American Heritage History of the 20's and 30's. New York: American Heritage, 1970.

Bakke, E. Wright. Citizens Without Work. New Haven, Conn.: Yale University Press, 1940.

_______. The Unemployed Worker. New Haven, Conn.: Yale University Press, 1940.

Beard, Charles A., and George H. Smith. The Future Comes. Westport, Conn.: Greenwood Press, 1972.

Bernstein, Irving. The Lean Years: A History of the American Worker, 1920-1933. Boston: Houghton Mifflin, 1960.

_______. The Turbulent Years: A History of the American Worker, 1933-1941. Boston: Houghton Mifflin, 1970.

Conkin, Paul. Tomorrow a New World. Ithaca, N.Y.: Cornell University Press, 1959.

Goldston, Robert C. The Great Depression: The United States in the Thirties. Indianapolis: Bobbs-Merrill, 1968.

Hatch, Alden. Franklin D. Roosevelt, An Informal Biography. New York: Holt, 1947.

Holland, Kenneth, and Frank Ernest Hill. Youth in the CCC. Washington, D.C.: Brookings Institution, 1942.

Leuchtenburg, William E. Franklin D. Roosevelt and the New Deal, 1932-1940. New York: Harper, 1963.

Moley, Raymond, with the assistance of Elliot A. Rosen. The First New Deal. New York: Harcourt, Brace, 1966.

Rawick, George. The New Deal and Youth (unpublished Ph.D. dissertation, University of Wisconsin, 1957).

Schnittkind, Henry Thomas. Franklin Delano Roosevelt. New York: Putnam, 1962.

Seldes, Gilbert. The Years of the Locust: America, 1929-1932. New York: Da Capo Press, 1973.

Shannon, David A. The Great Depression. Englewood Cliffs, N.J.: Prentice-Hall, 1960.

Terkel, Studs. Hard Times: An Oral History of the Great Depression in America. New York: Pantheon Books, 1970.

Wallace, Henry A. New Frontiers. Westport, Conn.: Greenwood Press, 1934.

Wecter, Dixon. The Age of the Great Depression, 1929-1941. New York: Macmillan, 1948.

THE ALL-AMERICAN SOAP BOX DERBY (1933-1976)

In the early 1930's Myron E. Scott, an American newspaper photographer, conceived the idea of conducting a contest between boys driving home-made gravity-propelled coaster wagons. Scott chose the name "soap box derby" for his contest in order to emphasize the use of such inexpensive materials as soap boxes in the construction of the vehicles.

The first Derby was held at Dayton, Ohio, in 1933. It drew 300 entries and was witnessed by a crowd of over 40,000. The idea of the Derby spread rapidly and similar contests were held in various cities throughout America. In subsequent years Derbies became annual occurrences and were held in a number of large cities. The event was sponsored by the Chevrolet Division of the General Motors Corporation plus local newspapers, radio and television stations, and fraternal and civic organizations. National headquarters of the Derby are in Detroit, Michigan; and Akron, Ohio became the permanent home of the Derby in 1935.

Derby Downs was opened in 1936. This is a specially designed race course for gravity-propelled vehicles. It consists of a three-lane sloping track 975.4 feet in length and with an average grade of six per cent. The track is surrounded by grandstands holding more than 60,000 seats. Racers have completed the course in as short a time as 27 seconds, attaining a speed of about 26 miles per hour from a standing start. The speeds of the contestants are measured by electronic timing devices and motion picture cameras are also used to check the order in which the racers finish.

The boys entering the Soap Box Derby must be at least eleven and not more than fifteen years of age. The rules require that the contestants design and build their own vehicles and that the component parts cost no more than fifteen dollars, exclusive of axles and wheels, which are furnished by local sponsors. Vehicles must not exceed a wheelbase of forty inches, a length of eighty inches, or a height of 28 inches. The total weight of vehicle and driver must be not more than 250 pounds.

Local races are held in various cities throughout the United states and abroad to determine the champion drivers in these areas. These local champions, totaling not more than 239, become eligible to enter the All-American Derby at Akron. All expenses of the competition in the Akron Derby are paid, and the top nine winners are rewarded with substantial scholarships. The top winner receives a $5,000 four-year college scholarship.

While the rules of the Derby specify that each boy entering shall design and build his own machine, there have, unfortunately, been instances where older people, such as the parents of an entrant, in their desire to see their sons win, have given them illegal assistance and have even equipped some vehicles with hidden devices which increase the vehicles' speed. When evidence of such fraudulence is discovered the car and driver are disqualified.

During the World War II years (1942-45) the running of the Soap Box Derby was suspended, but it was resumed in 1946.

Suggested Readings

"All-American Soap Box Derby Is Back," Industrial Arts and Vocational Education, April, 1946.
"Boy Speed Kings and the Racing Cars They Build," Scholastic, April 8, 1940.
"Design For Winning: Soap Box Racers Get Streamlined," Scholastic, February 3, 1941.
Gibson, G. "Watergate on Wheels: J. Gronen's Disqualification in the 1973 All-American Soap Box Derby," Ladies' Home Journal, August, 1974.
Horrell, B. "Soapbox Derby," Industrial Arts and Vocational Education, May, 1941.
Jackson, A. J., and W. Glover. "Derby Day in Burbank," Recreation, March, 1946.
"Pointers on Soap Box Racer Design," Popular Mechanics, April, 1947.
Radlauer, Edward. Soap Box Racing. Chicago: Children's Press, 1973.
"Soap Box Derby Launched," American City, March, 1946.
Telander, R. "Running the Gauntlet of Grownups: D. Abramovitz in the Oak Forest, Ill., Derby," Sports Illustrated, August 12, 1974.
"Tuning Up a Soap Box Racer," Popular Science, August, 1940.
Woodley, R. "How to Win the Soap Box Derby: Disqualification of James Gronen," Harper's Magazine, August, 1974.

THE FIRESIDE CHATS (1933-1945)

When, on March 4, 1933, Franklin Delano Roosevelt took office for his first term as President of the United States, the country was suffering from a major business and economic depression. Breadlines were commonplace, mortgage foreclosures were mounting, banks were failing, unemployed men were selling apples on street corners, and despair was widespread.

Roosevelt "created an immediate impression of action and initiative." His "New Deal," as his economic and social measures came to be called, included unemployment relief, farm mortgage aid, a civilian conservation corps, the Tennessee Valley Authority, the National Industrial Recovery Act and the Agricultural Adjustment Act. He declared a "bank holiday," closing banks for federal inspection and permitting those in good financial condition to reopen. He called Congress into special session and the "Hundred Days" between March 9 and June 16, 1933 saw more legislation passed than ever before.

It was Roosevelt's idea that the American public should be informed of his purposes, actions, plans, and objectives and be kept currently advised of their progress. In order to facilitate this he inaugurated his "fireside chats," radio broadcasts heard from coast to coast.

The first such broadcast was made on March 12, 1933. Speaking from the Diplomatic Reception Room of the White House in Washington, Roosevelt, in a friendly, informal manner, made millions of people feel that they were closer to their President than ever before in history. Invariably starting with the words, "My friends," a salutation he had used since his earliest days in politics, and delivered in a relaxed, intimate voice, he counseled, "Let us unite in banishing fear," emphasizing that the problems of the day were the people's no less than his to meet and solve.

These "fireside chats" became a regular feature of the Roosevelt administration. On December 29, 1940, he indicated that the United States must extend full industrial aid to the British war effort. He stated, "We must be the great arsenal of democracy."

Suggested Readings

Allen, Frederick Lewis. Since Yesterday: The Nineteen-Thirties in America, September 3, 1929-September 3, 1939. New York: Harper, 1940.

Andrist, Ralph K., ed. in charge. The American Heritage History of the 20's and 30's. New York: American Heritage, 1970.

Bernstein, Irving. The Turbulent Years: A History of the American Worker, 1933-1941. Boston: Houghton Mifflin, 1970.

Chase, Stuart. Rich Land, Poor Land. New York: AMS Press, 1969.

Divine, Robert A. Roosevelt and World War II. Baltimore: Johns Hopkins University Press, 1969.

Goldston, Robert C. The Great Depression: The United States in the Thirties. Indianapolis: Bobbs-Merrill, 1968.

Hatch, Alden. Franklin D. Roosevelt, An Informal Biography. New York: Holt, 1947.

Leuchtenburg, William E. Franklin D. Roosevelt and the New Deal, 1932-1940. New York: Harper, 1963.

Moley, Raymond, with the assistance of Elliot A. Rosen. The First New Deal. New York: Harcourt, Brace, 1966.

Peare, Catherine O. The FDR Story. New York: Crowell, 1962.

Rawick, George. The New Deal and Youth (unpublished Ph.D. dissertation, University of Wisconsin, 1957).

Roosevelt, James, and Sidney Shalett. Affectionately, F.D.R.; A Son's Story of a Lonely Man. London: Harrap, 1960.

Rosenman, Samuel, comp. The Public Papers and Addresses of Franklin D. Roosevelt. New York: Random House, 1938-1950.

Schlesinger, Arthur M., Jr. The Age of Roosevelt. Boston: Houghton Mifflin, 1957.

Schnittkind, Henry Thomas. Franklin Delano Roosevelt. New York: Putnam, 1962.

Seldes, Gilbert. The Years of the Locust: America, 1929-1932. New York: Da Capo Press, 1973.

Shannon, David A. The Great Depression. Englewood Cliffs, N.J.: Prentice-Hall, 1960.

Terkel, Studs. Hard Times: An Oral History of the Great

Depression in America. New York: Pantheon Books, 1970.
Wecter, Dixon. The Age of the Great Depression, 1929-1941. New York: Macmillan, 1948.
Weingast, David Elliott. Franklin D. Roosevelt, Man of Destiny. New York: Messner, 1952.

THE "MORRO CASTLE" DISASTER (1934)

One of the major marine disasters of the 20th century occurred on the morning of September 8, 1934, when the American Ward Line cruise ship "Morro Castle" burned off the New Jersey coast. Eighty-six passengers and forty-nine crew members died. The ship was a total loss.

The "Morro Castle" was launched at Newport News, Virginia, in 1930. She had been designed by Theodore E. Ferris, a prominent naval architect who combined speed, safety, and luxury in his plans. Special emphasis was placed on apparatus for detecting and fighting fires. Unfortunately, however, no fire detectors were installed in such places as the library, the writing room, the dining room, or the ballroom.

Following launching the "Morro Castle" was placed on the New York-Havana run, making one round trip each week. In addition to vacationing passengers she carried mail and general cargo.

On the night of Friday, September 7, 1934, she was completing her 174th trip, bound for New York. She was off the Delaware Capes and in the midst of a storm of near-hurricane force. Her captain, Robert R. Wilmott, had died in his cabin that evening of a heart attack and the ship was under the command of the former mate, acting captain William F. Warms. Sometime after midnight the smell of smoke was detected coming from a locker in the writing room. At 2:56 A.M. an officer opened the locker door and found the interior a mass of flames. He gave the alarm but by 3:30 the "Morro Castle" was burning wildly. The anchor had been dropped and she was three miles off Sea Girt, New Jersey. Passengers and crewmen jumped into the swirling sea. What lifeboats were lowered carried only a small percentage of their full capacities.

The burning ship was visible from Asbury Park, New Jersey, and throngs of people came to view it. Some bought admission to the Convention Hall, from which they could catch a glimpse of the wreck from second-story windows.

George Rogers, chief radio operator, sent out distress signals before the fire could put his transmitter out of commission. The ships "Andrea Luckenbach," "City of Savannah," and "Monarch of Bermuda" responded, rescuing some of the victims from the sea. The cutter "Tampa" and the sea-going tugs "Willett" and "Moran" approached the burning vessel. The anchor chain was cut and the "Tampa" took the "Morro Castle" in tow. However, the hawser connecting the two vessels snapped and the "Morro Castle" was driven to shore by the wind. There she grounded on a sand bank where she stayed until March 14, 1935, when she was towed away to a Baltimore shipyard to be cut up for scrap.

The investigation which followed the disaster disclosed that the "Morro Castle" was an anything but happy ship. She was understaffed with poorly paid, incompetent seamen. Fire drills were virtually nonexistent and few crew members had been instructed in their duties in case of fire. Captain Wilmott had delegated no authority to Mate Warms, whose functions under Wilmott's command had been little more than those of an order-carrier. Following the outbreak of the fire, passengers and crew panicked and no concerted leadership was evidenced by the ship's officers.

The cause of the fire has never been satisfactorily determined. Some attributed it to a carelessly tossed cigarette. Others felt that spontaneous combustion may have been the cause. Still others charged arson. It was shown that the wood paneling in the writing room and adjacent areas was highly flammable, which permitted the fire, once started, to spread rapidly through the ship.

Following the investigation and a hearing by the Steamboat Inspection Service, the United States Attorney's Office looked into the criminal aspects of the case. On December 3, 1934, a grand jury handed down indictments against Warms, certain other officers of the "Morro Castle," and her owners, the Atlantic, Gulf, and West Indies Steamship Company. The trial ended in January, 1936, with all defendants being found guilty. The steamship company was fined $10,000 and the defendants sentenced to fines and/or prison terms. In the following year the United States Court of Appeals reversed the convictions of Warms and the others but let the fines stand.

Some years later some circumstantial evidence suggesting that George Rogers, the chief radio operator, may have deliberately set the "Morro Castle" afire, came to light. Rogers' guilt, however, was never proven. He died in New Jersey State Prison on January 10, 1958, having been convicted of a first degree murder which had nothing to do with the ill-fated "Morro Castle."

Suggested Readings

Andrews, Mary Evans. "Hotel Fire at Sea," in Kartman, Ben, and Leonard Brown, eds. Disaster! New York: Pellegrini & Cudahy, 1948.

Andrist, Ralph K., ed. in charge. The American Heritage History of the 20's and 30's. New York: American Heritage, 1970.

Armstrong, Warren. "The Morro Castle," in his Fire Down Below. New York: John Day, 1968.

Burton, Hal. The Morro Castle. New York: Viking Press, 1973.

Clevely, Hugh. Famous Fires. New York: John Day, 1958.

Gallagher, Thomas. Fire at Sea: The Story of the "Morro Castle." New York: Rinehart, 1959.

Gordon, Thomas, and Max Morgan Witts. Shipwreck: The Strange Fate of the "Morro Castle." New York: Stein, 1972.

Hoehling, Adolph A. Great Ship Disasters. New York: Cowles,

1971.
McFee, William. "The Peculiar Fate of the 'Morro Castle'," in Leighton, Isabel, ed. The Aspirin Age, 1919-1941. New York: Simon & Schuster, 1949.
Mielke, Otto. Disaster at Sea. New York: Fleet, 1958.

THE "SHARE-OUR-WEALTH" PROGRAM AND THE TOWNSEND RECOVERY PLAN (1934-1939)

One of the basic concepts of President Franklin Delano Roosevelt's "New Deal" program was that of achieving a more equitable distribution of wealth. On June 19, 1935, Roosevelt presented a tax-on-wealth message to Congress, and in August of that year Congress passed a new tax act. This was a compromise measure. Roosevelt had requested heavy inheritance taxes which were not included. Surtaxes were scheduled to begin at $100,000 rather than $50,000. The Act did include gift and estate taxes, a graduated corporation tax and capital stock and excess profits taxes.

The 1935 legislation was practically forced by public opinion, shaped by the sufferings of many persons due directly to the depression. Several ideas involving the redistribution of wealth were, for a while, in vogue, one of these being Senator Huey P. Long's "Share-Our-Wealth" Program.

On February 5, 1934, Long had outlined his program to members of the Senate. The program, calculated by its sponsor to end poverty, provided for the limiting of individual incomes to $1,000,000 a year, the giving of old-age pensions of $30 a month to those over 65 possessing less than $10,000 in cash, and for giving "every deserving family" not less than $5,000 a year. It also stipulated that personal fortunes be limited to $3,000,000, a minimum annual wage of not less than $2,500, a shorter work week and government-paid college educations for "proven" youths.

Long's "Share-Our-Wealth" program attracted large groups of middle-class people, particularly in the rural midwest and south, and over 200,000 "Share-Our-Wealth" clubs were organized.

Like other Utopian ideas, Long's was economically unsound and, following his assassination by Dr. Carl A. Weiss in September, 1935, the program lost its sponsor and was abandoned.

Another scheme, popularly called the "Townsend Recovery Plan," was proposed by Dr. Francis Everett Townsend, an American physician and amateur economist. Dr. Townsend's aim was to pay $200 a month to every citizen in the United States over the age of 60, provided he spent it in the month in which it was received. Townsend clubs were organized and held their first national convention in Chicago in October, 1935. Over 6,000 delegates attended from all parts of the United States. The plan was sponsored in Congress by Representative John Steven McGroarty of California. It differed from the original Old Age Revolving Pensions, Ltd. plan in certain details and provided for a pension "not to exceed $200 a month." The money was to be derived from a national 2% transaction tax.

Unfortunately for the Townsendites it was not realized that the McGroarty Bill would not produce a sum sufficient to support the $200 monthly payment. The Townsend Recovery Plan, like so many others, "looked good on paper" but was impractical and unworkable. It was rejected by the House of Representatives on June 1, 1939.

Still another Utopian scheme by which the economic misery of the American people could be eliminated was Upton Sinclair's "Epic" (End Poverty in California) Plan. This met with no more success than those proposed by Long and Townsend. However, the basic idea of providing for the so-called senior citizens and others finds a counterpart in today's social security program, instituted in 1935 and amended in succeeding years. State laws meeting certain federal standards and designed to assist persons in need have also been passed.

Suggested Readings

Allen, Frederick Lewis. Big Change: America Transforms Itself, 1900-1950. New York: Harper, 1952.

________. Since Yesterday: The Nineteen-Thirties in America, September 3, 1929-September 3, 1939. New York: Harper, 1940.

Andrist, Ralph K., ed. in charge. The American Heritage History of the 20's and 30's. New York: American Heritage, 1970.

Beals, Carleton. The Story of Huey P. Long. Westport, Conn.: Greenwood Press, 1971.

Carter, Hodding. "Huey Long, American Dictator," in Leighton, Isabel, ed. The Aspirin Age, 1919-1941. New York: Simon & Schuster, 1949.

Goldston, Robert C. The Great Depression: The United States in the Thirties. Indianapolis: Bobbs-Merrill, 1968.

Holtzman, Abraham. The Townsend Movement: A Study in Old Age Pressure Politics (unpublished Ph.D. dissertation, Harvard University, 1952).

Long, Huey P. Every Man a King. New York: Quadrangle/New York Times Co., 1964.

________. My First Days in the White House. New York: Da Capo Press, 1972.

Meltzer, Milton. Brother, Can You Spare a Dime? The Great Depression, 1929-1933. New York: Knopf, 1969.

Mitchell, Broadus. Depression Decade, 1929-1941. New York: Holt, Rinehart, 1947.

Phillips, Cabell. New York "Times" Chronicle of American Life: From the Crash to the Blitz, 1929-1939. New York: Macmillan, 1969.

Schlesinger, Arthur M., Jr. The Age of Roosevelt. Boston: Houghton Mifflin, 1957.

Sinclair, Upton. The Autobiography of Upton Sinclair. New York: Harcourt, Brace, 1962.

Sindler, Allan. Huey Long's Louisiana. Baltimore: Johns Hopkins University Press, 1956.
Sperling, John G. Great Depressions; 1837-1844, 1893-1898, 1929-1939. Chicago: Scott, Foresman, 1966.
Terkel, Studs. Hard Times: An Oral History of the Great Depression in America. New York: Pantheon Books, 1970.
The Unofficial Observer, pseud. American Messiahs. New York: Da Capo Press, 1975.
______. The New Dealers. New York: Da Capo Press, 1975.
Wecter, Dixon. The Age of the Great Depression. New York: Macmillan, 1948.
Williams, T. Harry. "The Gentleman from Louisiana: Demagogue or Democrat?" Journal of Southern History, XXVI, 1960.
Wilson, Edmund. The American Jitters: A Year of the Slump. Plainview, N.Y.: Books for Libraries, 1932.

MURDER, INC. (1934-1941)

In the 1920's and 1930's various groups of racketeers and gangsters flourished in the major cities of the United States. These "mobs" were involved in such profitable, though illegal, activities as prostitution, bootlegging, dope peddling, extortion, bookmaking, "loan sharking," the "numbers racket," labor racketeering, selling "protection" to businessmen, slot machines, and other types of gambling. Competition for territories in which to operate was keen, and gang leaders often found it necessary to stifle competition by having members of other similar organizations murdered.

This situation was particularly acute in New York City, Brooklyn, and adjacent East Coast cities, although it was by no means confined to them alone, the city of Chicago being notorious as the home of such gangs as those led by Alphonse Capone, George "Bugs" Moran, and the Genna brothers.

By 1930 prohibition was coming to an end and the Great Depression was reducing the profits made by racketeers. In 1935 certain public officials started investigating the activities of the mobsters, particularly the "numbers racket," at that time dominated, in Manhattan, by Dutch Schultz. Then it was learned, after the investigation began to lag, that Schultz had contributed $30,000 to the election campaign of District Attorney William C. Dodge, who was pressing the investigation. The grand jury demanded that a special prosecutor be appointed and Governor Herbert H. Lehmann assigned Thomas E. Dewey, who later became governor of New York State, to the position.

In 1939 William O'Dwyer, district attorney of King's County (Brooklyn), appointed Burton B. Turkus to the homicide division of the police department. Turkus' force turned up information on the murder of a small-time racketeer named Red Alpert. Involved in this killing were three mobsters: Dukey Maffetore, Buggsy Goldstein, and Abe "Kid Twist" Reles. These three turned informer and it was then that the existence of an organization popularly known as "Murder, Inc." became known. Reles, the most knowledgeable of the three, disclosed that in 1934 the leaders of the New York criminal gangs had banded together in what was called the "Organization." The objective of the "Organization" was to achieve cooperation between the various gangs and present a united front against the law.

Frank Costello, a notorious gambler and racketeer, was the head of the "Organization," and Joe Adonis was his assistant. Louis "Lepke" Buchalter, known also as "Judge Louis," presided at a kangaroo court which sat in judgment on those who acted contrary to the decisions of the "Organization." Albert "The Boss" Anastasia handled the enforcement division, known as "Murder, Inc." Abe Reles declared that he was the "field commander" of this last division, which was ultimately charged with 63 killings.

The "Organization" had a board of directors which was chaired by Costello, and each criminal gang had its representative on the board. Legal assistance was provided for racketeers who needed it, territories were assigned, and disputes between rival gangs were arbitrated. No murders were to be committed unless previously approved by the board of directors. When a murder was so sanctioned, arrangements to have it committed "scientifically" would be made. Anastasia, according to Reles, received the decisions of the board of directors and arranged for them to be carried out.

At first confined only to organized crime in the New York area, the "Organization" expanded its activities to other cities, coast to coast. Reles, having been granted immunity from prosecution in return for his disclosures, implicated a number of high-ranking criminals. In August, 1939, Buchalter surrendered by arrangement to gossip columnist Walter Winchell at Madison Square in New York City. Winchell turned him over to J. Edgar Hoover, head of the Federal Bureau of Investigation. Buchalter was later executed on evidence furnished by Reles, as were seven other gangsters. Twice that number were sent to prison.

District Attorney O'Dwyer realized that it would be necessary to protect Reles from the vengeance of the criminal element and placed him under guard in a room on the sixth floor of the Half Moon Hotel in Coney Island. Reles had given him evidence concerning the killing of Pete Panto and Morris Diamond, both honest trade unionists, evidence which involved Anastasia. O'Dwyer announced that he had the "perfect murder case" against Anastasia; he, however, had gone into hiding and could not be found. Reles had also given O'Dwyer information incriminating the Purple Gang in Detroit, the Bug and Meyer mob in California, and other similar organizations. Consequently, the district attorney had put Reles and three other material witnesses under guard, as mentioned above.

On the morning of November 12, 1941, Abe Reles' dead body was found six floors below his hotel room window. Whether he jumped or was pushed to his death has never been determined.

Although Reles had given O'Dwyer information implicating Anastasia in the murders of Panto and Diamond almost nineteen months before his death, O'Dwyer never brought Anastasia before a grand jury to secure an indictment. O'Dwyer's excuse was that "with Anastasia in hiding nothing could be done."

In 1945 George J. Beldock, the newly appointed district attorney of King's County, stated that "the waterfront rackets inquiry had been deliberately abandoned [when O'Dwyer was district attorney]." Further, referring to Anastasia who, with two accomplices, Dandy Jack Parisi and Tony Romeo, had come out of hiding, he said, "The perpetrators of said crimes have thereby and because of the Statute of Limitations now secured complete immunity from prosecution and conviction."

Suggested Readings

Churchill, Allen. A Pictorial History of American Crime. New York: Holt, Rinehart, 1964.

Cipes, Robert M. The Crime War. New York: New American Library, 1968.

Corey, Herbert. Farewell, Mr. Gangster! America's War on Crime. New York: Appleton, 1936.

De Leeuw, Hendrik. Underworld Story: The Rise of Organized Crime and Vice-Racket in the U.S.A. London: N. Spearman, 1955.

Demaris, Ovid. The Lucky Luciano Story. New York: Tower, 1972.

Dewey, Thomas E. Twenty Against the Underworld. Edited by Rodney Campbell. Garden City, N.Y.: Doubleday, 1974.

Feder, Sid, and Joachim Joesten. The Luciano Story. New York: David McKay, 1954.

Fraenkel, Jack R., comp. Crime and Criminals: What Should Be Done About Them? Englewood Cliffs, N.J.: Prentice-Hall, 1970.

Hamilton, Charles. Men of the Underworld: The Professional Criminals' Own Story. New York: Macmillan, 1952.

Hanna, David. Bugsy Siegel: The Man Who Invented Murder, Inc. New York: Belmont-Tower, 1974.

Kunstler, William M. "Murder for Silence: The People of the State of New York versus Louis Buchalter," in his First Degree. New York: Oceana Publications, 1960.

Roberts, Carl E. B. The New World of Crime. London: Hylton, 1933.

Sterling, Hank, pseud. "The Muted Aria of the Singing Canary," in his Ten Perfect Crimes. New York: Stravon Publishers, 1954.

Symons, Julian. A Pictorial History of Crime. New York: Crown Publishers, 1966.

Turkus, Burton B., and Sid Feder. Murder, Inc.: The Story of "The Syndicate." New York: Farrar, Straus, 1951.

HUMPHREY'S EXECUTOR VS. UNITED STATES (1935)

The Supreme Court decision of 1935 in the case of Humphrey's Executor vs. United States involved the power of the President to remove an appointed official from office.

The facts of the case were these: On December 10, 1931, William E. Humphrey was nominated by President Herbert Hoover to succeed himself as a member of the Federal Trade Commission. The appointment was confirmed by the Senate and Humphrey was commissioned for a term of seven years. He took the required oath and entered upon his duties.

On July 25, 1933, President Franklin Delano Roosevelt requested Humphrey's resignation, stating in a letter that he felt that personnel of his own selection could carry on the work of the Commission more effectively than Humphrey. No reflections on Humphrey personally or on his manner of performing his duties were contained in Roosevelt's letter.

Humphrey replied, asking for time in which to consider the matter. Roosevelt wrote him again on October 7, 1933, this letter saying, "Effective as of this date you are hereby removed from the office of Commissioner of the Federal Trade Commission."

Humphrey did not acquiesce to Roosevelt's action and insisted that he was still a member of the Commission, entitled to perform his duties and to receive his $10,000 annual salary as provided by law. Suit was brought and eventually the case came before the Supreme Court of the United States.

Associate Justice George Sutherland read the Court's unanimous decision. The questions to be answered were: first, did the Federal Trade Commission Act restrict or limit the power of the President to remove a member of an independent quasi-legislative or quasi-judicial commission for reasons other than "inefficiency, neglect of duty, or malfeasance in office"? Second, if the answer to the first question is "yes," "Is such a restriction or limitation valid under the Constitution of the United States?"

Justice Sutherland stated that such organizations as the Federal Trade Commission and the Interstate Commerce Commission must be non-partisan, act with entire impartiality, and be free from "political domination and control--not subject to the orders of the President." He contrasted the duties of a postmaster, being an executive officer restricted to the performance of executive functions, as in the case of Myers vs. United States, with those of "an officer who occupies no place in the executive department and who exercises no part of the executive power vested by the Constitution in the President." "It is quite evident," he said, "that one who holds his office only during the pleasure of another cannot be depended upon to maintain an attitude of independence against the latter's will."

In summary, Sutherland stated that purely executive officers were subject to dismissal by the President. Those, however, who, in connection with their duties of administering quasi-legislative or quasi-judicial agencies, must act independently of executive control could not be removed during the prescribed term of office except for cause.

Suggested Readings

Binkley, Wilfred E. The Powers of the President. New York: Russell & Russell, 1973.

Carson, Hampton L. The History of the Supreme Court. New York: Burt Franklin, 1972.

Commager, Henry Steele, ed. "Humphrey's Executor v. United States," (Doc. No. 498) in his Documents of American History, 8th edition. New York: Appleton, 1968.

_______. "Myers v. United States," (Doc. No. 459) in his Documents of American History, 8th edition. New York: Appleton, 1968.

Cope, Alfred Haines, ed. Franklin D. Roosevelt and the Supreme Court. Boston: Heath, 1952.

Corwin, Edward S. The President: Office and Powers. New York: New York University Press, 1957.

Hatch, Alden. Franklin D. Roosevelt, An Informal Biography. New York: Holt, 1947.

Johnson, Gerald White. The Supreme Court. New York: Morrow, 1962.

Laski, Harold J. The American Presidency: An Interpretation. Westport, Conn.: Greenwood Press, 1972.

McCloskey, Robert G. The American Supreme Court. Chicago: University of Chicago Press, 1960.

Salomon, Leon I., ed. The Supreme Court. New York: Wilson, 1961.

Schnittkind, Henry Thomas. Franklin Delano Roosevelt. New York: Putnam, 1962.

Weingast, David Elliott. Franklin D. Roosevelt, Man of Destiny. New York: Messner, 1952.

THE WORKS PROGRESS ADMINISTRATION (1935-1938)

One of the objectives of President Franklin Delano Roosevelt's New Deal program was to relieve the nation's unemployment and at the same time avoid a policy of full-scale deficit spending. A Public Works Administration was created by the National Industrial Recovery Act which was passed in June, 1933. This, operating under Harold L. Ickes, Secretary of the Interior, was authorized to spend over three billion dollars to relieve unemployment. The P.W.A., however, proved ineffective.

Harry L. Hopkins, a former social worker, was appointed to administer the Federal Emergency Relief Administration which was established in May, 1933. Half a billion dollars was appropriated for this agency, which was to make direct grants to the states for relief services.

Hopkins, with President Roosevelt's assistance, was able to set up a temporary Civil Works Administration in October, 1933. This agency's primary purpose was to provide employment to some four million men in a complete federal make-work project. Such tasks as the construction of roads and playgrounds and the making of school building repairs were assigned to the workers. By March, 1934, the C. W. A. had employed over three million men and spent over $900,000,000. Roosevelt, alarmed at such large expenditures, terminated this agency and replaced it by the F. E. R. A., which resumed the responsibility for relief.

Roosevelt was of the opinion that further relief assistance was needed. Unemployment was increasing. Both he and Hopkins wished to keep the federal government out of direct relief, preferring instead to create a public employment plan which would provide, not a dole, but honest jobs for those who needed them. In January, 1935, the President called for a new emergency relief appropriation. The Emergency Relief Appropriation Act of April 8, 1935 provided five billion dollars.

Both Hopkins and Ickes were considered as administrators for this new program. Both desired the job and Roosevelt brought in Frank Walker to act as mediator between the two men. Hopkins managed to dominate the new program. His new agency, the Works Progress Administration, popularly known as the W. P. A., expanded quickly. Originally intended to furnish employment for unskilled laborers, it eventually aided actors, writers, students and artists. Hallie Flanagan of Vassar College's Little Theater administered the Federal Theater Project. Henry Alsberg, formerly a director of the Provincetown Theater, employed newspapermen, college professors and other literary craftsmen in the Federal Writers' Project which recorded local and regional history. Holger Cahill directed the Federal Art Project which produced pictures now on view in post offices and public libraries.

The W. P. A. was effective as an emergency measure, but problems, both political and financial, arose. Some corruption was inevitable and Hopkins was forced to change his approach when Congress gained control of W. P. A. appropriations and he refused to use the agency for patronage purposes.

It is generally agreed that the W. P. A. was by and large successful in solving a gigantic unemployment problem and establishing guidelines for coping with similar economic depressions in the future.

Suggested Readings

Allen, Frederick Lewis. Big Change: America Transforms Itself, 1900-1950. New York: Harper, 1952.

________. Since Yesterday: The Nineteen-Thirties in America, September 3, 1929-September 3, 1939. New York: Harper, 1940.

Andrist, Ralph K., ed. in charge. The American Heritage History of the 20's and 30's. New York: American Heritage, 1970.

Beard, Charles A., and Mary R. Beard. A Basic History of the

United States. New York: Doubleday, 1944.
Bernstein, Irving. The Turbulent Years: A History of the American Worker, 1933-1941. Boston: Houghton Mifflin, 1970.
Goldston, Robert C. The Great Depression: The United States in the Thirties. Indianapolis: Bobbs-Merrill, 1968.
Hatch, Alden. Franklin D. Roosevelt, An Informal Biography. New York: Holt, 1947.
Hopkins, Harry L. Spending to Save: The Complete Story of Relief. New York: Norton, 1936.
Howard, Donald S. The W. P. A. and Federal Relief Policy. New York: Russell Sage Foundation, 1943.
Ickes, Harold L. Back to Work. New York: Da Capo Press, 1935.
________. The Secret Diary of Harold L. Ickes. New York: Simon & Schuster, 1953-1954.
Leuchtenburg, William E. Franklin D. Roosevelt and the New Deal. New York: Harper, 1963.
Lynd, Robert S., and Helen M. Lynd. Middletown in Transition: A Study in Cultural Conflicts. New York: Harcourt, Brace, 1937.
Moley, Raymond, with the assistance of Elliot A. Rosen. The First New Deal. New York: Harcourt, Brace, 1966.
Peare, Catherine O. The FDR Story. New York: Crowell, 1962.
Schlesinger, Arthur M., Jr. The Age of Roosevelt. Boston: Houghton Mifflin, 1957.
Schnittkind, Henry Thomas. Franklin Delano Roosevelt. New York: Putnam, 1962.
Searle, Charles. Harry L. Hopkins, New Deal Administrator, 1933-38 (unpublished Ph.D. dissertation, University of Illinois, 1953).
Shannon, David A. The Great Depression. Englewood Cliffs, N.J.: Prentice-Hall, 1960.
Sherwood, Robert E. Roosevelt and Hopkins: An Intimate History. New York: Harper, 1950.
Wecter, Dixon. The Age of the Great Depression, 1929-1941. New York: Macmillan, 1948.
Weingast, David Elliott. Franklin D. Roosevelt, Man of Destiny. New York: Messner, 1952.
Witte, Edwin E. The Development of the Social Security Act. Madison, Wis.: University of Wisconsin Press, 1962.

THE "HINDENBURG" DISASTER (1937)

The "Hindenburg" was a large, German-built dirigible balloon which, on May 6, 1937, caught fire and burned while being fastened to its mooring mast at Lakehurst, New Jersey, after a voyage from Frankfort, Germany. She had been completely overhauled and inspected early in 1937. Prior to her arrival at Lakehurst on the first of a series of projected commercial flights over the North Atlantic she had made several European flights and a round trip to Rio de Janeiro, Brazil. These voyages had all been completed without incident and the "Hindenburg" had worked perfectly.

On what was to be her final flight the giant zeppelin carried a total of 97 persons: a crew of 61 which included a number of trainees, and 36 passengers. Commanded by Captain Max Pruss, the "Hindenburg" left the Frankfort Rhein-Main World Airport at 8:00 P.M. on May 3, 1937. It passed over Cologne and then proceeded by way of Holland to the North Sea and on out over the Atlantic Ocean. The North American coast near Newfoundland was sighted on the afternoon of May 5, and Boston, Massachusetts was reached about noon the following day. The ship passed over New York City and headed for Lakehurst, where she was to dock at the naval station mooring mast. Captain Pruss indicated that he planned to land at 6:00 P.M.

The landing was delayed beyond the appointed time by a thunderstorm which moved in from the West. The "Hindenburg" circled in a holding pattern until the storm cleared and at 7:21 P.M. dropped her landing ropes to the waiting ground crew below. The ground crew connected the landing ropes with corresponding ground lines and prepared to draw them taut. The steel mooring cable with which the ship was to be pulled to the mooring mast made its appearance at the nose of the giant dirigible.

At 7:25 P.M. a burst of flame appeared at the top of the ship just forward of where the upper vertical fin attached to the hull. The "Hindenburg," inflated with highly flammable hydrogen, was doomed.

The flames spread rapidly and the ship literally exploded in the air. She settled to the ground stern first, a mass of glowing, twisted wreckage. Herb Morrison, a radio announcer, on hand to report the arrival of the dirigible, described the holocaust to a coast-to-coast audience. "It's burst into flames! It's falling on the mooring mast! It's one of the worst catastrophes in the world!"

Within a minute after the appearance of the fire the ship was done for, but the fire raged for three hours more due to the large quantities of fuel oil aboard. Of the 36 passengers carried, thirteen died, either in the fire or later from injuries received. Twenty-two of the 61 crew members perished and one member of the ground crew died of burns. Thus, of the 97 persons aboard, 62 managed to survive.

Experts discussing the catastrophe are generally of the opinion that the fire was caused, not through human error or mechanical failure, but by a small gas leak. It was said that had the "Hindenburg" been inflated with non-flammable helium rather than the highly combustible hydrogen which filled its gas cells, no fire could have occurred.

Suggested Readings

Andrist, Ralph K., ed. in charge. The American Heritage History of the 20's and 30's. New York: American Heritage, 1970.

Clevely, Hugh. Famous Fires. New York: John Day, 1958.

Hoehling, Adolph A. "LZ-129 'Hindenburg'--Last Flight," in his Disaster: Major American Catastrophies. New York: Hawthorn Books, 1973.
_______. Who Betrayed the "Hindenburg?" Boston: Little, Brown, 1962.
Mather, M. G. "I Was on the 'Hindenburg'," Harper's Magazine, November, 1937.
Mooney, Michael Macdonald. The Hindenburg. New York: Dodd, Mead, 1972.
Robinson, Douglas H. Famous Aircraft: The LZ-129 "Hindenburg." Dallas: Morgan, 1964.
Rosendahl, Commander Charles E. "The Last Flight of the 'Hindenburg'," in Kartman, Ben, and Leonard Brown, eds. Disaster! New York: Pellegrini & Cudahy, 1948.
_______. Zeppelin, the Story of Lighter-Than-Air Craft. New York: Longmans, Green, 1937.
Wallechinsky, David, and Irving Wallace. "The 'Hindenburg'," in their The People's Almanac. Garden City, N.Y.: Doubleday, 1975.

THE LAST FLIGHT OF AMELIA EARHART PUTNAM (1937)

Amelia Earhart Putnam, better known by her maiden name, was one of the great pioneer flyers. As early as June, 1928, she achieved fame in this field when, as the first woman to fly across the Atlantic Ocean she, with Louis Gordon and Wilmer Stultz, made the trip from Newfoundland to England in a little over twenty hours.

She set another record in May, 1932, flying solo from Newfoundland to Ireland, the first woman to make the trip by air alone. She made the crossing in thirteen hours and thirty minutes, a new record. In August of the same year she became the first woman to make a transcontinental nonstop flight. Starting in Los Angeles, she landed at Newark, New Jersey, covering approximately 2,600 miles in nineteen hours and five minutes.

Her next attempt at record setting was her trip from Wheeler Field, Honolulu, to Oakland, California, made alone in 1935. She arrived safely after 18-1/4 hours in the air, the first woman to make this flight alone. Again, in 1935, she set a new speed record by flying nonstop from Mexico City to New York in fourteen hours and nineteen minutes.

Looking for new worlds to conquer, Amelia Earhart Putnam left the airport at Oakland, California, on May 20, 1937, bound for Miami. She was accompanied by her navigator, Frederick Noonan, the two planning a trip around the world. Departing from Miami on June 1, they ultimately arrived at Lae, New Guinea, from where they took off for Howland Island in mid-Pacific. On July 3, somewhere east of the Gilbert Islands, the plane carrying the two flyers disappeared.

An intensive search for Noonan and Mrs. Putnam was inaugurated. Ships and planes of the United States navy scoured the area but no trace of the two was ever found and their fate remains an unsolved mystery. It has been conjectured that their disappearance was deliberately contrived so that American naval units might search the Pacific, ostensibly looking for traces of the flyers but actually seeking information on Japanese activities there. This is mere speculation and the generally-held opinion is that their aircraft, suffering a malfunction or engine failure, was unavoidably lost in the ocean.

In 1937, shortly after her disappearance, George Palmer Putnam, the flyer's husband, edited and published Last Flight. This book consists largely of the diary of her tragic last journey, transmitted from her various stopping-places along the way.

Suggested Readings

Barker, Ralph. Great Mysteries of the Air. New York: Macmillan, 1967.

Bolton, Sarah K. Lives of Girls Who Became Famous. New York: Crowell, 1942.

Briand, Paul L., Jr. Daughter of the Sky: The Story of Amelia Earhart. New York: Holt, Rinehart, 1960.

De Leeuw, Adèle Louise. The Story of Amelia Earhart. New York: Grosset & Dunlap, 1955.

Earhart, Amelia. The Fun of It. Detroit: Gale Research Co., 1975.

________. Last Flight. Edited by George Palmer Putnam. New York: Putnam, 1937.

Earhart, Garst Shannon. Amelia Earhart, Heroine of the Skies. New York: Messner, 1947.

Goerner, Fred G. The Search for Amelia Earhart. New York: Doubleday, 1966.

Hyde, Margaret Oldroyd. Flight Today and Tomorrow. New York: Whittlesey House, 1962.

James, Edward T., ed. "Amelia Earhart," in his Notable American Women. Cambridge, Mass.: The Belknap Press of Harvard University Press, 1971.

Lewellen, John Bryan and Irwin Shapiro. The Story of Flight. New York: Golden Press, 1959.

Maizlish, I. L. Wonderful Wings: A Story of Aviation. Evanston, Ill.: Row, 1941.

May, Julian. Amelia Earhart, Pioneer of Aviation. Chicago: Children's Press, 1973.

Parlin, John. Amelia Earhart, Pioneer in the Sky. Champaign, Ill.: Garrard, 1962.

Putnam, George Palmer. Soaring Wings. New York: Manor Books, 1972.

Roberts, Joseph, and Paul L. Briand. The Sound of Wings. New York: Holt, Rinehart, 1957.

THE "INVASION FROM MARS" (1938)

One of the better pioneer radio drama programs of the 1930's was the Mercury Theater, hosted by Orson Welles, who also wrote, produced and directed many of the dramas as well as starring in them. Various classics, such as Dracula, Jane Eyre, Rebecca, and Frankenstein, were dramatized and broadcast. Such actors and actresses as Joseph Cotten, Everett Sloan, Ray Collins, and Agnes Moorehead were featured on the programs.

On the evening of October 30, 1938, Welles and his company of actors presented over the Columbia Broadcasting System Howard Koch's adaptation of the H. G. Wells' story, The War of the Worlds. This became one of the most famous radio broadcasts of all time and is remembered today as The Invasion from Mars.

The broadcast began with a clear explanation that the program to follow was an adaptation of Wells' science-fiction classic. Then the radio audience heard what appeared to be a remote control broadcast of the music of Ramón Raquello and his orchestra, playing in the Meridian Room of the Park Plaza Hotel, New York City. The music was suddenly interrupted by a "news bulletin" from Intercontinental Radio News reporting "several explosions of incadescent gas, occurring at regular intervals on the planet Mars."

The "remote broadcast" switched to the Hotel Martinet, Brooklyn, New York, where listeners heard Bobby Millette and his orchestra. As the music program progressed it was interrupted by further "news reports." "Professor Richard Pierson, an astronomer at Princeton University," played by Orson Welles, in an "interview" by "Carl Phillips, a radio commentator," and broadcasting from the "Wilmuth Farm, Grovers Mill, New Jersey," stated that several strange metallic cylinders from Mars had landed on Earth. Martians emerged from the cylinders.

The "news bulletins" continued, with realistic accounts of battles between the Martians and the U. S. Army, which had no chance against the poison gas and death rays used by the invaders.

Thousands of listeners, taking the drama at face value, panicked. Many ran into the streets, seeking safety of some sort. Members of the National Guard reported to headquarters to aid in the battles against the Martians. Hysterical people tied up telephone lines. Hospitals and police stations were besieged by terror-stricken crowds. In Michigan a woman planned to commit suicide rather than risk an encounter with one of the alien army. A number of people sought refuge in the hills, and for weeks after the broadcast teams of rescuers were going about the country assuring these people that it was safe to return home.

Orson Welles' broadcast established the power and effectiveness of radio drama, which flourished for the next quarter century and which now has its equally effective counterpart on television.

Suggested Readings

Cantril, Hadley, and Associates. The Invasion from Mars. Princeton, N.J.: Princeton University Press, 1940.
Harmon, Jim. The Great Radio Heroes. New York: Doubleday, 1967.
Houseman, John. "The Men from Mars," Harper's Magazine, December, 1948. Reprinted in Knowles, Horace, ed. Gentlemen, Scholars and Scoundrels. New York: Harper, 1959.
Jackson, Charles. "The Night the Martians Came," in Leighton, Isabel, ed. The Aspirin Age, 1919-1941. New York: Simon & Schuster, 1949.
Koch, Howard. The Night the Martians Landed. New York: Avon Books, 1975.
________. Panic Broadcast: Portrait of an Event. Boston: Little Brown, 1970.
McBride, Joseph. Orson Welles. New York: Viking Press, 1972.
Noble, Peter. The Fabulous Orson Welles. London: Hutchinson, 1956.
Sharp, Harold S., and Marjorie Z. Sharp. Index to Characters in the Performing Arts, Part IV: Radio and Television. Metuchen, N.J.: Scarecrow Press, 1973.
Wallechinsky, David, and Irving Wallace. "The Science Fiction Observatory," in their The People's Almanac. Garden City, N.Y.: Doubleday, 1975.
Wells, H. G. The War of the Worlds (science fiction). New York: Berkley, 1975.

THE DIES COMMITTEE (1938-1940)

In the 1930's the world was suffering the consequences of a gigantic economic depression. Unemployment was widespread and many people and business enterprises were in serious financial trouble. Although President Franklin Delano Roosevelt, through his "New Deal" program, made great efforts to restore the country to "a wider distribution of the wealth and property of the nation," progress towards this goal was exceedingly slow.

Since 1919, when Adolf Hitler and his followers founded the National Socialist German Workers Party, or Nazi Party, dictatorship had gradually attained progressively greater strength throughout Europe. It was thought that a similar dictatorship, aided by depression-generated public discontent, might seek to control the United States, as depicted in Sinclair Lewis' novel, It Can't Happen Here, published in 1935. Lewis described an America overwhelmed by the elected President Berzelius "Buzz" Windrip, who seized the reins of power and established himself as dictator.

Such organizations as the German-American Bund, headed by Fritz Kuhn, the Silver Shirt Legion of America, founded by William Dudley Pelley, as well as the Christian Front, the Social Justice Society and the Knights of the White Camelia--all subversive societies--were operating within the continental limits of the United States. It was feared that dictatorship, if unchecked, might well undermine democracy. This would have a damaging effect for centuries on the future of human liberties.

On May 26, 1938, the Dies Committee to Investigate un-American Activities was appointed. The attorney and legislator Martin Dies was made chairman of this body, which came to be known as the "Dies Committee." After almost a year and a half of investigation the Committee made its report to the House. During its existence it had been subject to severe criticism, in spite of which it was granted an extension of life in late 1939.

Dies made his report on January 3, 1940. In his opening remarks he declared that the struggle between dictatorship and democracy posed a serious dilemma. "It is of primary importance," he said, "to prevent the growth or influence of a group which seeks to undermine democracy and substitute dictatorship for it. But it is at least equally important that in combating subversive groups of this character nothing be done which would undermine the fundamental structure of constitutional liberty itself."

He then stated that the American public should be informed, from time to time, "of the activities of any such organization in their nation." He discussed the labor movement in the United States and declared that "the Committee's work should result in freeing the progressive and labor movements from Communist control or domination and in preventing sincere conservatives from temporizing with essentially Fascist or Nazi groups or philosophies." Further, he indicated that the Committee's investigation had disclosed evidence of Communist leadership within some labor unions and stated that "the American labor movement must, and will, as speedily as possible, free itself of Communist leadership and control wherever it exists."

Dies discussed the existence of various subversive organizations, identifying some of them by name, and declared that these organizations worked together to achieve their objectives. He mentioned the accomplishments of the Committee, stating that Fritz Kuhn had been convicted of embezzlement of his organization's funds and sentenced to prison; that Earl Browder and William Weiner, members of the Communist party, had been indicted on a charge of falsifying passports; that Nicholas Dozenberg had been charged with counterfeiting American money on orders from the Communist International; that William Dudley Pelley was a fugitive from justice; that Arno Rissi and Mrs. Leslie Fry, Nazi and Fascist leaders, had "fled the country for good and sufficient reason"; and that "many bills of a corrective nature had been introduced and passed by the House as a result of the testimony produced by the Committee."

In conclusion, Dies reported that several of the subversive organizations he had named were losing prestige and leadership; he said that "the time may speedily come when they cease to be seriously regarded by anyone in the United States."

Suggested Readings

Brandt, Raymond. "The Dies Committee: An Appraisal," Atlantic Monthly, CLXV, 1940.
Commager, Henry Steele, ed. "Report of the Dies Committee on UnAmerican Activities," (Doc. No. 528) in his Documents of American History, 8th edition. New York: Appleton, 1968.
Dimock, Marshall E. Congressional Investigating Committees. New York: AMS Press, 1929.
Eberling, Ernest J. Congressional Investigations. New York: Octagon Books, 1972.
Gellermann, William. Martin Dies. New York: Da Capo Press, 1944.
Greer, Thomas. What Roosevelt Thought. East Lansing, Mich.: Michigan State College Press, 1958.
Howe, Irving, Lewis Coser and Julius Jacobson. The American Communist Party. New York: Praeger, 1962.
Kempton, Murray. Part of Our Time. New York: Simon & Schuster, 1955.
Leuchtenburg, William E. Franklin D. Roosevelt and the New Deal, 1932-1940. New York: Harper, 1963.
Lewis, Sinclair. It Can't Happen Here (fiction). Garden City, N.Y.: Doubleday, 1935.
McGeary, M. Nelson. The Developments of Congressional Investigative Power. New York: Octagon Books, 1966.
Ogden, August Raymond. The Dies Committee. Washington, D.C.: Murray & Heister, 1944.
Saunders, D. A. "The Dies Committee: First Phase," Public Opinion Quarterly, III, 1939.
Shannon, David A. The Decline of American Communism. New York: Harcourt, Brace, 1945.
________. The Socialist Party of America. New York: Macmillan, 1955.
Wechsler, James Arthur. The Age of Suspicion. New York: Random House, 1953.

THE SINKING OF THE "SQUALUS" (1939)

On the morning of May 23, 1939, the United States navy submarine "Squalus" was making practice dives in the ocean off Portsmouth, New Hampshire. At 7:40 a radio message from the vessel was received at the navy yard. The message read, "Preparing to descend for one hour." By 10:20, no further word having been received, an attempt to contact the "Squalus" was made. This being unsuccessful, the submarine's sister ship "Sculpin" was dispatched to search for her.

At 12:10 P.M. the "Sculpin's" lookout spotted a yellow telephone buoy, of the type released by undersea craft in emergencies. Through it the commander of the "Sculpin" talked with Lieutenant Oliver Naquin, commander of the "Squalus." Naquin reported that his vessel was resting on the bottom, 240 feet below. He reported further that the high induction valve was open and that some compartments were full of water. Before further information could be obtained the telephone cable snapped. It was later learned that as the "Squalus" sank to the bottom, five men dashed to the safety of the control room, leaving 26 behind in the flooded aft section. The bulkhead door between the two compartments was slammed shut and bolted tight.

The next morning divers from the submarine tender "Falcon" landed on the deck of the "Squalus." A ten-ton diving bell was then lowered and guided to the sunken vessel by divers, working in shifts. The diving bell was set in place over the forward escape hatch which was then opened, and 33 officers and men were rescued. It required four trips of the diving bell to transport the trapped men to the surface, and on the last trip the bell's cable became tangled. It required four hours to correct it.

When the bell was sent to the rear escape hatch, the rear compartment, when opened, was found to be completely filled with water. Any hope of rescuing the remaining 26 crewmen vanished.

The navy then commenced operations to salvage the sunken submarine. Commander Allen McCann and A. I. McKee, both experts in such operations, decided that the "Squalus" could best be raised by pumping the water from the flooded sections and floating her to the surface with pontoons attached to the hull. One hundred and thirteen days later the pontoon-buoyed craft was brought to the surface and towed to the navy yard at Portsmouth. Twenty-five dead bodies were removed. The 26th, that of Robert P. Thompson, the cook, was missing and it was decided that it had probably been swept through a hatch during the salvage operation.

Lieutenant Naquin, one of the survivors, testified before the board of inquiry that his submarine sank because a high induction valve had failed to close when the vessel submerged, permitting water to stream in. He stressed the need for an apparatus which would insure that a submarine would not submerge unless all valves were properly shut.

The board, inspecting the undersea craft in drydock following the tragedy, watched the testing of the two high induction valves. One functioned perfectly. The other, bearing out Lieutenant Naquin's contention, failed to close.

Suggested Readings

"Dead Dogfish: The 'Squalus' Disaster and Rescue Apparatus," Time, June 5, 1939.

Ellsberg, Commander Edward. Men Under the Sea. New York: Dodd, Mead, 1939.

________. On the Bottom. New York: Dodd, Mead, 1929.

________. "Safety for Our Submarines," World's Work, March, 1928.

"Forty Fathoms Down," in Kartman, Ben, and Leonard Brown, eds. Disaster! New York: Pellegrini & Cudahy, 1948.

"Forty Fathoms Down; Epic of Submarine 'Squalus' Poses New

Problem for Navy," Newsweek, June 5, 1939.
Mielke, Otto. Disaster at Sea. New York: Fleet, 1958.
O'Donnell, J. "How Men Act in a Sunken Submarine," Collier's, August 7, 1926.
"The Risks Men Face in Subs," Popular Mechanics, March, 1928.
"The Sinking of the 'Squalus'," Commonweal, June 9, 1939.
Stephens, Edward. Submarines; the Story of Underwater Craft from the Diving Bell of 300 B.C. to Nuclear-powered Ships. New York: Golden Press, 1962.
"Who Is to Blame for Our Submarine Disasters?," Literary Digest, January 7, 1928.
"The Whole Truth: Inquiry Into the Sinking of the 'Squalus'," Time, July 3, 1939.
Zim, Herbert Spencer. Submarines, the Story of Undersea Boats. New York: Harcourt, Brace, 1942.

THE MANHATTAN PROJECT (1939-1945)

The "Manhattan Project" was the secret code name by which the design and manufacture of an atomic bomb, or A-bomb, to be used by the United States in World War II, was known.

In August, 1939, Dr. Albert Einstein advised President Franklin Delano Roosevelt that German scientists were attempting to find a method for manufacturing bombs using the power generated by nuclear fission rather than by the rapid burning or decomposition of some chemical compound. He urged the President to initiate a similar research project in the United States. Such a program was started. This was the "Manhattan Project," and it was to cost over two billion dollars.

The work was taken over by a small group of scientists, including Enrico Fermi, Leo Szilard, Edward Teller and Eugene P. Wigner. On December 12, 1942, a self-sustaining nuclear chain reaction was achieved for the first time. The scientists, working in secrecy in a laboratory below the football stadium at the University of Chicago, notified associates in other parts of the country by code message, "The Italian navigator has landed; the natives are friendly."

Plants for the production of fissionable material, costing over a billion dollars, were set up at Oak Ridge, Tennessee. These plants were designed for the continuous separation of the isotope U^{235} from natural uranium. The scientists in charge of this project also worked on the production of plutonium as an alternative bomb material.

At 5:30 on the morning of July 16, 1945, the world's first A-bomb was successfully detonated at Alamogordo Air Base in the desert of New Mexico. Later that month the bombs to be used on the Japanese cities of Hiroshima and Nagasaki were carried to Saipan on the United States steamship "Indianapolis."

On August 4, 1945, the first A-bomb was dropped on Hiroshima. The casualties were horrendous: killed, 78,150; injured,

37,425; and missing, 13,083. The second bomb was dropped on Nagasaki on August 5. More than half the city was wiped out and 18,000 buildings were completely demolished. Deaths totaled approximately 40,000 and injuries exceeded 80,000. Following this second bombing Japan surrendered.

In addressing the House of Commons on August 16, Prime Minister Winston Churchill stated that the atomic bomb, though inflicting many deaths and injuries on the Japanese, saved the lives of 1,000,000 American and 250,000 British soldiers by eliminating the necessity of an invasion of Japan.

Suggested Readings

Clark, Ronald William. The Birth of the Bomb. London: Phoenix House, 1961.

Groueff, Stephane. Manhattan Project: The Untold Story of the Making of the Atomic Bomb. Boston: Little, Brown, 1967.

Groves, Richard Leslie. Now It Can Be Told: The Story of the Manhattan Project. New York: Harper, 1962.

Hersey, John R. Hiroshima. New York: Knopf, 1946.

Hirschfeld, Burt. Cloud Over Hiroshima: The Story of the Atomic Bomb. New York: Messner, 1967.

Jungk, Robert. Brighter Than a Thousand Suns. Translated by James Cleugh. New York: Harcourt, Brace, 1958.

Kugelmass, J. Alvin. J. Robert Oppenheimer and the Atomic Story. New York: Messner, 1953.

Lamont, Lansing. Day of Trinity. New York: Atheneum, 1965.

Laurence, William Leonard. Men and Atoms: The Discovery, the Uses and the Future of Atomic Energy. New York: Simon & Schuster, 1959.

Moorehead, Alan. The Traitors. New York: Scribner's, 1952.

Purcell, John Francis. The Best-Kept Secret: The Story of the Atomic Bomb. New York: Vanguard Press, 1963.

Sherwin, Martin Jay. World Destroyed: The Atomic Bomb and the Grand Alliance. New York: Knopf, 1975.

Shute, Nevil, pseud. On the Beach (fiction). New York: Perennial Library, 1957.

Stimson, Henry L. "The Decision to Use the Atomic Bomb," Harper's Magazine, February, 1947. Reprinted in Knowles, Horace, ed. Gentlemen, Scholars and Scoundrels. New York: Harper, 1959.

THE SIXTY-FOUR-DOLLAR QUESTION AND THE TELEVISION QUIZ SHOW SCANDALS (1940 and 1958)

The phrase, "the sixty-four dollar question," which became a part of the American idiom, originated on a radio program,

"Take It or Leave It," first broadcast from New York City on April 21, 1940.

The program, later known as "The Sixty-Four-Dollar Question," was one of the first audience-participation quiz-type broadcasts. At various times the master of ceremonies, known also as the "quizmaster," was Bob Hawk, Garry Moore, Jack Paar, and Phil Baker. The last, with "Bottle," his English butler, and "Beetle," his invisible stooge and heckler, not to mention his inevitable accordion, is today best remembered as being associated with the program.

Contestants, after a brief interview by the quizmaster, were asked a series of six questions. For answering the first question correctly they were paid one dollar, which they could either keep or double by answering the second question correctly. For each question in the series for which the contestant gave the right answer the reward was doubled until the "sixty-four-dollar question" was answered properly. Any question missed meant that the contestant forfeited all he had won up to that point.

Members of the studio audience frequently called out, "You'll be sorry!" when a contestant elected to continue the double-or-nothing game rather than "quit when he was ahead."

Often guest "contestants," such as Jack Benny and Humphrey Bogart, appeared on the program. In 1944 a motion picture featuring Baker in his role as quizmaster was produced.

When television largely replaced radio as a broadcast entertainment medium "The Sixty-Four-Dollar Question" became "The Sixty-Four-Thousand-Dollar Question" with Hal March as quizmaster and Lynn Dollar as his assistant. On this television program the double-or-nothing formula of radio was retained but the financial rewards to successful contestants were much greater and the questions far more difficult than those asked on the radio broadcasts.

Many other quiz programs appeared, both on radio and television, following the success of those conducted by Baker and March. Groucho Marx, with his "You Bet Your Life," was highly successful on television and is still appearing on many stations in rerun form. "Dotto," "Tic-Tac-Dough," "The Sixty-Four Thousand Dollar Challenge" and "Twenty One" were all popular television quiz programs until it was disclosed that some such contests were rigged, with selected contestants being advised in advance what questions would be asked them. The quiz scandals were exposed in the fall of 1958 by Herbert Stempel, a disillusioned contestant on "Twenty One." An investigation followed, with over 150 witnesses appearing before the New York County grand jury. Some contestants, such as Dr. Joyce Brothers, who had won a large sum of money by answering correctly questions having to do with prize fighting, were able to show that they had not been involved in rigged questions. Some programs, such as "You Bet Your Life," were exonerated. Teddy Nadler, who won $252,000 on "The Sixty-Four Thousand Dollar Challenge," stated that "no one had ever given him any answers." Others, such as contestant Charles Lincoln Van Doren, a former college instructor who had starred on "Twenty One," winning $129,000, admitted that he had been "handed questions, answers and sometimes entire scripts before entering the isolation booth." His defeat of the "champion" Herbert Stempel had, he said, been "arranged," as was his loss to Mrs. Vivienne Nearing. He implicated producer Dan Enright and associate producer Albert Freedman in the affair.

Since then the quiz program has declined in popularity, being replaced with other types of radio and television entertainment, although such programs as "Password," in which various celebrities match wits with contestants, and certain types of game shows where prizes are given are currently viewed by large audiences.

Suggested Readings

"Blame for TV Fixes," Life, October 26, 1959.
Buxton, Frank, and Bill Owen. Radio's Golden Age. New York: Easton Valley Press, 1966.
Cook, F. J. "Corrupt Society," Nation, June 1, 1963.
"Dress Rehearsals Complete with Answers? TV Quiz Shows," U.S. News and World Report, October 19, 1959.
Evans, J. "The Sad Success of Teddy Nadler," Redbook, March, 1962.
"Hal March--He Got a $64,000 Break," Look, September 18, 1956.
Harmon, Jim. Great Radio Comedians. Garden City, N.Y.: Doubleday, 1970.
Lackmann, Ron. Remember Radio. New York: Putnam, 1970.
Langman, A. W. "Television's Rigged Honesty," Nation, December 26, 1959.
"Notes and Comment: Contestants in the Hoax," New Yorker, October 24, 1959.
Sharp, Harold S., and Marjorie Z. Sharp. Index to Characters in the Performing Arts. Part IV: Radio and Television. Metuchen, N.J.: Scarecrow Press, 1973.
"A Tawdry Hoax," Newsweek, June 22, 1959.
TV Personalities: Biographical Sketch Book. St. Louis: TV Personalities, 1954-1955-1956.
Van Doren, Charles. "I Was Involved in a Deception," Time, November 16, 1959.
Welles, C. "No Rig, No Fix, No Quiz," Life, October 18, 1963.

THE "MAD BOMBER" AND THE "SUNDAY BOMBER" (1940-1957 and 1960)

George Metesky was a $37.50 per week generator wiper and tool maker employed by the Consolidated Edison Company of New York, the utility which sells electricity and gas to consumers in New York City and Winchester County.

In 1931 a boiler backfired and Metesky was severely injured. For two hours he lay gasping helplessly before regaining sufficient strength to return home. The incident left an almost ineradicable mark on his mind and he became consumed by hate.

Nine years later he planted the first of his homemade bombs. Over the next seventeen years he placed a total of 32 bombs in various public buildings in the New York area. Some of his bombs failed to explode, but those which did injured some fifteen persons. One bomb which did not go off was left in the Paramount Theater in December, 1956.

In January, 1957, an open letter addressed to the "Mad Bomber" was published in the New York Journal-American. Metesky responded to this letter and his answer was published in the newspaper. Then Alice Kelly, an employee of Consolidated Edison, noticed that the letter written by the "Mad Bomber" contained phrases and characteristics similar to one in her company's correspondence files written by George Metesky. She notified the authorities and Metesky was arrested. Psychiatrists examined him and declared him mentally unfit to stand trial. He was committed to Matteawan State Hospital for the Criminal Insane at Beacon, New York, from which he was released in 1974.

One of the unexplained mysteries of New York City was the affair of the "Sunday Bomber" who, in October and November, 1960, exploded bombs in various public places in and around the area. These bombs were all detonated on Sundays and were placed where they were extremely likely to kill or injure many people and do a maximum amount of property damage.

The first bomb exploded without warning in Times Square on October 2. Six people were injured. A week later a second bomb was set off in front of the New York Public Library. Still another bomb was placed on the Staten Island ferryboat "Knickerbocker." The boat was severely damaged but there were no casualties. Although several passengers on the boat at the time of the explosion were questioned, none could furnish any information of use to the police.

In all, six bombs were planted, the final one being detonated in a car of the New York subway. On this occasion one woman was killed and twelve people were injured. Since then there have been no further bombings and the identity of the "Sunday Bomber" has never been determined. George Metesky, the "Mad Bomber," was eliminated as a suspect as he was confined in the Matteawan State Hospital during the period of the "Sunday Bomber's" activities.

Suggested Readings

"Bomb Plot," Time, April 25, 1955.
"Bombers Everywhere," Newsweek, January 14, 1957.
"George Did It," Time, February 4, 1957.
"Mad Bomber Case Closed by Police," New York Times, September 10, 1957, p. 35.
"The Mad Bomber in New York," Time, January 7, 1957.
"Manhattan's Mad Bomber," Newsweek, January 7, 1957.
"Metesky Indicted on Bomb Charges," New York Times, January 31, 1957, p. 29.

"Part-Time Sleuths," U.S. News and World Report, February 1, 1957.
"Proven Profile," Newsweek, February 4, 1957.
"Quiet Man With a Quirk and a Quarrel," Life, February 4, 1957.
"Suspect is Held as 'Mad Bomber'; He Admits Role," New York Times, January 22, 1957, p. 1.
Symons, Julian. A Pictorial History of Crime. New York: Crown Publishers, 1966.

THE "DAY OF INFAMY" (1941)

The phrase, "Yesterday, December 7, 1941--a date which will live in infamy," was used by President Franklin Delano Roosevelt in his message to Congress asking for a declaration of war against Japan following the air and submarine attack on the Pearl Harbor Naval Base located on the south shore of the island of Oahu, Hawaii.

War had been raging in Europe since 1939 when Adolf Hitler's armies invaded and seized Poland. Japan had determined to obtain military domination and economic control over Eastern Asia and the Western Pacific Islands. On July 23, 1941, the Japanese occupied French Indo-China, where they seized ten million dollars' worth of American-owned supplies originally destined for China. In May the British and American governments, in an effort to keep Japan out of the war, authorized the renewal of petroleum contracts to Japan by oil companies operating in the Netherlands Indies. This diplomatic gesture, however, was ineffective.

In May, 1941, the United States had stopped the exporting of raw materials from the Philippines to Japan. Diplomatic negotiations between the United States and Japan became strained and President Roosevelt's administration froze Japanese assets in the United States, a move followed by Great Britain and the Netherlands Indies government.

Japan had to choose between withdrawing from China and ceasing to interfere in the internal affairs of other countries, or going to war against the United States. It decided on the latter course of action, hoping for an Axis victory over Soviet Russia and an alliance with Germany.

On the morning of Sunday, December 7, 1941, while negotiations between Japanese and American diplomats for the preservation of peace were being conducted in Washington, carrier-based Japanese planes appeared at Pearl Harbor and, without warning, bombarded and severely crippled the American fleet. Further damage was inflicted by Japanese submarines. Eight American battleships and ten other naval vessels were either sunk or badly damaged. Other nearby military installations were also attacked. Simultaneous attacks were made on Malaya, Hong Kong, Guam, Wake Island, and the Philippines. The next day Midway Island was invaded.

This "Day of Infamy" resulted in an American declaration of war on Japan and launched America's active participation in World War II, which finally came to an end in 1945.

Suggested Readings

Beard, Charles A. President Roosevelt and the Coming of the War. New Haven, Conn.: Yale University Press, 1948.

Bliven, Bruce. From Pearl Harbor to Okinawa. New York: Random House, 1960.

Commager, Henry Steele, ed. "President Roosevelt's Broadcast on the War with Japan," (Doc. No. 541) in his Documents of American History, 8th edition. New York: Appleton, 1968.

________. "President Roosevelt's Message Asking for War Against Japan," (Doc. No. 540) in his Documents of American History, 8th edition. New York: Appleton, 1968.

Corbett, Edmund V., ed. "Pearl Harbour," in his Great True Stories of Tragedy and Disaster. New York: Archer House, 1963.

Daniels, Jonathan. "Pearl Harbor Sunday: The End of an Era," in Leighton, Isabel, ed. The Aspirin Age, 1919-1941. New York: Simon & Schuster, 1949.

Divine, Robert A. Roosevelt and World War II. Baltimore: Johns Hopkins University Press, 1969.

Farago, Ladislas. The Broken Seal: "Operation Magic" and the Secret Road to Pearl Harbor. New York: Random House, 1967.

Feis, Herbert. The Road to Pearl Harbor. Princeton, N.J.: Princeton University Press, 1950.

Hatch, Alden. Franklin D. Roosevelt, An Informal Biography. New York: Holt, 1947.

Hersey, John R. Hiroshima. New York: Knopf, 1946.

Kimmel, Husband E. Admiral Kimmel's Story. Chicago: Henry Regnery, 1955.

Lord, Walter. Day of Infamy. New York: Holt, 1957.

Millis, Walter. This Is Pearl: The United States and Japan. New York: Morrow, 1947.

Peare, Catherine O. The FDR Story. New York: Crowell, 1962.

Schnittkind, Henry Thomas. Franklin Delano Roosevelt. New York: Putnam, 1962.

Wallechinsky, David, and Irving Wallace. "The 8-Eyed Spy," in their The People's Almanac. Garden City, N.Y.: Doubleday, 1975.

Weingast, David Elliott. Franklin D. Roosevelt, Man of Destiny. New York: Messner, 1952.

Wohlstetter, Roberta. Pearl Harbor: Warning and Decision. Stanford, Calif.: Stanford University Press, 1962.

THE COCOANUT GROVE FIRE (1942)

The Cocoanut Grove, Boston's oldest nightclub, was gutted by fire on the night of Saturday, November 28, 1942, in one of the most tragic and disastrous blazes of modern times. A crowd of more than 800 patrons thronged the dance floor and the Melody Lounge. The bar was crowded and every available table was occupied. Some patrons were celebrating--or bewailing--the outcome of the afternoon's football game between Boston University and the College of the Holy Cross. Holy Cross had won.

Guests at the Cocoanut Grove included soldiers and sailors and their companions, a wedding party, a number of celebrities, including Charles "Buck" Jones, the western motion picture star, and the "usual crowd of night club habitués."

The first floor show of the evening was scheduled to begin at 10:00 P.M. Music was provided by Mickey Alpert and his orchestra; Bill Payne was a featured vocalist. Just as the entertainment was to start a girl ran across the now-cleared dance floor screaming "Fire!" Her hair was ablaze.

The papier-maché palm trees and gaudy, flammable decorations quickly caught fire. The curtains and draperies fell flaming from their supports. Smoke billowed through the hallways.

The merrymakers stampeded. People fell over one another, and the stairway leading up from the Melody Lounge was blocked with fallen bodies. Some people were tramped to death; others died from smoke or fire inhalation. The revolving door at the street entrance was blocked with swarms of people attempting to leave the burning nightclub. Another street door was found to be locked and remained so.

Not everyone lost his head. Bill Payne led twenty people safely through the basement to a cellar exit. Marshall Cook, a dancer in the floor show, conducted some 35 people to a second floor dressing room. From here they were able to escape through a window to the roof of the building next door, from which they descended by ladder to the safety of the ground. Others escaped by crawling through windows and by kicking out a glass brick wall.

The Cocoanut Grove fire was soon extinguished but 498 people had died in the holocaust, "Buck" Jones among them. More than 100 others were hospitalized with burns, broken bones, smoke inhalation, and other injuries. Only 100 persons escaped without injury.

Temporary morgues were set up to handle the dead bodies. Every available ambulance was pressed into service, as were private cars, taxicabs, trucks, and even a moving van. All the hospitals in the area were crowded with dead and dying victims.

The investigation of the Cocoanut Grove fire disclosed that it had started when S. F. Tomaszewski, a bus boy, while changing an electric light bulb, struck a match which inadvertently ignited a tinder-dry imitation palm tree. It also disclosed that, aside from the high concentration of highly flammable decorations, such vitally necessary safety fixtures as exit markers and sprinkler systems were conspicuously absent. Ironically enough, the premises had been inspected and found to be in compliance with the fire safety code only two weeks before. The inspection had been conducted by the Boston fire department.

Suggested Readings

"Boston Fire Death Toll 440: Night Club Holocaust Laid to Bus Boy's Lighted Match," New York Times, November 30, 1942, p. 1.

"Boston Fire: Record Death Toll Reaches 492 as 119 Survivors Struggle for Life," Life, December 14, 1942.

"Boston Holocaust," Life, December 7, 1942.

"Boston's Worst," Time, December 7, 1942.

"Catastrophe: Cocoanut Grove, Boston's Oldest Night Club," Newsweek, December 7, 1942.

Clevely, Hugh. Famous Fires. New York: John Day, 1958.

Kartman, Ben. "Boston's Trial by Fire," in Kartman, Ben and Leonard Brown, eds. Disaster! New York: Pellegrini & Cudahy, 1948.

Moulton, R. S. "Lessons from the Boston Night Club Fire," American City, December, 1942.

"300 Killed by Fire, Smoke and Panic in Boston Resort," New York Times, November 29, 1942, p. 1.

Washburn, E. S. "Lessons of the Cocoanut Grove Fire; Operation of the Master File," Survey Midmonthly, February, 1943.

Weeks, F. "After Cocoanut Grove," Atlantic Monthly, March, 1943.

THE "GRANDCAMP" EXPLOSION (1947)

One of the most devastating explosions of modern times occurred in the harbor of Texas City, Texas, on the morning of April 16, 1947, when the cargo of ammonium nitrate carried by the French freighter "Grandcamp" blew up.

The "Grandcamp" was docked near the Monsanto Company's chemical plant. It had caught fire, probably from a carelessly discarded cigarette. Fire fighters were attempting to extinguish the blaze and were considering the advisability of moving the vessel out into Galveston Bay. These men were well aware of the possibility of ammonium nitrate exploding and felt that the vessel was too close to the Monsanto plant for safety. Before the ship could be moved, the cargo detonated, virtually blowing the "Grandcamp" apart.

The sudden explosion caused a gigantic tidal wave which, rolling in from the bay, inundated the docks and wharves and flung a huge steel barge over a hundred yards inland. The Monsanto plant and several other waterfront structures then blew up. The roof of the Monsanto plant caved in, burying 800 workers. Oil and gasoline storage tanks in nearby refineries went up in flames and a rushing, invisible force was generated by the series of explosions. Doors and windows of buildings as much as a mile away were blown in. Houses were knocked down like rows of dominoes.

Many persons were killed or severely injured. The high school gymnasium was used as a temporary morgue and an emergency hospital was set up in the city hall. Dead bodies were removed from the wreckage. Some people stayed to help and others fled the town. Inevitably looters appeared on the scene and scavenged what they could.

At 1:11 the following morning the "High Flyer," a nitrate-laden sister ship of the "Grandcamp," exploded. She had apparently caught fire when the "Grandcamp" went off and had suffered a similar fate. This second explosion was of such intensity that it was recorded by a seismograph in Denver, Colorado. Coming on the heels of the first, it caused a primitive terror among the people of Texas City. A wild evacuation followed and all roads leading away from town were jammed with people salvaging what personal belongings they might.

The "Grandcamp" explosion resulted in 433 deaths, over 1,000 injuries, and 128 missing persons. Property damage exceeded $67,000,000, most of which was fortunately covered by insurance.

Suggested Readings

Armstrong, Warren. "The Texas City Disaster and the 'Seistan'," in his _Fire Down Below_. New York: John Day, 1968.
Clevely, Hugh. _Famous Fires_. New York: John Day, 1958.
"Disaster Report, Texas City," _New Republic_, June 9, 1947.
"Disaster: Texas City Diary," _Newsweek_, April 28, 1947.
Hearings, Texas City Disaster, U.S. House of Representatives, Judiciary Committee. Washington, D.C.: Government Printing Office, 1954.
Hoehling, Adolph A. "Explosion in Texas City, 1947," in his _Disaster: Major American Catastrophes_. New York: Hawthorn Books, 1973.
"Monsanto Dealing with a Disaster," _Nation's Business_, January, 1971.
National Board of Fire Underwriters. _Texas City Disaster_ (pamphlet). New York: 1947.
"Pluperfect Hell at Texas City," in Kartman, Ben, and Leonard Brown, eds. _Disaster!_ New York: Pellegrini & Cudahy, 1948.
Robinson, Donald B. _The Face of Disaster_. Garden City, N.Y.: Doubleday, 1959.
U.S. Coast Guard, Board of Investigation. _Texas City Explosion._ Galveston, 1947.
Wheaton, Elizabeth. _Texas City Remembers._ San Antonio: Naylor, 1947.

THE "BLACK DAHLIA" MURDER (1947)

Elizabeth Short, a 22-year-old girl nicknamed the "Black Dahlia" because of her fondness for black clothes and negligees, was found brutally murdered in a Los Angeles, California lot on Wednesday, January 15, 1947. She had been tortured before being killed. Her nude body was mutilated and had been skillfully bisected at the waist. The perpetrator of this grisly act has never been found.

In her brief life Elizabeth Short had drifted from coast to coast and had known many men intimately. She was born in Medford, Massachusetts. Her parents separated when she was six and her mother, struggling to support herself and her daughters, had little time to spare for them.

When Elizabeth was seventeen she left school and traveled to Miami, Florida. There she worked as a waitress and dated and had affairs with countless servicemen. From Miami she traveled to Vallejo, California, where her father was living. She stayed with him, dating servicemen in spite of his protests and, following a brush with the police which resulted in her being booked as a juvenile delinquent, was sent home to her mother in Medford. Instead of returning to Massachusetts, however, she went to Camp Cooke near Santa Barbara, California, where, as a civilian employee at the post exchange, she continued her associations with a series of soldiers.

It was at Camp Cooke that she met Major Matt Gordon, Jr., an air force officer from Pueblo, Colorado. She fell in love with him but he received his shipping orders and departed for India. She also met Lieutenant Joseph George Fickling, likewise an air force officer, in whom she became interested.

After a short visit to Medford she went to Los Angeles, in response to a letter from Lieutenant Fickling. He, however, had been sent to a field in Texas and was soon to be discharged from the service. He later wrote her to say, in essence, that he had had a change of heart and was not contemplating marriage.

Elizabeth Short led an aimless life in Los Angeles, posing nude for photographers and picking up men in cafes, hotel lobbies, and wherever else she could. In August, 1946, she received a telegram from Major Gordon's mother telling her that he had been killed in an air crash in India.

She then lived briefly with Mark Hansen, a theater owner, in San Carlos, California, but continued to date servicemen. Her address book, discovered after her murder, contained the names of nearly a hundred men.

Leaving Hansen's apartment, she lived with some other girls for a few weeks, then made another trip to Medford. From there she returned to Los Angeles where she worked as a waitress, did more posing for photographers, and continued to pick up men.

On December 6, 1946, she moved to San Diego, California, where she lived with Mrs. Vera French, a woman she had met in an all-night motion picture theater. She then met a man she called "Red," who was later found to be Robert Manley, a salesman and ex-serviceman. Though married, he and his wife were having a "misunderstanding" about their marriage.

"Red" Manley picked Elizabeth up at Mrs. French's home on the evening of January 8, 1947. She had packed her belongings and when Manley arrived, ordered him to drive her to Los Angeles. He declined, saying he had some business calls to make in San Diego the following day. She then indicated that she would take the bus.

It was one week later, on January 15, that the girl's nude, bisected body was found in a vacant lot on South Norton Avenue, Los Angeles, by Mrs. Betty Bersinger, who was out for a walk with her three-year-old child.

The police investigation failed to disclose any clues leading to the killer or any motive for the murder. From Marjorie Graham, a friend of the slain girl, it was learned that her trunk was in storage at the Los Angeles Union Station. When opened the trunk was found to contain, among other things, a letter from Lieutenant Fickling and an album of pictures of soldiers, from privates to a lieutenant general.

Fickling, when questioned in Charlotte, North Carolina, where he, then a civilian, was employed as a co-pilot by an airline, proved that he had not seen the "Black Dahlia" since leaving Los Angeles many months previously. Her father reported that he had not seen her for several years.

Other friends and acquaintances of the girl testified to seeing her at various times but none could shed any light on the murder. Edward Glen Thorpe, a cowboy, was arrested on suspicion but was able to prove his innocence.

Robert "Red" Manley was located. He said that after leaving Mrs. French's home in San Diego on the evening of January 8 he and the girl had spent the night together in a motel. The following morning, in spite of his business appointments in San Diego, they drove to Los Angeles. He took the girl to the Biltmore Hotel, where she said she was to meet her sister. The sister had not arrived when he left Elizabeth in the lobby at 6:30. Manley convinced the police that on the night of the murder he was at home with his wife.

A number of persons "confessed" to the crime. These confessions, when checked out, were found to be spurious. On January 29 an unsealed envelope was found in the Los Angeles post office. This contained Elizabeth Short's birth certificate and some other of her personal items. Pasted to the envelope were letters cut from magazine advertisements reading "To the Los Angeles Examiner and other papers. Here is Dahlia's belongings. Letter to follow." The police were unable to lift any finger prints from the envelope or its contents.

Two more untraceable communications were received. The "Black Dahlia's" killer seems to be still at large.

Suggested Readings

Abrahamsen, David, M.D. The Murdering Mind. New York: Harper & Row, 1973.
Banks, L. "I Killed Her; Black Dahlia Murder Case Has Produced a Rash of Psychopaths," Life, March 24, 1947.
Bromberg, Walter. Mold of Murder: A Psychiatric Study of Homicide. New York: Grune & Stratton, 1961.
Catton, Joseph. Behind the Scenes of Murder. New York: Norton, 1940.
"Dumais Off Murder Suspect List," New York Times, February 12, 1947, p. 21.
"GI Says It's Possible He Murdered 'Dahlia'," New York Times, February 9, 1947, p. 51.
Gribble, Leonard. "The Black Dahlia," in his They Had a Way With Women. New York: Roy Publishers, 1967.
Jesse, F. Tennyson. Murder and Its Motives. London: Harrap, 1952.
Sterling, Hank, pseud. "The Black Dahlia," in his Ten Perfect Crimes. New York: Stravon Publishers, 1954.
Wolfgang, Marvin E., comp. Studies in Homicide. New York: Harper & Row, 1967.

THE KINSEY REPORTS (1948 and 1953)

Two of the most widely read books of the 20th century were published by Dr. Alfred Charles Kinsey. They were Sexual Behavior in the Human Male (1948) and Sexual Behavior in the Human Female (1953). Their popularity has been attributed to the titillating nature of their titles but the books were highly technical and directed towards professional sociologists and physicians rather than to the general reading public.

Dr. Kinsey was an American zoologist and educator who, as a member of the faculty of Indiana University at Bloomington, conducted several important research projects. These included studies in insect taxonomy, particularly of gall wasps.

In 1938 Kinsey and his associates began a scientific investigation of human sexual behavior, later receiving financial assistance from Indiana University, the Rockefeller Foundation and the National Research Council. The funds furnished were to underwrite a fifteen-year survey in which confidential interviews were to be conducted with 100,000 men and women on all aspects of sexual behavior.

The "Kinsey Reports," by which name they were popularly known, aroused much controversy. Many critics of the books felt that the wide dissemination of such information as they contained would have a deleterious effect on American morals. Mathematicians and statisticians attacked the methodology of the survey, which precipitated a spate of books, technical reports, and periodical articles both attacking and defending it.

Suggested Readings

Christensen, Cornelia V. Kinsey, A Biography. Bloomington, Ind.: Indiana University Press, 1971.
Deutsch, Albert. "How Kinsey Studies the Sexual Behavior of American Women," Woman's Home Companion, August, 1953.
_______. Sex Habits of American Men: A Symposium on the Kinsey Report. New York: Prentice Hall, 1948.
Hall, C. W. "The Man Who Made a Mountain Out of Sex," American Magazine, October, 1953.
Himelhoch, Jerome, and S. F. Fava, eds. Sexual Behavior in American Society: An Appraisal of the First Two Kinsey Reports. New York: Norton, 1955.
Kinsey, Alfred Charles, et al. Sexual Behavior in the Human Female. Philadelphia: Saunders, 1953.
_______. Sexual Behavior in the Human Male. Philadelphia: Saunders, 1948.
Pomeroy, Wardell Baxter. Dr. Kinsey and the Institute for Sex Research. New York: Harper, 1972.
"Talk of the Town," New Yorker, March 27, 1948.
Wallechinsky, David, and Irving Wallace. "Sex Surveys," in their The People's Almanac. Garden City, N.Y.: Doubleday, 1975.
Wickware, F. S. "Report on Kinsey," Life, August 2, 1948.

THE BRINK'S ROBBERY (1950)

On January 17, 1950, shortly after seven o'clock in the evening, seven men wearing plastic halloween masks and pea jackets held up the Brink's Armored Car Service in Boston, Massachusetts, and escaped with $1,000,000 in cash and $500,000 in checks. This was the largest cash robbery in the nation's history up to that time.

The men entered the Prince Street side of the building in which the Brink's office was located, climbed the stairs and stepped into the area known as the counting room. A wire cage opening to the vault was inside this room. Brandishing a revolver, one of the masked men ordered the gate to the wire cage opened. Thomas B. Lloyd, head cashier, seeing that resistance was useless, promptly complied. The men entered, forced the six Brink's employees to lie face down on the floor, and tied them securely with cords they had brought with them. They also placed adhesive tape over the employees' mouths. Then, working swiftly, they collected their booty, carried it to the door, and departed.

Half an hour later one of the Brink's employees managed to wriggle free of his bonds and send in an alarm to the American District Telegraph Protective Agency. Police prowl cars were immediately alerted and hurried to the scene. Police Commissioner Sullivan was advised of the crime.

Roadblocks were set up across Boston's bridges, and vehicles were stopped and searched. Within an hour virtually every on-duty Boston policeman was hunting for the robbers. Roadblocks were set up in surrounding states.

The Brink's employees were questioned by the police. Lloyd stated that the robbers apparently had a pass key and that only one of them spoke during the holdup. Jim Allen, assistant cashier, the only person who could then tell how much money had been taken, had had his glasses stolen along with the money and checks and so was unable to indicate the amount of the loss. A Brink's guard named William Manter, supposedly on duty, was eating his supper in another part of the building and so saw nothing of the robbery. Other security measures at Brink's were found to be surprisingly lax.

Police detectives found a witness who stated she had observed six men get out of a black sedan parked near the Brink's office. These men had entered the building. A search for the black sedan failed to locate it. A police roundup and questioning of known hoodlums, criminals and racketeers led nowhere.

When the weight of the stolen money and checks had been calculated at 350 pounds, it was realized that a truck would be needed to haul it away. It was then established that a green Ford truck had been parked in the area on the evening of January 17.

Two boys, playing near the Mystic River at Somerville, Massachusetts, found two unloaded revolvers. One of these was identified as having been taken from the Brink's counting room. The other had been discarded by the finders before it could be traced.

The remains of a Ford truck which had been cut to pieces with an acetylene torch were found in the city dump at Stoughton, Massachusetts. These were traced to the Back Bay Ford Agency which reported that such a truck had been stolen from them some months before. An attempt to locate the person who had used the acetylene torch was unsuccessful.

Another lead involved the body of John T. Murphy which was found floating in Penelope Lake near Peekskill, New York. A pistol permit found on Murphy's body listed him as an employee of Brink's Boston branch. The branch denied that he had ever been employed there.

In May, 1950, a jewelry thief named Alfred Gagnon, awaiting trial at the Rhode Island State Prison, stated that he, Carleton M. O'Brien, and Joseph F. McGinnis had master-minded the Brink's holdup. This "confession" was regarded by the Boston police as spurious and nothing more than an effort on Gagnon's part to win a reduction in sentence for his part in a jewelry robbery. Shortly after Gagnon made his statement, which had been publicized in the newspapers, O'Brien was shot to death by an unknown assailant. McGinnis was picked up by the police, questioned and released. At the time of the Brink's robbery he had been chatting with a Boston police lieutenant. Gagnon was found to have been in Florida when the robbery was committed.

The case dragged on and new rumors made headlines. Then, in December, 1952, Mrs. Mary A. Hooley, her husband Paul, her brother Donald O'Keefe and a bookmaker named John H. Carlson were brought before Federal Judge William R. McCarthy for questioning. These people were not charged with any crime, but were asked to answer "certain questions concerning others." When asked specifically, "at any time did you see $70,000 in cash?" they refused to answer on grounds of self-incrimination.

The possible connection of these four with the Brink's robbery was shown when Joseph "Specs" O'Keefe, another of Mrs. Hooley's brothers, was charged with being one of the bandits. Sixty thousand dollars in small bills had been hidden in the home of Mrs. Hooley and her husband, and it was charged that the money had been transported from the Hooley home to that of "Specs" O'Keefe. "Specs," however, was then in jail in Pennsylvania, serving a three-year sentence for the illegal possession of firearms.

Carlson, Donald O'Keefe and the Hooleys refused to answer further questions put to them by the court, for which they were cited for contempt and sentenced to jail.

Eventually the federal statute of limitations ran out and the federal grand jury was no longer able to return an indictment. However, the statute of limitations in the state of Massachusetts did not terminate until 1956. In that year the authorities apprehended the criminals. Of the stolen funds, less than $100,000 was recovered. Eight men, including the "master-minds" who planned the robbery but did not participate in it, were sentenced to life imprisonment.

Suggested Readings

"Boston Dough Party," Newsweek, January 30, 1950.
"Brink's Jinx," Newsweek, September 8, 1952.
"Cool Million in Boston," Time, January 30, 1950.
Dinneen, Joseph F. "The Great Brink's Robbery," Collier's, January 13, 1951.
"FBI Breaks Brink's," Newsweek, December 29, 1952.
Feder, Sid, completed by Joseph F. Dinneen. The Great Brink's Holdup. Garden City, N.Y.: Doubleday, 1961.
"Million Dollar Robbery: Brink's Boston Branch," Life, January 30, 1950.
O'Keefe, Joseph James. The Men Who Robbed Brink's. New York: Random House, 1961.
Roberts, Carl E. B. The New World of Crime. London: Hylton, 1933.
Sterling, Hank, pseud. "The Two Million Dollar Brink's Robbery," in his Ten Perfect Crimes. New York: Stravon Publishers, 1954.
Symons, Julian. A Pictorial History of Crime. New York: Crown Publishers, 1966.

THE McCARTHY ACCUSATIONS (1950-1954)

Joseph Raymond McCarthy, an American lawyer, soldier, and United States senator, is remembered for his bitter attacks on high-ranking government officials during the administrations of Presidents Harry S. Truman and Dwight D. Eisenhower. McCarthy's roughshod method of making often unsubstantiated charges of subversive activities and Communist party membership against his political enemies became popularly known as "McCarthyism." This was defined by Truman in 1953 as "the corruption of truth, the abandonment of ... fair play and 'due process' of law ... the use of the 'big lie' and the unfounded accusation ... in the name of Americanism and security." "McCarthyism" was defined by the senator's supporters as "relentless determination to expose Communist conspiracy, particularly in the government."

McCarthy first came to national attention in February, 1950, when he charged that 57 employees of the Department of State were members of the Communist party. The charge was never proved. Over the next three years he made a number of accusations against high-ranking State Department officials, maintaining that they were guilty of involvement in subversive activities; in the process McCarthy made many political enemies and also won a wide popular following.

In January, 1953, McCarthy, having been reelected to the Senate, was appointed chairman of the Committee on Government Operations. As head of its Permanent Subcommittee of Investigations he looked into alleged Communist activities in the Central Intelligence Agency, the United States Information Service, the United Nations, the Government Printing Office, and the United States Army. Charles E. Wilson, Secretary of Defense, charged McCarthy and members of his Committee with threatening army officials in connection with the drafting of a former consultant of the Subcommittee. McCarthy, in turn, countercharged that the army had attempted to "blackmail" him in an effort to halt the Subcommittee's investigation.

McCarthy gave up his chairmanship of the Subcommittee temporarily, but that body continued to hold open hearings on the contradictory claims from April to June of 1954. These hearings, which were widely publicized in the press and on radio and television, attracted great interest throughout the country. Both sides offered much testimony but little substantiation to support their accusations. On August 31 the Permanent Subcommittee, unable to agree on the matter, issued majority and minority reports. In the Republican report all concerned were mildly criticized; the Democratic report, in turn, was sharply critical of the McCarthy faction and only mildly critical of the military.

On August 2 the Senate voted to set up a special committee to consider censuring McCarthy for conduct unbecoming a senator. On December 2 the Senate voted to condemn him "for his conduct toward two senatorial committees."

From then until his death in 1957 McCarthy's political influence diminished steadily.

Suggested Readings

Anderson, Jack, and Roland W. May. McCarthy: The Man, The Senator, The 'Ism.'' Boston: Beacon Press, 1952.
Biddle, Francis. The Fear of Freedom. Garden City, N.Y.: Doubleday, 1951.
Bontecou, Eleanor. The Federal Loyalty and Security Program. Ithaca, N.Y.: Cornell University Press, 1953.
Buckley, William F., and L. Brent Bozell. McCarthy and His Enemies: The Record and Its Meaning. Chicago: Henry Regnery, 1954.
Cochran, Bert. Harry Truman and the Crisis Presidency. New York: Funk & Wagnalls, 1973.
Coser, Lewis A., and Irving Howe. The American Communist Party. Boston: Beacon Press, 1957.
Draper, Theodore. The Roots of American Communism. New York: Viking Press, 1957.
Howe, Irving, Lewis Coser and Julius Jacobson. The American Communist Party. New York: Praeger, 1962.
Latham, Earl. The Communist Controversy in Washington: From the New Deal to McCarthy. Cambridge, Mass.: Harvard University Press, 1966.
_______. The Meaning of McCarthyism. Boston: Heath, 1965.
Lattimore, Owen. Ordeal by Slander. London: MacGibbon & Kee, 1952.
McCarthy, Abigail. Private Faces/Public Places. Garden City, N.Y.: Doubleday, 1972.
Mandelbaum, Seymour J. The Social Setting of Intolerance. Chicago: Scott, Foresman, 1964.
Rorty, James, and Moshe Decter. McCarthy and the Communists. Boston: Beacon Press, 1954.
Rovere, Richard H. Senator Joe McCarthy. New York: Harcourt, Brace, 1959.
Shannon, David A. The Decline of American Communism. New York: Harcourt, Brace, 1945.
Thomas, Lately. When the Angels Wept: The Senator Joe McCarthy Affair, a Story Without a Hero. New York: Morrow, 1973.

THE ALGER HISS CASE (1950)

The Alger Hiss case first came to public attention on August 3, 1948, when Whittaker Chambers, a senior editor of Time magazine and an admitted ex-Communist, named Hiss, a former State Department official, as a one-time key member of the Communist underground operating in Washington, D.C.

Some nine years previously Chambers had called the attention of Adolph A. Berle, Assistant Secretary of State, to the nature of Communistic penetration within the United States government. Hiss, among others, was investigated under the Hatch Act of 1941 and given a clean bill of health. He rose rapidly in the State Department. In 1945, after the discovery of espionage rings operating in the United States and Canada, the rumors against him were revived. He was investigated again, and a third investigation was made in May, 1947. James F. Byrnes, then Secretary of State, seemed satisfied with Hiss as a security risk, as did the Federal Bureau of Investigation.

In 1948 Chambers testified before the House Un-American Activities Committee that Alger Hiss had been cell leader of a Communist group before he, Chambers, broke with the Communist party in 1937. On August 5, 1948, Hiss denied the charge, telling the Committee that he did not know Chambers by name. He could not, however, state positively, when shown a photograph of him, that he had not seen Chambers at some time in the past.

Two days later Chambers reappeared before the Committee and furnished information to support his charges. Mention was made of an automobile which Hiss had allegedly donated to the Communists. Chambers demonstrated his familiarity with certain of Hiss' hobbies, including bird-watching. It appeared that Chambers had known Hiss in the past, and subsequently Hiss stated that he might have known Chambers under the name of George Crosley, one of a number of aliases which Chambers had used.

Richard M. Nixon, later 37th President of the United States, was active in pressing the Committee's investigation. On August 17, 1948, Hiss and Chambers confronted each other before a Subcommittee at the Hotel Commodore in New York. It was at this meeting that Hiss identified Chambers as George Crosley. Chambers, in turn, identified Hiss as a member of the Communist party. Hiss emphatically denied such membership.

Two days later, on a "Meet the Press" radio program, Chambers charged that "Alger Hiss was a Communist and may be one now." His statements made before the Committee were privileged but those made over the radio were not. Hiss, however, though he had defied Chambers to make his charges out of the presence of the Committee, so that he might bring suit for libel, failed to bring any such suit following the broadcast until a month had elapsed.

The libel suit was brought in Baltimore, Maryland, federal court. Hiss asked $75,000 for defamation. At a pre-trial hearing Chambers produced some official documents in Hiss' handwriting; these documents, he stated, had been received from Hiss and hidden by Chambers and were typical of documents which Hiss had copied and turned over to the Communists. This resulted in the matter being referred to the grand jury. On December 2 Chambers produced the "Pumpkin Papers." These were on five rolls of microfilm and were copies of secret state papers which, Chambers said, he had hidden in a hollowed-out pumpkin on his Maryland farm. This he did to foil Communist agents who might search his farmhouse in his absence.

Alger Hiss was charged with two counts of perjury by the grand jury. It was stated that he lied under oath when he denied giving state documents to Chambers and that he had not seen Chambers after January 1, 1947. He pleaded not guilty to each count.

His first trial began on May 31, 1949. The presiding judge was Samuel H. Kaufman. Hiss' team of defense lawyers was headed by Lloyd Paul Stryker and the prosecution was handled by Assistant United States Attorney Thomas F. Murphy.

Whittaker Chambers was essentially the only witness of importance to appear against Alger Hiss, and the issue became one of believing one man or the other. The jury, having heard evidence from Ramos C. Feehan, an F. B. I. typewriter expert, concerning documents allegedly typed on a Woodstock machine by Priscilla Hiss, Alger Hiss' wife, and from other witnesses in the case, could not come to a decision. Alger Hiss' second trial began on November 17, 1949. Like the first, it was held in New York. Judge Henry W. Goddard was on the bench. Lloyd Paul Stryker had been replaced by Claude B. Cross as chief defense counsel, and Murphy again headed the prosecution. Again Hiss pleaded not guilty.

On January 21, 1950, the jury announced its decision: guilty of perjury as charged. Hiss was sentenced to five years' imprisonment, of which he served three years and eight months at the Lewisburg Penitentiary in Pennsylvania.

Hiss was disbarred for the felony conviction and worked as a salesman for many years. In 1975 he was reinstated as a lawyer by the supreme court in his home state of Massachusetts.

Suggested Readings

Busch, Francis X. Guilty or Not Guilty? Indianapolis: Bobbs-Merrill, 1952.

Chambers, Whittaker. Witness. New York: Random House, 1952.

Cook, Fred J. The Unfinished Story of Alger Hiss. New York: Morrow, 1958.

Cooke, Alistair. A Generation on Trial: U. S. A. v. Alger Hiss. New York: Penguin Books, 1950.

De Toledano, Ralph, and Victor Lasky. Seeds of Treason: The True Story of the Hiss-Chambers Tragedy. New York: Funk & Wagnalls, 1950.

Hiss, Alger. In the Court of Public Opinion. New York: Knopf, 1957.

"The Hiss Case Ends at Last," Life, April 2, 1951.

Howe, Irving, Lewis Coser, and Julius Jacobson. The American Communist Party. New York: Praeger, 1962.

Kunstler, William. "The Alger Hiss Perjury Case," in Rubenstein, Richard E., ed. Great Courtroom Battles. Chicago: Playboy Press, 1973.

________. "A Traitor from Harvard," in his And Justice for All. New York: Oceana Publications, 1963.

Morris, Richard B. "The Case of Alger Hiss," in his Fair Trial. New York: Knopf, 1952.

Nixon, Richard M. Six Crises. New York: Doubleday, 1962.

Reuben, William A. Honorable Mr. Nixon and the Alger Hiss Case. New York: Action Books, 1956.

"Reviving the Hiss Case," Newsweek, March 29, 1976.

Seth, Ronald. The Sleeping Truth: The Hiss-Chambers Affair Reappraised. New York: Holt, 1968.
Smith, John Chabot. Alger Hiss: The True Story. New York: Holt, Rinehart & Winston, 1976.
Stripling, Robert E. Red Plot Against America. New York: Bell Publishing Co., 1949.
Wallechinsky, David, and Irving Wallace. "Alger Hiss," in their The People's Almanac. Garden City, N.Y.: Doubleday, 1975.
Younger, Irving. "Was Alger Hiss Guilty?," Commentary Magazine, August, 1975.
Zeligs, Meyer A. Friendship and Fratricide: An Analysis of Whittaker Chambers and Alger Hiss. New York: Viking Press, 1967.

THE IMPLOSION CONSPIRACY (1950-1953)

The "Implosion Conspiracy" involved the "nuclear spies" Julius Rosenberg and his wife Ethel Greenglass Rosenberg and their accomplice, Martin Sobel. These persons were convicted of betraying secret information concerning atomic energy to agents of the Soviet Union.

Following the atomic bombing of Hiroshima and Nagasaki in August, 1945, World War II quickly terminated. The bombs which wrecked these Japanese cities were designed and built in the United States but the scientists who developed them came from many countries. The international backgrounds of these scientists made the hope that the United States could be the sole possessor of the "secret of the atom" a slim one. This became obvious when, in 1949, the Russians detonated their first atomic bomb.

Atomic spying by the Russians was well established as early as 1945, and on March 4, 1946, Dr. Allan Nunn May, a British physicist and espionage agent, was arrested. He was ultimately tried and sentenced to ten years' imprisonment. Dr. Klaus Fuchs, a German-born naturalized British physicist, had also been found guilty by a British court of disclosing classified information to Russia. He was sentenced to fourteen years in prison. This was in March, 1950.

One of Fuchs' alleged accomplices was Harry Gold, a Swiss-born naturalized American biochemist who was involved in an espionage ring dealing with atomic information. Gold was arrested, and other arrests followed. David Greenglass, an employee at the Atomic Research Center at Los Alamos, New Mexico, who had given drawings of the implosion lens to Gold, was taken into custody and arraigned in early June, 1950.

Greenglass, under interrogation by F. B. I. agents, implicated his sister's husband, Julius Rosenberg, who was arrested in July. On August 11 Ethel Greenglass Rosenberg was taken into custody, as was Martin Sobel, who had been implicated by Julius Rosenberg.

The trial of Sobel and the Rosenbergs began in the Federal Court House, New York City, on March 6, 1951. Named in the indictment were Anatoli Yakovlev, a Russian consular official who had departed for the Soviet Union; David Greenglass, who had pleaded "guilty"; Harry Gold, and Ruth Greenglass, David's wife. The Rosenbergs pleaded "not guilty."

The prosecution witness against Sobel was Max Elitcher, who testified that Rosenberg, in attempting to recruit him as a member of the espionage ring, had implicated Sobel, Rosenberg's neighbor and friend. No connection was shown between Sobel and atomic spyings, but it was brought out that he had Communist associations.

The case against the Rosenbergs rested on the testimony of David Greenglass. Under oath he stated essentially that Julius Rosenberg had induced him to compile classified material on the atomic bomb, material which was typed by Ethel Rosenberg. Greenglass was then to turn the information over to a secret agent, who proved to be Harry Gold.

Gold, on the witness stand, speaking little of the Rosenbergs and not at all of Sobel, described his own espionage activities and dealings with Yakovlev. His testimony, however, corroborated many of Greenglass' sworn statements. The Rosenbergs, husband and wife, each took the stand and steadfastly denied all charges.

On March 29, 1951, the jury found Sobel and the Rosenbergs guilty. On April 5 Judge Irving R. Kaufman sentenced the Rosenbergs to death and Sobel to thirty years' confinement. Sobel was removed to prison and on June 19, 1953, the Rosenbergs were executed at Sing Sing Prison, Ossining, New York. They were the first convicted spies ever executed in the United States on the orders of a civil court.

Suggested Readings

Aymar, Brandt, and Edward Sagarin. "Julius and Ethel Rosenberg and Morton Sobel," in their _A Pictorial History of the World's Great Trials_. New York: Crown Publishers, 1967.

Fineberg, Andhil S. _The Rosenberg Case: Fact and Fiction._ New York: Oceana Publications, 1953.

Franklin, Charles. _The Great Spies_. New York: Hart, 1967.

Goldstein, Alvin H. _The Unquiet Death of Julius and Ethel Rosenberg_. New York: Lawrence Hill, 1975.

Howe, Irving, Lewis Coser and Julius Jacobson. _The American Communist Party_. New York: Praeger, 1962.

Laurence, John, _pseud_. _A History of Capital Punishment_. New York: Citadel Press, 1960.

Meeropol, Robert, and Michael Meeropol. _We Are Your Sons: The Legacy of Ethel and Julius Rosenberg_. Boston: Houghton Mifflin, 1975.

Moorehead, Alan. _The Traitors_. New York: Scribner's, 1952.

Nizer, Louis. _My Life in Court_. New York: Doubleday, 1961.

"The Rosenberg Case: Some Reflections on Federal Criminal Law," _Columbia Law Review_, February, 1954.

Schneir, Walter, and Miriam Schneir. Invitation to an Inquest. Garden City, N.Y.: Doubleday, 1965.
Sharp, Malcolm. Was Justice Done? The Rosenberg-Sobel Case. New York: Monthly Review Press, 1956.
Symons, Julian. A Pictorial History of Crime. New York: Crown Publishers, 1966.
The Testament of Ethel and Julius Rosenberg. New York: Cameron & Kahn, 1965.
U.S. vs. Rosenbergs, Sobel, Yakovlev and David Greenglass. Transcript of trial, published by Committee to Secure Justice for Morton Sobel. New York: 1951.
Wexley, John. The Judgment of Julius and Ethel Rosenberg. New York: Cameron & Kahn, 1955.

THE TRUMAN-MacARTHUR CONFRONTATION (1951)

General Douglas MacArthur was in Japan when Communist North Korea attacked South Korea on June 25, 1950. In spite of the fact that he had no responsibility for South Korea, he was ordered by President Harry S. Truman to assume command of a small American military mission stationed there. He was also to provide American air and sea support for the army of South Korea.

On June 28 MacArthur flew to Korea to make a reconnaissance. Following this he recommended that American ground forces be used to assist the South Koreans, to which recommendation Truman agreed. The United States was able to have the United Nations assume nominal responsibility for the defense of South Korea, and on July 8 MacArthur was placed in command of a unified international force set up to repel the invading North Korean Communists.

The first American troops arrived on July 1, 1950. By this time the North Koreans had defeated the South Korean troops, captured the city of Seoul and were advancing southward, meeting little opposition. In early September they had driven the defending South Koreans to the Pusan perimeter, a small beachhead near the southern tip of the Korean peninsula.

On September 15 MacArthur landed troops in a flanking movement at Inchon. Simultaneously the American Eighth Army launched a gigantic counterattack from the Pusan perimeter. The North Koreans fled toward the 38th parallel which marked the boundary between North and South Korea. Most of South Korea had been recaptured by the United Nations forces by the end of September.

MacArthur felt that total victory was within his grasp. He was aware that the Communist Chinese were massing in large forces north of the Yalu River which marked the boundary between North Korea and Communist China. On October 15 MacArthur and Truman met on Wake Island in the Pacific. The General believed that the Chinese would not attack and told Truman as much. With this assurance, Truman authorized MacArthur, with some qualifications, to invade North Korea with the object of destroying its military forces. United Nations troops crossed the 38th parallel on October 8 and advanced toward the Yalu River.

The Chinese attacked in force on November 24, deflecting MacArthur's army and driving it back over the 38th parallel. The General began a counterattack in February, 1951, which proved indecisive and found the Korean war settled into a stalemate.

Following China's intervention MacArthur stated that the Korean situation consituted a "new war" and proposed bombing Manchurian supply centers and "unleashing" Chinese Nationalists on Formosa against the Mainland, held by the Communists. President Truman, fearing that MacArthur's proposed tactics would cause a much larger war, rejected the General's proposals.

MacArthur's outspokeness about Formosa caused his first dispute with the White House. Truman, furious, had demanded that MacArthur withdraw his remarks, and the latter did so.

On April 5, 1951, Representative Joseph W. Martin, Jr. made known a letter written by the General. This letter, read to Congress, defended MacArthur's position and said, among other things, "there is no substitute for victory." On April 11 MacArthur was ordered to return to the United States, and President Truman informed the press that it was his intention to dismiss him from the Service. "By his act," Truman said, "MacArthur left me no choice. I could no longer tolerate his insubordination."

General MacArthur returned home a hero. On April 19, 1951, he addressed a joint session of Congress, at which he made his famous farewell address. This was a stirring defense of his views on Korea. It ended with a quotation from an old barrack ballad: "Old soldiers never die; they just fade away."

Following his dismissal from the army, MacArthur accepted a high level position in industry. He died in 1964.

Suggested Readings

Cochran, Bert. Harry Truman and the Crisis Presidency. New York: Funk & Wagnalls, 1973.

Daniels, Jonathan. The Man of Independence. New York: Lippincott, 1950.

Gunther, John. The Riddle of MacArthur: Japan, Korea, and the Far East. New York: Harper, 1951.

Hoare, Wilbur, Jr. "Truman," in May, Ernest R., ed. The Ultimate Decision: The President as Commander-in-Chief. New York: Braziller, 1960.

Hunt, Frazier. The Untold Story of Douglas MacArthur. New York: Devin-Adair, 1955.

MacArthur, General Douglas. Reminiscences. New York: McGraw-Hill, 1964.

Osgood, Robert E. Limited War: The Challenge to American Strategy. Chicago: University of Chicago Press, 1957.

Rees, David. Korea: The Limited War. New York: St. Martin's Press, 1964.

Rovere, Richard H., and Arthur M. Schlesinger, Jr. The General

and the President, and the Future of American Foreign Policy. New York: Farrar, Straus, 1951.
Spanier, John W. The Truman-MacArthur Controversy and the Korean War. New York: Norton, 1965.
Steinberg, Alfred. The Man from Missouri: The Life and Times of Harry S. Truman. New York: Putnam, 1962.
Truman, Harry S. Memoirs of Harry S. Truman. Garden City, N.Y.: Doubleday, 1956.
Whitney, Courtney. MacArthur: His Rendezvous With History. New York: Knopf, 1956.
Willoughby, Charles A., and John Chamberlain. MacArthur: 1941-1951. New York: McGraw-Hill, 1954.

THE ARNOLD SCHUSTER MURDER (1952)

In March, 1950, Willie "The Actor" Sutton, a notorious bank robber, together with three fellow robbers, staged a holdup at the Manufacturer's Trust Company in New York, making their escape with some $64,000. "Wanted" circulars showing Sutton's picture were printed and distributed. One such circular was placed in Mac's Clothes Shop in Brooklyn where it was seen by Arnold Schuster, the 24-year-old son of the proprietor.

On April 16, 1952, Schuster saw Sutton on the street and turned him in to the police. Sutton was arrested and held for trial.

Schuster claimed the reward which had been offered for Sutton's capture, and in so doing incurred the animosity of a number of misguided people who regarded his act as that of a "fink" or "rat." Threatening letters and telephone calls were received by him and his family, and on March 8, 1952, a letter was slipped under the door of his father's shop. The letter stated that "this would be Arnold Schuster's last day on earth."

At about 8:15 that evening young Schuster, who was employed in the shop, locked up and started home. He boarded a bus where he met and chatted with a neighbor woman friend. The woman left the bus a block before Schuster. When questioned later by the police she remembered that a man had boarded the bus when Schuster did and sat behind him. Who this man was and whether or not he had any connection with subsequent happenings has never been determined.

Schuster's family was watching a television program that evening. The program was interrupted by a spot news announcement that Arnold Schuster had been shot and killed. His body, with four .38 caliber bullets in it, had been discovered in a driveway near his home by Mrs. Muriel Geller, a neighbor. She reported her find to Dr. S. M. Fialka whose house was next to the driveway. Dr. Fialka promptly informed the police.

There seemed to be no clues leading to the murder except the four bullets found in Schuster's body. There were no witnesses to the killing. Dr. Fialka stated that he had heard a noise which he imagined to be the backfire of an automobile and immediately afterwards the sound of a car being driven rapidly away.

The motive for the killing remained a mystery. Willie Sutton was in jail and so could not have played an active part in the murder. Even though some demented underworld friend of Sutton's might have killed Schuster as an informer, such an act would damage Sutton's case when he was brought to trial. In spite of this Sutton's known friends and criminal associates were rounded up, questioned, and released for want of evidence. One of these, Frederick J. Tenuto, a murderer and 1947 escapee from Holmesburg State Prison in Pennsylvania, could not be found and is still at large.

Rewards totaling $39,000 were offered for evidence leading to the arrest and conviction of the killer. These rewards were never collected. The police eventually found a .38 Smith & Wesson revolver in a lot some five blocks from the scene of the murder and near the Brooklyn waterfront. This revolver was found to be one of fourteen stolen from a shipment at a Brooklyn pier on February 8, 1952. By questioning longshoremen at the pier the police ultimately traced the revolvers to John "Choppy" Mazziotta, a gangster who had purchased them from the longshoremen who had originally stolen them. Before he could be arrested and questioned, Mazziotta disappeared and has never been found. It was rumored but not proved that he had been killed by gangland associates.

To this day the murder of Arnold Schuster remains an unsolved mystery. One theory is that members of a criminal organization such as Murder, Inc. may have felt that if other private citizens were to inform the authorities of underworld activities when they learned of them, this would be bad for illegal business, and thus decided to have Schuster killed as an example and deterrent to others.

Suggested Readings

Abrahamsen, David, M.D. The Murdering Mind. New York: Harper & Row, 1973.
Bromberg, Walter. Mold of Murder: A Psychiatric Study of Homicide. New York: Grune & Stratton, 1961.
Catton, Joseph. Behind the Scenes of Murder. New York: Norton, 1940.
Cohen, Bruce J. Crime in America. Itasca, Ill.: Peacock, 1970.
"Death for an Informer," Newsweek, March 17, 1952.
"Good Citizen," Time, March 17, 1952.
Hibbert, Christopher. Roots of Evil: A Social History of Crime and Punishment. New York: Funk & Wagnalls, 1968.
Jesse, F. Tennyson. Murder and Its Motives. London: Harrap, 1952.
"Murderer Haunts Brooklyn," Life, March 24, 1952.
Reynolds, Quentin. I, Willie Sutton. New York: Farrar & Rinehart, 1953.
"Schuster Stymie," Newsweek, March 24, 1952.
Sterling, Hank, pseud. "Murder of a Good Citizen," in his Ten Perfect Crimes. New York: Stravon Publishers, 1954.
Sutherland, Edwin H. On Analyzing Crime. Edited by Karl Schussler. Chicago: University of Chicago Press, 1972.

"Underworld Challenge?," Senior Scholastic, March 19, 1952.
Winslow, Robert W. Crime in a Free Society. Encino, Calif.: Dickenson, 1968.
Wolfgang, Marvin E., comp. Studies in Homicide. New York: Harper & Row, 1967.

THE SAM SHEPPARD MURDER CASE (1954)

On the evening of Saturday, July 3, 1954, Dr. Samuel H. Sheppard and his wife Marilyn entertained their friends, Don and Nancy Ahern, at dinner in their Bay Village, Ohio, home on the shore of Lake Erie, near Cleveland. The Aherns left the Sheppard home around 12:30 A.M. Dr. Sheppard, according to his testimony, fell asleep on the sofa while his wife went upstairs to bed. He was, he said, awakened by a scream and ran upstairs. There he saw "misty figures grappling together," and then was struck down from behind "and apparently knocked out."

When Sheppard came to, he took his wife's pulse and "felt that she was gone." He then checked the condition of his seven-year-old son Samuel H. Sheppard, Jr., nicknamed "Chip." He found nothing amiss. He then thought he heard a noise below. Running downstairs he saw a form speeding toward the lake. Pursuing this form to the beach house, he was again attacked and rendered unconscious. When he revived a second time he found himself in water close to shore. He returned to his house and, after again examining his wife, telephoned his friend J. Spencer Houk, mayor of Bay Village. Houk and his wife promptly drove to the Sheppard residence. Sam Sheppard lay down and Houk, after determining that Marilyn Sheppard was dead, called the police, an ambulance, and Sam's brother, Dr. Richard Sheppard.

Marilyn Sheppard had been beaten to death with some object which was never found. Sam Sheppard went to the hospital for treatment of facial and neck injuries. During the subsequent trial he wore an orthopedic collar.

Following the inquest Sheppard was indicted by the grand jury on Tuesday, August 17, 1954, for first degree murder. The trial was scheduled to begin on October 18, with Judge Edward Blythin presiding and attorneys William J. Corrigan and Fred Garmone acting as defense counsel.

Sheppard was tried in an atmosphere of extreme hostility. The trial, one of the most sensational of the 1950's, was given nationwide publicity by newspaper, radio, and television. It developed that Marilyn Sheppard had been pregnant when she was murdered, that there had been some talk of divorce, and that Sam Sheppard had had an affair with Susan Hayes, an attractive laboratory technician.

The jury found Sheppard guilty of murder in the second degree and on December 21, 1954, Judge Blythin sentenced him to life imprisonment.

The sentence was appealed. The Court of Appeals of the Eighth Ohio District in two instances refused to reverse the decision of the lower court. It was then considered by the Supreme Court of Ohio, which, in a 5 to 2 decision, upheld the conviction.

Sam Sheppard was paroled in 1966. His medical career wrecked, he became a professional wrestler. That year he married Ariane Tebbenjohanns, who had written him when he was in prison, expressing her belief in his innocence. They were later divorced and Sheppard married again, this time to Colleen Strickland, the nineteen-year-old daughter of his wrestling manager.

In June, 1966, the United States Supreme Court reversed the conviction. This was the result of an appeal by F. Lee Bailey, a criminal attorney who had become convinced of Sheppard's innocence by Erle Stanley Gardner, detective story writer, attorney, and member of the Court of Last Resort. This is a non-profit organization which gives its services free in cases of possible injustice.

Sheppard never resumed his medical practice. He died in 1970.

Suggested Readings

Bailey, F. Lee, and Harvey Aronson. _The Defense Never Rests._ New York: New American Library, 1973.

________, and Henry B. Rothblatt. _Crimes of Violence, Homicide and Assault._ Rochester: Lawyers Cooperative, 1973.

________, and John Greenya. _For the Defense._ New York: Atheneum Publishers, 1975.

"Forget-Me-Not," _Newsweek_, December 31, 1962.

Holmes, Paul Allen. _Retrial: Murder and Dr. Sam Sheppard._ New York: Bantam Books, 1966.

______. _The Sheppard Murder Case._ New York: David McKay, 1961.

"How Sheppard Won," _Time_, November 25, 1966.

Kilgallen, Dorothy. "When Justice Took the Day Off," in her _Murder One._ New York: Random House, 1967.

Kunstler, William H. "Murder in a Triangle: The State of Ohio versus Sam H. Sheppard," in his _First Degree._ New York: Oceana Publications, 1960.

Pollack, Jack Harrison. _Dr. Sam: An American Tragedy._ Chicago: Henry Regnery Co., 1972.

Sheppard, Samuel H. _Endure and Conquer._ Cleveland: World, 1966.

Sheppard, Stephen A., with Paul Allen Holmes. _My Brother's Keeper._ New York: David McKay, 1964.

Symons, Julian. _A Pictorial History of Crime._ New York: Crown Publishers, 1966.

Wells, C. "The Sheppard Case Reopened," _Life_, July 31, 1964.

THE U-2 AFFAIR (1960)

On May 1, 1960, Lieutenant Francis Gary Powers, pilot of an American U-2 high-altitude jet reconnaissance plane, who had been taking aerial photographs of Soviet territory, was brought down near Sverdvolsk by a Soviet missile. Powers was captured unharmed and was charged by the Russians with participating in a military intelligence mission. President Dwight D. Eisenhower at first denied the charge but later found himself obliged to admit its truth and accept responsibility for the episode.

On May 5 Soviet Premier Nikita Khrushchev, in an angry announcement, declared that Powers was working for the United States Central Intelligence Agency. The Agency admitted that Powers was missing but claimed that he was working on a project involving weather measurement. On May 7 Khrushchev announced further that Powers had confessed to being engaged in an intelligence mission for the C.I.A. when he was shot down, and would be tried in Moscow. In rebuttal the United States State Department charged that it is "no secret" that all countries, including the Soviet Union, are engaged in "intelligence collection activities." Secretary of State Christian A. Herter, on May 9, stated that the United States "would be derelict to its responsibilities" if it failed to take intelligence measures to "overcome this danger of surprise attack" by the U.S.S.R.

Premier Khrushchev, on May 16, made certain demands as a result of the U-2 incident. These were, first, that all military flights over Russia be terminated by the United States; second, that the United States apologize for "past aggressions"; and, third, that the United States punish appropriately those responsible for such military flights. President Eisenhower had given his assurance that intelligence flights had been ended and no further such flights would be made. In spite of this Khrushchev refused to participate in the Big Four Summit Conference of heads of state (United States, Great Britain, France, and Russia) until his conditions were met, and the Conference collapsed. This Khrushchev blamed on the United States and President Eisenhower. He threatened dire consequences should intelligence flights over Russia be repeated. This warning applied not only to the United States but also to other nations permitting the use of their airfields for such flights.

On May 26, 1960, Eisenhower made a nationwide radio and television broadcast in which he stated that his goal was to reduce friction with the Soviet Union, and said also that it was necessary to maintain effective systems for gathering information about the military capabilities of other powerful nations. He took personal responsibility for "approving all the ... programs undertaken by our government to secure and evaluate military intelligence."

Lieutenant Powers was tried by a three-man tribunal in Moscow in August, 1960. After three days of hearings the tribunal, on August 19, found him guilty of "having collected information of a strategic significance which constitutes a state ... secret of the Soviet Union." The flyer was sentenced to ten years' confinement. On February 10, 1962 he was released in exchange for

Rudolf Abel, a Soviet spy who was reputed to be the director of a Russian spy network in the United States. Abel had been in American custody since 1957.

Suggested Readings

Commager, Henry Steele, ed. "The U-2 Affair," (Doc. No. 638) in his Documents of American History, 8th edition. New York: Appleton, 1968.
Demaris, Ovid. "Going to See Gary," Esquire, May, 1966.
"Flight and Capture of the U-2 in the Words of Pilot Powers," U.S. News and World Report, March 19, 1962.
Franklin, Charles. The Great Spies. New York: Hart, 1967.
Lamparski, Richard. "Francis Gary Powers," in his Whatever Became of ...? New York: Crown Publishers, 1974.
"Pilot Powers--All Clear," New Republic, March 19, 1962.
Powers, Barbara Moore, and W. W. Diehl. Spy Wife. New York: Pyramid Books, 1965.
Powers, Francis Gary. "Shot Down Over Russia," in Parker, W., ed. Men of Courage. Chicago: Playboy Press, 1972.
________, with Curt Gentry. Operation Overflight. New York: Holt, 1970.
"U-2 Spy: Questions Without Answers," Newsweek, February 26, 1962.
Wallechinsky, David, and Irving Wallace. "Francis Gary Powers," in their The People's Almanac. Garden City, N.Y.: Doubleday, 1975.
Wise, David, and Thomas B. Ross. The U-2 Affair. New York: Random House, 1962.

THE BAY OF PIGS INVASION (1961)

On April 17, 1961, shortly before dawn, an invading force of anti-Castro Cuban rebels landed in the area of Bahia de Cochinas (Bay of Pigs) in the southern Las Villas province of Cuba. The attack, directed by the U.S.-based Cuban National Revolutionary Council, ended in disaster for the invaders. Overwhelmingly defeated by Fidel Castro's forces, they surrendered two days later.

Premier Nikita Khrushchev of Russia was highly incensed over the incident and angry Russians demonstrated before the American embassy in Moscow. Khrushchev charged, in a message to President John F. Kennedy, that the "armed bands [which had] invaded Cuba" had been "trained, equipped, and armed" by the United States, and promised that Russia would assist Cuba to repel the "armed attack."

Kennedy issued a statement in which he assumed "sole responsibility for the events of the past days."

On May 17 Castro offered to exchange most of the prisoners captured in the invasion for 500 American heavy duty tractors. In

April, 1962, 1,179 prisoners captured by Castro's forces were tried in Havana by a military court. They were found guilty of "crimes committed against the nation in connivance with a foreign power." In June several Americans formed a committee to raise $62,000,000 with which to ransom the convicted prisoners. The ransom was arranged, the money being supplied by private donors. On December 23, 1962, the first group of ransomed prisoners arrived in Miami, Florida, having been flown from Cuba by a series of air lifts. In all, 1,113 prisoners were released, for which Castro received $50,000,000 in food, medicine, and cash.

On December 29 President Kennedy flew to Florida to welcome the freed prisoners. He stated at a rally attended by some 40,000 Cuban exiles that "on behalf of my government and my country I welcome you to the United States. I bring you my nation's respect for your courage and for your cause...."

Responsibility for the failure of the Bay of Pigs invasion is still a subject of debate.

Suggested Readings

Commager, Henry Steele, ed. "The Bay of Pigs; Ambassador Stevenson's Statement," (Doc. No. 645) in his Documents of American History, 8th edition. New York: Appleton, 1968.

Fay, Paul B. The Pleasure of His Company. New York: Harper, 1966.

Flammonde, Paris. The Kennedy Conspiracy. New York: Meredith Press, 1969.

Johnson, Haynes. The Bay of Pigs. New York: Norton, 1964.

Reidy, John P. The True Story of John Fitzgerald Kennedy. Chicago: Children's Press, 1967.

Rivero, Nicolas. Castro's Cuba: An American Dilemma. Washington, D.C.: Luce, 1962.

Salinger, Pierre. With Kennedy. Garden City, N.Y.: Doubleday, 1966.

Schlesinger, Arthur M., Jr. A Thousand Days: John F. Kennedy in the White House. Boston: Houghton Mifflin, 1965.

Sorenson, Theodore C. Kennedy. New York: Harper, 1965.

Winks, Robin W. The Cold War from Yalta to Cuba. New York: Macmillan, 1964.

Wood, James Playsted, and the editors of Country Beautiful Magazine. The Life and Words of John F. Kennedy. Garden City, N.Y.: Doubleday, 1964.

THE "FREEDOM RIDERS" (1961)

The so-called "Freedom Riders" were groups of people, both white and colored, who rode on common carrier buses traveling interstate to test segregation barriers in buses and bus terminals. The first of these was a biracial group of thirteen persons who set out by bus from Washington, D.C. on May 4, 1961. Their destination was New Orleans, Louisiana.

The concept of such freedom rides was triggered on December 1, 1955, at Montgomery, Alabama. On that date Mrs. Rosa Parks, a Negro seamstress, was arrested when she refused to relinquish her seat in the front section of a bus in order that a white man might occupy it.

The arrest of Mrs. Parks precipitated a general boycott of bus transportation by the Negroes of Montgomery. They resolved to stay off buses until such time as they could be seated on an equal basis with white passengers. The strategy for the boycott, which lasted for over a year, was planned by the Reverend Martin Luther King, Jr. and his associates at King's church. It was then that he first achieved prominence as a nationally known civil rights leader.

Other demonstrations followed. Negroes began occupying seats at lunch counters supposedly reserved for white patrons only. Civil rights marches, dealing with the Negro's right to vote, were held, notably one from Selma, Alabama to Montgomery, led by King in 1965. "Freedom riders" tested racial discrimination policies on buses. On September 22, 1961, the Interstate Commerce Commission ordered that such racial discrimination against bus travelers be discontinued.

Suggested Readings

Bennett, Lerone, Jr. What Manner of Man: A Biography of Martin Luther King, Jr. Chicago: Johnson, 1968.

Clayton, Edward Taylor. Martin Luther King: The Peaceful Warrior. Englewood Cliffs, N.J.: Prentice-Hall, 1968.

Fribourg, Marjorie G. Bill of Rights: Its Impact on the American People. Philadelphia: Smith, 1967.

Harris, Janet. The Long Freedom Road: The Civil Rights Story. New York: McGraw-Hill, 1967.

King, Martin Luther. Measure of a Man. Philadelphia: Pilgrim Press, 1968.

_______. Why We Can't Wait. New York: Harper, 1964.

Lewis, Anthony, and the New York Times. Portrait of a Decade: The Second American Revolution. New York: Random House, 1964.

Lewis, David L. King: A Critical Biography. New York: Frederick A. Praeger, 1970.

Lomax, Louis E. The Negro Revolt. New York: Harper & Row, 1962.

Miller, William Robert. Martin Luther King, Jr. New York: Weybright and Talley, 1968.

Myrdal, Gunnar. An American Dilemma: The Negro Problem and American Democracy. New York: Harper, 1962.

Peck, James. Freedom Ride. New York: Simon & Schuster, 1962.

Saunders, Doris, ed. The Day They Marched. Chicago: Johnson, 1963.

Schwartzmann, Ruth, and Joseph Stein. The Law of Personal Liberties. New York: Oceana Publications, 1955.
Spike, Robert W. The Freedom Revolution and the Churches. New York: Association, 1965.
Young, Margaret B. A Picture Life of Martin Luther King, Jr. New York: Watts, 1968.

THE BILLIE SOL ESTES SWINDLE (1962)

One of the most sensational business swindles of the 1960's was engineered by Billie Sol Estes, a Texas entrepreneur. Before his conviction following state and federal indictments for mail fraud and conspiracy, Estes created a $150,000,000 empire on paper and attempted to corner the grain storage business in West Texas. His downfall came when he induced farmers in eleven West Texas counties to sign mortgages on nonexistent storage tanks of anhydrous ammonia, a fertilizer which became a household word in the area when Estes' financial empire crumbled and he went bankrupt. The federal government charged that he sold the mortgages to finance companies for $24,000,000, using the proceeds to finance his business activities.

On November 7, 1962, Estes was tried for fraud in Tyler, Texas. Found guilty, he was sentenced to prison. In March of the following year a federal jury in El Paso, Texas, convicted him on four counts of mail fraud and one of conspiracy. On February 19, 1965, the American Bar Association asked the Supreme Court to reverse the 1962 conviction on the grounds that television coverage of his trial, despite his repeated objections, had violated his constitutional rights. By a 5 to 4 decision, handed down on June 7, 1965, the Supreme Court reversed the swindling conviction of the Tyler, Texas, court on the grounds specified by the Bar Association. This ruling, however, did not affect Estes' other convictions for mail fraud and conspiracy.

After six years of confinement Estes was paroled on the condition that he avoid self-employment and promotional activities. He became a truck driver for a butane company and a part-time cattle raiser on his brother's Texas ranch. Today the man who a federal judge said committed "as brazen an exhibition of perjury as I have ever seen in the forty years I have been going to the courtroom" lives quietly in Abilene, Texas, with his wife. He attends church regularly and refuses to talk with newsmen. His typical response to inquiries is, "I have no comment to make to you, but thank you for calling." The parole board also refuses to discuss his business activities.

When released from prison at Fort Leavenworth, Kansas, he said, "Business and money are no longer my gods. I am not on anyone's side any more except God's. He works in strange ways."

Suggested Readings

"Behind the Billie Sol Mess," Newsweek, May 28, 1962.
Duscha, Julius. "How Billie Sol Used the Farm Program," The Reporter, June 21, 1962.
______. Taxpayers' Hayride: The Farm Problem from the New Deal to the Billie Sol Estes Case. Boston: Little, Brown, 1964.
"Estes: Three-sided Country Slicker," Fortune, July, 1962.
Griffin, O. "How We Exposed Billie Sol Estes," Saturday Evening Post, June 23, 1962.
Hamilton, Donald. "Who Murdered Henry Marshall?," Argosy, November, 1962.
Holland, Cecil. "Decline and Fall," Time, May 25, 1962.
"The House of Cards Estes Built," Business Week, June 2, 1962.
"Investigations: The Estes Scandal," Time, June 15, 1962.
"The Strange Story of Too Much Grain," U.S. News and World Report, May 28, 1962.
Symons, Julian. A Pictorial History of Crime. New York: Crown Publishers, 1966.
"West Texas Stance: Estes on Trial," Newsweek, November 5, 1962.

THE CUBAN MISSILE CRISIS (1962)

As early as July, 1960, Premier Nikita Khrushchev of the Soviet Union threatened the United States with attack by Soviet rockets if the United States attempted to oust the Castro regime from the island of Cuba. Fidel Castro had led a successful revolution against President Fulgencio Batista in 1959 and had become Premier. His political leanings were toward those of Communist Russia.

On September 10, 1962, when western diplomats warned that the Soviets were planning to base nuclear submarines at Havana harbor, Castro indicated that he had merely granted permission to the Russians to use the harbor for a base for a fishing fleet.

President John F. Kennedy, on October 22, ordered a "quarantine" of Cuba by the American air and naval forces, having come to the conclusion that Castro, with the assistance of the Russians, was building missile bases there. Kennedy, in a nationwide radio and television broadcast from the White House, said, "This government has maintained the closest surveillance on the Soviet military buildup on the Island of Cuba. Within the past week unmistakable evidence has established the fact that a series of offensive missile sites is now in preparation on that imprisoned island. The purpose of these bases can be none other than to provide a nuclear striking capability against the Western Hemisphere." Following the broadcast the Organization of American States unanimously voted in Washington to stand behind the United States, despite the fact that both Cuba and the Soviet Union were enraged over Kennedy's "quarantine" of Cuba.

After considerable reflection Khrushchev, on October 28, reversed his position. He informed President Kennedy that he had ordered Soviet missiles to be withdrawn from Cuba, and promised that, under United Nations inspection, Soviet missile bases on the island would be dismantled. Kennedy, on November 2, announced that the bases in Cuba were being removed and on November 20 he lifted the "quarantine."

Suggested Readings

Abel, Elie. The Missile Crisis. Philadelphia: Lippincott, 1966.

Commager, Henry Steele, ed. "The Cuban Missile Crisis," (Doc. No. 650) in his Documents of American History, 8th edition. New York: Appleton, 1968.

________. "Kennedy's Speech at American University," (Doc. No. 651) in his Documents of American History, 8th edition. New York: Appleton, 1968.

________. "The Nuclear Test Ban Treaty," (Doc. No. 653) in his Documents of American History, 8th edition. New York: Appleton, 1968.

Fay, Paul B. The Pleasure of His Company. New York: Harper, 1966.

Hilsman, Roger. To Move a Nation. Garden City, N.Y.: Doubleday, 1967.

Reidy, John P. The True Story of John Fitzgerald Kennedy. Chicago: Children's Press, 1967.

Rivero, Nicolas. Castro's Cuba: An American Dilemma. Washington, D.C.: Luce, 1962.

Salinger, Pierre. With Kennedy. Garden City, N.Y.: Doubleday, 1966.

Schlesinger, Arthur M., Jr. A Thousand Days: John F. Kennedy in the White House. Boston: Houghton Mifflin, 1965.

Sorenson, Theodore C. Kennedy. New York: Harper, 1965.

Stebbins, Richard P. The United States in World Affairs. New York: Harper, 1963.

Winks, Robin W. The Cold War from Yalta to Cuba. New York: Macmillan, 1964.

Wood, James Playsted, and the editors of Country Beautiful Magazine. The Life and Words of John F. Kennedy. Garden City, N.Y.: Doubleday, 1964.

THE BOSTON STRANGLER (1962-1964)

On Thursday, June 14, 1962, at 7:45 in the evening, the first of thirteen victims of a man who came to be known as the "Boston Strangler" was found choked to death with the cord from her bathrobe. The dead woman was Mrs. Anna Slesers, a 55-

year-old divorcee who lived alone in an apartment at 77 Gainsborough Street, Boston. The body was discovered by her son Juris, a research engineer, who came to her apartment that evening to drive her to a church service in Roxbury. Her death was at first thought to be suicide but later it was determined that she had been murdered. She had been sexually molested and her apartment ransacked, although apparently nothing had been stolen. The initial investigation made by the police turned up no clues as to the identity of the killer.

Between June 14, 1962 and January 4, 1964, twelve other women in the Boston area were killed by an unknown assailant, precipitating the most gigantic homicide investigation in the history of Massachusetts.

The murders had certain characteristics in common. This led the police to believe that they all had probably been committed by the same person. The victims, all single women, ranged in age from 19 to 85. They were, so far as could be determined, inconspicuous, modest persons of blameless lives. Each woman had been strangled in her apartment with an article of wearing apparel which had been tied about her neck with a peculiar knot. Each victim had been sexually molested or assaulted. No evidence of forcible entry was found and no discernible motive for the murders was apparent. The apartments had all been searched by the killer but nothing had been taken from them. It seemed obvious that the murderer was insane.

The Boston police checked known sex offenders, patients recently released from mental hospitals, and other possible suspects, but without result. A free-lance advertising copy writer named Paul Gordon, who believed he had extrasensory perception, entered the case and gave his impressions to the police. He accused one Arnold Wallace of the murders. A woman, suspecting a neighbor of being the Boston Strangler, began to list his comings and goings in a notebook. Psychiatrists attempted to analyze the Strangler's characteristics. Plain-clothes men rode buses and the subways used by some of the women who had been murdered, hoping to find a lead to the killer. Boston was in a state of panic. Rewards for information leading to the apprehension and conviction of the Strangler were offered. Suspects were arrested, interrogated, and released.

Peter Hurkos, a Dutch mystic, was brought into the investigation. He had reportedly helped solve 27 murders in 17 countries. His expenses were to be paid by a Boston industrialist who wished to remain anonymous. Although skeptical, Assistant Attorney General John Bottomly, who was in charge of the investigation, felt that there was nothing to be lost by consulting Hurkos. In due course the Dutch mystic demonstrated a number of mind-reading feats but did not identify anyone positively as the Boston Strangler. He did designate a shoe salesman named Thomas O'Brien as the guilty man but O'Brien, while a psychotic and chronic paranoid, proved not to be the murderer of the thirteen women.

On Friday, March 17, 1961, a 29-year-old man named Albert Henry De Salvo was arrested for breaking into and entering apartments. He was found to have a long record of juvenile delinquencies and was currently engaged in posing as an agent for a model agency. In this capacity he had measured various women with a tape measure, for which he became to be known as the "Measuring Man." He had, while in the army during World War II, married a German girl by whom he had two children.

On May 4, 1961, De Salvo was sentenced on charges of assault and battery brought by some of the women he had measured. He served eleven months of an eighteen-month sentence and was paroled in April, 1962.

In November, 1964, De Salvo was again arrested, this time for assaulting a woman in her Cambridge, Massachusetts, apartment. Her description of the assailant matched that of the Measuring Man. It was found that he had similarly assaulted a number of other women following his release from prison.

De Salvo was interrogated, identified, subjected to psychiatric examination, found to be competent to stand trial, and committed to the hospital at Bridgewater. Here he met George Nassar, a fellow-inmate. He confessed to Nassar that he was the Boston Strangler. Nassar informed his attorney, F. Lee Bailey, of De Salvo's confession. Bailey agreed to defend De Salvo in the trial that seemed inevitable.

Bailey realized that, prior to trial, De Salvo would be subjected to further psychiatric examination. He had, after extensive investigation, become convinced that his client was, indeed, the Strangler. He knew that if De Salvo was found mentally competent to stand trial, he would undoubtedly be found guilty of murder and sentenced to death in the electric chair. Consequently, he arranged for a hearing to determine his client's competence to stand trial on indictments charging armed robbery, assault, and indecent and lascivious acts, but not murder.

Dr. Ames Robey, testifying as a psychiatrist, stated that it was his professional opinion that De Salvo was not competent to stand trial. Ten days later Judge Horace J. Cahill found De Salvo competent. On July 11, 1966, he was brought before Judge George G. Ponte. He pleaded not guilty. The district attorney recommended that De Salvo be remanded without bail to Bridgewater for trial at a later date. Bailey, as his attorney, agreed.

At this writing De Salvo is still at Bridgewater. It is unlikely that he will ever stand trial as the Boston Strangler or that he will ever be free again.

Suggested Readings

Bailey, F. Lee, and Harvey Aronson. The Defense Never Rests. New York: New American Library, 1973.

________, and Henry B. Rothblatt. Crimes of Violence, Homicide and Assault. Rochester: Lawyers Cooperative, 1973.

________, and John Greenya. For the Defense. New York: Atheneum Publishers, 1975.

Bromberg, Walter. Mold of Murder: A Psychiatric Study of Homicide. New York: Grune & Stratton, 1961.

Frank, Gerold. The Boston Strangler. New York: New American

Library, 1966.
_______. "The Boston Strangler" (condensation), Ladies' Home Journal, September, 1966.
Gardner, Erle Stanley. "The Mad Strangler of Boston," Atlantic Monthly, May, 1964.
Guttmacher, M. S. The Mind of the Murderer. Freeport, N.Y.: Books for Libraries, 1960.
Jesse, F. Tennyson. Murder and Its Motives. London: Harrap, 1952.
Lebowitz, M. "Making the Violent Scene," Nation, January 9, 1967.
Reinhardt, James Melvin. The Psychology of Strange Killers. Springfield, Ill.: Thomas, 1962.
"The Uncontrollable Vegetable," Newsweek, January 23, 1967.
"The Word is Out About the Boston Strangler," Life, October 7, 1966.

THE ASSASSINATION OF PRESIDENT KENNEDY (1963)

On the morning of November 22, 1963, John F. Kennedy, 35th President of the United States, landed at Love Field, Dallas, Texas, in the Presidential jet plane "Air Force One." He had made the trip for the purpose of strengthening the political situation in Texas. That state's Democratic party was riven by factionalism, with Governor John B. Connally, Jr. and Senator Ralph Yarborough at each other's throats. Kennedy and Vice President Lyndon B. Johnson thought to demonstrate an apparent reconciliation between Connally and Yarborough by appearing with them in the five major cities of Texas. They feared that if they did not, the Democratic candidate from that state would fail to win in the forthcoming Presidential election.

Arriving at Love Field, Kennedy left the plane, shook hands with members of the crowd which had assembled to greet him, and then proceeded to the Lincoln convertible in which he was scheduled to ride in a motorcade.

At 11:55 A.M., Dallas time, the procession started. The Presidential car, known as SS100X, contained six passengers. Secret service man Bill Greer drove. Beside him sat Roy Kellerman, another secret service agent. In the jump seats were the Connallys, John sitting on the right of his wife Nellie. John and Jacqueline Kennedy occupied the rear seats, the President on the right. A bouquet of roses, given to Mrs. Kennedy, lay on the seat between them. Vice President Johnson and his wife Lady Bird rode in another car, which also contained Senator Yarborough.

The motorcade proceeded from Love Field to Lemmon Avenue and thence to Stemmons Freeway by way of Turtle Creek Boulevard, Cedar Springs Road, Harwood Street, and Main Street. Kennedy ordered a brief halt at the intersection of Lemmon Avenue and Lomo Alto Drive in order to shake hands with a group of children assembled there. Shortly thereafter a second halt was made so that Kennedy might greet a group of nuns.

At 12:29 P.M. the motorcade turned right from Main to Houston Street. At the intersection of Houston and Elm Streets, one block north of Main, the procession turned to the left, to proceed along Elm. As it passed the Texas School Book Depository on the motorcade's right, a rifle bullet fired from the sixth-floor corner window of the Book Depository struck Kennedy in the neck. A second shot struck the President in the top of the head. Rushed to Parkland Memorial Hospital, he was pronounced dead at 1:00 P.M.

It later developed that a young man named Arnold Rowland had seen a man, later identified as Lee Harvey Oswald, silhouetted in the window of the Book Depository seconds before Kennedy was shot. The man was holding a rifle. A pipefitter named Howard L. Brennan also saw Oswald in the window and saw him fire the second shot. Brennan described Oswald to the police and his description was broadcast on the police radio. The Book Depository was checked and Oswald's rifle was found. Oswald had left by the front entrance and gone to his rooming house by bus and taxicab. There he obtained a pistol and then departed.

Less than half a mile from the rooming house Oswald was stopped by Police Officer J. D. Tippitt, whom he shot dead before at least nine witnesses. Then, carrying his pistol, he ran several blocks and entered the Texas Theater without paying. Shortly afterwards he was seized and arrested after a scuffle.

Lee Harvey Oswald was taken to police headquarters and interrogated. The following day he was formally charged with the murder of President Kennedy. As he was being taken, handcuffed to Detective James Leavelle, through the basement garage of the Dallas jail to be transported to the county jail, Jack Ruby, a nightclub proprietor, suddenly shot him with a .38 caliber revolver. This murder was witnessed by literally millions of people as it was televised by the National Broadcasting Company.

Oswald was taken to Parkland Memorial Hospital where he was pronounced dead at 1:07 P.M.

Jack Ruby was arrested and in due course was tried for the murder of Lee Harvey Oswald. On March 14, 1964, he was found guilty and sentenced to death. The conviction was reversed by the Texas Court of Appeals in October, 1966, and a new trial was ordered. Ruby died of cancer in January, 1967.

The assassination of John F. Kennedy was investigated by a commission headed by Earl Warren and reported in a 26-volume report of findings popularly known as the "Warren Report." Jim Garrison, District Attorney of New Orleans, charged that Clay Shaw, a local businessman, was involved in the murder. Others have urged that the assassination be re-investigated, especially in terms of seeking a possible connection between the shooting of Kennedy and those of his brother Robert and the Reverend Martin Luther King, Jr. This seems to be pure conjecture and at this writing no such re-investigation is contemplated.

Suggested Readings

Bell, Daniel, ed. The Radical Right. New York: Doubleday, 1963.

Bishop, Jim. The Day Kennedy Was Shot. New York: Funk & Wagnalls, 1968.

Breslin, Jimmy. "A Death in Emergency Room No. One," Saturday Evening Post, December 14, 1963.

Buchanan, James G. Who Killed Kennedy? New York: McFadden-Bartell, 1965.

Commager, Henry Steele, ed. "The Warren Report," (Doc. No. 662) in his Documents of American History, 8th edition. New York: Appleton, 1968.

Dudman, Richard. Men of the Far Right. New York: Pyramid Books, 1962.

Epstein, Edward J. Inquest. New York: Viking Press, 1966.

________. The Tangled Web. New York: Viking Press, 1968.

Flammonde, Paris. The Kennedy Conspiracy. New York: Meredith Press, 1969.

Ford, Gerald R., and J. R. Stiles. Portrait of the Assassin. New York: Simon & Schuster, 1965.

Forster, Arnold, and Benjamin R. Epstein. Danger on the Right. New York: Random House, 1964.

Fox, Sylvan. The Unanswered Questions About President Kennedy's Assassination. New York: Award Books, 1965.

Grosvenor, Melville Bell. "The Last Full Measure," National Geographic, March, 1964.

Hearings Before the President's Commission on the Assassination of President Kennedy (The Warren Report). 26 vols. Washington, D. C.: Government Printing Office, 1964.

James, Rosemary, and Jack Wardlaw. Plot or Politics? New Orleans: Pelican Publishing Co., 1967.

Joesten, Joachim. Marina Oswald. London: Peter Dawnay, Ltd., 1967.

________. Oswald: Assassin or Fall Guy? New York: Marzani & Munsell, 1964.

________. Oswald, the Truth. London: Peter Dawnay, Ltd., 1967.

Lane, Mark. A Citizen's Dissent. New York: Holt, Rinehart, 1968.

________. Rush to Judgment. New York: Holt, Rinehart, 1966.

Lawrence, Lincoln. Were We Controlled? New York: University Books, 1967.

Leslie, Warren. Dallas Public and Private. New York: Grossman Publishers, 1964.

Lewis, Richard Warren, and Lawrence Schiller. The Scavengers and Critics of the Warren Report. New York: Dell, 1967.

McGrory, Mary. In Memoriam: John Fitzgerald Kennedy. Washington, D. C.: Washington Star Newspaper Co., 1963.

Manchester, William. The Death of a President. New York: Harper, 1967.

________. Portrait of a President: John F. Kennedy in Profile. Boston: Little, Brown, 1967.

Marks, Stalley. Murder Most Foul. Los Angeles: Bureau of International Affairs, 1967.
Meagher, Sylvia. Accessories After the Fact. New York: Bobbs-Merrill, 1967.
Morin, Relman. Assassination: The Death of President Kennedy. New York: New American Library, 1968.
Murray, Norbert. Legacy of an Assassination. New York: Pro-People Press, 1964.
National Broadcasting Company, Inc. Seventy Hours and Thirty Minutes, as Broadcast on the NBC Television Network by NBC News. New York: Random House, 1966.
Popkin, Richard H. The Second Oswald. New York: Avon Books, 1966.
Potter, John Mason. Plots Against the Presidents. New York: Astor, 1968.
Rajski, Raymond B., comp. A Nation Grieved: The Kennedy Assassination in Editorial Cartoons. Rutland, Vt.: Tuttle, 1967.
Roberts, Charles. The Truth About the Assassination. New York: Grosset & Dunlap, 1967.
Rossiter, Clinton. The American Presidency. New York: Harcourt, Brace, 1960.
Sauvage, Leo. The Oswald Affair. Cleveland: World, 1966.
Schlesinger, Arthur M., Jr. A Thousand Days: John F. Kennedy in the White House. Boston: Houghton Mifflin, 1965.
Sparrow, John. After the Assassination: A Positive Appraisal of the Warren Report. New York: Chillmark Press, 1968.
Stafford, Jean. A Mother in History. New York: Farrar, Straus, 1966.
Symons, Julian. A Pictorial History of Crime. New York: Crown Publishers, 1966.
Thayer, George. The Farther Shores of Politics. New York: Simon & Schuster, 1967.
Thompson, Josiah. Six Seconds in Dallas. New York: Bernard Geis, 1967.
Wallechinsky, David, and Irving Wallace. "Assassinations--John F. Kennedy," in their The People's Almanac. Garden City, N.Y.: Doubleday, 1975.
Weisberg, Harold. Oswald in New Orleans. New York: Canyon Books, 1967.
________. Photographic Whitewash. Hyattstown, Md.: Privately published, 1966.
________. Whitewash: The Report on the Warren Report. Hyattstown, Md.: Privately published, 1965.
________. Whitewash II: The FBI-Secret Service Cover-Up. New York: Dell, 1966.
West, John R., et al. Death of the President. Covina, Calif.: Collectors Publications, and Hollywood, Calif.: Associated Professional Services, 1967.
Wicker, Tom. "Wicker Describes That Day in Dallas," Times Talk, December, 1963.

THE SALAD OIL SWINDLE (1963-1964)

The mastermind of the "Salad Oil Swindle" which rocked the American financial world and caused the ruin of Ira Haupt & Company, one of the most respected brokerage houses on Wall Street, was Anthony "Tino" De Angelis, an Italian-American entrepreneur. As a former hog butcher and meat processing executive he had, prior to his gigantic operations in edible vegetable oils, been engaged in a series of shady business transactions and deals.

On November 14, 1955, De Angelis organized the Allied Crude Vegetable Oil Refining Company which was to process crude vegetable oil into salad oil for export. He obtained financial backing from such blue ribbon export firms as Continental Grain Company and Bunge Corporation, both based in New York City. He bought crude oil in the midwest, shipped it to his plant at Bayonne, New Jersey, where it was stored in huge tanks until it could be refined and turned over to his customer-backers at a nominal markup. Although he paid "excessive prices" for his crude oil and sold to his customers "at a bargain," he seemed to prosper. Such suppliers as Central Soya Company of Fort Wayne, Indiana, and A. E. Staley Manufacturing Company of Decatur, Illinois, stored quantities of oil at the tank farm operated by De Angelis at Bayonne.

Because of De Angelis' unconventional business practices and shady background the larger New York banks refused to do business with Allied or any of the other companies he controlled, but were willing, if not anxious, to lend money indirectly to Allied through other customers. One such other customer was the American Express Field Warehousing Corporation, a subsidiary of the American Express Company. Donald K. Miller, then president of Field Warehousing, was under pressure from Howard L. Clark, president of the parent company, to generate a profit of at least $500,000 annually from his operation. When De Angelis' cousin Michael proposed to Miller that he open an American Express Field warehouse at Allied's Bayonne tank farm, Miller, after inspecting the facility, agreed. It was understood that American Express Field Warehousing would issue warehouse receipts against oil stored in the Bayonne tanks. Exporters would accept such receipts from American Express, use them as collateral for bank loans, and turn the money over to Allied.

The stage was set. When news of the salad oil swindle became public in the fall of 1963 it was learned that, through chicanery, De Angelis had contrived to issue dozens of warehouse receipts covering oil which did not exist, either in his Bayonne tanks or anywhere else. These receipts he had hypothecated with various brokers and exporters for cash.

The brokerage firm of Ira Haupt & Company, which specialized in stock transactions and had little to do with commodity operations, had agreed to trade for Allied and had accepted warehouse receipts as collateral, as had some other New York brokerage houses. Although Fred Barton, Haupt's senior specialist in commodity trading, had previously refused to do business with De Angelis, other Haupt executives, greedy for the commissions which the business would generate, overrode him. The warehouse receipts accepted by Haupt were, it turned out, forged and did not represent oil held in Allied's Bayonne tanks.

De Angelis also engaged in a check-kiting scheme in which the Bunge Corporation, a grain exporting firm, was victimized. Further, he began buying oil futures on the New York Produce Exchange and the Chicago Board of Trade at a fantastic rate. These purchases he financed by converting into cash the warehouse receipts stolen from Bunge through his check-kiting scheme and from the proceeds of the worthless receipts he produced from a pad of such receipts stolen from the American Express office.

Eventually De Angelis was the owner of a large percentage of the vegetable oil produced in the United States for the current crop year. He hoped that the price of oil would advance and he would be able to sell his holdings at a profit.

When, in the fall of 1963, the Commodity Exchange Authority informed De Angelis that his purchases of oil futures were being investigated, he realized that his luck had run out. He applied for voluntary bankruptcy for the Allied Company. A selling wave hit the commodity markets and De Angelis was faced with the necessity of paying some $20,000,000 in margin calls. He did not have the money.

Officials at Ira Haupt & Company learned to their horror that, with Allied's bankruptcy, they were no longer solvent, due to the invalid warehouse receipts they had accepted from De Angelis. Bunge Corporation learned that it had been swindled of approximately $3,040,000 due to the check-kiting maneuver. Other brokers, financial corporations and banks throughout the world found that they had lost heavily because of De Angelis' speculations. His cottonseed oil futures were liquidated at prices which cost the brokers more than $12,700. The liquidation of the soybean oil contracts meant a loss of $20,000,000 for the brokers. The assassination of President John F. Kennedy on November 22 precipitated a huge drop in stock prices on the various exchanges, further complicating the financial situations of all concerned.

The Haupt firm was liquidated. De Angelis was brought to trial. The total shortage for which he was responsible was in excess of $175,000,000, represented by spurious warehouse receipts which were in the hands of 51 companies and banks. De Angelis was induced to plead guilty to four charges of the indictment. Federal Judge Reynier J. Wortendyke sentenced him to prison. He served part of his sentence, was then paroled, and returned to his earlier occupation of hog butcher.

Suggested Readings

Baer, Jules B., and Olin Glenn Saxon. Commodity Exchanges and Futures Trading: Principles and Operating Methods. New York: Harper, 1949.

"Judgment Day for Tino," Newsweek, June 7, 1965.

Kempton, Murray. "The Salad Oil Mystery," New Republic, July 24, 1965.

"The Man Who Fooled Everybody," Time, June 4, 1965.
Mandel, P. "$150 Million in Oil-All Gone," Life, April 3, 1964.
Miller, Norman C. The Great Salad Oil Swindle. New York: Coward-McCann, 1965.
_______. "Tino's Bottomless Tanks of Oil," Saturday Evening Post, April 25, 1964.
"Salad Oil King Thinks Prison Saved His Life," Life, July 21, 1972.
"Twenty Years for Tino," Newsweek, August 30, 1965.
"Where Genius Went Wrong," Business Week, April 8, 1964.

THE BERKELEY STUDENT REVOLT (1964)

An area of sidewalk, 26 by 60 feet, at the Berkeley campus of the University of California, had been traditionally used by students for fund-raising campaigns and recruitment for off-campus politics. On September 14, 1964, Katherine Toule, Dean of Students, informed all student organizations that they would no longer be permitted to use the sidewalk area for such purposes. Her action promptly triggered a gigantic student revolt, and a united campus front called the Free Speech Movement rose in opposition to Dean Toule's mandate.

Five students were arrested for violating the new regulation. Four hundred additional students signed statements claiming equal guilt and demanding equal punishment.

The five students were placed on indefinite suspension by the university, and three other students, leaders of the protest movement which was organized on campus, were also suspended. On October 1, 1964, an attempt was made to arrest Jack Weinberg, a student who had set up and operated a table in behalf of the Free Speech Movement on the steps of the Administration Building. A spontaneous student sit-in surrounded and immobilized a police car for 32 hours. The Free Speech Movement, after a self-imposed moratorium, again set up tables. Once more hundreds of students signed statements of equal responsibility when some of their number were charged with violating university policy.

In November various university officials pressed for the suspension of Mario Savio, a student leader in the Free Speech Movement. Student rallies were held protesting this move by the university, these rallies being attended by hundreds of university enrollees. In early December over 800 students remained overnight in the Administration Building. Governor Edmund G. "Pat" Brown of California dispatched police who arrested the students and a week-long "strike" was proclaimed.

The university's academic senate met on December 8. The students' cause was voted upon, the vote favoring it 824 to 115. As a result of this Chancellor Edward Strong was deposed and significant changes in the university administration were made.

Other American universities found their policies and internal structures challenged by their student bodies. Massive demonstrations were held on campuses as far apart as those of the University of Hawaii, and Yale, Kent State and South Carolina.

Suggested Readings

Cohen, Mitchell, and Dennis Hale, eds. The New Student Left. Boston: Beacon Press, 1966.

Draper, Hal. Berkeley: The New Student Revolt. New York: Grove Press, 1965.

Ehrenreich, Barbara, and John Ehrenreich. Long March, Short Spring: The Student Uprising at Home and Abroad. New York: Monthly Review Press, 1969.

Esler, Anthony. Bombs, Beards and Barricades: 150 Years of Youth in Revolt. New York: Stein, 1971.

Heaps, Willard Allison. Riots, U.S.A., 1765-1970. New York: Seabury Press, 1970.

Keniston, Kenneth. Young Radicals: Notes on Committed Youth. New York: Harcourt, Brace, 1968.

McEvoy, James, and Abraham Miller, eds. Black Power and the Student Rebellion: Conflict on the American Campus. Belmont, Calif.: Wadsworth Publishing Co., 1969.

Miller, Michael V., and Susan Gilmore, eds. Revolution at Berkeley. New York: Dial Press, 1965.

Morrison, Robert. The Contemporary University: U.S.A. Boston: Houghton Mifflin, 1966.

Ruch, Floyd L., Gordon N. Mackenzie and Margaret McClean. People Are Important. Chicago: Scott, Foresman, 1941.

University of California Academic Senate (Select Committee on Education). Education at Berkeley. Berkeley, Calif.: University of California Press, 1968.

THE CIVIL RIGHTS PROTEST MARCH (1965)

In the mid-1960's the civil rights of Negroes were being disregarded in many parts of the American South and particularly in the state of Alabama. Qualified Negroes who sought to vote were prevented from doing so by technicalities and intimidation. An attempted civil rights march to Montgomery in protest of this situation came to a violent end on March 7, 1965, in the town of Selma when a sheriff's posse reinforced by state troopers set upon and scattered the marchers. The attackers used clubs, tear gas, and whips against the unarmed protesters.

Two days later Reverend James J. Reeb, a white minister from Boston who had participated in the aborted march, was attacked by white racists. He died shortly thereafter.

On March 13 Governor George C. Wallace of Alabama flew to Washington, D.C. to confer with President Lyndon B. Johnson. The President gave the segregation-minded Wallace to understand that "whether the Governor likes it or not," all persons in Alabama would be protected, by federal troops if necessary. Following this meeting Johnson, in a televised speech to the nation, called the Selma incident "an American tragedy," saying, "the blows that were received, the blood that was shed ... must strengthen the determination of each of us to bring full and equal justice to all of our people.... "

In another televised speech given before a specially called night session of Congress on March 15, 1965, President Johnson demanded immediate action on new legislation that would remove the barriers which had been set up to prevent American citizens from registering and voting. He was interrupted 36 times by applause and two standing ovations.

A federal judge, petitioned by Reverend Martin Luther King, Jr. and his civil rights supporters to authorize a protest march from Selma to Montgomery, gave his approval. On March 19 Governor Wallace advised President Johnson that it would be necessary to call out the state national guard to protect the marchers, an expense that the state of Alabama could not afford. Johnson then federalized 1,800 Alabama guardsmen and also sent two battalions of army military police to the state.

On March 21 King led his marchers out of Selma. They arrived in Montgomery on March 24 and converged on the state capitol. In addressing the 25,000 Negroes and white sympathizers participating in the march, King said, prior to their departure from Selma, 'Walk together, children. Don't you get weary, and it will lead to the promised land. And Alabama will be a new Alabama and America will be a new America."

Suggested Readings

Bennett, Lerone, Jr. What Manner of Man: A Biography of Martin Luther King, Jr. Chicago: Johnson, 1968.

Clayton, Edward Taylor. Martin Luther King: The Peaceful Warrior. Englewood Cliffs, N.J.: Prentice-Hall, 1968.

Fribourg, Marjorie G. Bill of Rights: Its Impact on the American People. Philadelphia: Smith, 1967.

Harris, Janet. The Long Freedom Road: The Civil Rights Story. New York: McGraw-Hill, 1967.

Heaps, Willard Allison. Riots, U.S.A., 1765-1970. New York: Seabury Press, 1970.

King, Martin Luther. Measure of a Man. Philadelphia: Pilgrim Press, 1968.

_______. Stride Toward Freedom: The Montgomery Story. New York: Harper, 1958.

_______. Why We Can't Wait. New York: Harper, 1964.

Lewis, Anthony, and The New York Times. Portrait of a Decade: The Second American Revolution. New York: Random House, 1964.

Lewis, David L. King: A Critical Biography. New York: Frederick A. Praeger, 1970.

Lomax, Louis E. The Negro Revolt. New York: Harper & Row, 1962.

Miller, William Robert. Martin Luther King, Jr. New York: Weybright and Talley, 1968.
Myrdal, Gunnar. An American Dilemma: The Negro Problem and American Democracy. New York: Harper, 1962.
Schwartzmann, Ruth, and Joseph Stein. The Law of Personal Liberties. New York: Oceana Publications, 1955.
Spike, Robert W. The Freedom Revolution and the Churches. New York: Association, 1965.
Young, Margaret B. A Picture Life of Martin Luther King, Jr. New York: Watts, 1968.

THE WATTS RIOT (1965)

Watts, California, the scene of the first of several outbreaks of racial violence in metropolitan areas, is situated on the southeast side of Los Angeles. It is populated largely by Negroes whose lives reflect the ghetto conditions under which they live. In August, 1965, more than a third of the Watts residents were unemployed and some fourteen per cent of the population was functionally illiterate. The Watts riot was the culmination of the racial tensions which had existed for decades between the blacks and whites in American society.

The outbreak of racial violence was precipitated on the night of August 11, 1965, when two young Negroes were arrested by a white California Highway Patrol officer for reckless driving. The officer attempted to administer a sobriety test. A hostile crowd gathered and it became necessary to make further arrests.

Feeling flared high and the patrolmen retreated under a hail of rocks, bottles and other such objects thrown by angry Negroes. Rumors of police brutality circulated rapidly through the Watts ghetto, and shortly before midnight a full-scale riot was under way. Motorists passing through the area were pelted with rocks. Although the police entered the district at once, sporadic outbursts of violence continued until dawn.

On the evening of August 12 the rioters returned to the Watts streets and resumed their activities. Buildings were set on fire and firemen who attempted to quell the blazes were attacked with rocks and other missiles. Others were fired upon. It was estimated that some 8,000 rioters were involved in the evening's uprising. The police were unable to prevent the widespread burnings and the looting of stores and houses.

On the night of August 13 the violence reached its peak. Arsonists set fires in an area comprising some fifty square blocks. Snipers, strategically placed, fired on policemen and firemen. The looting continued and crowds of angry Negroes surged through the community.

Early on the morning of August 14 the police began to make some progress towards regaining control of the streets. Some 14,000 National Guard troops were called in to assist the Los Angeles police officers in quelling the rioters. Burning and looting continued intermittently, but the presence of the fully armed soldiers eventually restored quiet to the Watts ghetto.

On the night of Saturday, August 14, a dusk-to-dawn curfew was ordered and went into effect. This was lifted three days later and most of the National Guard troops left the city.

The six-day riot received nationwide publicity. More than three thousand arrests were made and 34 deaths were reported. More than $40,000,000 worth of property damage was incurred.

The Watts riot was followed by similar upheavals in such large American cities as Detroit, Michigan, Newark, New Jersey, Washington, D.C., and elsewhere.

Suggested Readings

Brown, Claude. Manchild in the Promised Land. New York: Macmillan, 1965.

Cohen, Jerry, and William S. Murphy. Burn, Baby, Burn: The Los Angeles Race Riot, August, 1965. New York: Dutton, 1966.

Canot, Robert. Rivers of Blood, Years of Darkness: The Unforgettable Classic Account of the Watts Riot. New York: Bantam Books, 1967.

Fogelson, Robert. The Fragmented Metropolis: Los Angeles. Cambridge, Mass.: Harvard University Press, 1967.

Fribourg, Marjorie G. Bill of Rights: Its Impact on the American People. Philadelphia: Smith, 1967.

Harris, Janet. The Long Freedom Road: The Civil Rights Story. New York: McGraw-Hill, 1967.

Heaps, Willard Allison. Riots, U.S.A., 1765-1970. New York: Seabury Press, 1970.

Jacobs, Paul. Prelude to Riot: A View of Urban America from the Bottom. New York: Random House, 1968.

King, Martin Luther. Why We Can't Wait. New York: Harper, 1964.

Lewis, Anthony, and The New York Times. Portrait of a Decade: The Second American Revolution. New York: Random House, 1964.

Lomax, Louis E. The Negro Revolt. New York: Harper & Row, 1962.

Myrdal, Gunnar. An American Dilemma: The Negro Problem and American Democracy. New York: Harper, 1962.

O'Neill, William T. Coming Apart: An Informal History of America in the 1960's. Chicago: Quadrangle Books, 1971.

Ritchie, Barbara. Report of the National Advisory Commission of Civil Disorders. New York: Viking Press, 1969.

Schulberg, Budd, ed. From the Ashes: Voices of Watts. New York: New American Library, 1967.

Schwartzmann, Ruth, and Joseph Stein. The Law of Personal Liberties. New York: Oceana Publications, 1955.

Spike, Robert W. The Freedom Revolution and the Churches. New York: Association, 1965.

Tuttle, William M., Jr. Race Riot: Chicago in the Red Summer of 1919. New York: Atheneum Publishers, 1970.

THE ASSASSINATIONS OF MARTIN LUTHER KING, JR. AND ROBERT F. KENNEDY (1968)

Two important political figures were assassinated in mid-1968. The Reverend Martin Luther King, Jr. was shot to death in Memphis on April 4, and just two months later, on June 4, Robert F. Kennedy, United States senator and candidate for the Democratic Presidential nomination, was gunned down in Los Angeles.

King, an active civil rights leader and president of the Southern Christian Leadership Conference, was visiting Memphis. Early in 1968 he had announced a "Poor People's Campaign," to be held in Washington. His object was to dramatize the plight of America's poor, regardless of race. He had flown to Memphis to lead a demonstration of striking garbage collectors, most of whom were Negroes. On the evening of April 4, while he and his staff were gathered on the balcony of the Lorraine Motel, he was killed by a shot fired from across the street. He died almost instantly.

An intensive search for King's assailant spread to Canada and Great Britain. On June 8 James Earl Ray, traveling in Great Britain under the alias of Roman George Sneyd, was apprehended, arrested and returned to the United States for trial. In March, 1969, Ray, an ex-convict, pleaded guilty to the charge, was convicted and sentenced to 99 years imprisonment.

In the summer of 1968 Robert F. Kennedy was in Los Angeles. He had completed his campaign in the California Democratic Presidential primary and had just learned that he had defeated Senator Eugene McCarthy, winner of the May Oregon primary. Shortly after midnight on June 5, having addressed his supporters at a victory celebration at the Ambassador Hotel, he was leaving the building through the kitchen. A lone gunman approached him suddenly and fired three times. One bullet entered Kennedy's brain and five other persons, also struck by bullets, were less seriously injured. Kennedy died at the Good Samaritan Hospital on the morning of June 6.

Sirhan Bishara Sirhan, an Arab immigrant, was seized immediately after the shooting. Sirhan was indicted for first-degree murder, tried and convicted. In May, 1969, he was sentenced to death.

King's death precipitated widespread protest riots and fires in several American cities. The leadership of the SCLC was taken over by his long-time assistant, Ralph David Abernathy. Kennedy's death dramatically altered the 1968 Presidential campaign. In March President Lyndon Baines Johnson had announced that he would not seek reelection and the Democratic nomination became a three-way contest between Kennedy, McCarthy and Vice-President Hubert H. Humphrey. Humphrey was eventually nominated but lost the election to Richard M. Nixon, formerly Vice-President in the Eisenhower administration.

It has been suggested that the assassinations of Martin Luther King, Jr., Robert F. Kennedy and his brother John were somehow linked together in a gigantic conspiracy to do away with the three men, and possibly with others as well. At this writing, however, no such relationship among these murders has been established.

Suggested Readings

Bennett, Lerone, Jr. What Manner of Man: A Biography of Martin Luther King, Jr. Chicago: Johnson, 1968.

Clayton, Edward Taylor. Martin Luther King: The Peaceful Warrior. Englewood Cliffs, N.J.: Prentice-Hall, 1968.

Frank, Gerold. An American Death: The True Story of the Assassination of Dr. Martin Luther King, Jr., and the Greatest Manhunt of Our Time. Garden City, N.Y.: Doubleday, 1972.

Huie, William Bradford. He Slew the Dreamer: My Search for the Truth about James Earl Ray and the Murder of Martin Luther King. New York: Delacorte Press, 1970.

Kaiser, Robert Blair. RFK Must Die: The History of the Robert Kennedy Assassination and Its Aftermath. New York: Dutton, 1970.

King, Martin Luther. Measure of a Man. Philadelphia: Pilgrim Press, 1968.

Koch, Thilo. Fighters for a New World. New York: Putnam, 1969.

Lewis, David L. King: A Critical Biography. New York: Frederick A. Praeger, 1970.

Miller, William Robert. Martin Luther King, Jr. New York: Weybright & Talley, 1968.

Ross, Douglas. Robert F. Kennedy, Apostle of Change. New York: Trident, 1968.

Shannon, William V. Heir Apparent: Robert Kennedy and the Struggle for Power. New York: Macmillan, 1967.

United Press International Staff. Assassination: Robert F. Kennedy, 1925-1968. New York: Cowles Press, 1969.

Vanden Heuvel, William J., and Milton Gwirtzman. On His Own: Robert F. Kennedy, 1964-1968. Garden City, N.Y.: Doubleday, 1970.

Wallechinsky, David, and Irving Wallace. "Assassinations--Robert F. Kennedy," in their The People's Almanac. Garden City, N.Y.: Doubleday, 1975.

Witcover, Jules. 85 Days: The Last Campaign of Robert Kennedy. New York: Putnam, 1969.

Young, Margaret B. A Picture Life of Martin Luther King, Jr. New York: Watts, 1968.

THE "APOLLO" MOON LANDING (1969)

After thirteen years of competition between the Soviet Union and the United States in space exploration the United States succeeded, on July 20, 1969, in "putting a man on the moon." On that date Neil A. Armstrong, an American civilian astronaut, was watched by millions by way of their television screens as he climbed down from his lunar landing vehicle carrying a plaque inscribed "Here Men from the Planet Earth Set Foot on the Moon. We Came in Peace for All Mankind."

The first artificial earth satellite was the Russian "Sputnik," launched in October, 1957. This was the culmination of the efforts of such space scientists as Constantin Ziolokovsky, Hermann Oberth, Robert Goddard and Robert Esnault-Pelterie who, working independently, had done basic research in the matter of space exploration using high-powered rockets.

Various space travel organizations were formed in the United States, England and Germany during the 1920's and 1930's. These were largely hampered by a lack of funds with which to conduct experiments. The German army in 1932, however, became interested in the military possibilities of rockets and had established a rocket research group in the army weapons office. Here, under the direction of Colonel Walter Dornberger and the scientist Wernher von Braun, the German V-2 rocket was developed. This was used for bombing England in the closing months of World War II.

After the war, captured V-2 rockets were shipped to America and Russia for study and von Braun came to this country to continue his rocket research.

In February, 1949, the "WAC-Corporal" was developed. This was followed by the more sophisticated "Redstone" rocket. In July, 1955, President Dwight D. Eisenhower announced that the United States would put a small artificial satellite into orbit. This was the "Vanguard." Before it could be launched the Russians had successfully orbited the 184-pound "Sputnik."

Experiments continued, some successful, some failures. In March, 1958, the "Vanguard" launching was accomplished. In October of that year all non-military space programs were consolidated, to be administered by the National Aeronautics and Space Administration (NASA), and adequate funds and technical staffing were provided.

Russia launched a second "Sputnik," this carrying a live dog. Other "Sputniks" followed. On April 12, 1961, Colonel Yuri A. Gagarin, riding in the Russian capsule "Vostok I," made an almost complete orbit of the earth. In July, 1963, another Russian astronaut, Valery Fyodorovitch Bykovsky, made 81 orbits around the earth.

President John F. Kennedy, who believed that the United States should land a manned vehicle on the moon "before this decade is out," backed Project Mercury, America's first effort at manned space flight. Commander Alan B. Shepherd and others completed successful space flights. More flights followed under the Gemini series and these, in turn, were succeeded by the Apollo flights. The first Apollo effort ended in tragedy when, in January, 1967, a space cabin caught fire and astronauts Roger B. Chaffee, Virgil I. Grissom and Edward B. White were killed. This led to a twenty-month delay of the moon project while the Apollo program was subjected to a thorough investigation. Then, on October 11, 1968, "Apollo 7" successfully carried three astronauts around the earth in orbit, and in December of that year a three-man team orbited the moon.

The actual moon landing was made on July 20, 1969, as mentioned above. Armstrong was accompanied by Colonel Edwin E. Aldrin while Lieutenant Colonel Michael E. Collins remained aloft in the command module "Columbia." Three days later, on July 24, the three men splashed down in the Pacific Ocean where they were picked up by the carrier "Hornet."

Suggested Readings

Branley, Franklyn M. Experiments in the Principles of Space Travel. New York: Crowell, 1955.
Braun, Wernher von. Conquest of the Moon. Edited by Cornelius Ryan. New York: Viking Press, 1953.
_______. First Men to the Moon. New York: Holt, 1960.
Caidin, Martin. The Astronauts: The Story of Project Mercury, America's Man-in-Space Program. New York: Dutton, 1960.
Coggins, Jack, and Fletcher Pratt. Rockets, Satellites, and Space Travel. Edited by Willy Ley. New York: Random House, 1958.
Diamond, Edwin. The Rise and Fall of the Space Age. Garden City, N.Y.: Doubleday, 1964.
Emme, Eugene M. A History of Space Flight. New York: Holt, Rinehart, 1965.
Gallant, Roy A. Exploring the Moon. Garden City, N.Y.: Garden City Books, 1955.
Gaul, Albro T. The Complete Book of Space Travel. Cleveland: World, 1956.
Kennan, Erlend A., and Edmund H. Harvey, Jr. Mission to the Moon: A Critical Examination of NASA and the Space Program. New York: Morrow, 1969.
Ley, Willy. Events in Space. New York: David McKay, 1969.
_______. Rockets, Missiles, and Men in Space. New York: Viking Press, 1967.
Marcus, Abraham, and R. B. Marcus. Tomorrow the Moon! Planes, Missiles, Satellites, Space Travel. Englewood Cliffs, N.J.: Prentice-Hall, 1959.
Pierce, Philip N., and Karl Schuon. John H. Glenn, Astronaut. New York: Watts, 1962.
Rosholt, Robert L. An Administrative History of NASA, 1959-1963. Washington, D.C.: Government Printing Office, 1966.
Ross, Frank Xavier. Space Ships and Space Travel: The Scientifically Accurate Story of Man's Attempts and Plans to Travel Into Interplanetary Space. New York: Lothrop, 1954.

Shelton, William. Soviet Space Exploration: The First Decade. New York: Washington Square, 1968.
Sullivan, Walter, ed. America's Race for the Moon: The New York Times Story of Project Apollo. New York: Random House, 1962.
Swenson, Lloyd S., James M. Grimwood and Charles C. Alexander. This New Ocean: A History of Project Mercury. Washington, D.C.: Government Printing Office, 1966.
Thomas, Davis, ed. Moon: Man's Greatest Adventure. New York: Abrams, 1970.
Thomas, Shirley. Men of Space. Philadelphia: Chilton, 1960.
Verne, Jules. From the Earth to the Moon (science fiction). New York: Scholastic Book Service, 1972.

THE MANSON FAMILY MURDERS (1969)

One of the most gruesome mass murders of the twentieth century was committed shortly after midnight on the morning of August 9 and during the night of August 10, 1969, near Hollywood, California. On the first occasion five persons were shot and stabbed at the home of motion picture actress Sharon Tate Polanski, better known as Sharon Tate. Murdered in addition to the eight-month-pregnant actress were Abigail Folger, heiress to a coffee fortune; Boytek Frykowski, her lover; Jay Sebring, an internationally known hair stylist; and Stephen Earl Parent, a student. The bodies were discovered by Mrs. Winifred Chapman, a Negro maid employed by the actress, when she reported for work. She notified a neighbor who promptly reported the matter to the Los Angeles police.

William Garretson, caretaker at the Tate home, was arrested but later released when it was shown that he had no connection with the murders. The telephone wires outside the house had been cut. Blood from the many stab wounds inflicted by the murderers was literally everywhere and the word PIG had been written on the front door with blood.

The second murders were committed at the home of Leno La Bianca, owner of a chain of grocery stores. La Bianca and his wife Rosemary had been stabbed, as in the Tate murders, many times. The bodies were discovered by La Bianca's stepson Frank Struthers, Struthers's sister Susan, and Joe Dorgan, Susan's boy friend. DEATH TO PIGS and RISE had been written in blood on the walls and HEALTER SKELTER had been similarly inscribed on the refrigerator door. Robbery seemed ruled out as a motive for any of the brutal killings, as nothing of value had been taken from either the Tate or the La Bianca residence.

It was suspected that the two mass murders were connected. The police further recalled that on July 31 the body of Gary Hinman, a music teacher, had been found at his home in Malibu. Hinman had been stabbed to death and the words POLITICAL PIGGY had been printed in blood on his living room wall. Bobby Beausoleil, a young hippie musician, was ultimately arrested for the Hinman murder. It was learned that he had been living at Spahn's Movie Ranch near Chatsworth, California, with a group of other hippies led by a man named Charles Manson, who had apparently convinced them that he was Jesus Christ. Some 26 members of this group, known as the Family, were arrested on August 16 on a charge of auto theft but were released when it was discovered that the warrant was misdated. Manson then took his group to the Barker Ranch in Death Valley, California.

Various suspects were interrogated and released for lack of evidence. Then, in October, 1969, 24 members of the Family were arrested at Barker Ranch. Manson, 34 years old, an amateur philosopher, unsuccessful singer and guitarist and hater of the Establishment, had a long and checkered criminal history. He was among those taken. Others arrested included seventeen-year-old Kitty Lutesinger, a friend of Beausoleil. She told the police that Manson had sent Beausoleil and Susan Atkins, another Family member, to Hinman's home to get money from him, that a fight had ensued, and that Hinman had been killed.

Susan Atkins had been arrested at the Barker Ranch and was still in custody. Questioned by the police, she accused Beausoleil of stabbing Hinman but did not implicate Manson. She was booked for the Hinman murder and sent to Sybil Brand Institute in Los Angeles. Here she had as fellow prisoners two ex-call girls, Ronnie Howard and Virginia Graham.

In prison Susan told Virginia that she, Beausoleil and "another girl" had killed Gary Hinman. She also said that she was personally involved in the Tate murders. She stated, "we wanted to do a crime that would shock the world," that the victims had been picked at random and that they had acted on direct orders from Charles Manson. The Tate murders had been committed by herself and others, later identified as Charles "Tex" Watson, Linda Kasabian and Patricia Krenwinkel.

Virginia conferred with Ronnie Howard and the police were informed of what Susan had said. In the meantime Al Springer, a member of the Straight Satans motorcycle gang, had told the police that he had been at the Spahn Ranch and that Manson had bragged to him of a number of murders committed by himself and members of the Family. Danny DeCarlo, another motorcyclist, implicated Beausoleil, Susan Atkins, Manson, Mary Brunner and Bruce Davis, all Family members. He also connected Manson with the Tate murders and the killing of a ranch hand named Donald Shea at the Spahn Ranch.

Vincent T. Bugliosi, Deputy District Attorney, Los Angeles, prosecuted the case. Susan Atkins stated to her attorney in a taped interview that she, Watson, Krenwinkel and Kasabian had gone to the Tate residence and killed five people. She also stated that Manson, Watson and Leslie Van Houten had killed the La Biancas the following night. Susan Atkins had not entered the La Bianca home, she said, remaining with Krenwinkel and Steve Grogan in the automobile in which they had come. Bugliosi was permitted to listen to the tapes.

Linda Kasabian was taken into custody, having waived extradition from New Hampshire where she had gone. The matter of giving Susan Atkins immunity from prosecution in return for her testimony at the Grand Jury hearing and subsequent trial was considered and rejected. Patricia Krenwinkel was extradited from Alabama and indicted for murder. Manson, Watson, Kasabian, Van Houten and Atkins were also indicted.

Watson had fled to Texas but was eventually returned to California to stand trial after his efforts to avoid extradition had failed. He, Beausoleil and Grogan were tried and found guilty. Manson, Krenwinkel, Atkins and Van Houten were tried together before Judge Charles H. Older. Linda Kasabian, who had killed no one, although she had been at the scene of the Tate murders, agreed to testify for the prosecution in exchange for immunity. The defense was unable to get her to contradict herself during the eighteen days she was on the witness stand and the Supreme Court of California granted her immunity at the request of Prosecutor Bugliosi and Judge Older.

The four defendants were represented by a series of attorneys, the most spectacular of whom was Irving Kanarek, who defended Manson. The three girls attempted to take all blame for the murders and to absolve Manson. Ronald Hughes, attorney for Leslie Van Houten, refused to have anything to do with a proceeding "where he was forced to push a client out the window." For his refusal to assist Manson at the expense of his client he was later murdered by other members of the Family.

The trial of Manson and the three girls was characterized by many displays of unruliness on the part of the defendants, for which they were repeatedly ejected from the courtroom. Several attorneys were found in contempt of court and either fined or given short jail sentences. Manson and the other defendants cut X's on their foreheads and, after the verdicts were rendered, shaved their heads.

The proceedings lasted from late July, 1970 until the end of January of the following year. The defense put up no argument and called no witnesses. All defendants were found guilty of murder in the first degree and were sentenced to die in the gas chamber.

Charles "Tex" Watson was tried for murder in connection with both the Tate and La Bianca killings. He was found guilty and sentenced to die. Steve Grogan, found guilty of killing Donald Shea, received a like sentence, later reduced to life imprisonment. Bobby Beausoleil, tried for the Hinman murder, was also found guilty and sentenced to die.

The Spahn Ranch was destroyed in a forest fire which swept Southern California on September 26, 1970. On February 18, 1972, the California Supreme Court voted to abolish the death penalty in that state, which automatically reduced the sentences of all the defendants who had been sentenced to death to life imprisonment. They may apply for parole in 1978.

Following Manson's imprisonment, first at San Quentin, then at the maximum security prison at Folsom, California, the murders continued. These seem to be in retaliation for "wrongs" done various members of the Family. Lynette "Squeaky" Fromme, a hard-core follower of Manson, assumed leadership of the remaining members of the group after their leader was imprisoned. She was arrested several times on various charges but convicted only for minor offenses until 1975 when she attempted to assassinate President Gerald Ford. For this she was tried, convicted and sentenced to life imprisonment.

Suggested Readings

Atkins, Susan. The Killing of Sharon Tate. New York: New American Library, 1969.
Bishop, George Victor. Witness to Evil. Los Angeles: Nash Publishing Co., 1971.
Bugliosi, Vincent, with Curt Gentry. Helter Skelter: The True Story of the Manson Murders. New York: Norton, 1974.
"Death for the Family," Newsweek, April 12, 1971.
"The Demon of Death Valley," Time, December 12, 1969.
Gray, P. "Anatomy of an Outrage," Time, November 4, 1974.
Maas, Peter. "The Sharon Tate Murders," Ladies Home Journal, April, 1970.
Roberts, S. V. "Charles Manson: One Man's Family," New York Times Magazine, January 4, 1970, pp. 10-11.
Rubin, Jerry. We Are Everywhere. New York: Harper & Row, 1971.
Sanders, Ed. The Family: The Story of Charles Manson's Dune Buggy Attack Batallion. New York: Dutton, 1971.
Zamora, William. Trial by Your Peers. Secaucus, N.J.: Maurice Girodias, 1973.

THE CALLEY COURTMARTIAL (1970-1971)

The 1970/71 courtmartial of First Lieutenant William Laws Calley, Jr., for the alleged murder of 22 Vietnamese civilians at Mylai, involved the same question brought up at the Nuremburg trials of World War II criminals: Should a soldier carry out the orders of his superiors even though he believes them illegal, or should he refuse to obey such orders?

The circumstances leading up to the general courtmartial were quite involved. On March 16, 1968, Lieutenant Calley, as leader of the First Platoon of Charlie Company of the American army's Task Force Barker in South Vietnam, led his troops aboard a helicopter which was to take them to the village complex of Mylai. The troops had been briefed by Captain Ernest Medina, commander of Charlie Company, the day before. His instructions to them, reportedly as given him by higher headquarters, were to kill the Vietcong and destroy the village.

High-ranking officers of the task force observed the maneuver from helicopters flying overhead. Chief Warrant Officer Hugh

Thompson, flying an observation helicopter, saw native civilians in a ditch being shot by American troops. He landed his machine and rescued a few natives. Two days later Colonel Oran Henderson, commander of the Eleventh Brigade, who was one of the officers who had observed the maneuver from a helicopter, advised Captain Medina that an investigation was under way. Medina instructed his men to say nothing concerning the incident.

The investigation was apparently dropped. A year and a half later Robert Ridenhour, an American soldier who was not a member of Charlie Company but who had heard of the so-called massacre of Vietnamese civilians, interviewed some of the men who had participated in it. He then wrote letters to the White House and to Congress. Congressman L. Mendel Rivers, chairman of the House Armed Services Committee, called the matter to the attention of the Pentagon. The army started a secret investigation.

The army determined that 24 officers and enlisted men should be tried by military courtmartial for either covering up the matter or participating in it. Later charges were dropped against all but Colonel Henderson for the alleged cover-up and Captain Medina, Lieutenant Calley, and Sergeants Charles Hutto and David Mitchell for the killing.

Sergeant Paul Meadlo, a member of Charlie Company, told an anguished story on coast-to-coast television. Congress demanded the punishment of the men involved in the incident and President Richard M. Nixon stated that it appeared to him to be an unjustified massacre which could not be condoned.

The Calley courtmartial began at Fort Benning, Georgia, on November 12, 1970. Calley was charged with the premeditated murder of "not less than one hundred Oriental human beings." The prosecution was handled by Captain Aubrey M. Daniel III, and George Latimer acted as chief defense counsel. Colonel Clifford Ford was president of the court.

In his opening statement Captain Daniel indicated that Calley had ordered Private First Class Dennis I. Conti and Sergeant Paul Meadlo to "waste"--in other words, kill--the unarmed, unresisting native civilians, and that Conti and Meadlo did so, using their M-16 rifles. Meadlo, he said, obeyed unwillingly and burst into tears as he followed his officer's command. Other natives, said Daniel, were rounded up, pushed into the irrigation ditch, and shot.

Members of Calley's platoon, when called to testify for the prosecution, implicated him in the incident. Negro Specialist Fourth Class Robert Maples said under oath that Calley and Meadlo fired on Orientals in the ditch and that he, Maples, when ordered by Calley to do likewise, refused. "Meadlo was crying when he fired into the hole," the witness said.

Meadlo, when called to testify, refused at first but later appeared as a witness for the prosecution. Other members of Calley's platoon testified against him. In some cases the testimony of one witness contradicted that of another. Radio Operator Charles Sledge testified that Calley had shot a priest after hitting him with his rifle.

The defense's opening statement included the charge that Captain Medina had given orders that "every living thing in that hamlet should be killed," and had issued no instructions for the handling of civilians. The charge was also made that, though high-ranking officers, including Major General Samuel Koster, commander of the parent American Division, were observing ground action from helicopters, no cease fire order was given until four hours after the action started, and that was for a lunch break.

Defense witnesses testified that large numbers of Vietnamese had been killed by artillery or from gunships before the Americans arrived. This contradicted sworn statements by prosecution witnesses. It was shown that Mylai was in a hostile area. The defense's main argument was then put forth: that Calley was ordered by Medina to kill the civilians.

On the afternoon of February 23, 1971, Lieutenant Calley was placed in the witness chair. He testified under oath that Captain Medina had instructed the men and also the platoon leaders that everyone encountered at Mylai should be considered an enemy and "not left standing." "There are no civilians in the area," he said. Calley specifically denied firing into any group of people but did admit firing into the ditch. He also stated that he was following Medina's "order of the day" in telling Meadlo to "waste the Vietnamese if they couldn't be moved." He denied shooting a priest but admitted hitting him with the butt of his rifle. On cross-examination he contradicted himself, saying Medina had told him to "hang onto a few [Vietnamese] in case we hit a mine field." He then said, "Captain Medina rescinded his mine field order; he told me to waste them."

Captain Medina appeared as a prosecution witness on February 24. He denied instructing anyone at any briefing to kill women or children. On cross-examination he admitted trying to cover up the incident and also that he was so confused that he "might have commanded the shooting by others." He had, he testified, authorized Calley to "utilize prisoners to help lead his unit through the mine field."

Defense and prosecution then made their summations. Following this the court was closed and deliberations occupied the next thirteen days. When court reopened Colonel Ford announced that Lieutenant Calley had been found guilty of all charges and specifications, particularly of killing 22 Vietnamese civilians. He was sentenced to life imprisonment, which was later reduced to ten years.

The finding of the military court was appealed several times. After three years of legal maneuvering, during which Calley lived in his bachelor quarters at Fort Benning under house arrest, District Court Judge J. Robert Elliott, who had reversed Calley's conviction for the Mylai murders, freed him on bail and removed him from army custody. When Elliott took Calley's appeal under advisement, the latter was flown to the military prison at Fort Leavenworth, Kansas where, under minimum security, he worked as a clerk-typist.

In September, 1975, a thirteen-judge federal appeals court reversed the lower court's finding that pretrial publicity kept Calley from receiving a fair trial. Five appeal judges dissented, arguing that regardless of the publicity issue, Calley's trial was unfair because the defense was denied access to a House Committee's Mylai findings.

Lieutenant Calley's case was appealed to the Supreme Court which, in April, 1976, refused to review it, thus reinstating his conviction. He was promptly paroled by the army.

Suggested Readings

"Calley as Joshua," Time, October 7, 1974.
"Calley Loses," Time, September 22, 1975.
Calley, William Laws. Lieutenant Calley: His Own Story, as told to John Slack. New York: Viking Press, 1971.
"Closing the My Lai Case," Time, November 25, 1974.
DiMona, Joseph. Great Court Martial Cases. New York: Grosset & Dunlap, 1972.
________. "Mylai: The Court-Martial of Lt. Calley," in Rubenstein, Richard E., ed. Great Courtroom Battles. Chicago: Playboy Press, 1973.
Everett, Arthur, et al. Calley. New York: Dell Books, 1971.
Franco, Angelo. The Trial of Lt. William Calley. New York: Ledger-Enquirer Newspapers, 1971.
Greenhaw, Wayne. The Making of a Hero. Louisville: Touchstone, 1971.
Hammer, Richard. The Court-Martial of Lieut. Calley. New York: Coward-McCann, 1971.
Sheppard, R. Z. "Cog Ergo Sum," Time, October 22, 1973.
Sherrill, Robert. Military Justice Is to Justice as Military Music Is to Music. New York: Harper, 1969.
Tiede, Tom. Calley: Soldier or Killer? New York: Pinnacle Books, 1971.
Wallechinsky, David, and Irving Wallace. "The My Lai Massacre," in their The People's Almanac. Garden City, N.Y.: Doubleday, 1975.

THE MYSTERY SKYJACKER (1971)

The only person ever to "skyjack" a domestic airliner without being killed or brought to justice is "Dan Cooper" who, on November 24, 1971, successfully extorted $200,000 in ransom money and parachuted, apparently to safety, from a Northwest Airlines Boeing 727 jet plane.

Flight No. 305 was boarded at Portland, Oregon, International Airport by a man who called himself "Dan Cooper," who had purchased a one-way ticket to Seattle. The plane carried 36 passengers, three stewardesses and the cockpit crew. Captain William Scott, chief pilot, was in command.

Once the plane was aloft "Cooper" handed stewardess Tina Mucklow a note saying he had a bomb. Then, on his instructions, the girl sat beside him and wrote a note to the pilot, which

"Cooper" afterwards reclaimed. He demanded $200,000 in twenty-dollar bills and two parachutes. The money was to be delivered to him in a laundry sack. Failure to comply, he said, would cause him to blow up the plane.

When the aircraft reached Seattle the money and parachutes were ready. "Cooper" allowed the passengers and two of the stewardesses to disembark and then ordered the plane to be flown south to Reno, Nevada. The crew was to stay in the cockpit.

After takeoff a red light flashed in the cockpit, indicating that the plane's boarding ramp had been unlatched. As the plane crossed the Lewis River in Southwest Washington, Captain Scott called back over the intercom, "Anything we can do for you?" There was no answer.

Another light flashed, indicating that the ramp was fully extended. "Cooper's" voice then came back over the intercom: "No."

When the aircraft landed at Reno, Cooper, the sack of money and one parachute were gone. The ramp was still extended.

"Cooper's" drop zone was pinpointed at a place somewhere northeast of Woodland, Washington. When he left the plane, which was traveling at approximately 200 miles per hour, he was wearing a business suit and street shoes, poor equipment for a parachute jumper. A thunderstorm was raging and the temperature was seven degrees below zero.

The authorities, reasoning that no one could survive a parachute jump under such conditions, searched for a body and a laundry sack full of money. They used planes, helicopters, jeeps and track dogs. They found nothing.

Three years later "Dan Cooper" had still not been found. Julius Mattson, special agent in charge of the Portland, Oregon F.B.I. office, is still actively searching for him. "But," said Mattson, "I just wish we had something to go on. We don't have a thing--just a big zero."

In January, 1975, a skull, believed to be that of a Caucasian in his early thirties, was found in the Mount Hood National Forest ten miles east of Estacada, Oregon. Near the skull was a parachute hanging high in a fir tree. Dr. Larry Lewman of the Oregon state medical examiner's office, who studied the skull and checked with the F.B.I., expressed the opinion that it was not that of "Cooper," probably being that of a younger man, possibly a hunter who had died in the wilderness. The parachute was of a type used in connection with weather balloons.

Suggested Readings

Agrawala, S. K. Aircraft Highjacking and International Law. Dobbs Ferry, N.Y.: Oceana, 1973.

Arey, James A. Sky Pirates. New York: Scribner's, 1972.

"The Bandit Who Went Out into the Cold: Northwest Airlines Flight 305, Washington, D.C. to Seattle," Time, December 8, 1971.

"Bringing Skyjackers Down to Earth," Time, October 4, 1971.

Elten, J. A. "This Is a Highjacking!" (condensation). Reader's Digest, July, 1971.
"Highjacker Parachutes from Northwest 727," Aviation Week, December 6, 1971.
Hubbard, David Graham. The Skyjacker: His Flights of Fancy. New York: Macmillan, 1971.
"Progress in War on Skyjackers," U.S. News and World Report, August 9, 1971.
"Quantum Leap: Northwest Airlines Flight 305 to Seattle," Newsweek, December 6, 1971.
Rich, Elizabeth. Flying Scared: Why We Are Being Skyjacked and How to Put a Stop to It. New York: Stein, 1972.

THE WATERGATE BREAK-IN (1972)

The greatest national scandal of the 1970's, popularly called the "Watergate Break-in," culminated on the night of June 17, 1972, when five men acting for the Republican party broke into the Democratic National Committee Headquarters in the Watergate complex at Washington, D.C. These men were led by James W. McCord, Jr., one-time Central Intelligence Agency (CIA) man and security coordinator for the Committee to Re-elect the President (CREEP). He was assisted in the break-in by Bernard L. Barker, and by Virgilio R. Gonzales, Eugenio R. Martinez and Frank A. Sturgis, Cuban-Americans. The five were apprehended and arrested by Frank Wills, a guard. The building had previously been broken into on May 28, after two unsuccessful attempts, and wire-tapping equipment had been installed in two telephones. Alfred C. Baldwin III, a former FBI agent, who monitored the information received through the taps, came up with reports code-named "Gemstone." These were given to McCord, who passed them on to G. Gordon Liddy, General Counsel to CREEP.

The June 17 break-in was apparently made to repair a malfunctioning wire-tap. Found in the men's possession at the time of their arrest were lock picks, a telephone bugging device, camera equipment, rubber gloves and currency which represented Republican campaign funds.

The Watergate Break-in was part of an elaborate political espionage plan proposed by Liddy to White House Counsel John W. Dean III, CREEP Deputy Director Jeb Stuart Magruder and Attorney General John N. Mitchell in the latter's office on January 27, 1972. The plan involved extensive electronic bugging, political kidnaping and mugging, and the use of call girls to extract information from political enemies. Mitchell later testified that he was shocked by the proposal and ordered Liddy to "burn the plan and abandon any concept that such activities would be part of the campaign to reelect the President [Richard M. Nixon]."

On February 4 the four men met again and heard a revised proposal by Liddy. The cost of the operation had been scaled down from the original $1,000,000 to a comparatively modest $250,000, with only photography and wire-tapping plans remaining. According to testimony later given by Magruder before the seven-man Senate Select Committee on Presidential Campaign Activities, which had been created by a special Senate Resolution and was chaired by Senator Sam J. Erwin, Jr., Mitchell had, at a March 30 meeting at Key Biscayne, Florida, approved the final version of the plan "unethusiastically but unequivocally."

When McCord and his accomplices were caught by the guard Frank Wills, Magruder was in Los Angeles. He telephoned to Washington giving orders that incriminating "Gemstone" documents were to be destroyed and that a cover story was to be concocted. It was felt, according to Magruder's testimony before the Committee, that if the public were to learn that such high-ranking government officials as Mitchell were involved in the break-in, Nixon's reelection in the 1972 campaign would be severely jeopardized if not negated.

McCord, along with Liddy and E. Howard Hunt, Jr., former spy novelist, CIA agent and White House consultant, were tried in the U.S. District Court. They were found guilty and Judge John J. Sirica imposed stiff sentences on the defendants with the exception of McCord, whose sentence he postponed.

On March 23, 1973, Judge Sirica made public a letter he had received from McCord. This letter implicated higher-ups in the Watergate case and also charged perjury at the trial and stated that political pressure was put on some defendants.

In the course of his testimony before the Committee Magruder admitted his own complicity in the plot and implicated others. Mentioned were H. R. Haldeman, former White House Chief of Staff; Charles W. Colson, former special counsel to the President; Gordon C. Strachan, a Haldeman aide; Maurice H. Stans, Chairman of the Reelection Finance Committee; Herbert W. Kalmbach, former personal attorney to the President; Robert C. Mardian and Frederick C. LaRue, CREEP officials; Mitchell, Dean and others.

President Nixon was not specifically implicated by Magruder. Dean, who had been ousted from his position as White House counsel, had resolved not to be a scapegoat. On June 25 he appeared before the Committee and, under oath, charged Nixon with active participation in the Watergate bugging, burglary and subsequent cover-up.

The Watergate affair had all the elements of a complicated cloak-and-dagger melodrama. Witnesses before the Committee and the grand jury told conflicting stories, implicated each other and spoke of large sums of money, largely in cash, changing hands without being accounted for. Checks representing huge sums were "laundered" through Mexican banks. A State Department cable concerning the President was said to have been forged. Harvard law professor Archibald Cox was named special Watergate prosecutor. President Nixon was asked to furnish the Committee with tapes of conversations recorded in the White House. He refused to comply on the grounds of Presidential privilege. The Los Angeles office of Dr. Lewis Fielding, Daniel Ellsberg's psychiatrist, was broken into in an attempt to secure personal information concerning the latter. More charges and counter-charges were made. Rumors that Nixon contemplated resigning the Presidency were dismissed by him as "just plain poppycock."

On July 31, 1973, Representative Robert F. Drinan introduced an impeachment resolution against the President for "high crimes and misdemeanors." The California Bar Association began an inquiry into the conduct of six of its members tied to Watergate, including Nixon. A special grand jury was impaneled in Washington to investigate illegal campaign financing and other obstructions to justice.

On August 15, 1973, in a televised address, President Nixon denied any guilt, charging that the accusations made by Dean were uncorroborated. On August 22, at a news conference, he declared that Watergate was "water under the bridge," and that he was "going about the people's business rather than dwelling on the scandal."

Nixon, however, could not live the Watergate Affair down. Rather than suffer impeachment he resigned the Presidency on August 8, 1974, effective at noon the following day. He was succeeded by Gerald Ford, whom he had appointed Vice President following Spiro Agnew's resignation from that office. As of this writing the Watergate affair has not actually been settled.

Suggested Readings

Archer, Jules. *Watergate: America in Crisis.* New York: Crowell, 1975.

Bernstein, Carl, and Bob Woodward. *All the President's Men.* New York: Simon & Schuster, 1974.

Breslin, Jimmy. *How the Good Guys Finally Won.* New York: Viking Press, 1975.

Colson, Charles W. *Born Again.* New York: Chosen Books, 1975.

Dean, Maureen. *"Mo": A Woman's View of Watergate.* New York: Simon & Schuster, 1975.

Drew, Elizabeth. *Washington Journal: The Events of 1973-1974.* New York: Random House, 1975.

Evans, Les, and Allen Myers. *Watergate and the Myth of American Democracy.* New York: Pathfinder Press, 1974.

Friedman, Leon, comp. *United States v. Nixon.* New York: Chelsea House, 1974.

Higgins, George V. *Friends of Richard Nixon.* Boston: Little, Brown, 1975.

Hunt, E. Howard. *Undercover.* New York: Berkley, 1974.

Lucas, J. Anthony. *Nightmare: The Underside of the Nixon Years.* New York: Viking Press, 1975.

McCarthy, Mary T. *Mask of the State: Watergate Portraits.* New York: Harcourt, Brace, 1974.

McCord, James W., Jr. *A Piece of Tape: The Watergate Story, Fact and Fiction.* Washington, D.C.: Media Services, 1974.

Magnuson, Edward. "Post-Mortem: The Unmaking of a President," *Time*, May 12, 1975.

Magruder, Jeb Stuart. *American Life: One Man's Road to Watergate.* New York: Atheneum, 1974.

Mankiewicz, Frank. Perfectly Clear: Nixon from Whittier to Watergate. New York: Quadrangle, 1973.
_____. U. S. v. Richard M. Nixon: The Final Crisis. New York: Quadrangle, 1975.
Myerson, Michael. Watergate: The Crime in the Suites. New York: International, 1973.
New York Times. The Watergate Hearings: Break-in and Cover-up. Narrative by R. W. Apple, Jr.; chronology by Linder Amster; General editor, Gerald Gold. New York: Viking Press, 1973.
Nixon, Richard M. The Watergate Affair (phonotape). New York: Encyclopedia Americana/CBS News Audio Resource Library 04733, 1973.
Schell, Jonathan. The Time of Illusion. New York: Knopf, 1975.
Sorensen, Theodore C. Watchmen in the Night. Cambridge, Mass.: M.I.T. Press, 1975.
Tetick, Stanley. They Could Not Trust the King. New York: Collier Books, 1974.
Wallechinsky, David, and Irving Wallace. "Watergate," in their The People's Almanac. Garden City, N.Y.: Doubleday, 1975.
"Watergate Chronology," in Urdang, Laurence, editor-in-chief. The Official Associated Press Almanac 1974. Maplewood, N.J.: Hammond Almanac, Inc., 1973.
White, Theodore H. Breach of Faith: The Fall of Richard Nixon. New York: Atheneum-Reader's Digest Press, 1975.
Winter, R. K. Watergate and the Law. Washington, D.C.: American Enterprise Institute, 1974.
Wise, H. D. What Do We Tell the Children? New York: Braziller, 1974.

THE PATTY HEARST AFFAIR (1974-1976)

On February 4, 1974, Patricia "Patty" Campbell Hearst, the daughter of Randolph Hearst, editor and president of the San Francisco Examiner, was carried screaming from her Berkeley, California, apartment by a group of revolutionaries self-named the Symbionese Liberation Army. On February 7 the S.L.A. released a letter, claiming responsibility for the abduction. One of the kidnapers was identified as Donald De Freeze, a black criminal, ex-convict and leader of the group. Later, in tape recordings, Patty Hearst denounced her father, jilted her fiance Steven Weed, indicated that she had become a member of the S.L.A. and had taken the name "Tania."

Through tape recordings sent to a San Francisco television station the kidnapers stated that the price of Patty Hearst's freedom would be a donation of several million dollars' worth of food to the needy of California. Randolph Hearst arranged for a food distribution but his daughter did not return home after the food had been passed out. A subsequent food distribution demanded by the kidnapers was canceled when Patty was not released.

In April five members of the S. L. A. held up a branch of the Hibernia Bank in San Francisco. Cameras installed in the bank lobby photographed the robbers, later identified as the Hearst girl, Donald De Freeze, Patricia "Mizmoon" Soltysik, Mrs. Nancy Ling Perry and Camilla Hall.

On May 16 the group appeared in Los Angeles. A clumsy attempt at shoplifting a pair of shoes from a store in nearby Inglewood led to a shootout the following night when 500 policemen and F. B. I. agents laid siege to a house in South Los Angeles to which the group had been traced. Killed in the shooting were De Freeze, Mrs. Perry, Patricia Soltysik, William Wolfe, and an unidentified woman.

Following the shooting Patty Hearst dropped from sight. She was thought to be traveling with William T. Harris and his wife Emily, suspected S. L. A. members. In late August, 1974, her whereabouts were unknown and she was posted as "wanted" by the Federal Bureau of Investigation, No. 325,805, L 10.

Bill Walton, a professional basketball player, was questioned concerning the Hearst case but disclaimed all knowledge of it. He had been living with Jack Scott, a former athletic director, and Scott's wife Micki near Portland, Oregon. Published reports alleged that Scott had been in contact with the missing girl. It was rumored that Patty Hearst and the Harrises were living in Pennsylvania. It was also reported that they had been seen in Guatemala, Chicago, Costa Rica, Panama, Los Angeles and the San Francisco Bay district, among other places.

In September 1975, Patty Hearst, the Harrises and Wendy Yoshimura, a friend of the Hearst girl, were arrested by the F. B. I. in San Francisco. Patty Hearst was indicted for bank robbery and she, like William and Emily Harris, was also charged with the illegal possession and use of firearms to commit a felony. Her father engaged attorneys F. Lee Bailey and Albert H. Johnson to act as her defense counsel.

Patty's San Francisco trial in federal court made headlines. Her lawyers contended that, while she had participated in the bank robbery as charged, she had done so under duress. Nevertheless, in March of 1976 she was found guilty and on April 11 United States District Judge Oliver J. Carter imposed the maximum sentence of 35 years. He also ordered psychiatric examination, following which he said the sentence would be modified after he had received and considered the psychiatric report.

The girl was hospitalized shortly thereafter due to a collapsed lung. The Harrises were charged with Patty's kidnaping and other related crimes and were tried before Superior Court Judge Mark Brandler in Los Angeles. They and Patty Hearst fell out, with Patty being called "a turncoat and a liar." Patty refused to speak at a pre-trial hearing and Judge Brandler entered an innocent plea for her on each of the eleven state charges.

In mid-June Judge Carter died of a heart attack. Federal Judge William H. Orrick, Jr. was chosen by lot to replace him and to sentence Patty Hearst after evaluating the results of the psychiatric examination. The Harrises, tried together, were found guilty in a sensational hearing which included an alleged attempt to sabotage the trial by smuggling an old newspaper describing the Hearst kidnaping to the jury. How the newspaper was introduced into the heavily-guarded jury room has not been discovered.

At this writing the Patty Hearst Affair is not over. William and Emily Harris, convicted of kidnaping, assault and robbery charges, were each sentenced to serve eleven years in prison. On September 24, 1976, Patty Hearst appeared before Judge Orrick for sentencing. She declined to speak in her own behalf and stood quietly while the jurist said, "the violent nature of your conduct cannot be condoned," and sentenced her to seven years' confinement for bank robbery. In addition he decreed a two-year term, to run concurrently with the seven, for use of a firearm in the 1974 bank robbery. She will be eligible for parole in January 1978, the court giving her credit for 371 days of jail time served since her arrest in September 1975.

Patty Hearst is still in legal trouble. She faces trial on January 10, 1977 in Los Angeles on eleven state charges and is scheduled to appear before Superior Court Judge William Ritzi. Her parents declared Judge Orrick's sentence to be unfair. Randolph Hearst stated, "Judge Orrick can sleep soundly tonight knowing he has done his very best to keep our country safe from crimes committed by kidnap victims."

Suggested Readings

"After Hearst Arrest: Drive to Root Out U. S. Terror Gangs," U.S. News and World Report, October 6, 1975.
"All in the Family," Time, September 25, 1975.
Baker, Marilyn, with Sally Brompton. Exclusive! The Inside Story of Patricia Hearst and the S. L. A. New York: Macmillan, 1974.
"Battle Over Patty's Mind," Time, February 16, 1976.
"Brainwashing-Echoes of Korean War," U. S. News and World Report, February 23, 1976.
"Disturbed Young Woman," Time, October 13, 1975.
"Dragnet Spreads to Patty," Newsweek, June 3, 1974.
"Fiery End for Six of Patty's Captors," Time, May 27, 1974.
Gelman, D., and W. J. Cook. "Life Behind Bars," Newsweek, February 2, 1976.
"How P. O. W.'s Judge Tania," Time, March 8, 1976.
"Is Brainwashing an Excuse?" Time, March 23, 1976.
Mathews, T., and W. J. Cook. "Bailey Loses His Cool," Newsweek, March 15, 1976.
________. "End Game," Newsweek, March 22, 1976.
________. "Patty Tells Her Story," Newsweek, February 23, 1976.
________. "Three Faces of Patty," Newsweek, March 8, 1976.
________. "U.S. vs. P. Hearst," Newsweek, February 9, 1976.
Montagno, M. "Patty Hearst on the Stand," Newsweek, January 26, 1976.
"Patty Hearst Trial Heats Up," Time, March 24, 1975.
"Patty Hearst vs. Tania," Newsweek, February 23, 1976.
"Queen of the S.L.A?," Time, March 22, 1976.

"Some Questions for Patty," Newsweek, October 6, 1975.
"Stalking Patty Hearst," Newsweek, March 24, 1975.
"Tip Pays Off: Patty Hearst Is Captured," U.S. News and World Report, September 29, 1975.
Weed, Steven, and Scott Swanton. "My Life With Patty Hearst," McCall's, January, 1976.
________. My Search for Patty Hearst. New York: Crown Publishers, 1976.
West, Don, and Jerry Belcher. Patty/Tania. New York: Pyramid Books, 1975.

INDEX

For Marjorie's and my
good friends at Saint
Francis College -

Sincerely

Harold S. Sharp

March 26, 1977.